Gmelin Handbook of Inorganic Chemistry

8th Edition

Gmelin Handbook – Volumes on Platinum Metals

Ir	* Iridium Main Volume – 1939
	* Iridium Suppl. Vol. 1 (Metal, Alloys) – 1978
	° Iridium Suppl. Vol. 2 (Compounds) – 1978
Os	* Osmium Main Volume – 1939
	Osmium Suppl. Vol. 1 – 1980
Pd	* Palladium 1 (Element) – 1941
	* Palladium 2 (Compounds) – 1942
Pt	* Platin A 1 (History, Occurrence) – 1938
	* Platin A 2 (Occurrence) – 1939
	* Platin A 3 (Preparation of Platinum Metals) – 1939
	* Platin A 4 (Detection and Determination of the Platinum Metals) – 1940
	* Platin A 5 (Alloys of Platinum Metals: Ru, Rh, Pd) – 1949
	* Platin A 6 (Alloys of Platinum Metals: Os, Ir, Pt) – 1951
	* Platin B 1 (Physical Properties of the Metal) – 1939
	* Platin B 2 (Physical Properties of the Metal) – 1939
	* Platin B 3 (Electrochemical Behavior of the Metal) – 1939
	* Platin B 4 (Electrochemical Behavior and Chemical Reactions of the Metal) – 1942
	* Platin C 1 (Compounds up to Platinum and Bismuth) – 1939
	* Platin C 2 (Compounds up to Platinum and Caesium) – 1940
	* Platin C 3 (Compounds up to Platinum and Iridium) – 1940
	* Platin D (Complex Compounds of Platinum with Neutral Ligands) – 1957
Rh	* Rhodium Main Volume – 1938
	Rhodium Suppl. Vol. B 1 (Compounds) – 1982
	Rhodium Suppl. Vol. B 2 (Coordination Compounds) – 1984 (present volume)
Ru	* Ruthenium Main Volume – 1938
	* Ruthenium Suppl. Vol. – 1970

* Completely or ° partially in German

Gmelin Handbook of Inorganic Chemistry

8th Edition

Gmelin Handbuch der Anorganischen Chemie

Achte, völlig neu bearbeitete Auflage

Prepared and issued by Gmelin-Institut für Anorganische Chemie der Max-Planck-Gesellschaft zur Förderung der Wissenschaften

Director: Ekkehard Fluck

Founded by Leopold Gmelin

8th Edition: 8th Edition begun under the auspices of the Deutsche Chemische Gesellschaft by R. J. Meyer

Continued by E. H. E. Pietsch and A. Kotowski, and by Margot Becke-Goehring

Springer-Verlag Berlin · Heidelberg · New York · Tokyo 1984

Gmelin-Institut für Anorganische Chemie
der Max-Planck-Gesellschaft zur Förderung der Wissenschaften

ADVISORY BOARD

Dr. J. Schaafhausen, Vorsitzender (Hoechst AG, Frankfurt/Main-Höchst), Dr. G. Breil (Ruhrchemie AG, Oberhausen-Holten), Dr. G. Broja (Bayer AG, Leverkusen), Prof. Dr. G. Fritz (Universität Karlsruhe), Prof. Dr. N. N. Greenwood (University of Leeds), Prof. Dr. R. Hoppe (Universität Gießen), Prof. Dr. R. Lüst (Präsident der Max-Planck-Gesellschaft, München), Dr. H. Moell (BASF-Aktiengesellschaft, Ludwigshafen), Prof. Dr. E. L. Muetterties (University of California, Berkeley, California), Prof. Dr. H. Nöth (Universität München), Prof. Dr. A. Rabenau (Max-Planck-Institut für Festkörperforschung, Stuttgart), Prof. Dr. Dr. h.c. mult. G. Wilke (Max-Planck-Institut für Kohlenforschung, Mülheim/Ruhr)

DIRECTOR

Prof. Dr. Dr. h.c. Ekkehard Fluck

DEPUTY DIRECTOR

Dr. W. Lippert

CHIEF EDITORS

Dr. K.-C. Buschbeck – Dr. H. Bergmann, Dr. H. Bitterer, Dr. H. Katscher, Dr. R. Keim, Dipl.-Ing. G. Kirschstein, Dipl.-Phys. D. Koschel, Dr. U. Krüerke, Dr. H. K. Kugler, Dr. E. Schleitzer-Rust, Dr. A. Slawisch, Dr. K. Swars, Dr. B. v. Tschirschnitz-Geibler, Dr. R. Warncke

STAFF

A. Alraum, D. Barthel, Dr. N. Baumann, Dr. W. Behrendt, Dr. L. Berg, Dipl.-Chem. E. Best, P. Born, E. Brettschneider, Dipl.-Ing. V. A. Chavizon, E. Cloos, Dipl.-Phys. G. Czack, I. Deim, Dipl.-Chem. H. Demmer, R. Dombrowsky, R. Dowideit, Dipl.-Chem. A. Drechsler, Dipl.-Chem. M. Drößmar, M. Engels, Dr. H.-J. Fachmann, Dr. J. Faust, I. Fischer, Dr. R. Froböse, J. Füssel, Dipl.-Ing. N. Gagel, E. Gerhardt, Dr. U. W. Gerwarth, M.-L. Gerwien, Dipl.-Phys. D. Gras, Dr. V. Haase, K. Hartmann, H. Hartwig, B. Heibel, Dipl.-Min. H. Hein, G. Heinrich-Sterzel, H.-P. Hente, H. W. Herold, U. Hettwer, Dr. I. Hinz, Dr. W. Hoffmann, Dipl.-Chem. K. Holzapfel, Dr. S. Jäger, Dr. J. von Jouanne, Dipl.-Chem. W. Karl, H.-G. Karrenberg, Dipl.-Phys. H. Keller-Rudek, Dipl.-Phys. E. Koch, Dr. E. Koch, Dipl.-Chem. K. Koeber, Dipl.-Chem. H. Köttelwesch, R. Kolb, E. Kranz, Dipl.-Chem. I. Kreuzbichler, Dr. A. Kubny, Dr. P. Kuhn, Dr. W. Kurtz, M. Langer, Dr. A. Leonard, A. Leonhard, Dipl.-Chem. H. List, Prof. Dr. K. Maas, H. Mathis, E. Meinhard, Dr. P. Merlet, K. Meyer, M. Michel, Dipl.-Chem. R. Möller, K. Nöring, Dipl.-Min. U. Nohl, Dr. W. Petz, C. Pielenz, E. Preißer, I. Rangnow, Dipl.-Phys. H.-J. Richter-Ditten, Dipl.-Chem. H. Rieger, E. Rieth, Dr. J. F. Rounsaville, E. Rudolph, G. Rudolph, Dipl.-Chem. S. Ruprecht, Dr. R. C. Sangster, V. Schlicht, Dipl.-Chem. D. Schneider, Dr. F. Schröder, Dipl.-Min. P. Schubert, A. Schwärzel, Dipl.-Ing. H. M. Somer, E. Sommer, M. Teichmann, Dr. W. Töpper, Dipl.-Ing. H. Vanecek, Dipl.-Chem. P. Velić, Dipl.-Ing. U. Vetter, Dipl.-Phys. J. Wagner, R. Wagner, Dr. E. Warkentin, Dr. G. Weinberger, Dr. B. Wöbke, K. Wolff, U. Ziegler

CORRESPONDENT MEMBERS OF THE SCIENTIFIC STAFF

Dr. J. L. Grant, Dr. I. Kubach, Dr. D. I. Loewus, Dr. K. Rumpf, Dr. U. Trobisch

EMERITUS MEMBER OF THE INSTITUTE

Prof. Dr. Dr. E.h. Margot Becke

CORRESPONDENT MEMBERS OF THE INSTITUTE

Prof. Dr. Hans Bock
Prof. Dr. Dr. Alois Haas, Sc. D. (Cantab.)

Gmelin Handbook of Inorganic Chemistry

8th Edition

Rh
Rhodium

Supplement Volume B 2

Coordination Compounds with O- and N-Containing Ligands

With 24 illustrations

AUTHORS

William P. Griffith, Imperial College of Science and Technology, London, Great Britain

Jon A. McCleverty, University of Birmingham, Birmingham, Great Britain

Stephen D. Robinson, King's College, London, Great Britain

EDITORS

William P. Griffith, Imperial College of Science and Technology, London, Great Britain

Kurt Swars, Gmelin-Institut, Frankfurt am Main, Bundesrepublik Deutschland

System Number 64

Springer-Verlag Berlin · Heidelberg · New York · Tokyo 1984

AUTHORS OF THIS VOLUME

William P. Griffith	pp. 116/119, 124/125, 210/224, 300/301
Jon A. McCleverty	pp. 99/116, 125/200, 239/290
Stephen D. Robinson	pp. 1/98, 119/123, 200/210, 225/238, 291/299, 302/321

LITERATURE CLOSING DATE: 1982

Library of Congress Catalog Card Number: Agr 25-1383

ISBN 3-540-93496-0 Springer-Verlag, Berlin · Heidelberg · New York · Tokyo
ISBN 0-387-93496-0 Springer-Verlag, New York · Heidelberg · Berlin · Tokyo

Typesetting, printing, and bookbinding: LN-Druck Lübeck

Preface

The second volume of this trilogy is concerned with rhodium complexes of donors containing oxygen followed by those containing nitrogen. As a consequence of the traditional Gmelin arrangement of material some oxygen donors (oxo, hydroxo, aquo, nitrato, sulphato, carbonato and phosphato) have already been covered in Volume B 1. Amongst those dealt with here carboxylates are by far the most numerous, particularly those containing the $Rh_2(O_2CR)_4$ "lantern" unit. Since this is considered to be the dominant feature of its complexes all such species are considered in one section – thus, $Rh_2(O_2CCH_3)_4(P(C_6H_5)_3)_2$, which would normally appear under phosphorus in B 3, is dealt with in this volume. The other major section concerns β-diketonates, of which acetylacetonates are the most important group, and there is a small section on miscellaneous oxygen donor complexes.

The nitrogen donor section starts with nitrosyl complexes; as in previous recent Gmelin volumes all such species are grouped in one section. Sections on other simple ligands follow (NO_2^-, RCN, NH_2^-, N_2H_4), and then there is a large body of information on rhodium ammines. Complexes of saturated mono- and polyamines are considered next, including a large section on ethylenediamine complexes. Complexes of cyclic amines, saturated and unsaturated, are followed by those of unsaturated polydentate donors and then by mono- and polydentate heterocyclic ligands; prominent amongst these are the complexes of pyridine, 2,2′-bipyridyl and 1,10-phenanthroline. Finally, there are sections on miscellaneous nitrogen donors and miscellaneous nitrogen-oxygen donor complexes.

The final volume (B 3) in this series will be concerned principally with complexes containing sulphur, phosphorus, arsenic and antimony and also with heteronuclear metal-metal bonded species.

London, October 1983

William P. Griffith

Table of Contents

Page

Coordination Compounds of Rhodium 1

1 Complexes with Ligands Containing Oxygen 1

1.1 Carboxylates 1

1.1.1 Oxalato Complexes 1
Tris(oxalato) Complexes 2
Compounds with Nonmetallic Cations 2
Sodium Compounds, $Na_3[Rh(C_2O_4)_3] \cdot nH_2O$ 2
Potassium Compounds, $K_3[Rh(C_2O_4)_3] \cdot nH_2O$ 2
Heavy-Metal Compounds 6
Bis(oxalato) Complexes 7
Mono(oxalato) Complexes 8
Miscellaneous Oxalato Complexes 9

1.1.2 Rhodium(II) Carboxylates 9
Formates 10
Acetates 15
Unsubstituted Acetates 15
Methoxyacetate, $Rh_2(O_2CCH_2OCH_3)_4$ 37
Phenoxyacetate, $Rh_2(O_2CCH_2OC_6H_5)_4$ 37
Aminoacetate, $Rh_2(O_2CCH_2NH_2)_4\{(C_2H_5)_2O\}_2$ 37
Monofluoroacetates 38
Trifluoroacetates 38
Monochloroacetates 40
Dichloroacetates 40
Trichloroacetates 41
Chlorodifluoroacetates 42
Monobromoacetate, $Rh_2(O_2CCH_2Br)_4(C_2H_5OH)_2$ 42
Phenylacetate, $Rh_2(O_2CCH_2C_6H_5)_4$ 42
Propionates 42
Aminopropionates (Alaninates) 46
α-Alaninates 46
β-Alaninate, $[Rh_2(O_2CCH_2CH_2NH_3)_4(H_2O)_2](ClO_4)_4 \cdot 2H_2O$ 46
t-Butyloxycarbonylalaninate, $Rh_2(O_2CCH_2CH_2NHCOCOBu^t)_4$ 47
Chloropropionates 47
Butyrates 47
Perfluorobutyrates 48
Other Aliphatic Carboxylates 48
t-Pentanoates (Pivalates) 49
Benzoates 49
Salicylates 51
Thiosalicylates 51
Complexes with Substituted Benzoic Acids, $Rh_2(O_2CAr)_4$ 53

Page

Mandelates 53
Pyridine-2,6-dicarboxylates 53
Monothioacetates 53
Monothiobenzoates 55

1.1.3 Rhodium(II/III) Carboxylates 56
Formates 56
Acetates 56
Methoxyacetates, $[Rh_2(O_2CCH_2OCH_3)_4]^+$ 57
Phenoxyacetates, $[Rh_2(O_2CCH_2OC_6H_5)_4]^+$ 58
Trifluoroacetates 58
Propionates 58
n-Butyrates 59

1.1.4 Rhodium(III) Carboxylates 59
Acetates 59
Chloroacetates 61
Pyridine-2,6-dicarboxylates 62
Miscellaneous Carboxylates 62
Monothioacetate, $[Rh(OSCCH_3)(NH_3)_5]I_2$ 63

1.1.5 Aminocarboxylates 63
Rhodium(I) Leucinates 63
Rhodium(II) Aminocarboxylates 63
Rhodium(III) Aminocarboxylates 64
Aminoacetates 64
α-Aminopropionates, $Rh(\alpha\text{-ala})_3$ 67
β-Aminopropionates 68
α-Aminobutyrates 68
Aspartate, $Rh(aspH)_3 \cdot 2H_2O$ 69
Glutamates 69
Complexes with Substituted Glutamic Acids 69
Prolinates 70
Cysteinate, $\{Rh(L\text{-cys})_2Cl\}_2$ 70
Oxamate, $K_3(Rh(oxam)_3] \cdot 3H_2O$ 70
Complexes with Miscellaneous Amino Acids 70

1.1.6 Aminopolycarboxylic Acid Complexes 71
Iminodiacetic Acid Complexes 71
Methyliminodiacetic Acid Complexes 73
Nitrilotriacetic Acid Complexes 73
Diaminopolycarboxylic Acid Complexes 73
Ethylenediaminetetraacetic Acid Complexes 73
N-Hydroxyethylethylenediamine-N,N′,N′-triacetic Acid Complexes 78
(S,S)-Ethylenediaminedisuccinic Acid Complexes 78
Ethylenediamine-N,N′-diacetic Acid-N,N′-Dipropionic Acid Complexes 79

Page

1.3.11 Complexes with Oximidobenzotetronic Acid 96

1.3.12 Complexes with 8-Hydroxyquinoline N-Oxide 96

1.3.13 Complexes with 1,2,3-Triazole-4,5-dicarboxylic Acid 97

1.3.14 Complex with o-Hydroxycinnamic Acid (o-Coumaric Acid) 97

1.3.15 Complexes with Trimetaphosphimate Anion 97

2 Complexes with Ligands Containing Nitrogen 99

2.1 Monodentate Nitrogen Donors 99

2.1.1 Nitrosyl Complexes 99

Nitrosyl-nitrile Complexes, $[Rh(NO)(NCR)_4]X_2$ 99

Nitrosyl Halides 99

Nitrosyl-dithiocarbamato Complexes, $[Rh(NO)(S_2CNR_2)_3]X$ 100

Nitrosyl-porphyrin Complexes 100

Nitrosyl-phosphine Complexes 101

Complexes with PF_3 101

Cationic Nitrosyl Complexes, $[Rh(NO)_2(PPh_3)_2]^{n+}$ 101

Complexes $Rh(NO)(PR_3)_3$ 102

Nitrosyl-phosphine-halide Complexes, $Rh(NO)(PR_3)_2X_n$ 104

Monohalide Complexes 104

Dihalide Complexes 104

Trihalide Complexes, $Rh(NO)(PPh_3)_2X_3$ 106

Miscellaneous Phosphine Complexes 107

$Rh(NO)(PPh_3)_2XY$ Complexes 109

Tetraazadiene and Related Complexes 110

Carboxylato Complexes, $Rh(NO)(PPh_3)_2(OCOR)_2$ 110

Complexes Derived from Quinones, $Rh(NO)(PPh_3)_2Q$ 111

Nitrosyl Nitrato, Nitrito, Sulphato and Azido Complexes 111

Nitrosyl Complexes with Polydentate Phosphines 112

Nitrosyl-phosphite Complexes 113

Nitrosyl-tertiary-arsine Complexes 113

Nitrosyl-arsine-halide Complexes 114

Nitrosyl-tertiary-stibine Complex, $Rh(NO)(SbPh_3)_3$ 115

2.1.2 Thionitrosyl Complexes 115

2.1.3 Nitro Complexes 116

Unsubstituted Complexes 116

The $[Rh(NO_2)_6]^{3-}$ Ion 116

Salts of Alkali Metals and Ammonium 117

Salts of Heavy Metals 118

Substituted Complexes 119

2.1.4 Organonitrile Complexes 119

Complexes with Acetonitrile 119

Coordination Compounds of Rhodium

1 Complexes with Ligands Containing Oxygen

1.1 Carboxylates

In this large section we consider oxalato complexes first, since oxalate is unique as a carboxylate ligand. We then discuss the very large class of compounds formed by monobasic $RCOO^-$ ligands. Most of these compounds contain the fundamental $Rh_2(O_2CR)_4$ moiety with the carboxylate ligands in a bidentate bridging rôle spanning a short Rh-Rh bond to give the so-called "lantern" structure (see p. 9). Finally we consider aminocarboxylate and amino-polycarboxylate ligands in which, usually, both N and O donor atoms are coordinated to rhodium.

General Literature:

T. R. Felthouse, The Chemistry, Structure and Metal-Metal Bonding in Compounds of Rhodium(II), Progr. Inorg. Chem. **29** [1982] 73/166.

E. B. Boyar, S. D. Robinson, Rhodium(II) Carboxylato Complexes, Platinum Metals Rev. **26** [1982] 65/9.

I. B. Baranovskii, R. N. Shchelokov, Dinuclear Complexes of the Platinum Metals Containing a Metal-Metal Bond, Zh. Neorgan. Khim. **23** [1978] 3/17; Russ. J. Inorg. Chem. **23** [1978] 1/10.

A. Dobson, S. D. Robinson, Carboxylato Complexes of the Platinum Group Metals, Platinum Metals Rev. **20** [1976] 56/63.

In addition to the above, the chemistry of rhodium carboxylates has been discussed in review articles on metal carboxylates [1, 2], polynuclear metal carboxylates [3], and inorganic trifluoroacetates [4]. The infrared spectra of rhodium carboxylates have been reviewed at length [5].

References:

[1] C. Oldham (Progr. Inorg. Chem. **10** [1968] 223/58). – [2] S. Herzog, W. Kalies (Z. Chem. [Leipzig] **8** [1968] 81/92). – [3] J. Catterick, P. Thornton (Advan. Inorg. Chem. Radiochem. **20** [1977] 291/362). – [4] C. D. Garner, B. Hughes (Advan. Inorg. Chem. Radiochem. **17** [1975] 1/47, 20). – [5] Yu. Ya. Kharitonov, T. Ya. Mazo, N. A. Knyazeva (in: Yu. Ya. Kharitonov (Kolebatel'nye Spektry Neorganicheskoi Khimii, Akad. Nauk Inst. Obshch. Neorgan. Khim., Moskva 1971, pp. 314/41).

Rhodium(I) Carboxylates. For discussion of rhodium(I) carboxylato complexes containing phosphorus donor ligands see "Rhodium" Suppl. Vol. B3.

1.1.1 Oxalato Complexes

en = ethylenediamine, phen = 1,10-phenanthroline,
EDTA = ethylenediamine-tetraacetate, py = pyridine

General Literature:

K. V. Krishnamurty, G. M. Harris, The Chemistry of Metal Oxalato Complexes, Chem. Rev. **61** [1961] 213/46.

1.1.1.1 Tris(oxalato) Complexes

Most reports concerning the **tris(oxalato)rhodate ion** $[Rh(C_2O_4)_3]^{3-}$ (see "Rhodium" 1938, pp. 77/8, 83/6, 94, 95, 98, 101, 107) refer to the potassium salt, other salts mentioned briefly are those of strychnine, sodium, thallium(I); $K^+/[Ni(phen)_3]^{2+}$, $[Co(en)_3]^{3+}$, and silver(I).

Compounds with Nonmetallic Cations

$H_3[Rh(C_2O_4)_3]$. Electrophoresis studies indicate that the $H_3[Rh(C_2O_4)_3]$ obtained by heating $Rh(OH)_3$ with aqueous oxalic acid is contaminated with aquo-oxalato ions, mainly $[Rh(C_2O_4)_2(H_2O)_2]^-$ [1, 2].

$[C_{21}H_{23}O_2N_2]_3[Rh(C_2O_4)_3]$ ($C_{21}H_{23}O_2N_2$ = strychninium; see "Rhodium" 1938, p. 94) is obtained as a yellow precipitate by addition of aqueous strychnine nitrate to an aqueous solution of $K_3[Rh(C_2O_4)_3]$ and has been used to resolve the $[Rh(C_2O_4)_3]^{3-}$ ion [3, 4].

References:

[1] S. K. Shukla (J. Less-Common Metals **1** [1959] 333/42, 334). – [2] S. K. Shukla (Ann. Chim. [Paris] [13] **6** [1961] 1383/443, 1424). – [3] L. Damrauer, R. M. Milburn (J. Am. Chem. Soc. **93** [1971] 6481/6, 6481). – [4] A. L. Odell, R. W. Olliff, F. B. Seaton (J. Chem. Soc. **1965** 2280/1).

Sodium Compounds, $Na_3[Rh(C_2O_4)_3] \cdot nH_2O$ (see "Rhodium" 1938, pp. 77/8)

This is prepared as bright orange efflorescent crystals by dissolving hydrated rhodium oxide in aqueous $NaH(C_2O_4)$ and allowing the solution to crystallise. The degree of hydration is not specified. It has been resolved using the strychnine salt and the rate of acid-catalysed racemisation has been measured (see corresponding potassium salt, below).

A. L. Odell, R. W. Olliff, F. B. Seaton (J. Chem. Soc. **1965** 2280/1).

Potassium Compounds, $K_3[Rh(C_2O_4)_3] \cdot nH_2O$ (see "Rhodium" 1938, pp. 93/5)

Electrophoresis studies show that, to obtain complete conversion of $RhCl_3 \cdot aq$ to $K_3[Rh(C_2O_4)_3]$, the chloride must be refluxed with 4 to 5 times its own weight of $K_2C_2O_4$ for at least 6 h [1]. The hydrate $K_3[Rh(C_2O_4)_3] \cdot 4.5H_2O$ is obtained as orange-red crystals by dissolving freshly precipitated $Rh(OH)_3$ in aqueous $KH(C_2O_4)$ solution and allowing slow crystallisation to occur. On standing it partially dehydrates to form $K_3[Rh(C_2O_4)_3] \cdot 2.5H_2O$ and is fully dehydrated to $K_3[Rh(C_2O_4)_3]$ on heating at 184°C [2].

Resolution of the optically active $[Rh(C_2O_4)_3]^{3-}$ ion into *d* and *l* forms, both of which were isolated as potassium salts, is achieved by using *l*-$[Ni(phen)_3]I_2$ [3] and *d* or *l*-$[Co(en)_3]I_3$. Specific rotations are given as $[\alpha]_D = 0°$ and $[\alpha]_{546.1} = \pm 74°$ for a 0.1% solution at room temperature [4]. A value of $\pm 85°$ for $[\alpha]_D$ is also recorded [3].

The X-ray crystal and molecular structures of $K_3[Rh(C_2O_4)_3] \cdot 4.5H_2O$ [5] and *l*-$K_3[Rh(C_2O_4)_3] \cdot 2H_2O$ [6] have been determined.

The crystals of $K_3[Rh(C_2O_4)_3] \cdot 4.5H_2O$ are triclinic, space group $P\bar{1}-C_i^1$; Z = 2, a = 6.825, b = 10.457, c = 12.428 Å; $\alpha = 103.70°$, $\beta = 96.17°$, $\gamma = 85.40°$; density 2.195 g/cm^3 (crystallographic) and 2.152 g/cm^3 (pycnometric). Coordination about the rhodium, see **Fig. 1**, is distorted octahedral with three chelate oxalate ligands; Rh-O distances average 2.016 Å (range 2.000 to 2.046 Å). All potassium ions are 7-coordinate to water molecules and outer oxygen atoms of oxalate ligands [5].

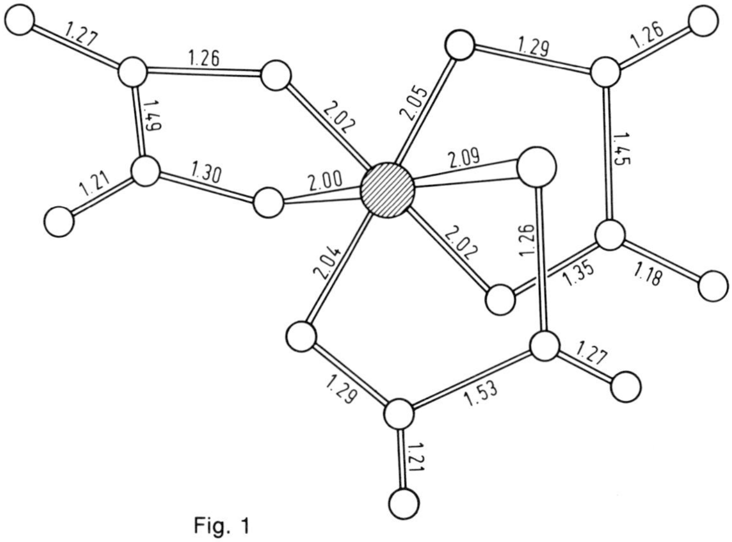

Fig. 1
Structure of tris(oxalate)rhodate anion.

The crystals of *l*-$K_3[Rh(C_2O_4)_3]\cdot 2H_2O$ are hexagonal, space group $P3_121\text{-}D_3^4$; Z = 6, a = 11.26, c = 20.06 Å; density 2.351 g/cm³ (crystallographic) and 2.345 g/cm³ (pycnometric). The compound is isomorphous with the corresponding iridium complex. Bond lengths were not reported [6].

These structure determinations appear to refute earlier suggestions (based on solid state broad line proton magnetic resonance spectra [7], thermogravimetric analysis (TGA) measurements [7, 8] and kinetic evidence [9, 10]) that some of the water of crystallisation in the complexes $K_3[Rh(C_2O_4)_3]\cdot nH_2O$ is constitutive and that the structures of $K_3[Rh(C_2O_4)_3]\cdot 4.5H_2O$ and $K_3[Rh(C_2O_4)_3]\cdot H_2O$ are more correctly represented as $K_6[Rh(C_2O_4)_3][Rh(C_2O_4)_2(HC_2O_4)(OH)]\cdot 8H_2O$ and $K_6[Rh(C_2O_4)_3][Rh(C_2O_4)_2(HC_2O_4)(OH)]\cdot H_2O$, respectively.

References:

[1] S. K. Shukla (J. Less-Common Metals **1** [1959] 333/42, 336). – [2] W. W. Wendlandt, T. D. George, K. V. Krishnamurty (J. Inorg. Nucl. Chem. **21** [1961] 69/76, 70). – [3] F. P. Dwyer, A. M. Sargeson (J. Phys. Chem. **60** [1956] 1331/2). – [4] J. W. Vaughn, V. E. Magnuson, G. J. Seiler (Inorg. Chem. **8** [1969] 1201/2). – [5] B. C. Dalzell, K. Eriks (J. Am. Chem. Soc. **93** [1971] 4298/300).

[6] P. Herpin (Bull. Soc. Franc. Mineral. Crist. **81** [1958] 201/19). – [7] A. L. Porte, H. S. Gutowsky, G. M. Harris (J. Chem. Phys. **34** [1961] 66/71). – [8]. R. D. Gillard, S. H. Laurie, P. R. Mitchell (J. Chem. Soc. A **1969** 3006/11). – [9] D. Barton, G. M. Harris (Inorg. Chem. **1** [1962] 251/5). – [10] R. H. Fenn, A. J. Graham, R. D. Gillard (Nature **213** [1967] 1012/3).

Vibrational Spectra. Numerous papers report vibrational spectra for various hydrates $K_3[Rh(C_2O_4)_3]\cdot nH_2O$. Low frequency (20 to 400 cm⁻¹) infrared spectra of microcrystalline samples reveal differences between the external vibrational frequencies of *dl*-$K_3[Rh(C_2O_4)_3]\cdot 4.5H_2O$

(49, 108, 127 and 187 cm^{-1}) and *l*-$K_3[Rh(C_2O_4)_3]\cdot 2H_2O$ (91 and 176 cm^{-1}) [1, 2]. Infrared (200 to 1800 cm^{-1}) and Raman spectra have been recorded using microcrystalline powders and aqueous solutions, respectively, for *dl*-$K_3[Rh(C_2O_4)_3]\cdot 4.5H_2O$ and *d*-$K_3[Rh(C_2O_4)_3]\cdot 2H_2O$ [3, 4]. The following assignments were made:

Compound	Frequencies in cm^{-1}			
	ν(C=O)	ν(C-O)	ν(Rh-O)	δ(O-C=O)
dl-$K_3[Rh(C_2O_4)_3]\cdot 4.5H_2O$	1700(s)	~1410(m)	550(s)	818(s)
	1670(s)	1380(s)		808(s)
	1655(s)	1260(m)		
		1233(s)		
d-$K_3[Rh(C_2O_4)_3]\cdot 2H_2O$	1700(s)	1385(s)	551(s)	818(s)
	1680(s)	~1255(sh)		806(s)
	1650(s)	1240(s)		

s = strong, m = medium, sh = shoulder

Vibrations were grouped into 3 irreducible representations (A_1, A_2 and E) of the point group D_3 [3]. Comparative infrared studies (800 to 1800 cm^{-1}) have been made for a range of tris(oxalato) complexes including $K_3[Rh(C_2O_4)_3]$ (degree of hydration not specified) [5], *dl*-$K_3[Rh(C_2O_4)_3]\cdot 4.5H_2O$ and *l*-$K_3[Rh(C_2O_4)_3]\cdot 2H_2O$ [6].

References:

[1] J. P. Mathieu (J. Raman Spectrosc. **1** [1973] 47/51, 49). – [2] J. P. Mathieu, J. Gouteron, J. Vaissermann (Compt. Rend. B **274** [1972] 880/2). – [3] J. Vaissermann (Compt. Rend. B **270** [1970] 948/51). – [4] J. Gouteron (J. Inorg. Nucl. Chem. **38** [1976] 55/61). – [5] F. Douvillé, C. Duval, J. Lecompte (Compt. Rend. **212** [1945] 697/700).

[6] J. P. Mathieu, H. Poulet (J. Chim. Phys. **59** [1962] 369/74).

Electronic Spectra. The electronic spectrum of $K_3[Rh(C_2O_4)_3]$ in aqueous potassium oxalate solution has been measured over the ranges 250 to 600 nm [1], 350 to 500 nm [2] (spectra reproduced in papers) and 300 to 600 nm [3]. The value of Δ (10 D_q) for the ion $[Rh(C_2O_4)_3]^{3-}$ is 26500 cm^{-1} [3] or 26400 cm^{-1} [4]; a value of 350 cm^{-1} was found for the parameter β [3]. The ORD (optical rotatory dispersion) spectrum of the $[Rh(C_2O_4)_3]^{3-}$ anion (potassium salt) has been recorded over the ranges 250 to 600 nm [5] and 350 to 600 nm [6] (spectra are reproduced in papers). The CD (circular dichroism) spectrum of the $[Rh(C_2O_4)_3]^{3-}$ ion has been recorded, λ_{max} = 400 nm, $(\varepsilon_l - \varepsilon_r)_{max}$ = +2.85 [7] and the spectrum has been published [8]. The effect of optically inactive counter cations (Li^+, Na^+, K^+, Rb^+, Cs^+, Mg^{2+}, Ca^{2+} and Ba^{2+}) on the ORD spectrum of the anion $[Rh(C_2O_4)_3]^{3-}$ has been studied [9]. Use of circular dichroism data to assign magnetic dipole transitions in optically active ions including $[Rh(C_2O_4)_3]^{3-}$ has been discussed [10].

X-Ray Photoelectron Spectroscopy. ESCA data (binding energies) for $K_3[Rh(C_2O_4)_3]\cdot 4.5H_2O$ are Rh $3d_{5/2}$ = 310.8, K $2p_{3/2}$ = 293.3 eV [11].

References:

[1] C. K. Jørgensen (Acta Chem. Scand. **10** [1956] 500/17, 508). – [2] K. V. Krishnamurty (Inorg. Chem. **1** [1962] 422/5). – [3] R. W. Olliff, A. L. Odell (J. Chem. Soc. **1964** 2417/21). – [4]

C. E. Schaffer, C. K. Jørgensen (J. Inorg. Nucl. Chem. **8** [1958] 143/8). – [5] M. Billardon (Compt. Rend. **251** [1960] 2320/2).

[6] M. Billardon, J. Badoz (Compt. Rend. **248** [1959] 2466/8). – [7] A. J. McCaffery, S. F. Mason, R. E. Ballard (J. Chem. Soc. **1965** 2883/92, 2884). – [8] J. Badoz, M. Billardon, J. P. Mathieu (Compt. Rend. **251** [1960] 1477/9). – [9] M. J. Albinak, D. C. Bhatnagar, S. Kirschner, A. J. Sonnessa (Can. J. Chem. **39** [1961] 2360/70, 2367). – [10] R. D. Gillard (J. Chem. Soc. **1963** 2092/4).

[11] V. I. Nefedov, M. A. Porai-Koshits (Mater. Res. Bull. **7** [1972] 1543/52, 1545).

Chemical Reactions. The hydrate $K_3[Rh(C_2O_4)_3]\cdot 4.5\,H_2O$ has been reported to effloresce at room temperature forming $K_3[Rh(C_2O_4)_3]\cdot H_2O$ [1] or $K_3[Rh(C_2O_4)_3]\cdot 2.5\,H_2O$ [2]. Thermogravimetric analysis (TGA) of the latter complex indicates stepwise loss of water to form $K_3[Rh(C_2O_4)_3]\cdot 0.5\,H_2O$ at 110°C and the anhydrous salt $K_3[Rh(C_2O_4)_3]$ at 184°C [2]. More recent TGA and EGA (evolved gas analysis) results show that $K_3[Rh(C_2O_4)_3]\cdot 4.5\,H_2O$ loses all water of crystallisation smoothly over the range 25 to 150°C and three molecules of CO_2 on heating from 225 to 260°C [3, 4, 5]. Enthalpy changes for dehydration of $K_3[Rh(C_2O_4)_3]\cdot 4.5\,H_2O$ and decomposition of $K_3[Rh(C_2O_4)_3]$ are given as 51 and 27.7 kJ/mol, respectively [4].

References:

[1] K. V. Krishnamurty (Inorg. Chem. **1** [1962] 422/5). – [2] W. W. Wendlandt, T. D. George, K. V. Krishnamurty (J. Inorg. Nucl. Chem. **21** [1961] 69/76). – [3] K. Nagase (Chem. Letters **1972** 205/6). – [4] K. Nagase (Bull. Chem. Soc. Japan **45** [1972] 2166/8). – [5] K. Nagase (Bull. Chem. Soc. Japan **46** [1973] 144/6).

Kinetic Studies. Resolution of *dl*-$K_3[Rh(C_2O_4)_3]$ has been noted above (p. 2). – Racemisation of the products in aqueous solution (pH = 7) is detectable only after a period of several months [1] but is catalysed by acid [2] and can be photolytically induced [3]. Kinetic data for the acid catalysed (1 M $HClO_4$) racemisation reaction are $E_A = 20$ kcal/mol = 83.7 kJ/mol, log A = 14.1 [2]; the quantum yield for the photochemical racemisation is given as 4.3×10^{-3} at 24°C using light of 546 nm [3].

Oxalate exchange and aquation reactions in aqueous acid solution (2 < pH < 8) are both reported to be first order in complex ion concentration. At pH > 8 the $[Rh(C_2O_4)_3]^{3-}$ ion decomposes into hydrated rhodium(III) oxide and oxalate [4]. The volume of activation for the acid (1 M) hydrolysis (aquation) of $K_3[Rh(C_2O_4)_3]$ to $K[Rh(C_2O_4)_2(H_2O)_2]$ at 60°C has been measured as -7.9 ± 0.6 cm³/mol, other activation parameters are $\Delta H^{\neq} = 82.7\pm2.1$ kJ/mol and $\Delta S^{\neq} = -90\pm6\ J\cdot K^{-1}\cdot mol^{-1}$. Plots of ln k versus pressure are linear (see original paper) [5].

A comparative kinetic study has been performed on the following processes:

a) exchange of oxalate (outer) oxygen atoms with water [6],
b) exchange of oxalate (inner) oxygen atoms with water [6],
c) racemisation [7],
d) aquation to form $[Rh(C_2O_4)_2(H_2O)_2]^-$ ion [7].

Processes (a) and (b) each follow a second order rate law of the form $R = k_2[H^+][Rh(C_2O_4)_3^{3-}]$, whereas processes (c) and (d) are governed by a two-term rate law $R = \{k_2[H^+] + k_3[H^+]^2\}$-$[Rh(C_2O_4)_3^{3-}]$.

Values of activation parameters (at ionic strength 0.54) [7]:

Reaction	Rate Constant in $mol^{-1} \cdot s^{-1}$ at 56.3°C	$\Delta H^{\neq}$ in kJ/mol	$\Delta S^{\neq}$ in $J \cdot K^{-1} \cdot mol^{-1}$
a) Outer oxygen exchange	$k_2 = 1.48 \times 10^{-3}$	70.7 ± 8	−83.7 ± 25
b) Inner oxygen exchange	$k_2 = 8.5 \times 10^{-5}$	98.0 ± 8	−26.4 ± 25
c) Racemisation	$k_2 = 4.2 \times 10^{-5}$	97.5 ± 6	−34.3 ± 17
	$k_3 = 2.4 \times 10^{-4}$	112.2 ± 6	+24.3 ± 17
d) Aquation	$k_2 = 1.67 \times 10^{-6}$	106.7 ± 8	−32.65 ± 25
	$k_3 = 1.16 \times 10^{-5}$	108.4 ± 8	−12.1 ± 25

Cerium(IV) oxidation of oxalate ligands in $K_3[Rh(C_2O_4)_3]$ under acid (1M H_2SO_4) conditions at 25°C involves an initial direct one-electron oxidation of $[Rh(C_2O_4)_3]^{3-}$ with a second-order rate constant $k_2 = (6.1 \pm 1.0) \times 10^{-4}\ M^{-1} \cdot s^{-1}$ [8].

Photolysis [9, 10] and radiolysis (γ-rays) [11] of $K_3[Rh(C_2O_4)_3] \cdot 4.5 H_2O$ at 77 K and 300 K in frozen matrices and in aqueous solutions gave products including formyl (HCO·) and hydrogen (H·) radicals which were detected by EPR and electronic spectra (700 to 328 nm).

References:

[1] M. Billardon (J. Chim. Phys. **61** [1964] 1070/5). – [2] A. L. Odell, R. W. Olliff, F. B. Seaton (J. Chem. Soc. **1965** 2280/1). – [3] S. T. Spees, A. W. Adamson (Inorg. Chem. **1** [1962] 531/9, 537). – [4] D. Barton, G. M. Harris (Inorg. Chem. **1** [1962] 251/5). – [5] D. A. Palmer, H. Kelm (J. Inorg. Nucl. Chem. **40** [1978] 1095/8).

[6] L. Damrauer, R. M. Milburn (J. Am. Chem. Soc. **90** [1968] 3884/5). – [7] L. Damrauer, R. M. Milburn (J. Am. Chem. Soc. **93** [1971] 6481/6). – [8] M. W. Hsu, H. G. Kruszyna, R. M. Milburn (Inorg. Chem. **8** [1969] 2201/7). – [9] L. A. Il'yukevich, N. I. Zotov, L. N. Neokladnova, Yu. V. Glazkov, V. V. Pansevich (Zh. Neorgan. Khim. **17** [1972] 2845/6; Russ. J. Inorg. Chem. **17** [1972] 1491/2). – [10] L. N. Neokladnova, L. A. Il'yukevich, N. I. Zotov, N. A. Reshetko (Prevrashch. Kompleks. Soedin. Deistviem Sveta Radiats. Temp. **1973** 30/8; C.A. **81** [1974] No. 31770).

[11] Yu. V. Glazkov, N. I. Zotov, L. A. Il'yukevich, L. N. Neokladnova (Khim. Vysokikh Energ. **9** [1975] 88/90; C.A. **82** [1975] No. 148395).

Heavy-Metal Compounds

en = ethylenediamine; phen = 1,10-phenanthroline

$Tl_3[Rh(C_2O_4)_3] \cdot 2H_2O$ is prepared as orange prisms by addition of Tl_2SO_4 to $K_3[Rh(C_2O_4)_3] \cdot 4.5 H_2O$ in aqueous solution. It is soluble in water and decomposes with liberation of oxalic acid in hot concentrated aqueous solution [1].

$K[Ni(phen)_3][Rh(C_2O_4)_3]$ is prepared by addition of an aqueous alcoholic solution of $[Ni(phen)_3]I_2$ to an aqueous solution of $K_3[Rh(C_2O_4)_3]$. Optical isomers of $[Ni(phen)_3]I_2$ have been used to resolve $[Rh(C_2O_4)_3]^{3-}$ [2].

$[Co(en)_3][Rh(C_2O_4)_3]$ is prepared by addition of aqueous $[Co(en)_3]I_3$ to aqueous $K_3[Rh(C_2O_4)_3]$. Optical isomers of $[Co(en)_3]I_3$ have been used as resolving agents for $[Rh(C_2O_4)_3]^{3-}$ [3].

$Ag_3[Rh(C_2O_4)_3]$ (see "Rhodium" 1938, p. 101) is prepared by addition of silver nitrate to aqueous solutions of $Na_3[Rh(C_2O_4)_3]$ and separated by filtration. It decomposes on heating to afford silver, rhodium, and CO_2 [4].

$K_6[Rh(C_2O_4)_3][Rh(C_2O_4)_2(C_2O_4H)(OH)]\cdot nH_2O$ (n = 1 or 8). These are alternative and largely discredited formulations of the salts $K_3[Rh(C_2O_4)_3]\cdot mH_2O$ (m = 1.0 or 4.5). For discussion see under $K_3[Rh(C_2O_4)_3]$, p. 3.

References:

[1] V. Riganti, T. Soldi (Ricerca. Sci. **30** [1960] 2157/8). – [2] F. P. Dwyer, A. M. Sargeson (J. Phys. Chem. **60** [1956] 1331/2). – [3] J. W. Vaughn, V. E. Magnuson, G. J. Seiler (Inorg. Chem. **8** [1969] 1201/2). – [4] L. Damrauer, R. M. Milburn (J. Am. Chem. Soc. **90** [1968] 3884/5).

1.1.1.2 Bis(oxalato) Complexes

en = ethylenediamine; phen = 1,10-phenanthroline

$K[Rh(C_2O_4)_2(H_2O)_2]$ mixtures of *cis* and *trans* forms are obtained as products in aquation studies on $K_3[Rh(C_2O_4)_3]$ [1 to 4]. *cis*-$K[Rh(C_2O_4)_2(H_2O)_2]\cdot nH_2O$ (n = 1 [6] or 1.75 [5]) is obtained by acid ($HClO_4$) aquation of $K_3[Rh(C_2O_4)_3]\cdot 1.5H_2O$ [5] or $K_3[Rh(C_2O_4)_3]\cdot H_2O$ (erroneously formulated in the paper) [6], and is separated from the less soluble *trans* isomer by fractional crystallisation [5] or ion exchange on Deacidite FF (chloride form) [6]. Alternatively the *cis* complex may be obtained by hydrolysis of $K_3[Rh(C_2O_4)_2Cl_2]$ followed by treatment of the solution with silver perchlorate and separated from the *trans* isomer by fractional crystallisation [5]. The complex forms dark orange crystals. The infrared spectrum (700 to 2000 cm^{-1}) is reproduced in one paper and the electronic spectrum is reported (λ_{max} = 394 nm) [5]. *trans*-$K[Rh(C_2O_4)_2(H_2O)_2]\cdot nH_2O$ (n = 2 [6] or 3 [5]) is prepared, and separated from the more soluble *cis* isomer, by the methods described above [5, 6]. It forms bright yellow crystals. The infrared spectrum (700 to 2000 cm^{-1}) is reproduced in one paper and the electronic spectrum is reported (λ_{max} = 387 nm) [5].

$K_3[Rh(C_2O_4)_2Cl_2]$ (see "Rhodium" 1938, p. 86). Both *cis* and *trans* isomers have been studied but have not always been assigned the correct stereochemistry. *cis*-$K_3[Rh(C_2O_4)_2Cl_2]$ is obtained as a green solid by heating a moist finely ground mixture of *cis*-$K[Rh(C_2O_4)_2(H_2O)_2]$ and KCl in the requisite proportions at 125°C [6]. An alternative preparation, based on a modification of Delépine's original procedure, gives the *trans* isomer and not, as claimed in the paper, the *cis* isomer [7]. The electronic spectrum has been measured [6, 3] (λ_{max} = 437 and 352 nm) and is reproduced in one paper [8]. *trans*-$K_3[Rh(C_2O_4)_2Cl_2]$ is obtained as a yellow solid by heating a finely ground mixture of *trans*-$K[Rh(C_2O_4)_2(H_2O)_2]$ and KCl in the requisite proportions at 125°C or by heating the *cis* isomer in aqueous solution [6]. The infrared spectrum has been recorded (700 to 4000 cm^{-1}) but the complex is erroneously formulated as the *cis* isomer in the paper [7]. The electronic spectrum has been measured [6] (λ_{max} = 465 and 400 nm) and has been reproduced in one paper [8].

References:

[1] D. Barton, G. M. Harris (Inorg. Chem. **1** [1962] 251/5). – [2] D. A. Palmer, H. Kelm (J. Inorg. Nucl. Chem. **40** [1978] 1095/8). – [3] L. Damrauer, R. M. Milburn (J. Am. Chem. Soc. **93** [1971] 6481/6). – [4] K. V. Krishnamurty (Inorg. Chem. **1** [1962] 422/5). – [5] N. S. Rowan, R. M. Milburn (Inorg. Chem. **11** [1972] 639/42).

[6] R. D. Gillard, G. Wilkinson (J. Chem. Soc. **1964** 870/3). – [7] J. P. Collman, H. F. Holtzclaw (J. Am. Chem. Soc. **80** [1958] 2054/6). – [8] C. K. Jørgensen (Acta Chem. Scand. **11** [1957] 151/65, 153).

1.1.1.3 Mono(oxalato) Complexes

For $[Rh(NH_3)_5(HC_2O_4)]^{2+}$ and $[Rh(NH_3)_5(C_2O_4)]^+$, see p. 134; for $[Rh(NH_3)_4(C_2O_4)]^+$, see p. 149.

***mer*-[Rh(C_2O_4)X(py)$_3$]** (X = Cl, Br, I) (see "Rhodium" 1938, p. 149). The **chloro complex** is prepared by heating *trans*-$[RhCl_2(py)_4]Cl$ with potassium oxalate in aqueous solution [1]; the reaction is catalysed by a trace of $NaBH_4$ [2]. The complex forms yellow needles [1, 2]. The electronic spectrum (λ_{max} = 396 nm) [1], infrared spectrum (200 to 1800 cm^{-1}) and Raman spectrum with ν = 220(s), 277(s), 317(s), 356(s) and 557(s) were recorded [2].

The **bromo complex** is similarly prepared from *trans*-$[RhBr_2(py)_4]Br$ as orange needles [1, 2]. The infrared spectrum is reported for 600 to 1800 cm^{-1} (selected bands) [1] and 190 to 1800 cm^{-1}. Raman bands are reported at 231 cm^{-1} and 199 cm^{-1} [2].

The **iodo complex** is prepared from the chloro complex by metathesis with sodium iodide in the presence of a trace of sodium borohydride. The infrared spectrum for 200 to 1800 cm^{-1} (selected bands) is recorded [2]. All three complexes are virtually insoluble in common solvents but the chloro and bromo complexes display a synergic solubility effect in water/pyridine solvent mixtures [1].

***mer*-[RhCl(C_2O_4)(4-pic)$_3$]·nH_2O** (n = 0 or 4; 4-pic = 4-picoline) are prepared by boiling the chloro complex $[RhCl_2(4\text{-pic})_4]Cl$ with potassium oxalate in aqueous ethanol (10% v/v) and crystallise on cooling. The anhydrate forms yellow platelets soluble in alcohols and chloroform [2].

***mer*-[RhCl(C_2O_4)(3,5-lut)$_3$]** (3,5-lut = 3,5-lutidine) is prepared by boiling the chloro complex $[RhCl_2(3,5\text{-lut})_4]Cl$ with potassium oxalate in aqueous ethanol (10% v/v) and crystallises on cooling. The complex forms yellow platelets soluble in alcohols and chloroform [2].

$\{(NH_3)_3Co(\mu\text{-}OH)_2[\mu^3\text{-}(C_2O_4)\text{-}Rh(NH_3)_5]Co(NH_3)_3\}[Y]_5 \cdot nH_2O$ (Y = ClO_4, n = 2; Y = Br, n = 3). The perchlorate is obtained as the dihydrate by treatment of $[Rh(HC_2O_4)(NH_3)_5][ClO_4]_2$ with $[(NH_3)_3Co(\mu\text{-}OH)_3Co(NH_3)_3][ClO_4]_3$. The cation is formulated as shown in **Fig. 2**.

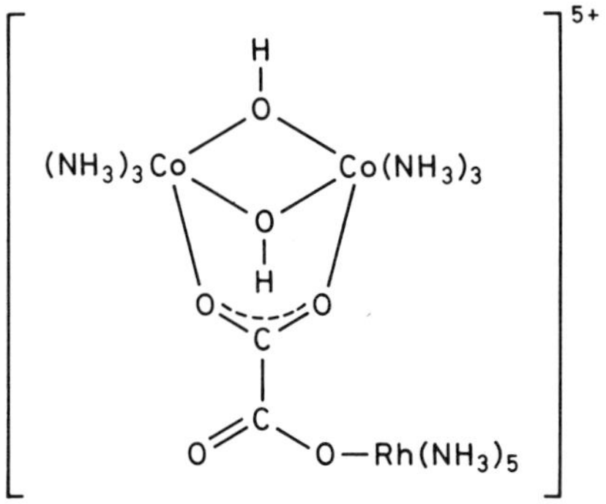

Fig. 2
Structure of $\{(NH_3)_3CO(\mu\text{-}OH)_2\text{-}\mu^3\text{-}(C_2O_4)\text{-}Rh(NH_3)_5CO(NH_3)_3\}^{5+}$ ion.

Treatment with sodium bromide affords the corresponding bromide salt as a trihydrate. Infrared bands for the latter complex are ν(C-O) = 1657(vs), 1593(vs), 1435(vw), 1306(vs) cm^{-1} [3].

References:

[1] R. D. Gillard, E. D. McKenzie, M. D. Ross (J. Inorg. Nucl. Chem. **28** [1966] 1429/34). – [2] A. W. Addison, R. D. Gillard, P. S. Sheridan, L. R. H. Tipping (J. Chem. Soc. Dalton Trans. **1974** 709/16). – [3] K. Wieghardt (Z. Naturforsch. **29b** [1974] 809/10).

1.1.1.4 Miscellaneous Oxalato Complexes

A series of ill-characterised, insoluble, polymeric amorphous oxalato complexes have been prepared by treatment of $K_3[RhCl_6]$ with neutral or weakly acid solutions containing excess hydrazine salts of oxalic acid. Products obtained are formulated as $Rh_4Cl_{4.5}(C_2O_4)_4$-$(N_2H_4)_{9.5}(N_2H_5)_{0.5}$, $Rh_4Cl_{2.5}(C_2O_4)_4(N_2H_3)_{1.5}(N_2H_4)_{10}$, and $Rh_4(C_2O_4)_5(N_2H_3)_2(N_2H_4)_7$.

L. Cambi, E. D. Paglia, G. Bargigia (Atti Accad. Nazl. Lincei Rend. Classe Sci. Fis. Mat. Nat. **30** [1961] 636/43 from C.A. **57** [1962] 6853).

1.1.2 Rhodium(II) Carboxylates

Gun = guanidine, $HN{=}C(NH_2)_2$

Rhodium(II) carboxylates, the first examples of which were reported in 1960 [1], are all binuclear and have the basic molecular formula $Rh_2(O_2CR)_4$. However, they are usually encountered as solvates, $Rh_2(O_2CR)_4(solv.)_2$, adducts $Rh_2(O_2CR)_4L_2$ (L = donor ligand), or as salts $M_2[Rh_2(O_2CR)_4X_2]$ (X = anion). They all possess the familiar "lantern" structure shown in **Fig. 3**, with a short rhodium-rhodium distance (~2.370 to 2.455 Å) indicative of a substantial metal-metal interaction. The presence of this interaction is confirmed by the diamagnetism of the complexes, but there has been much controversy concerning the Rh-Rh bond order. Rhodium-rhodium bond lengths, obtained from X-ray diffraction studies, are consistent with the presence of a Rh≡Rh triple bond. However, MO calculations favour a Rh-Rh single (σ) bond of exceptional strength, and are in good agreement with the electronic spectra [2]. The short Rh-Rh distance is attributed to mixing of higher-energy empty orbitals into the bonding scheme.

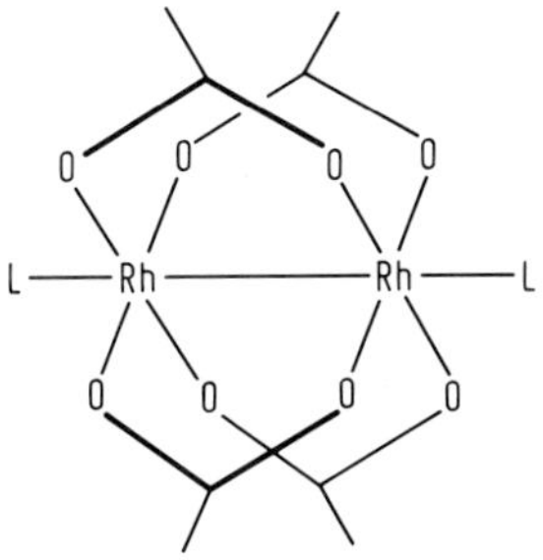

Fig. 3
"Lantern" structure of Rh^{II} carboxylates, $Rh_2(O_2CR)_4L_2$.

Several rhodium (II) carboxylates (in particular the propionate, butyrate and methoxyacetate and their solvates or adducts) show considerable potential as antitumour agents [3 to 12].

References:

[1] I. I. Chernyaev, E. V. Shenderetskaya, A. A. Karyagina (Zh. Neorgan. Khim. **5** [1960] 1163; Russ. J. Inorg. Chem. **5** [1960] 559). – [2] D. J. Cole-Hamilton (Coord. Chem. Rev. **35** [1981] 113/42, 131). – [3] J. L. Bear, H. B. Gray, L. Rainen, I. M. Chang, R. Howard, G. Serio, A. P. Kimball (Cancer Chemotherapy Rept. **59** [1975] 611/20). – [4] R. A. Howard, T. G. Spring, J. L. Bear (Cancer Res. **36** [1976] 4402/5). – [5] R. A. Howard, T. G. Spring, J. L. Bear (J. Clin. Hematol. Oncol. **7** [1977] 391/400).

[6] R. A. Howard, E. Sherwood, A. Erck, A. P. Kimball, J. L. Bear (J. Med. Chem. **20** [1977] 943/6). – [7] R. G. Hughes, J. L. Bear, A. P. Kimball (Proc. Am. Assoc. Cancer Res. **13** [1972] 120). – [8] A. Erck, L. Rainen, J. Whileyman, I. Chang, A. P. Kimball, J. L. Bear (Proc. Soc. Expt. Biol. Med. **145** [1974] 1278/83). – [9] J. L. Bear, R. A. Howard, A. M. Dennis (Current Chemotherapy [1978] 1321/3). – [10] R. A. Howard, A. P. Kimball, J. L. Bear (Cancer Res. **39** [1979] 2568/73).

[11] P. N. Rao, M. L. Smith, S. Pathak, R. A. Howard, J. L. Bear (J. Natl. Cancer Inst. **64** [1980] 905/11). – [12] I. Chang (Hanguk Saenghwa Hakhoe Chi **8** [1975] 37/43 from C. A. **84** [1976] No. 159616).

1.1.2.1 Formates

These were the first rhodium(II) carboxylates to be reported, but were initially formulated incorrectly as rhodium(I) species $H[Rh(O_2CH)_2L]$ (L = donor ligand) [1]. They were subsequently reformulated as rhodium(II) species $Rh_2(O_2CH)_4L_n$ (n = 0, 1 or 2) [2].

$Rh_2(O_2CH)_4$ is prepared by heating the dihydrate at 100°C [2]. Molecular orbital calculations have been used to interpret the electronic spectrum [3]. The infrared spectrum (4000 to 200 cm^{-1}) has been reported and assignments given [6].

$Rh_2(O_2CH)_4(H_2O)$ is obtained as dark green rhombs by heating aqueous $H_3[RhCl_6]$ with formic acid, and cooling the filtered solution. The complex is soluble in hot water [1, 4]. An X-ray diffraction study has been mentioned in reference [2] and reported at a conference [5]. The crystals are triclinic, a = 7.915, b = 9.371, c = 6.931 Å, α = 115°84′, β = 85°10′, γ = 97°10′; Z = 2. The formate-bridged binuclear units $Rh_2(O_2CH)_4$ have the familiar "lantern" structure (Fig. 3, p. 9) and are linked by further formate bridging. No other details are reported [5]. The infrared spectrum (4000 to 60 cm^{-1}) has been recorded and assignments made, ν(OCO) = 1589, 1541, 1529, 1340, 1318 cm^{-1} [6]. X-ray photoelectronic data have been reported (Rh $3d_{5/2}$ = 309.1 eV) [7].

$Rh_2(O_2CH)_4(H_2O)_2$ is obtained by treatment of the guanidine hydrochloride adduct $[GunH]_2[Rh_2(O_2CH)_4Cl_2]$ (Gun = guanidine, $HN{=}C(NH_2)_2$) with dilute sodium hydroxide and subsequent dissolution of the violet precipitate in aqueous sulphuric acid. It forms green water-soluble crystals [2]. Molecular orbital calculations for $Rh_2(O_2CH)_4(H_2O)_2$ have been used to interpret the electronic structure of rhodium(II) carboxylates [3, 16]. Infrared spectra covering the ranges 4000 to 60 cm^{-1} [6], 3600 to 400 cm^{-1} [8], and 600 to 60 cm^{-1} [9] have been reported for the normal and/or deuterated forms of the complex and assignments given [8, 9], ν(OCO) = 1600, 1544, 1533, 1343, 1319 cm^{-1} [6]. Force constants (in mdyn/Å) are K(Rh-O) = 2.66, K(Rh-Rh) = 0.77 [9].

$Rh_2(O_2CH)_4\{(NH_2)_2CO\}_2$ is obtained by addition of a saturated aqueous solution of urea to an aqueous suspension of $[GunH]_2[Rh_2(O_2CH)_4Cl_2]$ or $Na_2[Rh_2(O_2CH)_4(NO_2)_2]\cdot 4H_2O$, and deposits as blue rhombs. The complex is soluble in water and alcohol, and is a non-electrolyte, Λ^{25} = 9.65 $cm^2\cdot\Omega^{-1}\cdot mol^{-1}$ [10]. The infrared spectrum (4000 to 60 cm^{-1}) has been recorded and assignments given [6].

$Rh_2(O_2CH)_4(NH_3)_2$ is prepared by treatment of $Rh_2(O_2CH)_4(H_2O)$ with ammonia gas [1, 4] or by treatment of the salts $[GunH]_2[Rh_2(O_2CH)_4Cl_2]$ or $Na_2[Rh_2(O_2CH)_4(NO_2)_2]$ with ammonium carbonate in the presence of a little water [4]. It is a dark cherry-red solid [1, 4]. The infrared spectrum (4000 to 60 cm^{-1}) has been recorded and assignments given [6]. Molecular orbital calculations have been reported [16].

$Rh_2(O_2CH)_4(NH_2CH_3)_2$. The infrared spectrum (4000 to 60 cm^{-1}) of this adduct has been recorded and assignments made [6].

$Rh_2(O_2CH)_4(NH_2C_6H_5)_2$ is prepared by addition of aniline hydrate to an aqueous suspension of $[GunH]_2[Rh_2(O_2CH)_4Cl_2]$ and separates as a pink solid, insoluble in water and organic solvents [10].

$Rh_2(O_2CH)_4(NH_2C_{10}H_7)_2(H_2O)$. X-ray photoelectronic data have been recorded ($Rh\,3d_{5/2}$ = 309.5, N 1s = 400.0 eV) [11] but no preparative details are reported.

$Rh_2(O_2CH)_4(py)_2$ is prepared by dropwise addition of pyridine to a methanolic or aqueous solution of $Rh_2(O_2CH)_4(H_2O)_n$ (n = 1 or 2) [1, 4, 12, 13] and forms a dark pink, water-insoluble precipitate [1, 4, 13]. The infrared spectrum has been recorded (4000 to 60 cm^{-1}) and assignments given [6]. X-ray photoelectronic data have been recorded: $Rh\,3d_{5/2}$ = 309.1, N 1s = 399.7 eV [7].

$Rh_2(O_2CH)_4(quinoline)_2$ is prepared by dropwise addition of excess quinoline and alcohol to an aqueous suspension of the salt $[GunH]_2[Rh_2(O_2CH)_4Cl_2]$. The complex forms pink crystals, insoluble in water and organic solvents [10].

$Rh_2(O_2CH)_4\{o\text{-}C_6H_4(NH_2)CO_2H\}_2$ is obtained by addition of a saturated alcoholic solution of anthranilic acid ($o\text{-}C_6H_4(NH_2)CO_2H$) to an aqueous suspension of $[GunH]_2[Rh_2(O_2CH)_4Cl_2]$ and deposits as a fine violet-red crystalline solid [10]. X-ray photoelectronic data have been recorded: $Rh\,3d_{5/2}$ = 309.6, N 1s = 400.1 eV [11].

$Rh_2(O_2CH)_4(Gun)_2 \cdot xH_2O$. The anhydrous complex (x = 0) is prepared by treatment of green reduced $H_3[RhCl_6]$/formic acid solution with excess guanidinium carbonate at pH > 6 and is isolated as a dark violet water-insoluble solid [4]. The dihydrate (x = 2) is obtained as a violet solid by treatment of $Rh_2(O_2CH)_4(H_2O)_2$ with guanidine in aqueous alcohol solution, or by the action of sodium carbonate on $[GunH]_2[Rh_2(O_2CH)_4Cl_2]$ [14, 15]. The infrared spectrum (3600 to 400 cm^{-1}) of the dihydrate has been recorded and assignments given [14].

$Rh_2(O_2CH)_4(Gun \cdot CN)_2$. This violet adduct is obtained by treatment of $Rh_2(O_2CH)_4(H_2O)_2$ with cyanoguanidine in concentrated aqueous solution, and is sparingly soluble in water and alcohol. The infrared spectrum (3600 to 400 cm^{-1}) has been recorded and assignments given. Thermal decomposition data have been reported [14].

$Rh_2(O_2CH)_4(Gun \cdot NH_2)_2 \cdot H_2O$. This reddish violet adduct is prepared by treatment of $[Gun \cdot NH_2 \cdot H]_2[Rh_2(O_2CH)_4Cl_2]$ with aqueous sodium carbonate solution. The infrared spectrum (3600 to 400 cm^{-1}) has been recorded and assignments given. Thermal decomposition data have been reported [14].

References:

[1] I. I. Chernyaev, E. V. Shenderetskaya, A. A. Karyagina (Zh. Neorgan. Khim. **5** [1960] 1163; Russ. J. Inorg. Chem. **5** [1960] 559). – [2] I. I. Chernyaev, E. V. Shenderetskaya, A. G. Maiorova, A. A. Karyagina (Zh. Neorgan. Khim. **10** [1965] 537/8; Russ. J. Inorg. Chem. **10** [1965] 290/1). – [3] J. G. Norman, H. J. Kolari (J. Am. Chem. Soc. **100** [1978] 791/9). – [4] I. I. Chernyaev, E. V. Shenderetskaya, A. G. Maiorova, A. A. Karyagina (Zh. Neorgan. Khim. **11** [1966] 2575/82; Russ. J. Inorg. Chem. **11** [1966] 1383/7). – [5] A. S. Ancyškina (Acta Cryst. **21** [1966] A 135).

[6] G. Ya. Mazo, I. B. Baranovskii, R. N. Shchelokov (Zh. Neorgan. Khim. **24** [1979] 3330/6; Russ. J. Inorg. Chem. **24** [1979] 1855/7). – [7] V. I. Nefedov, Ya. V. Salyn, I. B. Baranovskii, A. G. Maiorova (Zh. Neorgan. Khim. **25** [1980] 216/25; Russ. J. Inorg. Chem. **25** [1980] 116/21). – [8] T. A. Mal'kova, V. N. Shafranskii, Yu. Ya. Kharitonov (Koord. Khim. **3** [1977] 1747/52; Soviet J. Coord. Chem. **3** [1977] 1371/6). – [9] Yu. Ya. Kharitonov, G. Ya. Mazo, N. A. Knyazeva (Zh. Neorgan. Khim. **15** [1970] 1440/1; Russ. J. Inorg. Chem. **15** [1970] 739/40). – [10] L. A. Nazarova, A. G. Maiorova (Zh. Neorgan. Khim. **21** [1976] 1070/4; Russ. J. Inorg. Chem. **21** [1976] 583/5).

[11] V. I. Nefedov, Ya. V. Salyn', A. G. Maiorova, L. A. Nazarova, I. B. Baranovskii (Zh. Neorgan. Khim. **19** [1974] 1353/7; Russ. J. Inorg. Chem. **19** [1974] 736/8). – [12] T. A. Stephenson, S. M. Morehouse, A. R. Powell, J. P. Heffer, G. Wilkinson (J. Chem. Soc. **1965** 3632/40,

3640). – [13] L. A. Nazarova, A. G. Maiorova (Zh. Neorgan. Khim. **18** [1973] 1710/2; Russ. J. Inorg. Chem. **18** [1973] 904/5). – [14] T. A. Veteva, V. N. Shafranskii (Zh. Obshch. Khim. **49** [1979] 488/93; J. Gen. Chem. [USSR] **49** [1979] 428/33). – [15] R. N. Shchelokov, A. G. Maiorova, O. M. Evstaf'eva, G. N. Emel'yanova (Zh. Neorgan. Khim. **22** [1977] 1414/6; Russ. J. Inorg. Chem. **22** [1977] 770/1).

[16] H. Nakatsuji, J. Ushio, K. Kanda, Y. Onishi, T. Kawamura, T. Yonezawa (Chem. Phys. Letters **79** [1981] 299/304).

$Na_2[Rh_2(O_2CH)_4(NO_2)_2]\cdot 4H_2O$. This salt is prepared by treatment of $Rh_2(O_2CH)_4(H_2O)$ with sodium nitrite in the solid state (moist slurry) or aqueous solution [1, 2]. The salt forms acicular dark orange crystals. The infrared spectrum (4000 to 60 cm^{-1}) has been recorded and assignments given [3]. X-ray photoelectronic data have been reported: Rh $3d_{5/2}$ = 309.5, N 1s = 403.5 eV [4].

$K_2[Rh_2(O_2CH)_4(NCO)_2]\cdot 2H_2O$. This salt is prepared as fine violet crystals by thoroughly grinding together $Rh_2(O_2CH)_4(H_2O)_2$ and potassium cyanate in water or a water/ethanol mixture. The infrared spectrum (3600 to 400 cm^{-1}) is reproduced in the paper and assignments are given; the cyanate ligand is coordinated through nitrogen. Thermal decomposition has been investigated [5].

$K[Rh_2(O_2CH)_4(NCS)]\cdot 2H_2O$. This salt is similarly prepared using potassium thiocyanate and is isolated as fine violet crystals. The infrared spectrum (3600 to 400 cm^{-1}) is reproduced and assignments are given; the thiocyanate ligand bridges between $Rh_2(O_2CH)_4$ units. Thermal decomposition has been investigated [5].

$K[Rh_2(O_2CH)_4(NCSe)]\cdot 2H_2O$. This salt is similarly prepared using KSeCN and is isolated as fine violet crystals. The infrared spectrum (3600 to 400 cm^{-1}) is reproduced and assignments are given, the selenocyanate ligand bridges between $Rh_2(O_2CH)_4$ units. Thermal decomposition has been investigated [5].

$[Gun H]_2[Rh_2(O_2CH)_4Cl_2]$. This salt, which is also formulated as $Rh_2(O_2CH)_4\cdot 2Gun\cdot HCl$, is obtained by addition of guanidinium carbonate or chloride to the green solution obtained by reduction of H_3RhCl_6 with formic acid, and is isolated from solution under acidic conditions (pH = 2 to 5) as a dark green, water-soluble precipitate [2]. The infrared spectrum (3600 to 400 cm^{-1}) has been recorded and assignments given [6].

$[Gun\cdot NH_2\cdot H]_2[Rh_2(O_2CH)_4Cl_2]$. This green crystalline salt is obtained by treatment of $Rh_2(O_2CH)_4(H_2O)_2$ with amino guanidine hydrochloride in alcoholic solution. It is very soluble in water. The infrared spectrum (3600 to 400 cm^{-1}) has been recorded and assignments given. Thermal decomposition data have been reported [6].

$Na_2[Rh_2(O_2CH)_4Cl_2]\cdot xH_2O$ (x = 0, 2 or 6). The salt is prepared as the **hexahydrate** by treatment of the guanidinium salt $[Gun H]_2[Rh_2(O_2CH)_4Cl_2]$ with aqueous alcoholic sodium chloride solution, and deposits as well-formed green platelets. The **tetrahydrate**, $Na_2[Rh_2(O_2CH)_4Cl_2]\cdot 4H_2O$ is obtained by acidification of $Na_2[Rh(O_2CH)_4(NO_2)_2]\cdot 4H_2O$ with hydrochloric acid. On heating the hydrates to 100°C the **anhydrous salt** $Na_2[Rh_2(O_2CH)_4Cl_2]$ is formed [2].

$Cs_2[Rh_2(O_2CH)_4Cl_2]$. No preparative details reported, but the infrared spectrum (4000 to 60 cm^{-1}) has been recorded and assignments given [3]. X-ray photoelectronic data have been reported: Rh $3d_{5/2}$ = 309.3, Cl 2p = 198.1 eV [4, 7].

$[Gun H]_2[Rh_2(O_2CH)_4Br_2]$. This salt, which has also been formulated as $Rh_2(O_2CH)_4\cdot 2Gun\cdot HBr$, is obtained by treatment of the corresponding hydrochloride with neat hydrogen bromide, and is isolated as bright green crystals. The salt is readily soluble in water, its conductance is reported to be 298 $cm^2 \cdot \Omega^{-1} \cdot mol^{-1}$ [2]. X-ray photoelectronic data have been recorded: $Rh\,3d_{5/2} = 309.2$, $Br\,3d = 69.0$ or 68.9 eV [4, 7].

$(NH_4)_2[Rh_2(O_2CH)_4Br_2]$. This salt, which has also been formulated as $Rh_2(O_2CH)_4 \cdot 2NH_4Br$, is obtained by treatment of $[GunH]_2[Rh_2(O_2CH)_4Br_2]$ with saturated aqueous ammonium bromide solution. It is isolated as a water-soluble dark green compound [2].

$Cs_2[Rh_2(O_2CH)_4Br_2]$. No preparative details have been recorded for this salt. However, the infrared spectrum (4000 to 60 cm^{-1}) has been recorded and assignments given [3]. X-ray photoelectronic data have also been reported: $Rh\,3d_{5/2} = 309.4$, $Br\,3d = 68.9$ eV [4, 7].

$Cs_2[Rh_2(O_2CH)_4Br(NO_2)]$. No preparative details have been recorded for this salt. However, X-ray photoelectronic data have been reported: $Rh\,3d_{5/2} = 309.7$, $N\,1s = 403.5$ eV [7].

$K_2[Rh_2(O_2CH)_4I_2] \cdot H_2O$. The infrared spectrum (4000 to 60 cm^{-1}) of this salt has been recorded and assignments given [3].

$K_2[Rh_2(O_2CH)_4(HSO_3)_2] \cdot 4H_2O$ is obtained as a red precipitate by treatment of $Rh_2(O_2CH)_4$ with saturated $K_2S_2O_5$ solution [8].

References:

[1] I. I. Chernyaev, E. V. Shenderetskaya, A. A. Karyagina (Zh. Neorgan. Khim. **5** [1960] 1163; Russ. J. Inorg. Chem. **5** [1960] 559). – [2] I. I. Chernyaev, E. V. Shenderetskaya, A. G. Maiorova, A. A. Karyagina (Zh. Neorgan. Khim. **11** [1966] 2575/82; Russ. J. Inorg. Chem. **11** [1966] 1383/7). – [3] G. Ya. Mazo, I. B. Baranovskii, R. N. Shchelokov (Zh. Neorgan. Khim. **24** [1979] 3330/6; Russ. J. Inorg. Chem. **24** [1979] 1855/7). – [4] V. I. Nefedov, Ya. V. Salyn', I. B. Baranovskii, A. G. Maiorova (Zh. Neorgan. Khim. **25** [1980] 216/25; Russ. J. Inorg. Chem. **25** [1980] 116/21). – [5] V. N. Shafranskii, T. A. Mal'kova, Yu. Ya. Kharitonov (Koord. Khim. **1** [1975] 375/83; Soviet J. Coord. Chem. **1** [1975] 297/303).

[6] T. A. Veteva, V. N. Shafranskii (Zh. Obshch. Khim. **49** [1979] 488/93; J. Gen. Chem. [USSR] **49** [1979] 428/33). – [7] V. I. Nefedov, Ya. V. Salyn', A. G. Maiorova, L. A. Nazarova, I. B. Baranovskii (Zh. Neorgan. Khim. **19** [1974] 1353/7; Russ. J. Inorg. Chem. **19** [1974] 736/8). – [8] I. B. Baranovskii, S. S. Abdullaev, G. Ya. Mazo, R. N. Shchelokov (Zh. Neorgan. Khim. **27** [1982] 536/8; Russ. J. Inorg. Chem. **27** [1982] 305/6).

$Rh_2(O_2CH)_4\{SC(NH_2)_2\}_2$. The infrared spectrum (4000 to 60 cm^{-1}) of this adduct has been recorded and assignments made but no preparative details are recorded [1]. X-ray photolelectronic data have been reported: $Rh\,3d_{5/2} = 309.2$, $S\,2p = 162.4$ eV [2].

$Rh_2(O_2CH)_4(CH_3CSNH_2)_2$. This adduct is prepared by addition of excess aqueous thioacetamide solution to an aqueous suspension of $[GunH]_2[Rh_2(O_2CH)_4Cl_2]$, and deposits as a red-violet solid [3].

$Rh_2(O_2CH)_4(NH_2CS \cdot NH \cdot NH_2)_2$. This adduct is prepared by addition of an aqueous solution of thiosemicarbazide hydrochloride ($NH_2CSNH \cdot NH_2 \cdot HCl$) to $[GunH]_2[Rh_2(O_2CH)_4Cl_2]$ or $Na_2[Rh_2(O_2CH)_4(NO_2)_2]$, and is isolated as an insoluble violet solid [3]. X-ray photoelectronic data have been recorded, $Rh\,3d_{5/2} = 309.8$, $C\,1s = 289.0$, $S\,2p_{3/2} = 163.1$ eV [4].

$Rh_2(O_2CH)_4(Me_2SO)(H_2O)_2$ (incorrectly formulated as $Rh_2(O_2CH)_2(Me_2SO)(H_2O)_2$ in paper). This adduct is obtained as green crystals by addition of dimethyl sulphoxide to an aqueous solution of $Rh_2(O_2CH)_4(H_2O)_2$ [5].

$Rh_2(O_2CH)_4(Me_2SO)_{1.5}$. This adduct precipitates as olive-green needles on addition of dimethyl sulphoxide to a concentrated alcoholic solution of $Rh_2(O_2CH)_4(H_2O)_2$. The infrared spectrum (2000 to 400 cm^{-1}) is reproduced and assignments given [5].

$Rh_2(O_2CH)_4(Me_2SO)_2$. This brown adduct is formed from $Rh_2(O_2CH)_4(Me_2SO)(H_2O)_2$ on standing. The infrared spectrum is reproduced (2000 to 400 cm^{-1}), recorded (4000 to 200 cm^{-1}) and assignments given [5]. Thermal properties have been reported [5] and X-ray photoelectronic data have been recorded, $Rh\,3d_{5/2} = 309.1$, $N\,1s = 399.7$ eV [2].

$Rh_2(O_2CH)_4(Me_2SO)_3$. This light brown adduct is prepared by grinding the diaquo complex with dimethyl sulphoxide. The infrared spectrum (2000 to 400 cm^{-1}) is reproduced and assignments given [5].

$Rh_2(O_2CH)_4(Et_2SO)_2$. This adduct is precipitated from an alcoholic solution of $[Rh_2(O_2CH)_4(H_2O)_2]$ by addition of diethyl sulphoxide. The infrared spectrum (2000 to 400 cm^{-1}) is reproduced and assignments given. Thermal properties have been reported [5].

$Rh_2(O_2CH)_4(PhCOSH)_2$. This adduct is prepared by treatment of $Rh_2(O_2CH)_4(H_2O)$ with thiobenzoic acid at room temperature, and is isolated as a reddish brown solid [6].

References:

[1] G. Ya. Mazo, I. B. Baranovskii, R. N. Shchelokov (Zh. Neorgan. Khim. **24** [1979] 3330/6; Russ. J. Inorg. Chem. **24** [1979] 1855/7). – [2] V. I. Nefedov, Ya. V. Salyn', I. B. Baranovskii, A. G. Maiorova (Zh. Neorgan. Khim. **25** [1980] 216/25; Russ. J. Inorg. Chem. **25** [1980] 116/21). – [3] L. A. Nazarova, A. G. Maiorova (Zh. Neorgan. Khim. **21** [1976] 1070/4; Russ. J. Inorg. Chem. **21** [1976] 583/5). – [4] V. I. Nefedov, Ya. V. Salyn', A. G. Maiorova, L. A. Nazarova, I. B. Baranovskii (Zh. Neorgan. Khim. **19** [1974] 1353/7; Russ. J. Inorg. Chem. **19** [1974] 736/8). – [5] T. A. Mal'kova, V. N. Shafranskii (Zh. Obshch. Khim. **47** [1977] 2592/6; J. Gen. Chem. [USSR] **47** [1977] 2365/9).

[6] I. B. Baranovskii, M. A. Golubnichaya, G. Ya. Mazo, R. N. Shchelokov (Koord. Khim. **1** [1975] 1573; Soviet J. Coord. Chem. **1** [1975] 1299).

$Rh_2(O_2CH)_4(PH_3)_2$. This compound has been used as a model for molecular orbital calculations on the electronic structure of rhodium(II) carboxylate-phosphine adducts [1, 2].

$Rh_2(O_2CH)_4(PPh_3)_2$. This adduct is prepared by addition of triphenylphosphine in diethyl ether to a cold methanolic solution of $Rh_2(O_2CH)_4(H_2O)$, and separates as orange crystals [3].

$K[Rh_2(O_2CH)_4(CN)(H_2O)]$. This violet adduct is formed by vigorously stirring $Rh_2(O_2CH)_4(Me_2SO)_2$ with potassium cyanide in aqueous alcohol for 24 h. The infrared spectrum (3500 to 500 cm^{-1}) is reproduced and assignments given, $\nu(CN) = 2138$ cm^{-1}. Cyanide bridges between $Rh_2(O_2CH)_4$ units are postulated [4].

$M_2[Rh_2(O_2CH)_2(O_2CO)_2]$ (M = K or Cs). These mixed formate/carbonate complexes have been mentioned as products isolated from the reaction of rhodium(II) formates with the appropriate metal carbonates. No experimental details are given [5].

$Rh_2(O_2CH)_2(o\text{-phen})_2Cl_2$. This complex is prepared by the treatment of $RhCl_2$(2-methylallyl) with formic acid and o-phenanthroline ($C_{12}H_8N_2$) in ethanol at 100°C [6, 7]. It forms deep violet-red crystals, soluble in water, slightly soluble in methanol but insoluble in other polar solvents. The X-ray crystal structure has been reported but few details given. The molecular structure is shown in **Fig. 4**. Bond lengths are as follows: Rh-Rh = 2.576 Å; Rh-O = 2.084, 2.085, 2.083 and 2.067 Å; Rh-N = 2.014, 2.034, 2.027 and 2.004 Å; Rh-Cl = 2.508 and 2.504 Å [7]. The infrared spectrum (4000 to 100 cm^{-1}) has been recorded [6, 7] and assignments made: ν(Rh-O) = 445, 433, 417, 397 cm^{-1}; ν(Rh-N) = 277, 260 cm^{-1} [6]. The electronic spectrum has been reported [7]. In aqueous solution the complex is a 1:2 electrolyte (10^{-4} M solution, $\Lambda = 215$ $cm^2 \cdot \Omega^{-1} \cdot mol^{-1}$) [7].

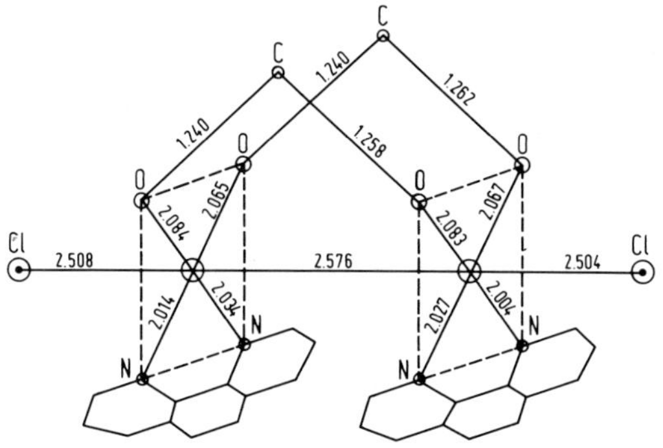

Fig. 4
Molecular structure of
$[Rh_2(O_2CH)_2(o\text{-phen})_2]Cl_2$.

$[Rh_2(O_2CH)_2(o\text{-phen})_2]Cl_2$. Adducts with Donor Ligands (L = PPh_3, $AsPh_3$, $^nBu_2S$, MeOH, py or H_2O). These adducts have been studied in methanol solution and their electronic spectra have been recorded [8].

$[Rh_2(O_2CH)_2(o\text{-phen})_2]Br_2$. This complex is prepared by treatment of $Rh_2Br_2(allyl)_4$ with anhydrous formic acid and o-phenanthroline in degassed ethanol at 100°C. The product is described as dark brown and its electronic spectrum has been reported [8]. Selected infrared bands in the region 4000 to 100 cm^{-1} are reported and assignments given [6].

$[Rh_2(O_2CH)_2(bipy)_2]Cl_2$. This complex is prepared from $Rh_2Br_2(allyl)_4$, anhydrous formic acid and 2,2'-bipyridyl at 100°C [6]. Selected infrared bands in the region 4000 to 100 cm^{-1} are reported and assignments given [6], the electronic spectrum has been recorded [8].

References:

[1] B. E. Bursten, F. A. Cotton (Inorg. Chem. **20** [1981] 3042/8). – [2] H. Nakatsuji, J. Ushio, K. Kanda, Y. Onishi, T. Kawamura, T. Yonezawa (Chem. Phys. Letters **79** [1981] 299/304). – [3] T. A. Stephenson, S. M. Morehouse, A. R. Powell, J. P. Heffer, G. Wilkinson (J. Chem. Soc. **1965** 3632/40, 3640). – [4] T. A. Mal'kova, V. N. Shafranskii, Yu. Ya. Kharitonov (Koord. Khim. **3** [1977] 1747/52; Soviet J. Coord. Chem. **3** [1977] 1371/6). – [5] R. N. Shchelokov, A. G. Maiorova, O. M. Evstaf'eva, G. N. Emel'yanova (Zh. Neorgan. Khim. **22** [1977] 1414/6; Russ. J. Inorg. Chem. **22** [1977] 770/1).

[6] F. Pruchnik, M. Zuber, H. Pasternak, K. Wajda (Spectrochim. Acta A **34** [1978] 1111/2). – [7] H. Pasternak, F. Pruchnik (Inorg. Nucl. Chem. Letters **12** [1976] 591/8). – [8] F. Pruchnik, M. Zuber (Roczniki Chem. **51** [1977] 1813/9).

1.1.2.2 Acetates

Rhodium(II) acetates have been particularly thoroughly studied and an extremely wide range of adducts has been reported.

Unsubstituted Acetates

$Rh_2(O_2CCH_3)_4$. The anhydrous complex is prepared by heating the dihydrate in vacuo at 80°C [1], by heating the methanol adduct in vacuo at 45°C [2, 3] or by heating the acetone adduct. The complex forms emerald-green crystals [2, 3] and is moderately soluble in many polar solvents including water, ethanol, acetone, acetic acid, acetonitrile, dimethyl sulphoxide, nitromethane, tetrahydrofuran and triethyl phosphate. With many of these solvents weak solvates are formed [4].

The infrared spectrum (4000 to 60 cm^{-1}) has been recorded and assignments given, ν(OCO) = 1584 and 1443 cm^{-1} [5]. A full normal coordinate analysis of all the vibrations of $Rh_2(O_2CCH_3)_4$ and $Rh_2(O_2CCD_3)_4$ has been reported and many assignments made [13]. The Raman active vibration ν(Rh-Rh) occurs at 351 cm^{-1} [6]. The electronic spectra of "$Rh_2(O_2CCH_3)_4$" in various polar solvents (C_2H_5OH, $(CH_3)_2CO$, THF, CH_3CO_2H, $(CH_3)_2SO$, CH_3CN) and aqueous salt solutions (NaX with X = Cl, Br, I, CNS, NO_2) have been measured. However, the species in solution are presumably solvates $Rh_2(O_2CCH_3)_4(solv.)_2$ and salts $Na_2[Rh_2(O_2CCH_3)_4X_2]$, respectively [4]. X-ray photoelectronic data have been recorded: Rh $3d_{5/2}$ = 308.5, Rh $3p_{1/2}$ = 522.7 eV and the X-ray photoelectronic spectrum between 280 and 320 eV has been reproduced [7].

The reaction of anhydrous $Rh_2(O_2CCH_3)_4$ with trifluoroacetic acid has been studied by NMR and mass spectroscopy, evidence for stepwise exchange of acetate and trifluoroacetate ligands was obtained but mixed ligand intermediates were not isolated [1]. Protonation of solutions of $Rh_2(O_2CCH_3)_4$ with aqueous fluoroboric acid followed by treatment with carbon monoxide at atmospheric pressure and 75°C affords $Rh_6(CO)_{16}$ in 85% yield [8]. Protonation of $Rh_2(O_2CCH_3)_4$ by non-complexing acids [$HClO_4$, HBF_4] affords cationic species originally formulated as "Rh_2^{4+}" [2, 9] but subsequently identified as $[Rh_2(O_2CCH_3)_3-(aq)]^+$ and $[Rh_2(O_2CCH_3)_2(aq)]^{2+}$ [10, 11]. Electronic spectra of the species $[Rh_2(O_2CCH_3)_3(aq)]^+$ and $[Rh_2(O_2CCH_3)_2(aq)]^{2+}$ have been recorded [10]. Treatment of $Rh_2(O_2CCH_3)_4$ with aqueous 2 M Na_2CO_3 solution at 100°C affords the carbonato complex anion $[Rh_2(CO_3)_4]^{4-}$ which can be isolated as a sodium, potassium or caesium salt [12]; see also "Rhodium" Suppl. Vol. B1, 1982, p. 183.

Treatment of $Rh_2(O_2CCH_3)_4$ with concentrated phosphoric acid at 150°C for a period of several hours affords dark green rhombic crystals of $Rh_2(H_2PO_4)_4(H_2O)_2$ [14] which also possess the familiar binuclear "lantern" structure [15].

References:

[1] J. L. Bear, J. Kitchens, M. R. Willcott (J. Inorg. Nucl. Chem. **33** [1971] 3479/86). – [2] P. Legzdins, R. W. Mitchell, G. L. Rempel, J. D. Ruddick, G. Wilkinson (J. Chem. Soc. A **1970** 3322/6). – [3] G. L. Rempel, P. Legzdins, H. Smith, G. Wilkinson (Inorg. Syn. **13** [1972] 90/1). – [4] S. A. Johnson, H. R. Hunt, H. M. Neumann (Inorg. Chem. **2** [1963] 960/2). – [5] G. Ya. Mazo, I. B. Baranovskii, R. N. Shchelokov (Zh. Neorgan. Khim. **24** [1979] 3330/6; Russ. J. Inorg. Chem. **24** [1979] 1855/7).

[6] A. P. Ketteringham, C. Oldham (J. Chem. Soc. Dalton Trans. **1973** 1067/70). – [7] A. M. Dennis, R. A. Howard, K. M. Kadish, J. L. Bear, J. Brace, N. Winograd (Inorg. Chim. Acta **44** [1980] L139/L141). – [8] B. R. James, G. L. Rempel, W. K. Teo (Inorg. Syn. **16** [1976] 49/51). – [9] P. Legzdins, G. L. Rempel, G. Wilkinson (Chem. Commun. **1969** 825). – [10] C. R. Wilson, H. Taube (Inorg. Chem. **14** [1975] 2276/9).

[11] T. J. Pinnavaia, R. Raythatha, J. G.-S. Lee, L. J. Halloran, J. F. Hoffman (J. Am. Chem. Soc. **101** [1979] 6891/7). – [12] C. R. Wilson, H. Taube (Inorg. Chem. **14** [1975] 405/9). – [13] I. K. Kireeva, G. Ya. Mazo, R. N. Shchelokov (Zh. Neorgan. Khim. **24** [1979] 396/407; Russ. J. Inorg. Chem. **24** [1979] 220/5). – [14] I. B. Baranovskii, S. S. Abdullaev, R. N. Shchelokov (Zh. Neorgan. Khim. **24** [1979] 3149; Russ. J. Inorg. Chem. **24** [1979] 1753). – [15] L. M. Dikareva, G. G. Sadikov, M. A. Porai-Koshits, I. B. Baranovskii, S. S. Abdullaev (Zh. Neorgan. Khim. **25** [1980] 875; Russ. J. Inorg. Chem. **25** [1980] 488/9).

$Rh_2(O_2CCH_3)_4(H_2O)_2$. This dihydrate is prepared by boiling a solution of $H_3[RhCl_6]$ or $(NH_4)_3[RhCl_6]$ in a mixture of acetic acid, water and ethanol. It separates from the solution as well-formed, green prismatic crystals on cooling [1, 2]. The same product is obtained by dropwise addition of water to finely divided $Rh_2(O_2CCH_3)_4$ [3, 4].

Two X-ray diffraction studies have been described. A preliminary report described the crystals as monoclinic, space group C2/c-C_{2h}^6 with a = 13.05, b = 8.60, c = 13.76 Å, β = 118° and Z = 4. The rhodium-rhodium bond length was given as 2.45 Å [5]. A more recent report describes the crystals as monoclinic, space group C2/c-C_{2h}^6 with a = 13.287, b = 8.608, c = 14.042 Å, β = 117°14′; Z = 4; D_{exp} = 2.23, D_{calc} = 2.24 g/cm³. The molecule has the "lantern" structure with axial water ligands, the rhodium-rhodium bond length is 2.3855(5) Å, the average Rh-O(acetate) distance is 2.039(8) Å and the Rh-O(water) distance is 2.310(3) Å [6, 7]. The infrared spectra of the dihydrate, 4000 to 60 cm^{-1} [8], 3600 to 500 cm^{-1} [9] and its deuterated analogue 4000 to 60 cm^{-1} [8] have been reported and assignments given: ν(OCO) = 1584, 1450 cm^{-1} [8]. The low frequency (600 to 60 cm^{-1}) infrared and Raman spectra of $Rh_2(O_2CCH_3)_4(H_2O)_2$ [10, 11] and $Rh_2(O_2CCD_3)_4(H_2O)_2$ [10] have been recorded. The frequency of ν(Rh-Rh) has been assigned as 320 cm^{-1} [11] and 155 cm^{-1} [10]. The force constant for the Rh-Rh bond is calculated to be 0.67 mdyn/Å [10]. There are numerous reports on the electronic spectrum of $Rh_2(O_2CCH_3)_4(H_2O)_2$; λ_{max} = 585, 440 nm [12, 13], ~600, ~450, 250, 218 nm [14], 587, 447 nm [3], 584, 441 nm [4], and 587, 447 nm [15]. The visible and near ultraviolet spectra of single crystals have been measured in the range 10 to 295 K [16]. Single-crystal polarised absorption spectra (645 to 333 nm) have been reported for the (101) face of $Rh_2(O_2CCH_3)_4(H_2O)_2$ at 15 and 300 K and are reproduced in the paper [17]. The electronic spectrum measured in aqueous or ethanolic solution is pressure dependent; shifts of 2 nm/1000 atm to higher energies are recorded for the band near 600 nm [18]. The electronic structure of $Rh_2(O_2CCH_3)_4(H_2O)_2$ has been discussed and assignments given for the electronic spectrum [19, 20]. X-ray photoelectronic data have been recorded, Rh $3d_{5/2}$ = 309.2 eV [21, 22]; the thermogravimetric analysis curve has been reproduced [23]. The complex is diamagnetic, μ_{eff} = 0.52 B.M. [23].

References:

[1] L. A. Nazarova, I. I. Chernyaev, A. S. Morozova (Zh. Neorgan. Khim. **10** [1965] 539/41; Russ. J. Inorg. Chem. **10** [1965] 291/2). – [2] L. A. Nazarova, I. I. Chernyaev, A. S. Morozova (Zh. Neorgan. Khim. **11** [1966] 2583/6; Russ. J. Inorg. Chem. **11** [1966] 1387/9). – [3] S. A. Johnson, H. R. Hunt, H. M. Neumann (Inorg. Chem. **2** [1963] 960/2). – [4] J. Kitchens, J. L. Bear (J. Inorg. Nucl. Chem. **31** [1969] 2415/21). – [5] M. A. Porai-Koshits, A. S. Antsyshkina (Dokl. Akad. Nauk SSSR **146** [1962] 1102/5; Proc. Acad. Sci. USSR Chem. Sect. **142/147** [1962] 902/5).

[6] F. A. Cotton, B. G. De Boer, M. D. LaPrade, J. R. Pipal, D. A. Ucko (J. Am. Chem. Soc. **92** [1970] 2926/7). – [7] F. A. Cotton, B. G. De Boer, M. D. LaPrade, J. R. Pipal, D. A. Ucko (Acta Cryst. B **27** [1971] 1664/71). – [8] G. Ya. Mazo, I. B. Baranovskii, R. N. Shchelokov (Zh. Neorgan. Khim. **24** [1979] 3330/6; Russ. J. Inorg. Chem. **24** [1979] 1855/7). – [9] T. A. Mal'kova, V. N. Shafranskii, Yu. Ya. Kharitonov (Koord. Khim. **3** [1977] 1747/52; Soviet J. Coord. Chem. **3** [1977] 1371/6). – [10] Yu. Ya. Kharitonov, G. Ya. Mazo, N. A. Knyazeva (Zh. Neorgan. Khim. **15** [1970] 1440/1; Russ. J. Inorg. Chem. **15** [1970] 739/40).

[11] A. P. Ketteringham, C. Oldham (J. Chem. Soc. Dalton Trans. **1973** 1067/70). – [12] C. R. Wilson, H. Taube (Inorg. Chem. **14** [1975] 405/9). – [13] C. R. Wilson, H. Taube (Inorg. Chem. **14** [1975] 2276/9). – [14] L. Dubicki, R. L. Martin (Inorg. Chem. **9** [1970] 673/5). – [15] J. J. Ziolkowski (Bull. Acad. Polon. Sci. Ser. Sci. Chim. **21** [1973] 125/9).

[16] G. Bienek, W. Tuszynski, G. Gliemann (Z. Naturforsch. **33b** [1978] 1095/8). – [17] D. S. Martin, T. R. Webb, G. A. Robbins, P. E. Fanwick (Inorg. Chem. **18** [1979] 475/8).– [18] E. Sinn

(Inorg. Nucl. Chem. Letters **11** [1975] 665/8). – [19] J. G. Norman, H. J. Kolari (J. Am. Chem. Soc. **100** [1978] 791/9). – [20] J. G. Norman, G. E. Renzoni, D. A. Case (J. Am. Chem. Soc. **101** [1979] 5256/67).

[21] V. I. Nefedov, Ya. V. Salyn', G. A. Maiorova, L. A. Nazarova, I. B. Baranovskii (Zh. Neorgan. Khim. **19** [1974] 1353/7; Russ. J. Inorg. Chem. **19** [1974] 736/8). – [22] V. I. Nefedov, Ya. V. Salyn', I. B. Baranovskii, A. G. Maiorova (Zh. Neorgan. Khim. **25** [1980] 216/25; Russ. J. Inorg. Chem. **25** [1980] 116/22). – [23] V. I. Belova, Z. S. Dergacheva (Zh. Neorgan. Khim. **16** [1971] 3065/70; Russ. J. Inorg. Chem. **16** [1971] 1626/9).

$Rh_2(O_2CCH_3)_4(CH_3OH)_2$. This adduct is prepared by refluxing a mixture of rhodium trichloride trihydrate, sodium acetate trihydrate and glacial acetic acid in ethanol for 1 h, then collecting the crystals which deposit on cooling and recrystallising them from methanol. The adduct forms blue-green crystals [1, 2]. The Raman spectrum has been reproduced (4000 to 0 cm^{-1}) [3], the value of ν(Rh-Rh) has been given as 170 cm^{-1} [3] and as 336 cm^{-1} [4]. The adduct reacts with hydrogen halides (HCl, HBr) to afford rhodium metal and rhodium(III) halides [5, 6].

$Rh_2(O_2CCH_3)_4(C_2H_5OH)_2$. This green adduct is prepared by heating a mixture of rhodium trichloride trihydrate, sodium acetate trihydrate and glacial acetic acid in absolute ethanol. The infrared spectrum (4000 to 700 cm^{-1}) has been recorded and selected assignments given [7].

$Rh_2(O_2CCH_3)_4\{(CH_3)_2CO\}_2$. This adduct is prepared by refluxing a suspension of $Rh(OH)_3 \cdot H_2O$ in glacial acetic acid for 18 h, evaporating the solution to dryness, extracting the residue with acetone, and cooling the concentrated acetone extract. The adduct forms dark green crystals which readily lose acetone on standing at room temperature [8].

$Rh_2(O_2CCH_3)_4(THF)_2$. This adduct is prepared by adding drops of tetrahydrofuran ($THF = C_4H_8O$) to finely powdered $[Rh_2(O_2CCH_3)_4]$ and allowing the excess to evaporate. It forms a green solid which loses THF on heating to 120°C. The electronic spectrum has been reported, $\lambda_{max} = 597$ and 441 nm in THF solution [8].

$Rh_2(O_2CCH_3)_4(CH_3CO_2H)_2$. This adduct is prepared by dropwise addition of acetic acid to finely powdered $Rh_2(O_2CCH_3)_4$. It is green, and slowly loses acetic acid even at room temperature. The electronic spectrum has been recorded, $\lambda_{max} = 592$ and 442 nm in acetic acid solution [8].

$NaH[Rh_2(O_2CCH_3)_4(O_2CCH_3)_2]$ is obtained in solution by treatment of $[Rh_2(H_2O)_8(H_2O)_2]^{4+}$ in aqueous 3 M $HClO_4$ (or 2 M H_2SO_4) solution with acetic acid and sodium acetate. The electronic spectrum has been reported: $\lambda_{max} = 587$, 446 nm [9].

References:

[1] P. Legzdins, R. W. Mitchell, G. L. Rempel, J. D. Ruddick, G. Wilkinson (J. Chem. Soc. A **1970** 3322/6). – [2] G. A. Rempel, P. Legzdins, H. Smith, G. Wilkinson (Inorg. Syn. **13** [1972] 90/1). – [3] J. San Filippo, H. J. Sniadoch (Inorg. Chem. **12** [1973] 2326/33, 2331). – [4] A. P. Ketteringham, C. Oldham (J. Chem. Soc. Dalton Trans. **1973** 1067/70). – [5] H. D. Glicksman, A. D. Hamer, T. J. Smith, R. A. Walton (Inorg. Chem. **15** [1976] 2205/9).

[6] H. D. Glicksman, R. A. Walton (Inorg. Chim. Acta **33** [1979] 255/9). – [7] G. Winkhaus, P. Ziegler (Z. Anorg. Allgem. Chem. **350** [1967] 51/61). – [8] S. A. Johnson, H. R. Hunt, H. M. Neumann (Inorg. Chem. **2** [1963] 960/2). – [9] J. J. Ziółkowski (Bull. Acad. Polon. Sci. Ser. Sci. Chim. **21** [1973] 119/24; 124/9).

$Rh_2(O_2CCH_3)_4(HCONH_2)_2$. This adduct is prepared by addition of formamide to a concentrated aqueous solution of $Rh_2(O_2CCH_3)_4(H_2O)_2$. It forms minute light violet crystals which are insoluble in water, alcohol, acetone or chloroform. The infrared spectrum (3600 to 500 cm^{-1}) and thermogravimetric analysis curves are reproduced in the paper. Coordination of the formamide ligand is thought to occur through the oxygen atom [1].

$Rh_2(O_2CCH_3)_4\{HCON(CH_3)_2\}_2$ is obtained by allowing dimethylformamide to react with $Rh_2(O_2CCH_3)_4(H_2O)_2$ over a period of several days. It forms dark blue prismatic crystals. The infrared spectrum (3600 to 500 cm^{-1}) and thermogravimetric analysis curves are reproduced in the paper. Coordination of the dimethylformamide ligands is thought to occur through oxygen [1].

$Rh_2(O_2CCH_3)_4(CH_3CONH_2)_2 \cdot 2H_2O$ is prepared by grinding together $Rh_2(O_2CCH_3)_4(H_2O)_2$, acetamide and a little water, then allowing the mixture to stand for 2 to 3 d [1, 2]. It forms a light green crystalline mass. The infrared spectrum (3600 to 500 cm^{-1}) and thermogravimetric analysis curves are reproduced in the paper. The acetamide ligands are thought to coordinate through oxygen [1].

$Rh_2(O_2CCH_3)_4\{CH_3CON(CH_3)_2\}_2 \cdot 2H_2O$ is prepared by allowing a concentrated solution of $Rh_2(O_2CCH_3)_4(H_2O)_2$ in dimethylacetamide to stand for 3 d. It forms green crystals sparingly soluble in water and ethanol. The infrared spectrum (3600 to 500 cm^{-1}) is reproduced in the paper. The dimethylacetamide ligands are thought to coordinate through oxygen [1].

$Rh_2(O_2CCH_3)_4\{(NH_2)_2CO\}_2$ is prepared by treatment of $[GunH]_2[Rh_2(O_2CCH_3)_4Cl_2]$ with urea in aqueous solution. It separates as a blue precipitate and is sparingly soluble in water and alcohol [3]. The infrared spectrum (4000 to 60 cm^{-1}) has been recorded and assignments given [4].

$Rh_2(O_2CCH_3)_4\{(NH_2)_2CO\}_2 \cdot 2H_2O$ is prepared by addition of excess urea to an aqueous solution of $Rh_2(O_2CCH_3)_4(H_2O)_2$. It separates slowly from solution as minute green crystals which are readily soluble in acetone, methanol and chloroform [1]. The infrared spectrum (3600 to 500 cm^{-1}) and thermogravimetric curves are reproduced in the paper [1]. The **mono-aquo complex** is diamagnetic, $\mu_{eff} = 0.51$ B.M. [6].

$Rh_2(O_2CCH_3)_4(C_6H_5CONH_2)_2$ is prepared by dropwise addition of an alcoholic solution of benzamide to an aqueous solution of $Rh_2(O_2CCH_3)_4(H_2O)_2$. It separates as light violet octahedra which are moderately soluble in water, ethanol and acetone. The infrared spectrum (3600 to 500 cm^{-1}) and thermogravimetric analysis curves are reproduced in the paper [1]. The benzamide ligand is thought to coordinate through oxygen [1, 5]. Infrared assignments have been given. The complex decomposes between 170 and 300°C [5].

$Rh_2(O_2CCH_3)_4\{(C_6H_5NHNH)_2CO\}_2$ is prepared by addition of concentrated aqueous solution of $Rh_2(O_2CCH_3)_4(H_2O)_2$ to an acetone solution of 1,5-diphenylcarbonohydrazide, $(C_6H_5NHNH)_2CO$. It forms long violet needles, which are insoluble in water but dissolve in methanol, ethanol, acetone and chloroform [1]. The infrared spectrum (3600 to 500 cm^{-1}) is reported [1].

$Rh_2(O_2CCH_3)_4(NH_2NHCONH_2)_2$ is prepared by neutralising an aqueous solution of semicarbazide hydrochloride with potassium hydroxide, and adding the mixture to an aqueous solution of $Rh_2(O_2CCH_3)_4(H_2O)_2$. The adduct slowly deposits from solution as minute red prisms which are soluble in water and methanol but insoluble in acetone, ether and chloroform. The infrared spectrum (3600 to 500 cm^{-1}) is reported [1].

$Rh_2(O_2CCH_3)_4(o\text{-}NH_2C_6H_4CO_2H)_2$ is prepared by the addition of a saturated alcoholic solution of anthranilic acid ($o\text{-}NH_2C_6H_4CO_2H$) to an aqueous solution of $[GunH]_2[Rh_2(O_2CCH_3)_4Cl_2]$, and separates as red lamellar crystals [3].

References:

[1] V. N. Shafranskii, T. A. Mal'kova (Zh. Obshch. Khim. **45** [1975] 1065/9; J. Gen. Chem. [USSR] **45** [1975] 1051/4). – [2] L. A. Nazarova, I. I. Chernyaev, A. S. Morozova (Zh. Neorgan. Khim. **10** [1965] 539/41; Russ. J. Inorg. Chem. **10** [1965] 291/2). – [3] L. A. Nazarova, A. G. Maiorova (Zh. Neorgan. Khim. **21** [1976] 1070/4; Russ. J. Inorg. Chem. **21** [1976] 583/5). – [4] G. Ya. Mazo, I. B. Baranovskii, R. N. Shchelokov (Zh. Neorgan. Khim. **24** [1979] 3330/6; Russ. J. Inorg. Chem. **24** [1979] 1855/7). – [5] T. A. Mal'kova, V. N. Shafranskii (Zh. Neorgan. Khim. **19** [1974] 2501/5; Russ. J. Inorg. Chem. **19** [1974] 1366/8).

[6] V. I. Belova, Z. S. Dergacheva (Zh. Neorgan. Khim. **16** [1971] 3065/70; Russ. J. Inorg. Chem. **16** [1971] 1626/9).

$Rh_2(O_2CCH_3)_4(NH_3)_2$ is prepared by passing ammonia gas over finely divided $Rh_2(O_2CCH_3)_4$ [1, 2], or by the action of ammonia on $Rh_2(O_2CCH_3)_4(H_2O)_2$ [3]. It forms crimson crystals which lose ammonia on heating to 120°C [1]. The electronic spectrum has been recorded, $\lambda_{max}=528$ and 442 nm [2], reproduced (320 to 200 nm) and interpreted [4]. X-ray photoelectronic data have been reported: Rh $3d_{5/2}=309.2$, N $1s=400.0$ eV [5]. Thermogravimetric analysis (TGA) results have been recorded [2] and the TGA curve is reproduced [6]. The complex is diamagnetic, $\mu_{eff}=0.51$ B.M. [6] and is a non-electrolyte ($\Lambda^{25}=28.34$ $cm^2 \cdot \Omega^{-1} \cdot mol^{-1}$) [7].

$Rh_2(O_2CCH_3)_4(N_2H_4)(H_2O)$ is prepared by treatment of $Rh_2(O_2CCH_3)_4(H_2O)_2$ with hydrazine hydrate and is isolated as red crystals which are sparingly soluble in water [7].

$Rh_2(O_2CCH_3)_4(NH_2CH_3)_2$ is prepared by passing dry methylamine vapour over powdered anhydrous $Rh_2(O_2CCH_3)_4$. It is a rose-red solid. The electronic spectrum has been recorded, $\lambda_{max}=519$ nm (shoulder). Thermogravimetric data are reported [2].

$Rh_2(O_2CCH_3)_4(NH_2C_2H_5)_2$ is prepared by adding a few drops of ethylamine to powdered anhydrous $Rh_2(O_2CCH_3)_4$. It is a rose-red solid. The electronic spectrum has been recorded, $\lambda_{max}=523$ nm (shoulder) [2].

$Rh_2(O_2CCH_3)_4\{NH(CH_3)_2\}_2$ is prepared by passing dry dimethylamine vapour over powdered anhydrous $Rh_2(O_2CCH_3)_4$. It is a rose-red solid. The electronic spectrum (750 to 350 nm) is reproduced. The endothermic heat of dissociation has been measured, $\Delta H=110.8$ kJ/mol. Thermogravimetric data are recorded and the thermogram curve is reproduced [2].

$Rh_2(O_2CCH_3)_4\{NH(C_2H_5)_2\}_2$ is prepared by adding a few drops of diethylamine to powdered anhydrous $Rh_2(O_2CCH_3)_4$, then removing the excess in vacuo. It is a rose-red solid [2].

The crystals are orthorhombic, space group Pbcn-D_{2h}^{14}, a=16.329(4), b=8.011(4), c=17.660(6) Å, Z=4; $D_{exp}=1.60$ g/cm³, $D_{calc}=1.691$ g/cm³. The molecule has the familiar "lantern" structure with a Rh-Rh bond length of 2.402(0) Å [8, 9]. Other bond lengths are Rh-O=2.046(3), 2.031(3), 2.034(3) and 2.038(3) Å, Rh-N=2.301 Å [8].

The electronic spectrum has been recorded, $\lambda_{max}=530$ nm (shoulder). The endothermic heat of dissociation of the adduct is $\Delta H=138.4$ kJ/mol [2].

$Rh_2(O_2CCH_3)_4\{N(CH_3)_3\}_2$ is prepared by treatment of powdered anhydrous $Rh_2(O_2CCH_3)_4$ with liquid [1] or gaseous [2] trimethylamine. It is described as a pink [1] or rose-red [2] solid. The electronic spectrum has been recorded, $\lambda_{max}=521$ nm (shoulder); the endothermic heat of dissociation is $\Delta H=103.2$ kJ/mol [2].

$Rh_2(O_2CCH_3)_4\{N(C_2H_5)_3\}_2$ is prepared by treatment of powdered anhydrous $Rh_2(O_2CCH_3)_4$ with triethylamine. It is described as a purple solid. The electronic spectrum has been recorded, $\lambda_{max}=540$, 458 nm; the endothermic heat of dissociation is $\Delta H=113.1$ kJ/mol [2].

$Rh_2(O_2CCH_3)_4(en)_2$. This adduct has been reported [1] but does not appear to have been fully characterised.

$Rh_2(O_2CCH_3)_4(en)(H_2O)$ is prepared by treatment of $Rh_2(O_2CCH_3)_4$ or $Rh_2(O_2CCH_3)_4(Gun)_2$ with ethylenediamine hydrate [7]. It is red in colour [7] and diamagnetic, $\mu_{eff} = 0.49$ B.M. [6]. An anhydrous form $Rh_2(O_2CCH_3)_4(en)$ is also mentioned, $\mu_{eff} = 0.50$ B.M. [6].

References:

[1] S. A. Johnson, H. R. Hunt, H. M. Neumann (Inorg. Chem. **2** [1963] 960/2). – [2] J. Kitchens, J. L. Bear (J. Inorg. Nucl. Chem. **31** [1969] 2415/21). – [3] L. A. Nazarova, I. I. Chernyaev, A. S. Morozova (Zh. Neorgan. Khim. **10** [1965] 539/41; Russ. J. Inorg. Chem. **10** [1965] 291/2). – [4] L. Dubicki, R. L. Martin (Inorg. Chem. **9** [1970] 673/5). – [5] V. I. Nefedov, Ya. V. Salyn', A. G. Maiorova, L. A. Nazarova, I. B. Baranovskii (Zh. Neorgan. Khim. **19** [1974] 1353/7; Russ. J. Inorg. Chem. **19** [1974] 736/8).

[6] V. I. Belova, Z. S. Dergacheva (Zh. Neorgan. Khim. **16** [1971] 3065/70; Russ. J. Inorg. Chem. **16** [1971] 1626/9). – [7] L. A. Nazarova, I. I. Chernyaev, A. S. Morozova (Zh. Neorgan. Khim. **11** [1966] 2583/6; Russ. J. Inorg. Chem. **11** [1966] 1387/9). – [8] Y. B. Koh, G. G. Christoph (Inorg. Chem. **18** [1979] 1122/8). – [9] G. G. Christoph, Y. B. Koh (J. Am. Chem. Soc. **101** [1979] 1422/34).

$Rh_2(O_2CCH_3)_4(C_6H_5NH_2)_2$ is prepared by treatment of $Rh_2(O_2CCH_3)_4(H_2O)_2$ [1] or $[GunH]_2$-$[Rh_2(O_2CCH_3)_4Cl_2]$ [2] with aniline in aqueous ethanol. It forms violet-red platelets which are sparingly soluble in water and organic solvents [2]. The infrared spectrum (4000 to 500 cm^{-1}) has been recorded and assignments given. A thermogravimetric analysis curve has been published [1].

$Rh_2(O_2CCH_3)_4(p\text{-}H_2NC_6H_4NO_2)_2$ is similarly prepared using p-nitroaniline. It is a violet solid, almost insoluble in water but soluble in many organic solvents. The infrared spectrum (4000 to 500 cm^{-1}) has been recorded and assignments given [1]; for thermal stability data see original publication.

$Rh_2(O_2CCH_3)_4(p\text{-}H_2NC_6H_4Cl)_2$ is prepared by dropwise addition of an aqueous solution of $Rh_2(O_2CCH_3)_4(H_2O)_2$ to a well-stirred alcoholic solution of p-chloroaniline. It is a violet solid. The infrared spectrum has been recorded (4000 to 500 cm^{-1}) and assignments given. The electronic spectrum (350 to 200 nm) is reproduced. The complex is almost insoluble in water but soluble in many organic solvents [1]; for thermal stability data see original publication.

$Rh_2(O_2CCH_3)_4(p\text{-}H_2NC_6H_4Br)_2$ is similarly prepared using p-bromoaniline. It is a violet solid, almost insoluble in water but soluble in many organic solvents. The infrared spectrum (4000 to 500 cm^{-1}) has been recorded and assignments are given. Thermal stability data are recorded [1].

$Rh_2(O_2CCH_3)_4(p\text{-}H_2NC_6H_4I)_2$ is similarly prepared using p-iodoaniline. It is a violet solid, almost insoluble in water but soluble in many organic solvents. The infrared spectrum (4000 to 500 cm^{-1}) has been recorded and assignments given [1]; for thermal stability data see original publication.

$Rh_2(O_2CCH_3)_4(p\text{-}H_2NC_6H_4CH_3)_2$ is similarly prepared using p-toluidine. It is a violet solid, almost insoluble in water but soluble in many organic solvents. The infrared spectrum (4000 to 500 cm^{-1}) has been recorded and assignments given. The thermal stability has been investigated and a thermogravimetric curve is reproduced [1].

$Rh_2(O_2CCH_3)_4(p\text{-}H_2NC_6H_4OCH_3)_2$ is similarly prepared using p-anisidine. It is a violet solid, almost insoluble in water but soluble in many organic solvents. The infrared spectrum (4000 to 500 cm^{-1}) has been recorded and assignments given [1]; for thermal stability data see original paper.

$Rh_2(O_2CCH_3)_4(p\text{-}H_2NC_6H_4OC_2H_5)_2$ is similarly prepared using p-phenitidine ($p\text{-}H_2NC_6H_4\text{-}OC_2H_5$). It is a violet solid, almost insoluble in water but soluble in many organic solvents. The infrared spectrum (4000 to 500 cm^{-1}) has been recorded and assignments given [1]; for thermal stability data see original paper.

References:

[1] T. A. Mal'kova, V. N. Shafranskii (Zh. Neorgan. Khim. **20** [1975] 1308/13; Russ. J. Inorg. Chem. **20** [1975] 735/8). – [2] L. A. Nazarova, A. G. Maiorova (Zh. Neorgan. Khim. **21** [1976] 1070/4; Russ. J. Inorg. Chem. **21** [1976] 583/5).

$Rh_2(O_2CCH_3)_4(C_5H_5N)_2$ is prepared by the addition of pyridine to a concentrated ethanolic solution of $Rh_2(O_2CCH_3)_4$ [1] or by treatment of $Rh_2(O_2CCH_3)_4(H_2O)_2$ [2, 3], or $Rh_2(O_2CCH_3)_4$ [4] with neat pyridine. It is an extremely insoluble rose-red solid [1].

The crystals are monoclinic, space group $C2/c\text{-}C_{2h}^6$ with a = 9.923(3), b = 17.009(6), c = 12.539(3) Å, β is variously given as 106.60°, 96.60° [5] and 83.40° [6]; D_{exp} = 1.85(1) g/cm³, D_{calc} = 1.896 g/cm³, Z = 4. The molecule has the familiar lantern structure (Fig. 3, p. 9) and the Rh-Rh bond length is given as 2.3963 (uncorrected) [5, 6] and 2.3994 Å (corrected) [5]. Other bond lengths are Rh-O = 2.035(2), 2.038(2), 2.042(2), 2.040(2) Å, Rh-N = 2.231(3), 2.223(2) Å [5]. The infrared spectrum (2000 to 500 cm^{-1}) has been recorded and assignments given [7, 8]. The electronic spectrum has been recorded, λ_{max} = 514 nm (shoulder) [4] and has been interpreted [9].

ESCA data have been recorded: Rh $3d_{5/2}$ = 308.5, Rh $3d_{3/2}$ = 313.3, N 1s = 399.9 eV [10]; Rh $3d_{5/2}$ = 309.2, N 1s = 399.8 eV [11]; N 1s = 399.8 eV [12]; Rh $3d_{5/2}$ = 309.2, N 1s = 399.8 eV [13]. The endothermic heat of dissociation has been given as 153.4 ± 4 kJ/mol [4]. The thermal stability has been investigated [7] and thermogravimetric curves are reproduced [14, 15]. The complex is diamagnetic, μ_{eff} = 0.50 B.M. [14].

$Rh_2(O_2CCH_3)_4(2\text{-}CH_3C_5H_4N)_2$ is prepared by addition of excess 2-picoline ($2\text{-}CH_3 \cdot C_5H_4N$) to an aqueous solution of $Rh_2(O_2CCH_3)_4(H_2O)_2$. It forms bright pink crystals, practically insoluble in water but soluble in chloroform and methanol. The infrared spectrum (1600 to 600 cm^{-1}) has been recorded and assignments given [7]; for thermal stability see original paper.

$Rh_2(O_2CCH_3)_4(3\text{-}CH_3C_5H_4N)_2$ is similarly prepared using 3-picoline ($3\text{-}CH_3C_5H_4N$) and forms bright pink crystals, practically insoluble in water but soluble in chloroform and methanol. The infrared spectrum (1600 to 600 cm^{-1}) has been recorded and assignments given [7]; for thermal stability data see original publication.

$Rh_2(O_2CCH_3)_4(4\text{-}CH_3C_5H_4N)_2$ is similarly prepared using 4-picoline ($4\text{-}CH_3C_5H_4N$). It forms bright pink crystals, practically insoluble in water but soluble in chloroform and methanol. The infrared spectrum (1600 to 600 cm^{-1}) is reported and assignments given [7]; for thermal stability data see original publication.

$Rh_2(O_2CCH_3)_4\{2,6\text{-}(CH_3)_2C_5H_3N\}_2$ is similarly prepared using 2,6-lutidine ($2,6\text{-}(CH_3)_2C_5H_3N$). It forms bright pink crystals, practically insoluble in water but soluble in chloroform and methanol. The infrared spectrum (1600 to 600 cm^{-1}) is reported and assignments given [7]; for thermal stability data see original publication.

$Rh_2(O_2CCH_3)_4\{2,4,6\text{-}(CH_3)_3C_5H_2N\}_2$ is similarly prepared using neat 2,4,6-collidine (2,4,6-$(CH_3)_3C_5H_2N$) as solvent. It forms bright pink crystals, insoluble in water but soluble in chloroform and methanol. The infrared spectrum (1600 to 600 cm^{-1}) has been recorded and assignments given [7]; for thermal stability data see original publication.

$Rh_2(O_2CCH_3)_4(C_9H_7N)_2$ is prepared by addition of quinoline to a solution of $Rh_2(O_2CCH_3)_4$-$(H_2O)_2$ in acetic acid [7] or by treatment of $[GunH]_2[Rh_2(O_2CCH_3)_4Cl_2]$ with quinoline in the minimum volume of water [16]. It forms bright pink crystals, practically insoluble in water but soluble in chloroform and methanol. The infrared spectrum (1600 to 600 cm^{-1}) has been recorded and assignments given [7]; for thermal stability data see original paper.

$Rh_2(O_2CCH_3)_4(C_{13}H_9N)_2$ is prepared by addition of acridine ($C_{13}H_9N$) to an alcoholic solution of $Rh_2(O_2CCH_3)_4(H_2O)_2$. It forms green crystals, insoluble in water but soluble in chloroform and methanol. The infrared spectrum (1600 to 600 cm^{-1}) has been recorded and assignments given [7]; for thermal stability data see original paper.

References:

[1] S. A. Johnson, H. R. Hunt, H. M. Neumann (Inorg. Chem. **2** [1963] 960/2). – [2] L. A. Nazarova, I. I. Chernyaev, A. S. Morozova (Zh. Neorgan. Khim. **10** [1965] 539/41; Russ. J. Inorg. Chem. **10** [1965] 291/2). – [3] T. A. Stephenson, S. M. Morehouse, A. R. Powell, J. P. Heffer, G. Wilkinson (J. Chem. Soc. **1965** 3632/40). – [4] J. Kitchens, J. L. Bear (J. Inorg. Nucl. Chem. **31** [1969] 2415/21). – [5] Y. B. Koh, G. G. Christoph (Inorg. Chem. **17** [1978] 2590/6).

[6] G. G. Christoph, Y. B. Koh (J. Am. Chem. Soc. **101** [1979] 1422/34). – [7] T. A. Mal'kova, V. N. Shafranskii (Zh. Obshch. Khim. **45** [1975] 631/5; J. Gen. Chem. [USSR] **45** [1975] 618/21). – [8] G. Ya. Mazo, I. B. Baranovskii, R. N. Shchelokov (Zh. Neorgan. Khim. **24** [1979] 3330/6; Russ. J. Inorg. Chem. **24** [1979] 1855/7). – [9] L. Dubicki, R. L. Martin (Inorg. Chem. **9** [1970] 673/5). – [10] A. D. Hamer, D. G. Tisley, R. A. Walton (J. Chem. Soc. Dalton Trans. **1973** 116/20).

[11] V. I. Nefedov, Ya. V. Salyn', I. B. Baranovskii, A. G. Maiorova (Zh. Neorgan. Khim. **25** [1980] 216/25; Russ. J. Inorg. Chem. **25** [1980] 116/22). – [12] V. I. Nefedov, Ya. V. Salyn', A. V. Shtemenko, A. S. Kotelnikova (Inorg. Chim. Acta **45** [1980] L49/L50). – [13] V. I. Nefedov, Ya. V. Salyn', A. G. Maiorova, L. A. Nazarova, I. B. Baranovskii (Zh. Neorgan. Khim. **19** [1974] 1353/7; Russ. J. Inorg. Chem. **19** [1974] 736/8). – [14] V. I. Belova, Z. S. Dergacheva (Zh. Neorgan. Khim. **16** [1971] 3065/70; Russ. J. Inorg. Chem. **16** [1971] 1626/9). – [15] R. A. Howard, A. M. Wynne, J. L. Bear, W. W. Wendlandt (J. Inorg. Nucl. Chem. **38** [1976] 1015/8).

[16] L. A. Nazarova, A. G. Maiorova (Zh. Neorgan. Khim. **21** [1976] 1070/4; Russ. J. Inorg. Chem. **21** [1976] 583/5).

Remark. In the following five adducts coordination of the nitrogen bases is thought to occur through the nitrogen atom in the heterocycle.

$Rh_2(O_2CCH_3)_4(2\text{-}NH_2\cdot C_5H_4N)_2$ is prepared from $Rh_2(O_2CCH_3)_4(H_2O)_2$ and 2-aminopyridine in aqueous solution. It forms pale pink crystals. The infrared spectrum (3600 to 500 cm^{-1}) has been recorded and assignments given. Thermal stability data are recorded [1].

$Rh_2(O_2CCH_3)_4(3\text{-}NH_2\cdot C_5H_4N)_2$ is similarly prepared using 3-aminopyridine, and is isolated as pale pink crystals. The infrared spectrum (3600 to 500 cm^{-1}) has been recorded and assignments given. Thermal stability data are recorded [1].

$Rh_2(O_2CCH_3)_4(4\text{-}NH_2\cdot C_5H_4N)_2$ is similarly prepared using 4-aminopyridine, and is isolated as pale pink crystals. The infrared spectrum (3600 to 500 cm^{-1}) has been recorded and assignments given. Thermal stability data are recorded [1].

$Rh_2(O_2CCH_3)_4(3\text{-}NH_2CO\cdot C_5H_4N)_2$ is similarly prepared using 3-nicotinamide and is isolated as pale pink crystals. The infrared spectrum (3600 to 500 cm^{-1}) is recorded and assignments given. Thermal stability data are recorded [1].

$Rh_2(O_2CCH_3)_4\{3\text{-}(C_2H_5)_2NCO\cdot C_5H_4N\}_2$ is similarly prepared using 3-diethylnicotinamide and is isolated as pale pink crystals. The infrared spectrum (3600 to 500 cm^{-1}) is recorded and assignments given. Thermal stability data are recorded [1].

$Rh_2(O_2CCH_3)_4(4\text{-}CN\cdot C_5H_4N)_n$ (n = 1, 2). The mono adduct (n = 1) is prepared by heating the bis adduct (n = 2) in vacuo at 150°C for 30 min. No information concerning the synthesis of the bis adduct is given. X-ray photoelectronic data for the mono adduct are reported: Rh $3d_{5/2}$ = 308.2, Rh $3p_{1/2}$ = 522.3, O 1s = 531.4 eV [2].

References:

[1] T. A. Mal'kova, V. N. Shafranskii (Zh. Neorgan. Khim. **19** [1974] 2501/5; Russ. J. Inorg. Chem. **19** [1974] 1366/8). – [2] A. M. Dennis, R. A. Howard, K. M. Kadish, J. L. Bear, J. Brace, N. Winograd (Inorg. Chim. Acta **44** [1980] L139/L141).

$Rh_2(O_2CCH_3)_4(o\text{-}C_{12}H_8N_2)$. A product of this apparent stoichiometry is isolated as an olive-green solid on treatment of $Rh_2(O_2CCH_3)_4$ with phenanthroline in acetone at 0°C [1].

$Rh_2(O_2CCH_3)_4(CH_3CN)_2$ is prepared by placing a few drops of methyl cyanide on some finely powdered $Rh_2(O_2CCH_3)_4$ and removing the excess under vacuum. It is a violet solid. The electronic spectrum has been recorded, λ_{max} = 552 and 447 nm (CH_3CN solution) [1].

$Rh_2(O_2CCH_3)_4(NO)_2$ is prepared by passage of nitric oxide over finely divided $Rh_2(O_2CCH_3)_4$ until constant weight is attained, and is isolated as a brown solid; ν(NO) = 1800, 1710 cm^{-1} [1].

$[Rh_2(O_2CCH_3)_4(HO\cdot N{=}CH{-}CH{=}N\cdot OH)]_n$ is obtained by treatment of $Rh_2(O_2CCH_3)_4$ with glyoxime, and isolated as a red solid. A polymeric structure with $Rh_2(O_2CCH_3)_4$ "lantern" units bridged by glyoxime molecules coordinated through nitrogen is proposed.

$[Rh_2(O_2CCH_3)_4\{HON{=}C(C_4H_3O){-}C(C_4H_3O){=}N\cdot OH\}]_n$ is similarly obtained using α-furildioxime, and is isolated as a red solid. A polymeric structure analogous to that described above is proposed [2].

$Rh_2(O_2CCH_3)_4(Gun\cdot NH_2)(H_2O)$ is prepared by treatment of $[Gun\cdot NH_2\cdot H]_2[Rh_2(O_2CCH_3)_4Cl_2]$ ($Gun\cdot NH_2$ is aminoguanidine) with Na_2CO_3 and is isolated as a reddish-violet solid which is sparingly soluble in water and alcohol. The infrared spectrum (3600 to 300 cm^{-1}) has been recorded and assignments made [3].

$Rh_2(O_2CCH_3)_4(Gun\cdot CN)(H_2O)$ is prepared by addition of cyanoguanidine to a concentrated aqueous solution of $Rh_2(O_2CCH_3)_4(H_2O)_2$, and is isolated as a violet solid. The infrared spectrum (3600 to 300 cm^{-1}) has been recorded and assignments made. The thermogravimetric analysis (TGA) curve is reproduced [3].

References:

[1] S. A. Johnson, H. R. Hunt, H. M. Neumann (Inorg. Chem. **2** [1963] 960/2). – [2] H. J. Keller, K. Seibold (Z. Naturforsch. **25b** [1970] 552/4). – [3] T. A. Veteva, V. N. Shafranskii (Zh. Obshch. Khim. **49** [1979] 488/93; J. Gen. Chem. [USSR] **49** [1979] 428/33).

$Rh_2(O_2CCH_3)_4$(adenine)(H_2O) is obtained by mixing aqueous solutions of $Rh_2(O_2CCH_3)_4$-$(H_2O)_2$ and adenine ($C_5H_4N_5$), and separates as pink crystals. NMR data indicate that the adenine coordinates through the N(7) site [1, 8].

$Rh_2(O_2CCH_3)_4$(9-methyladenine) is obtained by mixing aqueous solutions of $Rh_2(O_2CCH_3)_4$-$(CH_3OH)_2$ and 9-methyladenine ($C_6H_6N_5$), and separates as a pink solid. The infrared spectrum (3600 to 600 cm^{-1}) is reported and assignments given. The ^1H NMR spectrum has been recorded [2].

$Rh_2(O_2CCH_3)_4$(adenosine) is obtained by addition of adenosine to an aqueous solution of $Rh_2(O_2CCH_3)_4(H_2O)_2$ [1] or $Rh_2(O_2CCH_3)_4(CH_3OH)_2$ [2], and separates as a pink precipitate [1, 2]. The infrared spectrum (3600 to 600 cm^{-1}) has been recorded and assignments given. ^1H NMR data are reported. The electronic spectrum is given, λ_{max} = 559 nm [2]. The product of this reaction has also been formulated as $Rh_2(O_2CCH_3)_4$(adenosine)(H_2O)·H_2O [8].

$Rh_2(O_2CCH_3)_4$(triacetyladenosine) is prepared by stirring together an aqueous solution of $Rh_2(O_2CCH_3)_4(CH_3OH)_2$ and an aqueous suspension of triacetyladenosine. It is isolated as a pink precipitate. The infrared spectrum (3600 to 600 cm^{-1}) has been recorded and assignments given. ^1H NMR data are reported [2].

$Rh_2(O_2CCH_3)_4$(tetra-acetyladenosine)$_2$ is prepared by stirring together an aqueous mixture of $Rh_2(O_2CCH_3)_4(CH_3OH)_2$ and tetra-acetyladenosine. The product deposits as a pink precipitate. The infrared spectrum has been recorded and assignments given. ^1H NMR data are reported [2].

$Rh_2(O_2CCH_3)_4$(adenosine 5′-monophosphate) is prepared by addition of adenosine 5′-monophosphate to an aqueous suspension of $Rh_2(O_2CCH_3)_4(CH_3OH)_2$ followed by stirring until dissolution is complete, and is isolated by precipitation with acetone. The infrared spectrum (3600 to 600 cm^{-1}) has been recorded and assignments given. The electronic spectrum is reported, λ_{max} = 560 nm [2].

$Rh_2(O_2CCH_3)_4$(adenosine 5′-diphosphate) is similarly prepared using adenosine 5′-diphosphate. The infrared spectrum (3600 to 600 cm^{-1}) has been recorded and assignments given. The electronic spectrum is reported, λ_{max} = 555 nm [2].

$Rh_2(O_2CCH_3)_4$(adenosine 5′-triphosphate) is similarly prepared using adenosine 5′-triphosphate. The infrared spectrum (3600 to 600 cm^{-1}) has been recorded and assignments given. ^1H NMR data are reported [2].

Stepwise stability constants for the formation of mono or bis adducts of adenosine-5′-monophosphate, adenosine-5′-diphosphate and adenosine-5′-triphosphate in aqueous solution have been reported [3, 4]. The pink colour of the adducts formed and their pH dependence is taken to indicate that coordination to rhodium occurs through one of the adenine nitrogen atoms [3, 4]. Polymeric structures linked through the N(1) and N(7) sites on the adenine are proposed for the 1:1 adducts. Steric hindrance at the N(1) site on tetra-acetyladenosine leads to formation of a 1:2 adduct [2].

$Rh_2(O_2CCH_3)_4$(theophylline)$_2$2H_2O is prepared from $Rh_2(O_2CCH_3)_4(CH_3OH)_2$ and theophylline ($C_7H_8N_4O_2$) at pH≈5 (no other details recorded), and is isolated as violet red columnar crystals [5].

The X-ray crystal structure is monoclinic, space group C2/c-C^6_{2h}, a = 9.872(12), b = 23.760(28), c = 15.926(18) Å, β = 117.06(9)°, Z = 4; D_{exp} = 1.75, D_{calc} = 1.674 g/cm³. The molecule has the familiar "lantern" structure (Fig. 3, p. 9). The Rh-Rh distance is 2.412(6) Å and the average Rh-O(acetate) distance is 2.07(4) Å. The axial theophylline ligands are bound through N(9) with an average Rh-N distance of 2.23(3) Å [5].

$Rh_2(O_2CCH_3)_4$(caffeine)$_2$ is prepared from $Rh_2(O_2CCH_3)_4(CH_3OH)_2$ and caffeine ($C_8H_{10}N_4O_2$) at pH = 6 (no other details given), and is isolated as violet columnar crystals [5].

The X-ray crystal structure is triclinic, space group $P\bar{1}$-C_i^1, a = 9.132(5), b = 12.834(8), c = 12.623(8) Å, α = 143.55(4)°, β = 120.56(6)°, γ = 66.47(7)°, Z = 1; D_{exp} = 1.82, D_{calc} = 1.882 g/cm^3. The molecule has the familiar "lantern" structure (Fig. 3, p. 9). The Rh-Rh distance is 2.395(1) Å and the average Rh-O(acetate) distance is 2.036(10) Å. The axial caffeine ligands are bound through N(9) with an average Rh-N distance of 2.315(9) [5].

$Rh_2(O_2CCH_3)_4D(H_2O)$, $Rh_2(O_2CCH_3)_4D_2$ (D = picolinic acid, 2-$HO_2CC_5H_4N$; iso-nicotinic acid, 4-$HO_2CC_5H_4N$; niacin, 3-$HO_2CC_5H_4N$; pyridine, C_5H_5N; imidazole, $C_3H_4N_2$; histidine, $C_6H_9N_3O_2$).

These donor ligands (D) react with aqueous buffered solutions of $Rh_2(O_2CCH_3)_4$ to give a blue to pink colour change. Spectral data afford stability constants and forward and reverse rate constants for formation of mono and bis adducts [6].

Stepwise formation constants and thermodynamic data for formation of 1:1 and 1:2 imidazole adducts have been reported [3, 7] as well as kinetic and thermodynamic parameters for histidine adduct formation [3].

References:

[1] N. Farrell (Chem. Commun. **1980** 1014/6). – [2] G. Pneumatikakis, N. Hadjiliadis (J. Chem. Soc. Dalton Trans. **1979** 596/9). – [3] K. Das, E. L. Simmons, J. L. Bear (Inorg. Chem. **16** [1977] 1268/71). – [4] L. Rainen, R. A. Howard, A. P. Kimball, J. L. Bear (Inorg. Chem. **14** [1975] 2752/4). – [5] K. Aoki, H. Yamazaki (Chem. Commun. **1980** 186/8).

[6] J. L. Bear, R. A. Howard, J. E. Korn (Inorg. Chim. Acta **32** [1979] 123/6). – [7] K. Das, J. L. Bear (Inorg. Chem. **15** [1976] 2093/5). – [8] N. Farrell (J. Inorg. Biochem. **14** [1981] 261/5).

$K_2[Rh_2(O_2CCH_3)_4(NO_2)_2]$ is prepared by treatment of $Rh_2(O_2CCH_3)_4(H_2O)_2$ or $Rh_2(O_2CCH_3)_4$-(amine)$_2$ (amine not specified) with saturated potassium nitrite solution, and is isolated as water soluble brick-red polyhedral crystals [1]. The infrared spectrum (4000 to 60 cm^{-1}) has been recorded and assignments made [2]. X-ray photoelectronic data have been recorded, N1s = 403.6 eV [3]; Rh3d$_{5/2}$ = 309.3, N1s = 403.6 eV [4]; Rh3d$_{5/2}$ = 309.7, N1s = 403.7 eV [5]. A thermogravimetric analysis curve is reproduced [7].

$K_2[Rh_2(O_2CCH_3)_4(NCO)_2]\cdot 2H_2O$ is prepared by thoroughly grinding KNCO and $Rh_2(O_2CCH_3)_4$-$(H_2O)_2$ together in a little water or water/ethanol, filtering the mixture after 24 h, and washing the residue with small amounts of water and ethanol. The salt forms fine violet tetragonal platelets. The infrared spectrum (3600 to 400 cm^{-1}) is reproduced and assignments given [6, 8]. Thermal stability data are recorded [6, 8]; a thermogram and thermogravimetric curve are reproduced [8]. The cyanate ligand is coordinated through nitrogen [8].

$K[Rh_2(O_2CCH_3)_4(NCS)]\cdot 2H_2O$ is similarly prepared using KSCN but is separated after only 2 to 3 h as violet-green dichroic tetragonal platelets [6, 8]. The infrared spectrum (3600 to 400 cm^{-1}) is reproduced and assignments given [6, 8]. Thermal stability data are recorded [6, 8]; a thermogram and thermogravimetric curve are reproduced [8]. A polymeric structure with bridging thiocyanate ligands is proposed for the complex anion [6, 8].

$K[Rh_2(O_2CCH_3)_4(NCS)\{HCON(CH_3)_2\}]$ is obtained when potassium thiocyanate is mixed with $Rh_2(O_2CCH_3)_4(H_2O)_2$ in aqueous dimethylformamide solution. The infrared spectrum (3000 to 400 cm^{-1}) is reported and assignments given. Thermal stability data are recorded [6].

$K[Rh_2(O_2CCH_3)_4(NCSe)]\cdot 2H_2O$ is prepared by grinding together KSeCN and $Rh_2(O_2CCH_3)_4$-$(H_2O)_2$ in a little water or water/ethanol mixture [6, 8]. It is immediately separated as a violet-grey product (possibly contaminated with selenium or $Rh(SeCN)_x$) [6]. The infrared spectrum (3600 to 400 cm^{-1}) is reproduced and assignments given [6, 8]. Thermal stability data are recorded [6, 8]; a thermogram and thermogravimetric curve are reproduced [8]. A polymeric structure with bridging NCSe ligands is proposed for the complex anion [6, 8].

$K[Rh_2(O_2CCH_3)_4(NCSe)\{HCON(CH_3)_2\}]$. This salt is obtained when potassium selenocyanate is mixed with $Rh_2(O_2CCH_3)_4(H_2O)_2$ in aqueous N,N-dimethylformamide solution. The infrared spectrum (3000 to 400 cm^{-1}) is reported and assignments given. Thermal stability data are recorded [6].

References:

[1] L. A. Nazarova, I. I. Chernyaev, A. S. Morozova (Zh. Neorgan. Khim. **10** [1965] 539/41; Russ. J. Inorg. Chem. **10** [1965] 291/2). – [2] G. Ya. Mazo, I. B. Baranovskii, R. N. Shchelokov (Zh. Neorgan. Khim. **24** [1979] 3330/6; Russ. J. Inorg. Chem. **24** [1979] 1855/7). – [3] V. I. Nefedov, Ya. V. Salyn', A. V. Shtemenko, A. S. Kotelnikova (Inorg. Chim. Acta **45** [1980] L49/L50). – [4] V. I. Nefedov, Ya. V. Salyn', I. B. Baranovskii, A. G. Maiorova (Zh. Neorgan. Khim. **25** [1980] 216/25; Russ. J. Inorg. Chem. **25** [1980] 116/22). – [5] V. I. Nefedov, Ya. V. Salyn', A. G. Maiorova, L. A. Nazarova, I. B. Baranovskii (Zh. Neorgan. Khim. **19** [1974] 1353/7; Russ. J. Inorg. Chem. **19** [1974] 736/8).

[6] V. N. Shafranskii, T. A. Mal'kova, Yu. Ya. Kharitonov (Koord. Khim. **1** [1975] 375/83; Soviet J. Coord. Chem. **1** [1975] 297/303). – [7] V. I. Belova, Z. S. Dergacheva (Zh. Neorgan. Khim. **16** [1971] 3065/70; Russ. J. Inorg. Chem. **16** [1971] 1626/9). – [8] V. N. Shafranskii, T. A. Mal'kova, Yu. Ya. Kharitonov (Zh. Strukt. Khim. **16** [1975] 212/5; J. Struct. Chem. [USSR] **16** [1975] 195/8).

$M_2[Rh_2(O_2CCH_3)_4X_2]$, $M[Rh_2(O_2CCH_3)_4X]$ (M = organic cation or alkali metal cation, X = halide). A substantial range of salts with this general formula have been reported. X-ray and X-ray photoelectron spectra have been used to investigate the electronic structure of the anion $[Rh_2(O_2CCH_3)_4Cl_2]^{2-}$. The evidence points to a bond order close to one rather than three for the Rh-Rh interaction [1, 2].

$[NH_4]_2[Rh_2(O_2CCH_3)_4Cl_2]\cdot 2NH_4Cl$ is prepared by treatment of $Rh_2(O_2CCH_3)_4(H_2O)_2$ with a large excess of ammonium chloride [3].

$[enH_2][Rh_2(O_2CCH_3)_4Cl_2]\cdot 6H_2O$ is prepared by treatment of "various diacetato-rhodium compounds" with ethylenediamine dihydrochloride and is isolated as well-formed green polyhedral crystals [3].

$[NH_2CONH\cdot NH_3]_2[Rh_2(O_2CCH_3)_4Cl_2]$. This salt, which is also formulated as $Rh_2(O_2CCH_3)_4$-$(NH_2CONHNH_3Cl)_2$, is prepared by addition of a saturated aqueous solution of semicarbazide hydrochloride to $Rh_2(O_2CCH_3)_4(H_2O)_2$ in alcohol. The salt forms blue-green lamellar crystals [4].

$[GunH]_2[Rh_2(O_2CCH_3)_4Cl_2]$. This salt is prepared by addition of guanidine hydrochloride to an aqueous solution of $Rh_2(O_2CCH_3)_4(H_2O)_2$ and separates as green polyhedral crystals [5, 6].

The crystals are tetragonal, space group I4-C_4^5, a = 8.586(1), c = 14.006(2) Å, Z = 4; D_{calc} = 2.04 g/cm^3. The crystals consist of alternating guanidinium cations and binuclear $[Rh_2(O_2CCH_3)_4Cl_2]^{2-}$ anions. The anions possess the "lantern" structure (Fig. 3, p. 9). Bond distances are Rh-Rh = 2.397(2), Rh-O = 2.04(1) and 2.05(1) Å, Rh-Cl = 2.571(6) and 2.610(5) Å [7].

The infrared spectrum (4000 to 60 cm^{-1}) has been recorded and assignments given [6, 8]. X-ray photoelectronic data have been recorded, Rh$3d_{5/2}$ = 309.3, Cl2p = 198.2 eV [9, 10]. A thermogravimetric analysis curve has been reproduced [12].

$[GunNH_2 \cdot H]_2[Rh_2(O_2CCH_3)_4Cl_2]$ is obtained as water-soluble green crystals by treatment of $Rh_2(O_2CCH_3)_4(H_2O)_2$ with aminoguanidine hydrochloride in alcoholic solution. The infrared spectrum (3600 to 300 cm^{-1}) has been recorded and assignments given. Thermal decomposition data are reported [6].

$K_2[Rh_2(O_2CCH_3)_4Cl_2] \cdot 2KCl$ is prepared by treatment of $Rh_2(O_2CCH_3)_4(H_2O)_2$ with excess potassium chloride in aqueous solution, and separates as polyhedral green crystals [5]. The infrared spectrum (4000 to 60 cm^{-1}) has been recorded and assignments made [8]. X-ray photoelectronic data have been recorded, Rh$3d_{5/2}$ = 309.2, Cl$2p_{3/2}$ = 198.2 eV [9]. The thermogravimetric analysis curve has been reproduced [12].

$M_2[Rh_2(O_2CCH_3)_4Cl_2]$ (M = K or Cs). X-ray photoelectronic data have been recorded, Cl2p = 198.1 eV [11].

$NH_4[Rh_2(O_2CCH_3)_4Cl]$ is prepared by the action of a dilute solution of NH_4Cl on a concentrated solution of $Rh_2(O_2CCH_3)_4$ or its guanidine derivative [3].

$K[Rh_2(O_2CCH_3)_4Cl]$ is similarly prepared as fine green lamellar crystals using potassium chloride [3].

$[N_2H_5][Rh_2(O_2CCH_3)_4Cl(H_2O)]$. The thermal stability of this diamagnetic complex has been investigated; μ_{eff} = 0.5 B.M. [12].

$(NH_4)_2[Rh_2(O_2CCH_3)_4Br_2] \cdot 2NH_4Br$ is prepared by the action of excess NH_4Br on $Rh_2(O_2CCH_3)_4(H_2O)_2$ or $Rh_2(O_2CCH_3)_4(Gun)_2$ [3]. The complex is diamagnetic, μ_{eff} = 0.51 B.M. [12]. Preliminary crystallographic data have been reported; the crystals are orthorhombic, space group Pncm-D_{2h}^7, lattice parameters a = 9.193, b = 9.481, c = 14.533 Å, Z = 4; D = 2.22 g/cm^3 [13].

$[GunH]_2[Rh_2(O_2CCH_3)_4Br_2]$ is prepared by treatment of $[GunH]_2[Rh_2(O_2CCH_3)_4Cl_2]$ with potassium bromide, and is isolated as green lamellar crystals [3].

Preliminary crystallographic data have been reported. The crystals have a tetragonal unit cell, space group I4-C_4^5, lattice parameters a = 9.03, c = 14.00 Å, Z = 4; D = 2.043 g/cm^3. The anion has the familiar "lantern" structure (Fig. 3, p. 9) [13].

The infrared spectrum (4000 to 60 cm^{-1}) has been recorded and assignments given [8]. X-ray photoelectronic data have been recorded, Rh$3d_{5/2}$ = 309.2, Br3d = 68.9 eV [9, 10]. The complex is diamagnetic, μ_{eff} = 0.49 B.M. [12].

$K_2[Rh_2(O_2CCH_3)_4Br_2] \cdot 2KBr$ is prepared by the action of excess KBr on $Rh_2(O_2CCH_3)_4(H_2O)_2$ or $Rh_2(O_2CCH_3)_4(Gun)_2$ [3]. The infrared spectrum (4000 to 60 cm^{-1}) has been recorded and assignments given [8]. X-ray photoelectronic data have been reported, Rh$3d_{5/2}$ = 309.2, Br3d = 69.2 eV [9].

$M_2[Rh_2(O_2CCH_3)_4Br_2]$ (M = K, Cs). X-ray photoelectronic data have been recorded, Br3d = 68.9 eV [11].

$[GunH]_2[Rh_2(O_2CCH_3)_4I_2] \cdot 2H_2O$ is prepared by treatment of $[GunH]_2[Rh_2(O_2CCH_3)_4Cl_2]$ with potassium iodide, and is isolated as green lamellar crystals [3].

References:

[1] V. I. Nefedov, Ya. V. Salyn', A. P. Sadovskii (J. Electron Spectrosc. Relat. Phenomena **16** [1979] 299/305). – [2] V. I. Nefedov, A. P. Sadovskii, Ya. V. Salyn' (Koord. Khim. **5** [1979] 1204/8; Soviet J. Coord. Chem. **5** [1979] 949/53). – [3] L. A. Nazarova, I. I. Chernyaev, A. S. Morozova (Zh. Neorgan. Khim. **11** [1966] 2583/6; Russ. J. Inorg. Chem. **11** [1966] 1387/9). – [4] L. A. Nazarova, A. G. Maiorova (Zh. Neorgan. Khim. **21** [1976] 1070/4; Russ. J. Inorg. Chem. **21** [1976] 583/5). – [5] L. A. Nazarova, I. I. Chernyaev, A. S. Morozova (Zh. Neorgan. Khim. **10** [1965] 539/41; Russ. J. Inorg. Chem. **10** [1965] 291/2).

[6] T. A. Veteva, V. N. Shafranskii (Zh. Obshch. Khim. **49** [1979] 488/93; J. Gen. Chem. [USSR] **49** [1979] 428/33). – [7] L. M. Dikareva, G. G. Sadikov, I. B. Baranovskii, M. A. Porai-Koshits (Zh. Neorgan. Khim. **25** [1980] 3146/7; Russ. J. Inorg. Chem. **25** [1980] 1725). – [8] G. Ya. Mazo, I. B. Baranovskii, R. N. Shchelokov (Zh. Neorgan. Khim. **24** [1979] 3330/6; Russ. J. Inorg. Chem. **24** [1979] 1855/7). – [9] V. I. Nefedov, Ya. V. Salyn', A. G. Maiorova, L. A. Nazarova, I. B. Baranovskii (Zh. Neorgan. Khim. **19** [1974] 1353/7; Russ. J. Inorg. Chem. **19** [1974] 736/8). – [10] V. I. Nefedov, Ya. V. Salyn', I. B. Baranovskii, A. G. Maiorova (Zh. Neorgan. Khim. **25** [1980] 216/25; Russ. J. Inorg. Chem. **25** [1980] 116/22).

[11] V. I. Nefedov, Ya. V. Salyn', A. V. Shtemenko, A. S. Kotelnikova (Inorg. Chim. Acta **45** [1980] L49/L50). – [12] V. I. Belova, Z. S. Dergacheva (Zh. Neorgan. Khim. **16** [1971] 3065/70; Russ. J. Inorg. Chem. **16** [1971] 1626/9). – [13] L. M. Dikareva (Acta Cryst. **21** [1966] A140).

$\mathbf{Rh_2(O_2CCH_3)_4\{(C_2H_5)_2S\}_2}$ is prepared by placing a few drops of diethyl sulphide on finely divided anhydrous $Rh_2(O_2CCH_3)_4$, and removing excess ligand under vacuum. It is a burgundy solid [1]. The electronic spectrum has been measured, $\lambda_{max} = 541$ nm (shoulder) [1] and is reproduced (750 to 350 nm) [2]. The heat of dissociation is $\Delta H = 27.4$ kcal/mol [1] and thermogravimetric analysis curves are reproduced. The complex is diamagnetic, $\mu_{eff} = 0.21$ B.M. [2].

$\mathbf{Rh_2(O_2CCH_3)_4\{(C_2H_5)_2S\}}$ is prepared from the bis adduct by heating to 147°C, cooling the residue rapidly, washing with benzene and recrystallising from dichloromethane. It forms deep purple crystals. The electronic spectrum has been reported, $\lambda_{max} = 555$ nm (shoulder). The complex is diamagnetic, $\mu_{eff} = 0.28$ B.M. [2].

$\mathbf{Rh_2(O_2CCH_3)_4\{(NH_2)_2CS\}_2}$ is prepared by the action of thiourea on rhodium(II) acetate dihydrate [17] or "various" rhodium(II) acetate derivatives [3], and is described as fine violet lamellar crystals [3]. Coordination of the thiourea is thought to occur through sulphur [3, 17]. The infrared spectrum has been recorded (4000 to 60 cm^{-1}) [4] and has been reproduced (3600 to 400 cm^{-1}) [17], assignments are given [4, 17]. Thermal decomposition has been investigated and thermograms reproduced [17]. X-ray photoelectronic data have been recorded, $Rh\,3d_{5/2} = 308.4$, $Rh\,3d_{3/2} = 313.2$ eV [5].

$\mathbf{Rh_2(O_2CCH_3)_4(NH_2CSCH_3)_2}$ is prepared by moistening $[GunH]_2[Rh_2(O_2CCH_3)_4Cl_2]$ with a saturated solution of thioacetamide [13] or by treatment of $Rh_2(O_2CCH_3)_4(H_2O)_2$ with aqueous thioacetamide [17]. It is a violet-brown solid which is insoluble in water but reacts with donor solvents to form the corresponding adducts [13]. The infrared spectrum (3600 to 400 cm^{-1}) of the **dihydrate** is reproduced in a paper and assignments given. Thermal decomposition is investigated and a thermogram is reproduced. The thioacetamide ligand is thought to coordinate through sulphur [17].

$\mathbf{Rh_2(O_2CCH_3)_4(NH_2NHCSNH_2)_2 \cdot 2H_2O}$ is prepared by treatment of $Rh_2(O_2CCH_3)_4(H_2O)_2$ in alcoholic solution with aqueous thiosemicarbazide ($NH_2NHCSNH_2$), and is described as a dark red-brown or violet solid. The infrared spectrum (3600 to 400 cm^{-1}) has been reproduced and assignments given. Thermal decomposition has been investigated. The thiosemicarbazide ligand is thought to coordinate through sulphur [17].

$Rh_2(O_2CCH_3)_4(CH_3CONHCSNH_2)_2$ is obtained by treatment of $Rh_2(O_2CCH_3)_4(H_2O)_2$ with acetylthiourea in alcoholic solution, and is isolated as a red-brown or dark violet precipitate. The infrared spectrum (3600 to 400 cm^{-1}) has been reproduced and assignments given. Thermal decomposition has been investigated. The acetylthiourea ligand is thought to coordinate through sulphur [17].

$Rh_2(O_2CCH_3)_4(C_6H_5CSNH_2)_2$ is obtained by treatment of $Rh_2(O_2CCH_3)_4(H_2O)_2$ with thiobenzamide in alcoholic solution and is isolated as a red-brown or dark violet crystalline precipitate. The infrared spectrum (3600 to 400 cm^{-1}) has been reproduced and assignments given. Thermal decomposition has been investigated. The thiobenzamide ligand is thought to coordinate through sulphur [17].

$Rh_2(O_2CCH_3)_4(C_6H_5NHCSNHC_6H_5)_2$ is obtained by treatment of $Rh_2(O_2CCH_3)_4(H_2O)_2$ with diphenylthiourea in alcoholic solution, and is isolated as a red-brown or deep violet crystalline precipitate. The infrared spectrum (3600 to 400 cm^{-1}) is reproduced and assignments given. The diphenylthiourea ligand is thought to coordinate through sulphur [17].

$Rh_2(O_2CCH_3)_4(C_6H_5NHCSC_6H_5)_2$ is prepared by treatment of $Rh_2(O_2CCH_3)_4(H_2O)_2$ with phenylthiobenzamide in alcoholic solution, and is described as a red-brown or deep violet crystalline solid. The infrared spectrum (3600 to 400 cm^{-1}) is reproduced and assignments given. The N-phenylthiobenzamide ligand is thought to coordinate through sulphur [17].

$Rh_2(O_2CCH_3)_4(C_4H_8S)_2$ is prepared by suspending $Rh_2(O_2CCH_3)_4$ in benzene, adding tetrahydrothiophene (C_4H_8S) and allowing the solvent to evaporate. It forms large dark red crystals [6].

The crystals are orthorhombic, space group Pbcn-D_{2h}^{14}, a = 9.801(2), b = 16.087(3), c = 14.592(3) Å, Z = 4. The molecule has the familiar "lantern" structure (Fig. 3, p. 9), bond lengths are Rh-Rh = 2.413(1) Å, Rh-O = 2.035(3), 2.046(3), 2.036(3), 2.042(3) Å, Rh-S = 2.517 Å [6].

$Rh_2(O_2CCH_3)_4\{(CH_3)_2SO\}_2$ is prepared by adding a few drops of dimethyl sulphoxide to finely powdered $Rh_2(O_2CCH_3)_4$ then removing excess ligand in vacuo [1, 7, 8] or by slow evaporation of an aqueous solution of $Rh_2(O_2CCH_3)_4$ containing dimethylsulphoxide [6]. It forms orange crystals [1, 7, 8].

The X-ray crystal structure has been reported. The crystals are orthorhombic, space group Pbca-D_{2h}^{15}, a = 8.377(2), b = 16.726(2), c = 14.840(2) Å, Z = 4; D_{calc} = 1.91 g/cm^3. The molecules possess the usual "lantern" structure, the Rh-Rh bond length is 2.406(1) Å and the $(CH_3)_2SO$ ligands are coordinated through sulphur, Rh-S = 2.451 Å. Rhodium-oxygen bond lengths are 2.032(3), 2.043(3), 2.035(4), 2.035(3) Å [6].

The infrared spectrum has been recorded (4000 to 60 cm^{-1}) [4] and reproduced (2000 to 400 cm^{-1}) [8]; assignments have been given [4, 8]. A value of 311 cm^{-1} has been reported for ν(Rh-Rh) [9]. The electronic spectrum has been recorded, λ_{max} = 500 nm [7], and has been reproduced (700 to 350 nm) [1]. X-ray photoelectronic data have been reported, Rh $3d_{5/2}$ = 309.2, S 2p = 166.5 eV [10]; S 2p = 166.5 eV [11]; Rh $3d_{5/2}$ = 308.5, O 1s = 532.3 eV [5]. Thermogravimetric and differential scanning calorimetry curves are reproduced [1, 2, 12].

$Rh_2(O_2CCH_3)_4\{(C_2H_5)_2SO\}_2$ is precipitated from an alcoholic solution of $Rh_2(O_2CCH_3)_4(H_2O)_2$ by addition of diethyl sulphoxide. The infrared spectrum (2000 to 400 cm^{-1}) is reproduced and assignments given. Thermal properties are reported [8].

$Rh_2(O_2CCH_3)_4(NH_2NHCSNH_2 \cdot HCl)_2$ is prepared by dissolving $Rh_2(O_2CCH_3)_4(H_2O)_2$ in the minimum volume of warm water and adding excess thiosemicarbazide hydrochloride; it rapidly separates as dark violet lamellar crystals. The complex is sparingly soluble in water but dissolves in hot acetic acid to form a green solution. The violet colour and low solubility of the complex suggest that it is an adduct rather than a salt of formula $[NH_2NHCSNH_3]_2$-$[Rh_2(O_2CCH_3)_4Cl_2]$ [13].

$Na_2[Rh_2(O_2CCH_3)_4(O_2SC_6H_5)_2]$ is prepared by dissolving $Na[O_2SC_6H_5] \cdot 2H_2O$ in a methanolic solution of $Rh_2(O_2CCH_3)_4$ and concentrating the mixture under reduced pressure until precipitation occurs. It forms orange crystals. The infrared spectrum (3600 to 390 cm^{-1}) has been recorded. The axial $C_6H_5SO_2^-$ ligands, which are coordinated through sulphur, are very readily replaced by water, methanol, pyridine or dimethylsulphoxide [14].

$Na_2[Rh_2(O_2CCH_3)_4(O_2S \cdot C_6H_4CH_3\text{-}p)_2]$ is similarly prepared using $Na[O_2SC_6H_4CH_3\text{-}p] \cdot 2H_2O$ and forms orange crystals [14].

$Rh_2(O_2CCH_3)_4(BTD)_2$ (BTD is substituted 2,1,3-benzothiadiazole; substituents 4-CH_3, 5-CH_3, 4-CH_3O, 5-CH_3O, 5-C_2H_5O, 4-NH_2, 4-NO_2, 5-Br and 4,6-Cl_2). These adducts are prepared by addition of alcoholic solutions of the appropriate benzothiadiazole to an aqueous solution of $Rh_2(O_2CCH_3)_4(H_2O)_2$, and are isolated as brown, water-insoluble precipitates. Their infrared spectra have been recorded (1600 to 600 cm^{-1}, selected bands, assignments given). Differential thermal analysis curves are reproduced. The adducts are insoluble in ethanol, chloroform and ether but dissolve in dimethylformamide and dioxane with formation of green solutions [15].

$Rh_2(O_2CCH_3)_4\{CH_3SCH_2CH(NH_2)CO_2H\}_2$ is prepared by stirring an aqueous suspension of $Rh_2(O_2CCH_3)_4(CH_3OH)_2$ overnight with S-methyl cysteine, and is isolated as orange-violet crystals by concentration of the solution under reduced pressure, followed by dilution with acetone. The S-methyl cysteine ligands are coordinated through sulphur. Selected infrared bands are reported, and the electronic spectrum has been recorded, $\lambda_{max} = 558$ nm (aqueous solution) [16].

$Rh_2(O_2CCH_3)_4\{C_2H_5SCH_2CH(NH_2)CO_2H\}_2$ is similarly prepared using S-ethyl cysteine, and is isolated as an orange-violet solid. Coordination of the S-ethyl cysteine occurs through sulphur. Selected infrared bands are reported and assignments given, the electronic spectrum has been recorded, $\lambda_{max} = 560$ nm (aqueous solution) [16].

$Rh_2(O_2CCH_3)_4\{CH_3SCH_2CH_2CH(NH_2)CO_2H\}_2$ is similarly prepared using 1-methionine, and is isolated as an orange-violet solid. Coordination of the 1-methionine occurs through sulphur. Selected infrared bands are reported and assignments given, the electronic spectrum has been recorded, $\lambda_{max} = 562$ nm (aqueous solution) [16].

$K_4[Rh_2(O_2CCH_3)_4(SO_3)_2] \cdot 4H_2O$ is prepared by treatment of $Rh_2(O_2CCH_3)_4$ with saturated aqueous K_2SO_3. The red product turns green slowly but again becomes red on addition of water [18].

References:

[1] J. Kitchens, J. L. Bear (J. Inorg. Nucl. Chem. **31** [1969] 2415/21). – [2] J. Kitchens, J. L. Bear (J. Inorg. Nucl. Chem. **32** [1970] 49/58). – [3] L. A. Nazarova, I. I. Chernyaev, A. S. Morozova (Zh. Neorgan. Khim. **11** [1966] 2583/6; Russ. J. Inorg. Chem. **11** [1966] 1387/9). – [4] G. Ya. Mazo, I. B. Baranovskii, R. N. Shchelokov (Zh. Neorgan. Khim. **24** [1979] 3330/6; Russ. J. Inorg. Chem. **24** [1979] 1855/7). – [5] A. D. Hamer, D. G. Tisley, R. A. Walton (J. Chem. Soc. Dalton Trans. **1973** 116/20).

[6] F. A. Cotton, T. R. Felthouse (Inorg. Chem. **19** [1980] 323/8). – [7] S. A. Johnson, H. R. Hunt, H. M. Neumann (Inorg. Chem. **2** [1963] 960/2). – [8] T. A. Mal'kova, V. N. Shafranskii (Zh.

Obshch. Khim. **47** [1977] 2592/6; J. Gen. Chem. [USSR] **47** [1977] 2365/9). – [9] A. P. Ketteringham, C. Oldham (J. Chem. Soc. Dalton Trans. **1973** 1067/70). – [10] V. I. Nefedov, Ya. V. Salyn', I. B. Baranovskii, A. G. Maiorova (Zh. Neorgan. Khim. **25** [1980] 216/25; Russ. J. Inorg. Chem. **25** [1980] 116/22).

[11] V. I. Nefedov, Ya. V. Salyn', A. V. Shtemenko, A. S. Kotelnikova (Inorg. Chim. Acta **45** [1980] L49/L50). – [12] R. A. Howard, A. M. Wynne, J. L. Bear, W. W. Wendlandt (J. Inorg. Nucl. Chem. **38** [1976] 1015/8). – [13] L. A. Nazarova, A. G. Maiorova (Zh. Neorgan. Khim. **21** [1976] 1070/4; Russ. J. Inorg. Chem. **21** [1976] 583/5). – [14] J. G. Norman, E. O. Fey (J. Chem. Soc. Dalton Trans. **1976** 765/7). – [15] Yu. N. Kukushkin, S. A. Simanova, V. K. Krylov, S. I. Bakhireva, I. A. Belen'kaya (Zh. Obshch. Khim. **46** [1976] 888/92; J. Gen. Chem. [USSR] **46** [1976] 885/8).

[16] G. Pneumatikakis, P. Psaroulis (Inorg. Chim. Acta **46** [1980] 97/100). – [17] T. A. Mal'kova, V. N. Shafranskii (Zh. Fiz. Khim. **49** [1975] 2805/9; Russ. J. Phys. Chem. **49** [1975] 1653/6). – [18] I. B. Baranovskii, S. S. Abdullaev, G. Ya. Mazo, R. N. Shchelokov (Zh. Neorgan. Khim. **27** [1982] 536/8; Russ. J. Inorg. Chem. **27** [1982] 305/6).

$Rh_2(O_2CCH_3)_4\{P(C_6H_5)_3\}_2$ is prepared by addition of triphenylphosphine in diethyl ether to a cold, methanolic solution of $Rh_2(O_2CCH_3)_4$, and separates as an orange powder [1]. The crystals are triclinic, space group $P\bar{1}-C_i^1$, lattice parameters a = 9.56(1), b = 9.19(1), c = 12.85(1) Å, α = 110.09(5)°, β = 102.56(5)°, γ = 88.85(5)°, Z = 1; D_{calc} = 1.55, D_{obs} = 1.56(2) g/cm³. The Rh-Rh bond length is 2.4505(2) Å and the average Rh-P distance is 2.4771(5) Å [8]. The infrared spectrum (4000 to 60 cm^{-1}) has been recorded and assignments given [2]. Raman and far-infrared data have been reported; ν(Rh-Rh) = 289 cm^{-1}, ν(Rh-O) = 320, 374 and 379 cm^{-1} (shoulder) [3]. X-ray photoelectronic data have been recorded, Rh $3d_{3/2}$ = 312.9, Rh $3d_{5/2}$ = 308.0 eV [4]; Rh $3d_{5/2}$ = 309.1 eV [5].

$Rh_2(O_2CCH_3)_4\{P(C_6H_5)_3\}_2 \cdot 2C_2H_5OH$ has been mentioned but no preparative details given [6]. The complex is diamagnetic, μ_{eff} = 0.55 B.M. Thermogravimetric analysis curves are reproduced in the paper [6].

$Rh_2(O_2CCH_3)_4\{P(OCH_3)_3\}_2$ is prepared by treatment of $Rh_2(O_2CCH_3)_4$ with the stoichiometric amount of trimethylphosphite in an organic solvent; no other details are recorded [7].

The crystals are monoclinic, space group $P2_1/c-C_{2h}^5$, a = 14.133(3), b = 15.799(4), c = 8.309(2) Å, β = 138.65°, Z = 2; D_{exp} = 1.85(1), D_{calc} = 1.877 g/cm³. The Rh-Rh bond distance is 2.4555(3) Å and the average Rh-P bond distance is 2.437(5) Å [7].

$Rh_2(O_2CCH_3)_4\{P(OC_6H_5)_3\}_2$ is prepared by addition of a stoichiometric amount of triphenylphosphite to a solution of $Rh_2(O_2CCH_3)_4$ in an organic solvent, no other details are recorded [7].

The crystals are monoclinic, space group $C2/c-C_{2h}^6$, a = 26.134(9), b = 9.951(2), c = 22.512(7) Å, β = 61.45°, Z = 4; D_{exp} = 1.42(1), D_{calc} = 1.372 g/cm³. The Rh-Rh bond length is 2.445(1) Å and the average Rh-P bond length is 2.418(3) Å [7].

A later paper gives revised data: β = 118.55(2)°, D_{calc} = 1.489 g/cm³, Rh-Rh = 2.4434(6), Rh-P = 2.412(1) Å [8].

$Rh_2(O_2CCH_3)_4(PF_3)_2$ is prepared by passing a stoichiometric amount of phosphorus trifluoride into a solution of $Rh_2(O_2CCH_3)_4$ in an organic solvent, no other details are recorded [7].

The crystals are cubic, space group $I23-T^3$, a = 14.339(2) Å, Z = 6; D_{calc} = 2.087 g/cm³. The Rh-Rh bond length is 2.430(3) Å and the average Rh-P bond length is 2.42(1) Å [7].

$Rh_2(O_2CCH_3)_4\{As(C_2H_5)_3\}_2$ is prepared by adding a drop of triethyl arsine to a saturated solution of $Rh_2(O_2CCH_3)_4$ in diethyl ether, decanting off the solution and allowing the solvent to evaporate. It is a red-orange solid. The electronic spectrum has been recorded, $\lambda_{max} = 512$ nm (reflectance spectrum) [9].

$Rh_2(O_2CCH_3)_4\{As(C_6H_5)_3\}_2$ is obtained by treating $Rh_2(O_2CCH_3)_4$ or its methanol adduct with $As(C_6H_5)_3$ in boiling methanol, and precipitates rapidly as maroon crystals, melting point 267 to 268°C [10].

References:

[1] T. A. Stephenson, S. M. Morehouse, A. R. Powell, J. P. Heffer, G. Wilkinson (J. Chem. Soc. **1965** 3632/40). – [2] G. Ya. Mazo, I. B. Baranovskii, R. N. Shchelokov (Zh. Neorgan. Khim. **24** [1979] 3330/6; Russ. J. Inorg. Chem. **24** [1979] 1855/7). – [3] A. P. Ketteringham, C. Oldham (J. Chem. Soc. Dalton Trans. **1973** 1067/70). – [4] A. D. Hamer, D. G. Tisley, R. A. Walton (J. Chem. Soc. Dalton Trans. **1973** 116/20). – [5] V. I. Nefedov, Ya. V. Salyn', I. B. Baranovskii, A. G. Maiorova (Zh. Neorgan. Khim. **25** [1980] 216/25; Russ. J. Inorg. Chem. **25** [1980] 116/22).

[6] V. I. Belova, Z. S. Dergacheva (Zh. Neorgan. Khim. **16** [1971] 3065/70; Russ. J. Inorg. Chem. **16** [1971] 1626/9). – [7] G. G. Christoph, Y.-B. Koh (J. Am. Chem. Soc. **101** [1979] 1422/34). – [8] G. G. Christoph, J. Halpern, G. P. Khare, Y. B. Koh, C. Romanowski (Inorg. Chem. **20** [1981] 3029/37). – [9] J. Kitchens, J. L. Bear (J. Inorg. Nucl. Chem. **31** [1969] 2415/21). – [10] R. W. Mitchell, J. D. Ruddick, G. Wilkinson (J. Chem. Soc. A **1971** 3224/30, 3228).

$Rh_2(O_2CCH_3)_4(CO)_2$ is prepared from carbon monoxide and $Rh_2(O_2CCH_3)_4$ in organic solvents; no further details are recorded [1].

The crystals are orthorhombic, space group Pbca-D_{2h}^{15}, a = 14.368(8), b = 12.151(5), c = 8.800(3) Å, Z = 4; $D_{calc} = 2.153$ g/cm^3. The Rh-Rh distance is 2.4196(4) Å and the average Rh-CO distance is 2.092(4) Å [1].

$K_2[Rh_2(O_2CCH_3)_4(CN)_2]$. This orange salt is prepared by addition of aqueous potassium cyanide to $Rh_2(O_2CCH_3)_4(C_5H_5N)_2$ or to an ethanolic solution of $Rh_2(O_2CCH_3)_4(H_2O)_2$. The infrared spectrum is reproduced (3600 to 500 cm^{-1}) and assignments given, ν(CN) = 2100 cm^{-1} [2].

References:

[1] G. G. Christoph, Y.-B. Koh (J. Am. Chem. Soc. **101** [1979] 1422/34). – [2] T. A. Veteva, V. N. Shafranskii (Zh. Obshch. Khim. **49** [1979] 488/93; J. Gen. Chem. [USSR] **49** [1979] 428/33).

$Rh_2(O_2CCH_3)_2(O_2CC_6H_4OH\text{-}o)_2(H_2O)_2$. This mixed carboxylate complex is prepared by heating a mixture of $Rh_2(O_2CCH_3)_4(H_2O)_2$ or $[GunH]_2[Rh_2(O_2CCH_3)_4Cl_2]$ and a large excess of salicylic acid o-$C_6H_4(OH)(CO_2H)$, in aqueous alcoholic solution for 1.5 to 2 h. It separates from solution as green lamellar crystals, readily soluble in water but sparingly soluble in organic solvents [1]. X-ray photoelectronic data have been recorded, Rh $3d_{5/2} = 309.2$ eV [2].

$Rh_2(O_2CCH_3)_2(O_2CCH_2CH_2CH_2CO_2H)_2$ is prepared by heating a finely divided 1:1 mixture of $Rh_2(O_2CCH_3)_4$ and glutaric acid ($HO_2CCH_2CH_2CH_2CO_2H$) at 120°C for 1 h and is isolated as a green solid after washing with diethyl ether. Infrared data have been reported and assignments (in cm^{-1}) given: ν(OH) = 3200, 2700, ν(CO_2) = 1700, 1686 (noncoordinated), 1573 (coordinated), δ(CO_2) = 692, ν(Rh-O) = 388, 352 and 340 [3].

$Rh_2(O_2CCH_3)_2(O_2CC_6H_4OH\text{-}o)_2(NH_3)_2$ is obtained in poor yield by treatment of the corresponding aquo complex (see p. 33) with concentrated ammonia solution, and separates as red lamellar crystals [1].

$Rh_2(O_2CCH_3)_2(O_2CC_6H_4OH\text{-}o)_2(C_5H_5N)_2$ is similarly prepared using a small volume of pyridine, and separates from solution as red tablets, sparingly soluble in water [1]. X-ray photoelectronic data have been recorded, $Rh\,3d_{5/2} = 309.1$, $N\,1s = 399.8$ eV [2].

$Rh_2(O_2CCH_3)_2(O_2CC_6H_4OH\text{-}o)_2(CH_3CSNH_2)_2$ is prepared by treatment of the corresponding aquo adduct (see above) with a saturated aqueous solution of thioacetamide and 2 to 3 ml of alcohol. It forms fine violet tablets which are sparingly soluble in alcohol but soluble in water [1].

References:

[1] L. A. Nazarova, A. G. Maiorova (Zh. Neorgan. Khim. **18** [1973] 1710/2; Russ. J. Inorg. Chem. **18** [1973] 904/5). – [2] V. I. Nefedov, Ya. V. Salyn', A. G. Maiorova, L. A. Nazarova, I. B. Baranovskii (Zh. Neorgan. Khim. **19** [1974] 1353/7; Russ. J. Inorg. Chem. **19** [1974] 736/8). – [3] M. A. Golubnichaya, I. B. Baranovskii, G. Ya. Mazo, R. N. Shchelokov (Zh. Neorgan. Khim. **26** [1981] 2868/71; Russ. J. Inorg. Chem. **26** [1981] 1534/6).

$Rh_2(O_2CCH_3)_2(C_6H_6NO)_2(imidazole)$ (C_6H_6NO = 2-oxy-6-methylpyridine anion). This complex is obtained as small blue-red dichroic crystals by fusing $Rh_2(O_2CCH_3)_4$ with 2-hydroxy-6-methylpyridine at ~160°C and dissolving the melt in acetonitrile. The origin of the imidazole ($C_3H_4N_2$) ligand is unknown.

The crystals are monoclinic, space group C2/c-C_{2h}^6, a = 10.702(2), b = 13.138(3), c = 15.369(2) Å, β = 102.58(1)°, Z = 4; D_{calc} = 1.91 g/cm³. The molecules possess the familiar "lantern" structure, see **Fig. 5**, but with a **trans** pair of acetate ligands replaced by a pair of 2-oxy-6-methylpyridine anions; the imidazole ligand occupies the least hindered axial site. The Rh-Rh bond length is 2.388(2) Å.

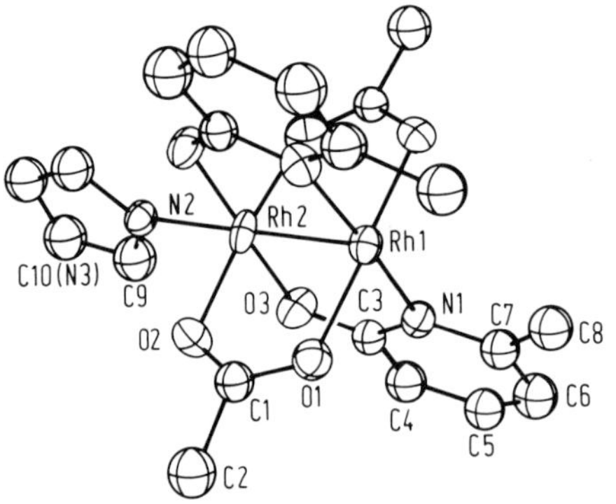

Fig. 5
Molecular structure of $Rh_2(O_2CCH_3)_2(C_6H_6NO)_2(imidazole)$.

$Rh_2(O_2CCH_3)_2(C_6H_6NO)_2(imidazole)\cdot 2CH_2Cl_2$. This solvate is obtained as blue-red prisms on recrystallising the preceding compound from dichloromethane solution.

The crystals are monoclinic, space group C2/c-C_{2h}^6, a = 21.488(10), b = 12.603(2), c = 11.086(2) Å, β = 112.21°, Z = 4; D_{calc} = 1.86 g/cm³. The dimeric rhodium(II) unit in this complex is essentially identical with that found in the preceding complex. The Rh-Rh bond length is the same (2.388 Å).

F. A. Cotton, T. R. Felthouse (Inorg. Chem. **20** [1981] 584/600).

$Rh_2(O_2CCH_3)_2(HOH\cdots O_2CCH_3)_2(NH_3)_2$ is prepared by the reduction of $\{Rh(O_2CCH_3)(HOH\cdot O_2CCH_3)(OH)(NH_3)\}_2$ (see p. 59) with ethyl alcohol in acetic acid at 50 to 60°C, and is isolated as a yellow powder. A modified "lantern" structure with the rhodium atoms bridged by two acetate anions and two $HOH\cdots OC(O)CH_3$ groups is proposed. Selected infrared bands are reported, the electronic spectrum is recorded, λ_{max} = 587, 395, 305, 255 and 193 nm. The complex is diamagnetic. Thermal decomposition data are recorded and a thermogravimetric analysis curve is reproduced [1].

$Rh_2(O_2CCH_3)_2(HOH\cdots O_2CCH_3)(OH)(H_2O)(NH_3)_2$ is prepared by evaporation of an aqueous solution of the preceding complex at ambient temperature. It is isolated as green plates [1].

The crystals are monoclinic, space group Cc-C_s^4 or C2/c-C_{2h}^6, a = 13.281 ± 0.01, b = 8.623 ± 0.009, c = 14.218 ± 0.01 Å, β = 118.71° ± 0.10°, Z = 4; D_{pykn} = 2.23 g/cm³ [2].

A molecular structure based on the familiar binuclear "lantern" arrangement is proposed. The rhodium atoms are thought to be bridged by two acetate anions and an $HOH\cdots OC(O)CH_3$ group; water and hydroxide ligands occupy equatorial sites on different rhodium atoms, and the ammonia ligands are located on the axial sites [1].

The infrared spectrum (4000 to 250 cm^{-1}) is reproduced and the electronic spectrum is reported to be essentially identical to that of the preceding complex. The complex is diamagnetic. Thermal decomposition data are recorded and a thermogravimetric analysis curve is reproduced. Structural diagrams for the preceding two complexes are transposed in the original paper [1].

References:

[1] G. Pannetier, J. Segall (J. Less-Common Metals **22** [1970] 305/13). – [2] A. Dereigne, J. M. Manoli, J. Segall (J. Less-Common Metals **22** [1970] 314/6).

$Rh_2(O_2CCH_3)_2(CH_3COCHCOCH_3)_2$ is obtained as an insoluble brown precipitate by heating together under reflux in aqueous solution $Rh_2(O_2CCH_3)_4(H_2O)_2$ and excess sodium acetyl acetonate. It is insoluble in common organic solvents but dissolves in pyridine to form an unstable adduct. The electronic spectrum (λ_{max} = 688 nm) and selected infrared bands, ν(CO), are reported.

$Rh_2(O_2CCH_3)_2(CF_3COCHCOCH_3)_2(H_2O)_2$ is obtained as an insoluble green precipitate by heating under reflux in aqueous solution $Rh_2(O_2CCH_3)_4(H_2O)_2$ and trifluoroacetylacetone. The electronic spectrum with λ_{max} = 682, 420 (shoulder), 360 nm and selected infrared bands, ν(CO), are reported.

$Rh_2(O_2CCH_3)_2(CF_3COCHCOCH_3)_2(C_5H_5N)_2$ is obtained as brick red crystals by evaporating a pyridine solution of the corresponding diaquo adduct and reprecipitating the residue from benzene/petroleum ether. The electronic spectrum with λ_{max} = 560, 420 (shoulder), 353 nm and selected infrared bands, ν(CO), are reported.

$Rh_2(O_2CCH_3)_2(CF_3COCHCOCH_3)_2(2\text{-}CH_3C_5H_4N)_2$ is similarly obtained as red-brown crystals using 2-picoline. Selected infrared bands, ν(CO), are reported.

$Rh_2(O_2CCH_3)_2(CF_3COCHCOCF_3)_2(H_2O)_2$ is obtained by repeated, prolonged reflux of $Rh_2(O_2CCH_3)_4(H_2O)_2$ with hexafluoroacetylacetonate in aqueous solution. The green-yellow complex is slightly soluble in methanol and ethanol, and gives red-brown adducts with donor solvents. The electronic spectrum with λ_{max} = 610 to 600, 450, 390 nm and selected infrared bands, ν(CO), are reported.

$Rh_2(O_2CCH_3)_2(CF_3COCHCOCF_3)_2(C_5H_5N)_2$ is obtained by dissolving the corresponding diaquo adduct in pyridine, then evaporating the solution to dryness and crystallising the residue from petroleum ether. The brick red complex is very soluble in ether and chlorinated solvents. The electronic spectrum with λ_{max} = 600, 445 and 440 nm and selected infrared bands, ν(CO), are recorded.

$Rh_2(O_2CCH_3)_2(CF_3COCHCOCF_3)_2(2\text{-}CH_3C_5H_4N)_2$ is similarly prepared using 2-picoline and is isolated as a deep red solid. Selected infrared bands, ν(CO), are recorded.

$Rh_2(O_2CCH_3)_2(CF_3COCHCOCF_3)_2(2\text{-}C_5H_4ClN)_2$ is similarly prepared using 2-chloropyridine and is isolated as a crystalline red-brown solid. Selected infrared bands, ν(CO), are reported.

The preceding eight compounds are diamagnetic and are thought to possess binuclear Rh-Rh bonded structures with bridging acetate ligands, chelate β-diketonate ligands and axial H_2O or N-donor groups.

S. Cenini, R. Ugo, F. Bonati (Inorg. Chim. Acta **1** [1967] 443/7).

$Rh_2(O_2CCH_3)_2(C_{10}H_8N_2)_2Cl_2$ is prepared by treatment of a rhodium(III) allyl complex $[Rh_2Cl_2(C_3H_5)_4$ or $RhCl_2(C_4H_7)]$ with acetic acid and bipyridyl in ethanol at 100°C. Far infrared data (450 to 150 cm^{-1}) are reported and assignments given [1]. The electronic spectrum has been recorded [2].

$Rh_2(O_2CCH_3)_2(o\text{-}C_{12}H_8N_2)_2Cl_2 \cdot 3\,CH_3CO_2H$ is similarly prepared using o-phenanthroline [1, 2]. Far infrared data (450 to 150 cm^{-1}) have been recorded and assignments given [1]. The electronic spectrum has been recorded [2]. The above complexes are thought to possess binuclear Rh-Rh bonded structures with bridging acetate and chelating bipyridyl or o-phenanthroline ligands (see corresponding formate complexes, pp. 14/5).

$Rh_2(O_2CCH_3)_2(C_4H_4N_2O_2)_2(H_2O)_2$ is obtained as a black precipitate by warming a solution of $Rh_2(O_2CCH_3)_4(H_2O)_2$ and dimethylglyoxime in methanol for 40 min [3].

$Rh_2(O_2CCH_3)_2(C_4H_4N_2O_2)_2\{P(C_6H_5)_3\}_2(H_2O)$ is prepared by slow evaporation of a methanol solution containing the above compound and triphenylphosphine. It forms deep red crystals.

The crystals are monoclinic, space group $P2_1/b\text{-}C^5_{2h}$ with a = 15.21(2), b = 28.88(3), c = 13.79(2) Å, β = 125.55(3)°, Z = 4. The molecules are binuclear with a Rh-Rh bond, 2.618(5) Å, bridging acetate ligands, chelating dimethylglyoxime ligands and axially located triphenylphosphine ligands as shown in **Fig. 6**. Metal-ligand distances are Rh-O = 2.081(7) to 2.094(6) Å, Rh-N = 1.952(4) to 1.996(6) Å, Rh-P = 2.476(9) to 2.494(9) Å [3].

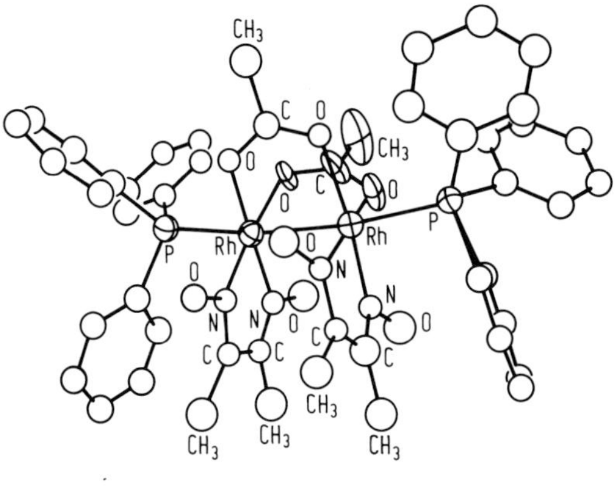

Fig. 6
Molecular structure of $Rh_2(O_2CCH_3)_2(C_4H_4N_2O_2)_2\{P(C_6H_5)_3\}_2$.

References:

[1] F. Pruchnik, M. Zuber, H. Pasternak, K. Wajda (Spectrochim. Acta A **34** [1978] 1111/2). – [2] F. Pruchnik, M. Zuber (Roczniki Chem. **51** [1977] 1813/9). – [3] J. Halpern, E. Kimura, J. Molin-Case, C. S. Wong (Chem. Commun. **1971** 1207/8).

Methoxyacetate, $Rh_2(O_2CCH_2OCH_3)_4$

This is prepared from $Rh_2(O_2CCH_3)_4$ by treatment with free methoxyacetic acid, no other details are given [1].

The anti-tumour activity of $Rh_2(O_2CCH_2OCH_3)_4$, coupled with its good solubility properties, has led to several studies of adduct formation with ligands of biological interest. Kinetic and/or thermodynamic data for adduct formation have been reported with niacin [2], isonicotinic acid [2], imidazole [3, 4, 5], pyridine [2, 4], L-histidine [4, 5], 5'-adenosine-mono-, di-, and triphosphates [4, 5, 6]. The interaction of $Rh_2(O_2CCH_2OCH_3)_4$ with numerous enzymes [7], with leukaemia L 1210 cells [8] and with Ehrlich ascites tumour [1] has also been investigated. Redox properties and anti-tumour activity have been examined [9].

References:

[1] R. A. Howard, E. Sherwood, A. Erck, A. P. Kimball, J. L. Bear (J. Med. Chem. **20** [1977] 943/6). – [2] J. L. Bear, R. A. Howard, J. E. Korn (Inorg. Chim. Acta **32** [1979] 123/6). – [3] K. Das, J. L. Bear (Inorg. Chem. **15** [1976] 2093/5). – [4] K. Das, E. L. Simmons, J. L. Bear (Inorg. Chem. **16** [1977] 1268/71). – [5] R. A. Howard, T. G. Spring, J. L. Bear (J. Clin. Hematol. Oncol. **7** [1977] 391/400).

[6] L. Rainen, R. A. Howard, A. P. Kimball, J. L. Bear (Inorg. Chem. **14** [1975] 2752/4). – [7] R. A. Howard, T. G. Spring, J. L. Bear (Cancer Res. **36** [1976] 4402/5) – [8] R. A. Howard, A. P. Kimball, J. L. Bear (Cancer Res. **39** [1979] 2568/73). – [9] K. M. Kadish, K. Das, R. Howard, A. Dennis, J. L. Bear (Bioelectrochem. Bioenergetics **5** [1978] 741/53).

Phenoxyacetate, $Rh_2(O_2CCH_2OC_6H_5)_4$

This is prepared from rhodium(II) acetate dimer and phenoxyacetic acid, or from rhodium trichloride trihydrate and an equimolar mixture of free phenoxyacetic acid and its sodium salt in alcohol.

The electrochemical oxidation-reduction has been investigated by polarography, cyclic voltammetry and controlled potential electrolysis.

K. Das, K. M. Kadish, J. L. Bear (Inorg. Chem. **17** [1978] 930/4).

Aminoacetate, $Rh_2(O_2CCH_2NH_2)_4\{(C_2H_5)_2O\}_2$

The preparative method employed to prepare this pale green compound is not clearly reported, and the characterisation is incomplete. The complex is apparently paramagnetic, $\mu_{eff} = 1.4$ B.M.

V. I. Belova, Z. S. Dergacheva (Zh. Neorgan. Khim. **16** [1971] 3065/70; Russ. J. Inorg. Chem. **16** [1971] 1626/9).

Monofluoroacetates

$Rh_2(O_2CCH_2F)_4(H_2O)_2$. Selected infrared bands (3600 to 400 cm^{-1}) have been recorded but no preparative details are reported.

$Rh_2(O_2CCH_2F)_4(C_5H_5N)_2$ is mentioned as a precursor but no details reported.

$K_2[Rh_2(O_2CCH_2F)_4(CN)_2]\cdot 0.5\,H_2O$ is obtained as an orange solid by treatment of an alcoholic solution of $Rh_2(O_2CCH_2F)_4(C_5H_5N)_2$ with aqueous potassium cyanide. The infrared spectrum (3600 to 500 cm^{-1}) has been reproduced and assignments given, $\nu(CN) = 2105\ cm^{-1}$.

T. A. Mal'kova, V. N. Shafranskii, Yu. Ya. Kharitonov (Koord. Khim. **3** [1977] 1747/52; Soviet J. Coord. Chem. **3** [1977] 1371/6).

Trifluoroacetates

$Rh_2(O_2CCF_3)_4$. This complex is prepared by the action of trifluoroacetic acid on hydrous rhodium(III) oxide, followed by extraction with dichloromethane and crystallisation from benzene [1]. It has also been obtained by repeated treatment of $Rh_2(O_2CCH_3)_4$ with trifluoroacetic acid, followed by evaporation of the solution to dryness and heating the residue at 110°C [2]. Heating of the bis-ethanol adduct $Rh_2(O_2CCF_3)_4(C_2H_5OH)_2$ at 120°C for 30 min also affords the anhydrous complex [3]. The complex forms blue-green crystals [1], but has been reported to be purple at 77 K [4]. The infrared spectrum (4000 to 60 cm^{-1}) has been recorded and assignments given [5]. The mass spectrum has been measured and has been reproduced; the highest observed m/e value (658) corresponds to the expected molecular weight of the intact dimer. The complex is diamagnetic [4]. The electrochemical oxidation-reduction of $Rh_2(O_2CCF_3)_4$ has been investigated by polarography, cyclic voltammetry and controlled potential electrolysis [6].

$Rh_2(O_2CCF_3)_4(H_2O)_2$ has been described as a blue solid [4] but no preparative details have been reported. The infrared spectrum (4000 to 60 cm^{-1}) has been recorded and assignments given, $\nu(OCO) = 1670$, 1650, 1460 cm^{-1} [5].

$Rh_2(O_2CCF_3)_4(H_2O)(C_2H_5OH)$. A thermogravimetric analysis curve has been reproduced for this complex but only a general reference to the method of synthesis is given [7]. The complex is diamagnetic, $\mu_{eff} \approx 0.5$ B.M. [7].

$Rh_2(O_2CCF_3)_4(C_2H_5OH)_2$. This adduct is prepared by heating a mixture of sodium trifluoroacetate and rhodium(III) trichloride trihydrate in ethanol at 55 to 60°C for 20 min. It forms blue needles [3].

The crystals are triclinic, space group P1-C_i^1, a = 10.131(4), b = 10.880(3), c = 13.333(4) Å, $\alpha = 65.86(3)°$, $\beta = 71.45(3)°$, $\gamma = 62.07(2)°$, Z = 2; $D_{exp} = 1.98(1)$, $D_{calc} = 2.115\ g/cm^3$. The molecule, which exists in two crystallographically independent forms, has the familiar "lantern" structure (Fig. 3, p. 9). Bond lengths are Rh-Rh = 2.396(2) Å, Rh-O = 2.02(1), 2.04(1), 2.03(1), 2.03(1) Å and Rh-$O_{C_2H_5OH}$ = 2.28(1) Å (form 1), and Rh-Rh = 2.409(2) Å, Rh-O = 2.04(1), 2.04(1), 2.04(1), 2.04(1) Å, Rh-$O_{C_2H_5OH}$ = 2.26(1) Å (form 2) [8]. The infrared spectrum (4000 to 60 cm^{-1}) has been recorded and assignments given [5], selected infrared frequencies are recorded by [3]. X-ray photoelectronic data have been reported, Rh $3d_{5/2}$ = 310.2 eV [9].

$Rh_2(O_2CCF_3)_4(TMPNO)$ (TMPNO = 2,2,6,6-tetramethylpiperidine-N-oxide). This 1:1 adduct is formed when $Rh_2(O_2CCF_3)_4$ and TMPNO (1:1 molar ratio) are dissolved together in toluene. The electron spin resonance (ESR) spectrum is reproduced, and a rhodium hyperfine coupling of 1.7 ± 0.2 G is reported [10].

$Rh_2(O_2CCF_3)_4\{(CH_3)_2SO\}_2$ is obtained by dissolving $Rh_2(O_2CCF_3)_4$ in a 1:1 benzene-chloroform solution of dimethylsulphoxide, then evaporating to dryness and recrystallising the residue from benzene-chloroform [11].

The crystals are triclinic, space group $P\bar{1}$-C_i^1, a = 8.621(2), b = 9.500(2), c = 8.766(2) Å, α = 112.68(1)°, β = 110.99(2)°, γ = 87.83(2)°, Z = 1; D_{calc} = 2.20 g/cm³. The molecule has the familiar "lantern" structure (Fig. 3, p. 9) with axial $(CH_3)_2SO$ ligands coordinated through oxygen. Bond lengths are Rh-Rh = 2.419(1) Å, Rh-O(CF_3CO_2) = 2.031(3), 2.022(3), 2.016(3) and 2.029(3) Å, Rh-O{$(CH_3)_2SO$} = 2.236(3) Å [11].

The infrared spectrum (4000 to 60 cm^{-1}) has been recorded and assignments given [5].

The X-ray structure of the d_6-$(CH_3)_2SO$ adduct has recently been determined and shown to display a structural deuterium isotope effect [14].

$Rh_2(O_2CCF_3)_4\{(CH_3)_2SO_2\}_2$ is prepared by dissolving the non-solvated complex in a 1:1 solution of benzene/chloroform and adding $(CH_3)_2SO_2$ in excess. It separates from the blue-green solution on evaporation as dark green needles and prisms [12].

The crystals are triclinic, space group $P\bar{1}$-C_i^1, lattice parameters a = 10.482(1), b = 14.004(2), c = 9.533(2) Å, α = 108.17(2)°, β = 101.64(2)°, γ = 82.27(1)°, Z = 2; D_{calc} = 2.16 g/cm³. The molecule has the familiar "lantern" structure with axial $(CH_3)_2SO_2$ ligands coordinated through oxygen. Bond lengths are Rh-Rh = 2.401(1), Rh-O(carboxylate) = 2.030(3) (mean), Rh-O$(CH_3)_2SO_2$ = 2.291(3) Å [12].

$Rh_2(O_2CCF_3)_4(C_5H_5N)_2$ is formed by dropwise addition of pyridine to a cold, ethanolic solution of $Rh_2(O_2CCF_3)_4$ and forms red crystals from dichloromethane [1]. It has also been described as a red-orange solid [4]. The infrared spectrum (4000 to 60 cm^{-1}) has been recorded and assignments made [5]. X-ray photoelectronic data have been reported, Rh $3d_{5/2}$ = 309.6 eV [9].

$Rh_2(O_2CCF_3)_4\{N(C_2H_5)_3\}_2$ is described as a rose-red solid, no preparative details are given [4].

$Rh_2(O_2CCF_3)_4\{S(C_2H_5)_2\}_2$ is described as a rose-red solid, no preparative details are given [4].

$Rh_2(O_2CCF_3)_4(PPh_3)_2$ is obtained by mixing stoichiometric amounts of $Rh_2(O_2CCF_3)_4$ and triphenylphosphine in methanol, and precipitates as a microcrystalline yellow-brown solid with a purple sheen [13]. It has also been described as a yellow-orange solid [4].

The crystals are triclinic, space group $P\bar{1}$-C_i^1, lattice parameters a = 9.974(1), b = 13.365(2), c = 9.154(2) Å, α = 105.24(1)°, β = 91.06(1)°, γ = 107.42(1)°, Z = 1; D_{calc} = 1.76 g/cm³. Important bond lengths are Rh-Rh = 2.486(1), Rh-P = 2.494(2), Rh-O = 2.045(5) Å (mean) [13].

ESCA data have been recorded; Rh $3d_{5/2}$ = 310.0 eV [9].

$Rh_2(O_2CCF_3)_4\{P(OPh)_3\}_2$ is prepared by mixing stoichiometric amounts of $Rh_2(O_2CCF_3)_4$ and triphenylphosphite in methanol, and is isolated as an orange-brown solid which forms thin yellow plates from methanol [13].

The crystals are triclinic, space group $P\bar{1}$-C_i^1, lattice parameters a = 9.772(1), b = 14.194(2), c = 9.565(2) Å, α = 103.76(1)°, β = 93.38(2)°, γ = 74.90(1)°, Z = 1; D_{calc} = 1.71 g/cm³. Important bond lengths are Rh-Rh = 2.470(1), Rh-P = 2.422(2), Rh-O = 2.042(5) Å (mean) [13].

$Rh_2(O_2CCF_3)_4(C_9H_{15}NO_2)_2$ ($C_9H_{15}NO_2$ = 2,2,6,6-tetramethyl-4-hydroxypiperidinyl-1-oxy). The preparation and X-ray crystal structure of this complex have recently been reported [15].

$Rh_2(O_2CCF_3)_4(H_2O)_2(C_8H_{18}NO)_2$ ($C_8H_{18}NO$ = di-tert-butyl-nitroxide). The preparation and X-ray crystal structure of this complex have recently been reported [15].

References:

[1] T. A. Stephenson, S. M. Morehouse, A. R. Powell, J. P. Heffer, G. Wilkinson (J. Chem. Soc. **1965** 3632/40). – [2] S. A. Johnson, H. R. Hunt, H. M. Neumann (Inorg. Chem. **2** [1963] 960/2). – [3] G. Winkhaus, P. Ziegler (Z. Anorg. Allgem. Chem. **350** [1967] 51/61). – [4] J. Kit-

chens, J. L. Bear (Thermochim. Acta **1** [1970] 537/44). – [5] G. Ya. Mazo, I. B. Baranovskii, R. N. Shchelokov (Zh. Neorgan. Khim. **24** [1979] 3330/6; Russ. J. Inorg. Chem. **24** [1979] 1855/7).

[6] K. Das, K. M. Kadish, J. L. Bear (Inorg. Chem. **17** [1978] 930/4). – [7] V. I. Belova, Z. S. Dergacheva (Zh. Neorgan. Khim. **16** [1971] 3065/70; Russ. J. Inorg. Chem. **16** [1971] 1626/9). – [8] M. A. Porai-Koshits, L. M. Dikareva, G. G. Sadikov, I. B. Baranovskii (Zh. Neorgan. Khim. **24** [1979] 1286/95; Russ. J. Inorg. Chem. **24** [1979] 716/21). – [9] V. I. Nefedov, Ya. V. Salyn', I. B. Baranovskii, A. G. Maiorova (Zh. Neorgan. Khim. **25** [1980] 216/25; Russ. J. Inorg. Chem. **25** [1980] 116/22). – [10] R. M. Richman, T. C. Kuechler, S. P. Tanner, R. S. Drago (J. Am. Chem. Soc. **99** [1977] 1055/8).

[11] F. A. Cotton, T. R. Felthouse (Inorg. Chem. **19** [1980] 2347/51). – [12] F. A. Cotton, T. R. Felthouse (Inorg. Chem. **20** [1981] 2703/8). – [13] F. A. Cotton, T. R. Felthouse, S. Klein (Inorg. Chem. **20** [1981] 3037/42). – [14] F. A. Cotton, T. R. Felthouse (Inorg. Chem. **21** [1982] 431/5). – [15] F. A. Cotton, T. R. Felthouse (Inorg. Chem. **21** [1982] 2667/75).

Monochloroacetates

$Rh_2(O_2CCH_2Cl)_4(H_2O)_2$. An elemental analysis has been reported for this adduct, but no preparative details are recorded [1]. The infrared spectrum (3600 to 400 cm^{-1}) has been recorded and selected assignments given, $\nu(OCO) = 1608$ and 1418 cm^{-1} [2, 4].

$Rh_2(O_2CCH_2Cl)_4(C_2H_5OH)_2$ is prepared by treatment of $RhCl_3 \cdot 3H_2O$ with sodium monochloroacetate in ethanol solution and is isolated as green needles. The infrared spectrum has been recorded and selected assignments given [3].

$K_2[Rh_2(O_2CCH_2Cl)_4(NCO)_2] \cdot H_2O$ is obtained as violet tetragonal platelets by thoroughly grinding $Rh_2(O_2CCH_2Cl)_4(H_2O)_2$ and excess KNCO in water/ethanol. The infrared spectrum (3600 to 500 cm^{-1}) is reproduced and assignments given. Thermal decomposition data are recorded [2].

$K_2[Rh_2(O_2CCH_2Cl)_4(NCS)_2] \cdot 2H_2O$ is similarly prepared from $Rh_2(O_2CCH_2Cl)_4(H_2O)_2$ using KNCS and is isolated as violet-green tetragonal platelets. The infrared spectrum (3600 to 500 cm^{-1}) is reproduced and assignments given. Thermal decomposition data are recorded [2].

$K[Rh_2(O_2CCH_2Cl)_4(NCSe)] \cdot 2H_2O$ is similarly prepared using KNCSe but could not be fully purified. The infrared spectrum (3600 to 500 cm^{-1}) is reproduced and assignments given. Thermal decomposition data are recorded [2].

$K_2[Rh_2(O_2CCH_2Cl)_4(CN)_2] \cdot 2H_2O$ is obtained as orange crystals by treatment of $Rh_2(O_2CCH_2Cl)_4(C_5H_5N)_2$ in alcohol solution with aqueous KCN solution. The infrared spectrum (3600 to 500 cm^{-1}) has been recorded and assignments given, $\nu(CN) = 2143$ cm^{-1} [4].

References:

[1] V. I. Belova, Z. S. Dergacheva (Zh. Neorgan. Khim. **16** [1971] 3065/70; Russ. J. Inorg. Chem. **16** [1971] 1626/9). – [2] V. N. Shafranskii, T. A. Mal'kova, Yu. Ya. Kharitonov (Koord. Khim. **1** [1975] 375/83; Soviet J. Coord. Chem. **1** [1975] 297/303). – [3] G. Winkhaus, P. Ziegler (Z. Anorg. Allgem. Chem. **350** [1967] 51/61). – [4] T. A. Mal'kova, V. N. Shafranskii, Yu. Ya. Kharitonov (Koord. Khim. **3** [1977] 1747/52; Soviet J. Coord. Chem. **3** [1977] 1371/6).

Dichloroacetates

$Rh_2(O_2CCHCl_2)_4$ is prepared by heating the ethanol adduct (see p. 41) at ~120°C for 30 min. The infrared spectrum has been recorded.

$Rh_2(O_2CCHCl_2)_4(C_2H_5OH)_2$ is prepared by treatment of $RhCl_3 \cdot 3H_2O$ with sodium dichloroacetate in ethanol at 55 to 60°C for 20 min, and is isolated as blue needles. The infrared spectrum has been recorded and selected assignments given, ν(OCO) = 1619, 1395 cm^{-1}.

$Rh_2(O_2CCHCl_2)_4(C_5H_5N)_2$ is prepared by treatment of the corresponding ethanol adduct with pyridine in ethanol, and is isolated as red crystals.

$Rh_2(O_2CCHCl_2)_4\{P(C_6H_5)_3\}_2$ is obtained as brown needles by treatment of the corresponding ethanol adduct with triphenylphosphine in ethanol.

G. Winkhaus, P. Ziegler (Z. Anorg. Allgem. Chem. **350** [1967] 51/61).

Trichloroacetates

$Rh_2(O_2CCCl_3)_4$ is prepared by heating the corresponding ethanol adduct (see below) at 120°C for 30 min [1]. The complex is reported to be green in colour [2]. Infrared data (1800 to 600 cm^{-1}) have been recorded [1]. The complex is diamagnetic, $\mu_{eff} = 0.41$ B.M. [2].

$Rh_2(O_2CCCl_3)_4(C_2H_5OH)_2$ is prepared by treatment of $RhCl_3 \cdot 3H_2O$ with sodium trichloroacetate in ethanol at 55 to 60°C for ~20 min. It forms blue needles from chloroform/light petroleum [1]. The infrared spectrum (3600 to 600 cm^{-1}) has been recorded and selected assignments given, ν(OCO) = 1645 and 1367 cm^{-1} [1]. X-ray photoelectronic data have been recorded, Rh $3d_{5/2} = 309.7$ eV [3].

$Rh_2(O_2CCCl_3)_4\{(CH_3)_2SO\}_2$ is reported to be blue in solution (solvent not specified) and presumably has O-bonded $(CH_3)_2SO$ ligands. A thermogravimetric analysis curve is reproduced [2].

$Rh_2(O_2CCCl_3)_4(C_5H_5N)_2$ is prepared by treatment of the corresponding ethanol adduct with pyridine [1] and is isolated as a rose-red solid [1, 2]. The infrared spectrum (3200 to 600 cm^{-1}) has been recorded [1]. X-ray photoelectronic data have been reported, Rh $3d_{5/2} = 309.4$ eV [3].

$Rh_2(O_2CCCl_3)_4\{N(C_2H_5)_3\}_2$ is described as a rose-red solid, no preparative details are given [2].

$Rh_2(O_2CCCl_3)_4\{S(C_2H_5)_2\}_2$ is described as a burgundy solid. No preparative details are given [2].

$Rh_2(O_2CCCl_3)_4(C_6H_5SH)_2$. Infrared data have been published but no preparative details are given [1].

$Rh_2(O_2CCCl_3)_4\{P(C_6H_5)_3\}_2$ is prepared by treatment of the corresponding ethanol adduct with triphenylphosphine in ethanol solution [1], and is isolated as brown [1] or orange [2] crystals. The infrared spectrum (3100 to 600 cm^{-1}) has been reported [1].

References:

[1] G. Winkhaus, P. Ziegler (Z. Anorg. Allgem. Chem. **350** [1967] 51/61). – [2] J. Kitchens, J. L. Bear (Thermochim. Acta **1** [1970] 537/44). – [3] V. I. Nefedov, Ya. V. Salyn', I. B. Baranovskii, A. G. Maiorova (Zh. Neorgan. Khim. **25** [1980] 216/25; Russ. J. Inorg. Chem. **25** [1980] 116/22).

Chlorodifluoroacetates

$Rh_2(O_2CCClF_2)_4$ is obtained as a green solid by heating the ethanol adduct (see below) at 120°C for 30 min. The infrared spectrum (1700 to 700 cm^{-1}) has been recorded [1].

$Rh_2(O_2CCClF_2)_4(C_2H_5OH)_2$ is obtained as blue needles by heating a mixture of $RhCl_3 \cdot 3H_2O$ and sodium chloro-difluoroacetate in ethanol at 55 to 60°C for ~20 min. The infrared spectrum (3000 to 700 cm^{-1}) has been recorded and selected assignments given, ν(OCO) = 1656 and 1428 cm^{-1}.

G. Winkhaus, P. Ziegler (Z. Anorg. Allgem. Chem. **350** [1967] 51/61).

Monobromoacetate, $Rh_2(O_2CCH_2Br)_4(C_2H_5OH)_2$

This is prepared by treatment of $RhCl_3 \cdot 3H_2O$ with sodium monobromoacetate in ethanol at 55 to 60°C for ~20 min, and is isolated as green rhombic crystals. The infrared spectrum (3000 to 700 cm^{-1}) has been recorded and selected assignments given, ν(OCO) = 1600 and 1412 cm^{-1}.

G. Winkhaus, P. Ziegler (Z. Anorg. Allgem. Chem. **350** [1967] 51/61).

Phenylacetate, $Rh_2(O_2CCH_2C_6H_5)_4$

A general reference to preparative methods is given but no details are recorded. Electrochemical oxidation-reduction has been studied by polarography, cyclic voltammetry and controlled potential electrolysis.

K. Das, K. M. Kadish, J. L. Bear (Inorg. Chem. **17** [1978] 930/4).

1.1.2.3 Propionates

$Rh_2(O_2CC_2H_5)_4$ is obtained, apparently in unsolvated form, by heating a mixture of hydrated rhodium oxide, excess propionic acid and ethanol under reflux [1]. It is also conveniently prepared by heating under reflux a mixture of $RhCl_3 \cdot nH_2O$ and sodium propionate [2]. It forms green solid [1] or green crystals from acetone solution [2]. The Raman and far infrared spectra have been recorded, ν(Rh-Rh) = 347 cm^{-1}, ν(Rh-O) = 354, 411 and 436 cm^{-1} [3]. The thermogravimetric analysis curve is reproduced in two papers [2, 4]. The electrochemical oxidation-reduction of $Rh_2(O_2CC_2H_5)_4$ has been investigated by polarography, cyclic voltammetry and controlled potential electrolysis, cyclic voltammograms and a polarogram are reproduced [5].

$Rh_2(O_2CC_2H_5)_4(H_2O)_2$ is described as a green solid, no preparative details are given [2]. The infrared spectrum (3600 to 300 cm^{-1}) has been recorded and assignments given, ν(OCO) = 1580, 1470 cm^{-1} [6].

$Rh_2(O_2CC_2H_5)_4\{N(C_2H_5)_3\}_2$ is described as a rose-violet solid, no preparative details are given [2].

$Rh_2(O_2CC_2H_5)_4(C_5H_5N)_2$ is prepared by dropwise addition of pyridine to a cold methanol solution of $Rh_2(O_2CC_2H_5)_4$, and is recrystallised from cyclohexane/light petroleum [1]. It is described as a rose-pink solid [2]. Thermal properties have been investigated and a thermogravimetric analysis curve is reproduced [4].

$Rh_2(O_2CC_2H_5)_4(C_5H_{11}N)_2$ is similarly prepared using piperidine, and is recrystallised from light petroleum as dark red crystals [1].

$Rh_2(O_2CC_2H_5)_4(C_9H_7N)_2$ is prepared by addition of isoquinoline in diethyl ether to a cold methanolic solution of $Rh_2(O_2CC_2H_5)_4$, and is recrystallised from cold benzene/light petroleum as dark red crystals [1].

$Rh_2(O_2CC_2H_5)_4(C_{13}H_9N)_2$ is prepared by addition of a slight excess of acridine to an acetonitrile solution of $Rh_2(O_2CC_2H_5)_4$ and is obtained as dark red-green dichroic crystals by slow evaporation [7].

The crystals are triclinic, space group $P\bar{1}$-C_i^1, a = 8.332(2), b = 10.008(2), c = 11.328(4) Å, α = 109.60(2)°, β = 100.15(3)°, γ = 97.21(2)°, Z = 1; D_{calc} = 1.66 g/cm³. The molecule has the familiar "lantern" structure with axial N-bonded acridine ligands. Bond distances are Rh-Rh = 2.417 Å, Rh-O = 2.040(3), 2.027(3), 2.041(3) and 2.039(3) Å, Rh-N = 2.413(3) Å (average) [7].

$Rh_2(O_2CC_2H_5)_4(C_7H_6N_2)_2$ is similarly prepared using 7-azaindole as ligand and benzene or methanol as solvent. It forms dark red prisms [7].

The crystals are orthorhombic, space group Pbca-D_{2h}^{15}, a = 20.110(2), b = 20.040(2), c = 14.224(2) Å, Z = 8; D_{calc} = 1.70 g/cm³. The molecule has the familiar "lantern" structure with axial azaindole ligands coordinated through the nitrogen of the six-membered ring. Bond distances are Rh-Rh = 2.403(1) Å, Rh-O = 2.047(5), 2.043(5), 2.022(5) and 2.043(5) Å, Rh-N = 2.275(6) Å (average) [7].

$[Rh_2(O_2CC_2H_5)_4(C_{12}H_8N_2)]_n$ is prepared by adding a benzene solution of phenazine to a suspension of $Rh_2(O_2CC_2H_5)_4$ in benzene and stirring the mixture for 3 h. It crystallises as thin red needles and plates from methanol [7].

The crystals are triclinic, space group $P\bar{1}$-C_i^1, a = 8.744(1), b = 9.890(4), c = 8.037(1) Å, α = 111.82(2)°, β = 90.14(1)°, γ = 82.24(2)°, Z = 1; D_{calc} = 1.76 g/cm³. The structure is polymeric with $Rh_2(O_2CC_2H_5)_4$ "lantern" units linked by bridging phenazine ligands. Bond distances are Rh-Rh = 2.409(1) Å, Rh-O = 2.046(3), 2.043(3), 2.033(3) and 2.038(3) Å, Rh-N = 2.362(4) Å [7].

$[Rh_2(O_2CC_2H_5)_4(C_{10}H_{16}N_2)]_n$ is prepared by allowing durenediamine ($C_{10}H_{16}N_2$) and $Rh_2(O_2CC_2H_5)_4$ solutions in methanol to mix slowly by diffusion, and is isolated as red-violet acicular crystals [7].

The X-ray crystal structure is triclinic, space group $P\bar{1}$-C_i^1, a = 8.922(3), b = 9.092(3), c = 8.316(2) Å, α = 104.44(2)°, β = 101.02(2)°, γ = 84.70(2)°, Z = 1; D_{calc} = 1.72 g/cm³. The structure is polymeric with $Rh_2(O_2CC_2H_5)_4$ "lantern" units linked by bridging durenediamine ligands. Bond lengths are Rh-Rh = 2.387(1) Å, Rh-O = 2.030(4), 2.028(4), 2.032(4) and 2.031(4) Å, Rh-N = 2.324(6) Å [7].

$Rh_2(O_2CC_2H_5)_4$(Gun·CN)2H_2O is obtained as a finely crystalline violet solid by treatment of $Rh_2(O_2CC_2H_5)_4(H_2O)_2$ with cyanoguanidine (Gun·CN = $C_2H_4N_4$) in aqueous solution. The infrared spectrum (4000 to 300 cm^{-1}) has been reported and assignments given, ν(CN) = 2224, 2170 cm^{-1} [8]. A polymeric structure with bridging cyanoguanidine ligands is proposed. Thermal decomposition data are recorded [8].

$Rh_2(O_2CC_2H_5)_4$(Gun·NH_2)(H_2O) is obtained as reddish-violet crystals by treatment of $[GunNH_2 \cdot H]_2[Rh_2(O_2CC_2H_5)_4Cl_2]$ with aqueous sodium carbonate (GunNH_2 = CH_6N_4). The infrared spectrum (3600 to 300 cm^{-1}) has been recorded and assignments given [8].

$Rh_2(O_2CC_2H_5)_4D(H_2O)$, $Rh_2(O_2CC_2H_5)_4D_2$. Kinetic and thermodynamic data for formation of complexes of the above stoichiometry in which D is niacin or isonicotinic acid [9], imidazole

[10, 11, 12], L-histidine [11, 12], 5′-AMP [11, 12, 13] and 5′-ATP [12, 13] have been reported (AMP, ATP = adenosine-mono- and -triphosphate).

$Rh_2(O_2CC_2H_5)_4$ with Other Biological Species. The interactions of $Rh_2(O_2CC_2H_5)_4$ with enzymes [12, 14], leukaemia L1210 cells [15] and Ehrlich ascites tumour [16] have been investigated.

References:

[1] T. A. Stephenson, S. M. Morehouse, A. R. Powell, J. P. Heffer, G. Wilkinson (J. Chem. Soc. **1965** 3632/40). – [2] J. Kitchens, J. L. Bear (Thermochim. Acta **1** [1970] 537/44). – [3] A. P. Ketteringham, C. Oldham (J. Chem. Soc. Dalton Trans. **1973** 1067/70). – [4] R. A. Howard, A. M. Wynne, J. L. Bear, W. W. Wendlandt (J. Inorg. Nucl. Chem. **38** [1976] 1015/8). – [5] K. Das, K. M. Kadish, J. L. Bear (Inorg. Chem. **17** [1978] 930/4).

[6] T. A. Mal'kova, V. N. Shafranskii, Yu. Ya. Kharitonov (Koord. Khim. **3** [1977] 1747/52; Soviet J. Coord. Chem. **3** [1977] 1371/6). – [7] F. A. Cotton, T. R. Felthouse (Inorg. Chem. **20** [1981] 600/8). – [8] T. A. Veteva, V. N. Shafranskii (Zh. Obshch. Khim. **49** [1979] 488/93; J. Gen. Chem. [USSR] **49** [1979] 428/33). – [9] J. L. Bear, R. A. Howard, J. E. Korn (Inorg. Chim. Acta **32** [1979] 123/6). – [10] K. Das, J. L. Bear (Inorg. Chem. **15** [1976] 2093/5).

[11] K. Das, E. L. Simmons, J. L. Bear (Inorg. Chem. **16** [1977] 1268/71). – [12] R. A. Howard, T. G. Spring, J. L. Bear (J. Clin. Hematol. Oncol. **7** [1977] 391/400). – [13] L. Rainen, R. A. Howard, A. P. Kimball, J. L. Bear (Inorg. Chem. **14** [1975] 2752/4). – [14] R. A. Howard, T. G. Spring, J. L. Bear (Cancer Res. **36** [1976] 4402/5). – [15] R. A. Howard, A. P. Kimball, J. L. Bear (Cancer Res. **39** [1979] 2568/73).

[16] J. L. Bear, H. B. Gray, L. Rainen, I. M. Chang, R. Howard, G. Serio, A. P. Kimball (Cancer Chemotherapy Rept. I **59** [1975] 611/20).

$K_2[Rh_2(O_2CC_2H_5)_4(NCO)_2]\cdot 2H_2O$ is obtained as fine violet crystals by thoroughly grinding together $Rh_2(O_2CC_2H_5)_4$ and KNCO in water or aqueous ethanol and filtering to remove excess potassium cyanate. The infrared spectrum (3600 to 400 cm^{-1}) has been reproduced and assignments given. The cyanate ligands are bound through nitrogen, ν(CN) = 2214 cm^{-1}. Thermal decomposition has been investigated [1].

$K[Rh_2(O_2CC_2H_5)_4(NCS)]\cdot 2H_2O$ is similarly prepared using KNCS, and is isolated as violet-green dichroic tetragonal platelets. The infrared spectrum (3600 to 400 cm^{-1}) has been reproduced and assignments given, ν(CN) = 2121 cm^{-1}. The thiocyanate ligands link $Rh_2(O_2CC_2H_5)_4$ units. Thermal decomposition was investigated [1].

$K[Rh_2(O_2CC_2H_5)_4(NCSe)]\cdot 2H_2O$ is similarly prepared using KNCSe, and is isolated as an impure grey-violet solid, possibly contaminated with selenium. The infrared spectrum (3600 to 400 cm^{-1}) has been reproduced and assignments given, ν(CN) = 2108 cm^{-1}. The selenocyanate ligands link $Rh_2(O_2CC_2H_5)_4$ units. Thermal decomposition data are reported [1].

$[Gun\,H]_2[Rh_2(O_2CC_2H_5)_4Cl_2]$ is obtained as green crystals by mixing aqueous solutions of $Rh_2(O_2CC_2H_5)_4(H_2O)_2$ and guanidine hydrochloride, and allowing the solution to evaporate slowly. The infrared spectrum (3600 to 500 cm^{-1}) has been recorded and assignments given [2].

$[Gun\cdot NH_2H]_2[Rh_2(O_2CC_2H_5)_4Cl_2]$ is obtained as green crystals by addition of crystalline aminoguanidine hydrochloride to an ethanolic solution of $Rh_2(O_2CC_2H_5)_4(H_2O)_2$. The infrared spectrum (3600 to 300 cm^{-1}) has been recorded and assignments given. Thermal decomposition data have been recorded [2].

References:

[1] V. N. Shafranskii, T. A. Mal'kova, Yu. Ya. Kharitonov (Koord. Khim. **1** [1975] 375/83; Soviet J. Coord. Chem. **1** [1975] 297/303). – [2] T. A. Veteva, V. N. Shafranskii (Zh. Obshch. Khim. **49** [1979] 488/93; J. Gen. Chem. [USSR] **49** [1979] 428/33).

$Rh_2(O_2CC_2H_5)_4\{(CH_3)_2SO\}_2$ is prepared by adding drops of dimethylsulphoxide to anhydrous $Rh_2(O_2CC_2H_5)_4$ [1] or $Rh_2(O_2CC_2H_5)_4(H_2O)_2$ [2] and removing excess under reduced pressure, or by dissolving $Rh_2(O_2CC_2H_5)_4(H_2O)_2$ in 1:1 $H_2O/(CH_3)_2SO$ and evaporating the orange solution formed [3]. It forms orange prisms [3, 4].

The crystals are monoclinic, space group $P2_1/m\text{-}C^2_{2h}$, a = 9.057(2), b = 15.709(4), c = 18.164(2) Å, β = 101.62(1)°, Z = 4; D_{calc} = 1.72 g/cm³. The molecule has the familiar "lantern" structure, the axial $(CH_3)_2SO$ ligands are coordinated through sulphur (see figure in original paper). Bond distances are Rh-Rh = 2.407 Å, Rh_I-O = 2.025(3), 2.026(3), 2.028(3), 2.027(3) Å, Rh_I-S = 2.453(1) Å, Rh_{II}-O = 2.025(3), 2.033(3), 2.019(4), 2.042(3) Å, Rh_{II}-S = 2.445(1) Å [3].

The infrared spectrum (2000 to 400 cm^{-1}) has been reproduced and assignments given [2]. Thermal properties have been investigated and a thermogravimetric analysis curve has been reproduced [1].

$Rh_2(O_2CC_2H_5)_4\{(C_2H_5)_2SO\}_2$ precipitates from alcoholic solution of $Rh_2(O_2CC_2H_5)_4(H_2O)_2$ on addition of diethylsulphoxide. The infrared spectrum (2000 to 400 cm^{-1}) has been reproduced and assignments given, thermal properties have been investigated [2].

$Rh_2(O_2CC_2H_5)_4\{S(C_2H_5)_2\}_2$ is described as a burgundy solid but no other information is recorded [4].

$Rh_2(O_2CC_2H_5)_4\{P(C_6H_5)_3\}_2$ is prepared by addition of a diethyl ether solution of triphenylphosphine to a cold methanolic solution of $Rh_2(O_2CC_2H_5)_4$, and is crystallised from cyclohexane [5]. It is reported to be yellow-orange in colour [4].

$Rh_2(O_2CC_2H_5)_4\{As(C_6H_5)_3\}_2$ is obtained by treating $Rh_2(O_2CC_2H_5)_4$ or its methanol adduct with $As(C_6H_5)_3$ in boiling methanol, and precipitates rapidly as maroon crystals, melting point 228 to 229°C [6].

References:

[1] R. A. Howard, A. M. Wynne, J. L. Bear, W. W. Wendlandt (J. Inorg. Nucl. Chem. **38** [1976] 1015/8). – [2] T. A. Mal'kova, V. N. Shafranskii (Zh. Obshch. Khim. **47** [1977] 2592/6; J. Gen. Chem. [USSR] **47** [1977] 2365/9). – [3] F. A. Cotton, T. R. Felthouse (Inorg. Chem. **19** [1980] 2347/51). – [4] J. Kitchens, J. L. Bear (Thermochim. Acta **1** [1970] 537/44). – [5] T. A. Stephenson, S. M. Morehouse, A. R. Powell, J. P. Heffer, G. Wilkinson (J. Chem. Soc. **1965** 3632/40).

[6] R. W. Mitchell, J. D. Ruddick, G. Wilkinson (J. Chem. Soc. A **1971** 3224/30, 3228).

$K[(H_2O)\{Rh_2(O_2CC_2H_5)_4\}CN\{Rh_2(O_2CC_2H_5)_4\}(H_2O)]\cdot 2H_2O$. This tetranuclear cyanide-bridged complex anion is obtained as its potassium salt by addition of crystalline KCN to a concentrated ethanolic solution of $Rh_2(O_2CC_2H_5)_4(H_2O)_2$. It forms violet crystals. The infrared spectrum (3600 to 500 cm^{-1}) is reproduced and assignments given, ν(CN) = 2127 cm^{-1} [1].

$K_3[(NC)\{Rh_2(O_2CC_2H_5)_4\}CN\{Rh_2(O_2CC_2H_5)_4\}(CN)]$. This tetranuclear cyanide-bridged complex anion is obtained as its potassium salt by addition of concentrated aqueous potassium cyanide to an alcoholic solution of $Rh_2(O_2CC_2H_5)_4(H_2O)_2$. It is an orange solid. The infrared spectrum (3600 to 500 cm^{-1}) is reproduced and assignments given, ν(CN) = 2140, 2102 cm^{-1} [1].

$K_2[Rh_2(O_2CC_2H_5)_4(CN)_2]$ is obtained as an orange solid by stirring a mixture of $Rh_2(O_2CC_2H_5)_4(C_5H_5N)_2$ or $Rh_2(O_2CC_2H_5)_4$(quinoline)$_2$ and KCN in a small volume of water. The infrared spectrum (3600 to 500 cm^{-1}) is reproduced and assignments given [1].

$Rh_2(O_2CC_2H_5)_4(CNC_6H_{11})_2$ is obtained as an orange solid by dropwise addition of cyclohexylisocyanide to a saturated alcoholic solution of $Rh_2(O_2CC_2H_5)_4(H_2O)_2$. The infrared spectrum (3000 to 400 cm^{-1}) has been recorded and assignments given, ν(CN) = 2172 cm^{-1}. Thermogravimetric curves are reproduced [2].

References:

[1] T. A. Mal'kova, V. N. Shafranskii, Yu. Ya. Kharitonov (Koord. Khim. **3** [1977] 1747/52; Soviet J. Coord. Chem. **3** [1977] 1371/6). – [2] V. N. Shafranskii, T. A. Mal'kova (Zh. Obshch. Khim. **46** [1976] 1197/200; J. Gen. Chem. [USSR] **46** [1976] 1181/4).

Aminopropionates (Alaninates)

α-Alaninates

$Rh_2\{O_2CCH(NH_2)CH_3\}_4Py_2$ is prepared by heating an aqueous mixture of $Rh_2(O_2CCH_3)_4$-$(H_2O)_2$ and D-α-alanine on a waterbath for 1 h, removing precipitated rhodium metal and excess D-α-alanine then adding excess pyridine. It deposits in poor yield (~5%) as a fine pink crystalline precipitate. At 160 to 170°C two pyridine ligands are lost and the colour changes from pink to green [1].

$Rh_2\{O_2CCH(NH_2)CH_3\}_4\{(CH_3)_2SO\}_2$ is similarly prepared using dimethyl sulphoxide in place of pyridine, and deposits in poor yield (~13%) as an orange solid. Infrared data have been reported: ν(NH_2) = 3245, 3150; ν(SO) = 1070; ν(Rh-O) = 392, 381, 351 cm^{-1}; the colour and the ν(SO) value indicate coordination of the $(CH_3)_2SO$ ligand through sulphur. The electronic spectrum has been recorded, λ_{max} = 625 and 391 nm. ESCA data are Rh $3d_{5/2}$ = 308.9, N 1s = 400.7, S 2p = 166.6 eV [1].

β-Alaninate, $[Rh_2(O_2CCH_2CH_2NH_3)_4(H_2O)_2](ClO_4)_4 \cdot 2H_2O$

This complex salt is prepared by treating $Rh_2(O_2CCH_2CH_2NHCOCOBu^t)_4$ with perchloric acid in methanol solution at 0°C, then allowing the mixture to warm slowly to 45°C and maintaining that temperature until precipitation is complete. The salt forms fine blue-green crystals.

The crystals are orthorhombic, space group Pccn-D_{2h}^{10}, a = 14.431(17), b = 19.176(5), c = 13.369(15) Å, Z = 4; D_{calc} = 1.788 g/cm^3. The cation has the familiar lantern structure (Fig. 3, p. 9). Bond lengths are Rh-Rh = 2.38 Å, Rh-O(carboxylate) = 2.05 Å (mean), Rh-O(H_2O) = 2.34 Å [2].

In a later paper the complex is reformulated as a tetrahydrate and revised X-ray data are reported. Values (where changed) are a = 14.435(7), b = 19.177(5), c = 13.385(5) Å; D_{calc} = 1.91 g/cm^3. Bond lengths Rh-Rh = 2.386(3), Rh-O(carboxylate) = 2.04(1) (mean), Rh-O(H_2O) = 2.33(1) Å [3].

The electronic spectrum of the complex in aqueous solution in the presence and absence of pyridine is reproduced. The ^1H NMR spectrum of the salt after exchange with D_2O is reproduced [2].

t-Butyloxycarbonylalaninate, $Rh_2(O_2CCH_2CH_2NHCOCOBu^t)_4$

The complex is prepared by adding $RhCl_3 \cdot 3H_2O$ to an ethanolic solution of t-butyloxycarbonyl, β-alanine and sodium ethoxide, then heating the mixture under gentle reflux for ~90 min. Isolation is achieved by evaporation, extraction with diethyl ether, washing with water, re-evaporation and, finally, chromatography on alumina using ethyl acetate. The complex is blue-green in solution [2].

References:

[1] M. A. Golubnichaya, I. B. Baranovskii, G. Ya. Mazo, R. N. Shchelokov (Zh. Neorgan. Khim. **26** [1981] 2868/71; Russ. J. Inorg. Chem. **26** [1981] 1534/6). – [2] A. M. Dennis, R. A. Howard, J. L. Bear, J. D. Korp, I. Bernal (Inorg. Chim. Acta **37** [1979] L561/L563). – [3] J. D. Korp, I. Bernal, J. L. Bear (Inorg. Chim. Acta **51** [1981] 1/7).

Chloropropionates

$Rh_2(O_2CCHClCH_3)_4$. The electrochemical oxidation-reduction of this complex has been investigated using polarography, cyclic voltammetry and controlled-potential electrolysis. Only a general reference to the methods of preparation is given [1].

$Rh_2(O_2CCH_2CHCl_2)_4(C_2H_5OH)_2$. This salt is obtained as blue needles by treatment of $RhCl_3 \cdot 3H_2O$ with the sodium salt of the carboxylic acid. The infrared spectrum (3000 to 600 cm^{-1}) has been recorded and assignments given, ν(OCO) = 1621, 1376, 1399, 1488 cm^{-1} [2].

References:

[1] K. Das, K. M. Kadish, J. L. Bear (Inorg. Chem. **17** [1978] 930/4). – [2] G. Winkhaus, P. Ziegler (Z. Anorg. Allgem. Chem. **350** [1967] 51/61).

1.1.2.4 Butyrates

$Rh_2(O_2CC_3H_7)_4$ is prepared by heating $RhCl_3 \cdot 3H_2O$ in a saturated alcoholic solution of anhydrous sodium butyrate (only a general reference to procedure is given) [1] or by refluxing $Rh_2(O_2CCH_3)_4$ in butyric acid [2]. Selected infrared bands have been recorded, ν(OCO) = 1574, 1455 cm^{-1} [3]. Temperatures and enthalpies of ligand evolution and cage breakdown in air and N_2 have been investigated, thermogravimetric and differential scanning calorimetry curves are reproduced [1].

X-ray photoelectronic data have been recorded, $Rh\,3d_{5/2}$ = 308.4, $Rh\,3p_{1/2}$ = 522.6 eV [4]. Electrochemical properties have been investigated [2, 5]; evidence for reversible formation of $[Rh_2(O_2CC_3H_7)_4]^+$ and irreversible formation of $[Rh_2(O_2CC_3H_7)_4]^-$ has been reported. The thermodynamics of adduct formation have been examined [2].

$Rh_2(O_2CC_3H_7)_4\{(CH_3)_2SO\}_2$ is prepared by mixing $Rh_2(O_2CC_3H_7)_4$ with excess dimethylsulphoxide, then evaporating the mixture to dryness under reduced pressure. Temperatures and enthalpies of ligand evolution and cage breakdown in air and N_2 have been investigated, thermogravimetric and differential scanning calorimetry curves are reproduced in the paper [1]. Enthalpies for formation of mono- and bis-dimethylsulphoxide adducts have been reported [2].

$Rh_2(O_2CC_3H_7)_4(C_5H_5N)_2$ is prepared by mixing anhydrous $Rh_2(O_2CC_3H_7)_4$ with excess pyridine, then evaporating the mixture to dryness under reduced pressure. Thermal properties have been investigated, thermogravimetric and differential scanning calorimetry curves are reproduced [1]. Thermodynamic data for formation of mono- and bis-pyridine adducts have been reported [2].

$Rh_2(O_2CC_3H_7)_4D$, $Rh_2(O_2CC_3H_7)_4D_2$. Thermodynamic data for formation of mono- and bis-adducts, $Rh_2(O_2CC_3H_7)_4D$ and $Rh_2(O_2CC_3H_7)_4D_2$ where D is methyl cyanide, pyridine, 1-methylimidazole, piperidine, diethyl sulphide, 4-picoline N-oxide and dimethylsulphoxide have been reported [2, 10]. The electrochemical oxidation-reduction of many of these systems has been investigated [2].

$K_2[Rh_2(O_2CC_3H_7)_4(CN)_2]$. This orange complex is obtained by addition of aqueous KCN to alcoholic $Rh_2(O_2CC_3H_7)_4(C_5H_5N)_2$ solution. The infrared spectrum (3600 to 500 cm^{-1}) has been reproduced and assignments given [3].

$Rh_2(O_2CC_3H_7)_4$ with Biological Species. The interactions of $Rh_2(O_2CC_3H_7)_4$ with Ehrlich ascites tumour cells [6, 7], P338 tumour cells [5] and leukaemia L1210 cells [8] have been investigated. Adducts formed with nucleic acid bases have been characterised by spectroscopic means [9].

Isobutyrates, $Rh_2\{O_2CCH(CH_3)_2\}_4$(9-ethyladenine)$_n$ (n not specified). The association of the rhodium(II)isobutyrate complex $Rh_2\{O_2CCH(CH_3)_2\}_4(H_2O)_2$ with 9-ethyladenine has been investigated by infrared spectroscopy. Adduct formation involves donation from the nitrogen bound to 6-C of the 9-ethyladenine [11, 12].

Perfluorobutyrates

$Rh_2(O_2CC_3F_7)_4$ is obtained as a yellow green solid by heating $Rh_2(O_2CCH_3)_4$ with $C_3F_7CO_2H$ and $(C_3F_7CO)_2O$, then drying the initial dark blue product over NaOH in vacuo at 100°C [13].

$Rh_2(O_2CC_3F_7)_4(H_2O)_2$ is obtained as a blue solid by addition of water to the preceding complex [13].

$Rh_2(O_2CC_3F_7)_4L$, $Rh_2(O_2CC_3F_7)_4L_2$ (L = CH_3CN, C_5H_5N, N-methylimidazole, $(CH_3)_2SO$ and $CH_3CON(CH_3)_2$). Thermodynamic and electronic spectral data have been reported for these adducts [13].

References:

[1] R. A. Howard, A. M. Wynne, J. L. Bear, W. W. Wendlandt (J. Inorg. Nucl. Chem. **38** [1976] 1015/8). – [2] R. S. Drago, S. P. Tanner, R. M. Richman, J. R. Long (J. Am. Chem. Soc. **101** [1979] 2897/903). – [3] T. A. Mal'kova, V. N. Shafranskii, Yu. Ya. Kharitonov (Koord. Khim. **3** [1977] 1747/52; Soviet J. Coord. Chem. **3** [1977] 1371/6). – [4] A. M. Dennis, R. A. Howard, K. M. Kadish, J. L. Bear, J. Brace, N. Winograd (Inorg. Chim. Acta **44** [1980] L139/L141). – [5] K. M. Kadish, K. Das, R. Howard, A. Dennis, J. L. Bear (Bioelectrochem. Bioenergetics **5** [1978] 741/53).

[6] R. A. Howard, E. Sherwood, A. Erck, A. P. Kimball, J. L. Bear (J. Med. Chem. **20** [1977] 943/6). – [7] J. L. Bear, H. B. Gray, L. Rainen, I. M. Chang, R. Howard, G. Serio, A. P. Kimball (Cancer Chemotherapy Rept. I **59** [1975] 611/20). – [8] R. A. Howard, A. P. Kimball, J. L. Bear (Cancer Res. **39** [1979] 2568/73). – [9] B. S. Yu, B.-K. Kim (Arch. Pharmacol. Res. **1** [1978] 1/6). – [10] R. S. Drago, J. R. Long, R. Cosmano (Inorg. Chem. **20** [1981] 2920/7).

[11] B. S. Yu, S. T. Choo, I. M. Chang (Yakhak Hoe Chi **19** [1975] 215/7 from C.A. **84** [1976] No. 113739). – [12] I. M. Chang (Soul Taehakkyo Saengyak Yonguso Opjukjip **15** [1976] 69/71 from C.A. **88** [1978] No. 68435). – [13] R. S. Drago, J. R. Long, R. Cosmano (Inorg. Chem. **21** [1982] 2196/202).

1.1.2.5 Other Aliphatic Carboxylates

n-Pentanoate, $Rh_2(O_2CC_4H_9)_4$ is prepared by refluxing $Rh(OH)_3 \cdot H_2O$ in pentanoic acid. No further details are recorded, but anti-tumour activity has been reported [1].

t-Pentanoates (Pivalates)

$Rh_2\{O_2CC(CH_3)_3\}_4(H_2O)_2$ is prepared by heating $Rh_2(O_2CCH_3)_4$ with pivalic acid at 130°C for 30 min [2, 3] or by treatment of $RhCl_3 \cdot 3H_2O$ with an equimolar mixture of pivalic acid and sodium pivalate in alcoholic solution [3]. It forms dark green crystals [2].

The crystals are triclinic, space group $P\bar{1}$-C_i^1, a = 9.408(1), b = 11.799(2), c = 6.984(1) Å, α = 105.98(1)°, β = 94.24(1)°, γ = 99.48(1)°, Z = 1; D_{calc} = 1.47 g/cm^3. The molecule has the familiar "lantern" structure (Fig. 3, p. 9). Bond distances are Rh-Rh = 2.371 Å, Rh-O(carboxylate) = 2.036(2), 2.036(2), 2.043(2), 2.044(2) Å, Rh-O(H_2O) = 2.295(2) Å [2].

$Rh_2\{O_2CC(CH_3)_3\}_4(C_2H_5OH)_2$. No preparative details given but X-ray photoelectronic data have been recorded, Rh $3d_{5/2}$ = 309.2 eV [4].

$Rh_2\{O_2CC(CH_3)_3\}_4(C_5H_5N)_2$. No preparative details given but X-ray photoelectronic data have been recorded, Rh $3d_{5/2}$ = 308.8 eV [4].

1,3-Glutarate, $Rh_2(O_2CCH_2CH_2CH_2CO_2H)_4$ is prepared by heating a finely divided mixture of $Rh_2(O_2CCH_3)_4$ and excess glutaric acid ($HO_2CCH_2CH_2CH_2CO_2H$) at 120°C for 1.5 to 2 h and is washed with diethyl ether to yield a green-blue solid. Water of crystallisation (amount not specified) is easily removed at 105°C. Infrared data have been reported and assignments given; frequencies in cm^{-1}: ν(OH) ≈ 3100, 2650, 2580 (shoulder); ν(CO_2) = 1691, 1683 (noncoordinate), 1567 (shoulder), ~1553 (coordinate); ν(RhO) ≈ 350 and 315 [5].

Hexanoate, $Rh_2(O_2CC_5H_{11})_4$. This is prepared by treatment of $Rh(OH)_3 \cdot H_2O$ with hexanoic acid; no further details are given. Anti-tumour activity has been investigated [1].

Cyclohexanoate, $Rh_2(O_2CC_5H_9)_4$. The electrochemical oxidation-reduction of this complex has been studied by polarography, cyclic voltammetry and controlled potential electrolysis. Only a general reference to the method of preparation is given [3].

References:

[1] R. A. Howard, E. Sherwood, A. Erck, A. P. Kimball, J. L. Bear (J. Med. Chem. **20** [1977] 943/6). – [2] F. A. Cotton, T. R. Felthouse (Inorg. Chem. **19** [1980] 323/8). – [3] K. Das, K. M. Kadish, J. L. Bear (Inorg. Chem. **17** [1978] 930/4). – [4] V. I. Nefedov, Ya. V. Salyn', I. B. Baranovskii, A. G. Maiorova (Zh. Neorgan. Khim. **25** [1980] 216/25; Russ. J. Inorg. Chem. **25** [1980] 116/22). – [5] M. A. Golubnichaya, I. B. Baranovskii, G. Ya. Mazo, R. N. Shchelokov (Zh. Neorgan. Khim. **26** [1981] 2868/71; Russ. J. Inorg. Chem. **26** [1981] 1534/6).

1.1.2.6 Benzoates

$Rh_2(O_2CC_6H_5)_4$ is prepared by dissolving $RhCl_3 \cdot 3H_2O$ in absolute ethanol saturated with anhydrous sodium benzoate and heating the solution under reflux. Crystallisation from acetone gives an adduct which is readily desolvated by heating in vacuo for 2 h. The complex is described as yellow-green in colour. The corrected molar susceptibility is -259.8×10^{-6} cgs units at 298 K, μ_{eff} = 0.28 B.M. The thermogravimetric curve is reproduced and the mass spectrum of the gaseous decomposition products has been recorded [1]. X-ray photoelectronic data have been reported, Rh $3d_{5/2}$ = 309.3 eV [2]. The use of rhodium(II) benzoate as a stationary phase for gas chromatographic analysis of sigma donor ligands and olefins has been described [3].

$Rh_2(O_2CC_6H_5)_4(H_2O)_2$ is prepared by treatment of $Rh_2(O_2CC_6H_5)_4(C_5H_5N)_2$ with excess dilute HCl [4]. It is described as a blue-green solid [1, 4]. The infrared spectrum (4000 to 30 cm^{-1}) has been recorded and assignments given, ν(OCO) = 1555, 1410 cm^{-1} [4].

$Rh_2(O_2CC_6H_5)_4\{(CH_3)_2CO\}_2$ is obtained as deep blue-green needles by crystallisation of the non-solvated complex from acetone [1]. Thermogravimetric studies have been reported [5].

$Rh_2(O_2CC_6H_5)_4\{(CH_3)_2SO\}_2$ is obtained by treatment of the non-solvated complex with neat dimethylsulphoxide [4]. It has been described elsewhere as an orange solid [1]. The infrared spectrum (4000 to 30 cm^{-1}) has been recorded and selected assignments given [4]. X-ray photoelectronic data have been reported, $Rh\,3d_{5/2}=309.1$, $S\,2p=166.5$ eV [2, 4].

$Rh_2(O_2CC_6H_5)_4\{HCON(CH_3)_2\}_2$. X-ray photoelectronic data have been recorded for this complex, $Rh\,3d_{5/2}=309.1$, $N\,1s=399.8$ eV [4], but no preparative details are recorded.

$Rh_2(O_2CC_6H_5)_4\{N(C_2H_5)_3\}_2$. This adduct has been described as a rose-violet solid but no preparative details have been recorded [1].

$Rh_2(O_2CC_6H_5)_4(C_5H_5N)_2$ is prepared by addition of pyridine to the green solution obtained by heating a mixture of rhodium(III) chloride trihydrate, benzoic acid and sodium benzoate in methanol at 60°C [6]. It has been described as raspberry-red crystals [6] or a rose-pink solid [1]. The infrared spectrum has been recorded (4000 to 30 cm^{-1}) and assignments made, the region 400 to 30 cm^{-1} is reproduced [4]. X-ray photoelectronic data have been recorded, $Rh\,3d_{5/2}=309.2$ eV [2], $Rh\,3d_{5/2}=309.0$, $N\,1s=399.7$ eV [4]. The X-ray crystal structure has recently been determined [9].

$Rh_2(O_2CC_6H_5)_4\{(NH_2)_2CS\}_2$ is prepared by warming a finely divided suspension of $Rh_2(O_2CC_6H_5)_4(H_2O)_2$ in saturated aqueous thiourea solution. No further details are recorded. The infrared spectrum (4000 to 30 cm^{-1}) has been recorded and selected assignments given [4]. X-ray photoelectronic data have been reported, $Rh\,3d_{5/2}=309.0$, $S\,2p=162.6$ eV [2], $Rh\,3d_{5/2}=309.0$, $S\,2p=162.6$, $N\,1s=400.0$ eV [4].

$Rh_2(O_2CC_6H_5)_4\{S(C_2H_5)_2\}_2$. This adduct has been described as a rose-red solid but no further information has been recorded [1].

$Rh_2(O_2CC_6H_5)_4\{P(C_6H_5)_3\}_2$ is prepared by addition of methanolic triphenylphosphine solution to the green solution obtained by heating a mixture of $RhCl_3\cdot3H_2O$, benzoic acid and sodium benzoate in methanol at 60°C [6]. It is variously described as a yellow-orange [1] or light-brown solid [6]. Raman and far infrared spectra have been recorded, ν(Rh-Rh) = 288 cm^{-1}, ν(Rh-O) = 334, 415, 440 cm^{-1} [7].

$Rh_2(O_2CC_6H_5)_4\cdot1.5KCN\cdot2H_2O$. This grey-violet product is obtained by treatment of $Rh_2(O_2CC_6H_5)_4$ with KCN in methanol solution. The infrared spectrum (3600 to 500 cm^{-1}) is reproduced and assignments given, ν(CN) = 2108 cm^{-1} [8]. The complex may involve cyanide bridges and be best formulated as $K_3[(NC)\{Rh_2(O_2CC_6H_5)_4\}CN\{Rh_2(O_2CC_6H_5)_4\}(CN)]\cdot4H_2O$.

References:

[1] J. Kitchens, J. L. Bear (Thermochim. Acta **1** [1970] 537/44). – [2] V. I. Nefedov, Ya. V. Salyn', I. B. Baranovskii, A. G. Maiorova (Zh. Neorgan. Khim. **25** [1980] 216/25; Russ. J. Inorg. Chem. **25** [1980] 116/22). – [3] V. Schurig, J. L. Bear, A. Zlatkis (Chromatographia **5** [1972] 301/4). – [4] I. B. Baranovskii, M. A. Golubnichaya, G. Ya. Mazo, V. I. Nefedov, Ya. V. Salyn', R. N. Shchelokov (Koord. Khim. **3** [1977] 743/9; Soviet J. Coord. Chem. **3** [1977] 576/82). – [5] G. A. Barclay, R. F. Broadbent, J. V. Kingston, G. R. Scollary (Thermochim. Acta **10** [1974] 73/83).

[6] P. Legzdins, R. W. Mitchell, G. L. Rempel, J. D. Ruddick, G. Wilkinson (J. Chem. Soc. A **1970** 3322/6). – [7] A. P. Ketteringham, C. Oldham (J. Chem. Soc. Dalton Trans. **1973** 1067/70). – [8] T. A. Mal'kova, V. N. Shafranskii, Yu. Ya. Kharitonov (Koord. Khim. **3** [1977] 1747/52; Soviet J. Coord. Chem. **3** [1977] 1371/6). – [9] G. Li, Y. Sun (Huaxue Xuebao [Hua Hsueh Hsueh Pao] **39** [1981] 945/50 from C.A. **97** [1982] No. 102117).

1.1.2.7 Salicylates

These compounds appear to possess the familiar binuclear "lantern" structure with a rhodium-rhodium bond, however, participation of the phenolic oxygen atom in metal-ligand bonding has been postulated on the basis of infrared evidence. In the absence of more definitive evidence the exact bonding mode of the salicylate residue remains uncertain.

$Rh_2(O_2CC_6H_4OH)_4(H_2O)_2$ is prepared by the treatment of $Rh_2(O_2CCH_3)_4(H_2O)_2$, $[Gun H]_2$-$[Rh_2(O_2CCH_3)_4Cl_2]$, $[Gun H]_2[Rh_2(O_2CH)_4(O_2CCH_3)_2]C_2H_5OH$ or $RhCl_3 \cdot 5H_2O$ in aqueous solution with crystalline salicylic acid at 70 to 80°C. The complex is a green solid, insoluble in water but readily soluble in ethanol, acetone and ether. The infrared spectrum (4000 to 400 cm^{-1}) is reproduced, and the electronic spectrum has been recorded and assigned. X-ray photoelectronic data have been reported, Rh $3d_{5/2}$ = 309.7 eV.

$Rh_2(O_2CC_6H_4OH)_4\{HCON(CH_3)_2\}_2$. The electronic spectrum of this blue-green complex has been recorded, λ_{max} = 581, 463 nm, but no other details are given.

$Rh_2(O_2CC_6H_4OH)_4(NH_3)_2$. This pink adduct is obtained by treatment of the diaquo complex with ammonia (no details given). The infrared spectrum (4000 to 400 cm^{-1}) is reproduced in the paper, and the electronic spectrum has been recorded, λ_{max} = 543, 454 nm. X-ray photoelectronic data are reported, Rh $3d_{5/2}$ = 309.6, N 1s = 400.3 eV.

$Rh_2(O_2CC_6H_4OH)_4(N_2H_4)$. This cherry-red adduct is prepared by treatment of the diaquo complex with hydrazine (no details given). The infrared spectrum (4000 to 400 cm^{-1}) is reproduced and the electronic spectrum has been recorded, λ_{max} = 543, 441 nm. X-ray photoelectronic data have been reported, Rh $3d_{5/2}$ = 309.7, N 1s = 401.3 eV.

$Rh_2(O_2CC_6H_4OH)_4(en)_2$. This cherry-red adduct is prepared by treatment of the diaquo complex with ethylenediamine (no details given). The electronic spectrum with λ_{max} = 539 nm and data Rh $3d_{5/2}$ = 309.7, N 1s = 400.4 eV have been recorded.

$Rh_2(O_2CC_6H_4OH)_4(C_6H_5NH_2)_2$. This pink adduct is prepared by treatment of the diaquo complex with aniline (no details given). The electronic spectrum with λ_{max} = 549, 441 nm and X-ray photoelectronic data Rh $3d_{5/2}$ = 309.4, N 1s = 400.2 eV have been recorded.

$Rh_2(O_2CC_6H_4OH)_4(C_5H_5N)_2$. This cherry-red adduct is obtained by treatment of the diaquo complex with pyridine (no details given). The electronic spectrum with λ_{max} = 532, 441 nm and X-ray photoelectronic data Rh $3d_{5/2}$ = 309.3, N 1s = 400.1 eV have been recorded.

$Rh_2(O_2CC_6H_4OH)_4(CH_3CSNH_2)_2$. This red-violet adduct is prepared by treatment of the diaquo adduct with thioacetamide in aqueous alcohol suspension. X-ray photoelectronic data have been recorded, Rh $3d_{5/2}$ = 309.3, N 1s = 400.2 eV.

$Rh_2(O_2CC_6H_4OH)_4\{(NH_2)_2CS\}_2$. This red-violet adduct is prepared by treatment of the diaquo complex with thiourea in aqueous alcohol suspension. The infrared spectrum (4000 to 400 cm^{-1}) is reproduced, and X-ray photoelectronic data have been recorded, Rh $3d_{5/2}$ = 309.3, N 1s = 400.2, S $2p_{3/2}$ = 162.9 eV.

R. N. Shchelokov, A. G. Maiorova, G. N. Kuznetsova, I. F. Golovaneva, O. N. Evstaf'eva (Zh. Neorgan. Khim. **25** [1980] 1891/7; Russ. J. Inorg. Chem. **25** [1980] 1049/52).

1.1.2.8 Thiosalicylates

$Rh_2(O_2CC_6H_4SH)_4 \cdot xH_2O$. The **octahydrate** (x = 8) is prepared by boiling an aqueous suspension of $Rh_2(O_2CC_6H_4SH)_4(HO_2CC_6H_4SH) \cdot 7H_2O$ (see p. 52) for 3 to 4 h. The water content is variable, heating at 100°C affords the **dihydrate** (x = 2), at 170°C all water is lost leaving the

anhydrous complex ($x=0$) [1]. The **dihydrate** is also obtained by heating $Rh_2(O_2CC_6H_4SH)_4(C_2H_5OH)\cdot 4H_2O$ at an unspecified temperature [1]. X-ray photoelectronic data have been recorded for a **hexahydrate** ($x=6$), $Rh\,3d_{5/2}=309.2$, $S\,2p_{3/2}=163.5$ eV [2].

$Rh_2(O_2CC_6H_4SH)_4(C_2H_5OH)\cdot xH_2O$. This orange product ($x=4$) is prepared by treatment of $Rh_2(O_2CCH_3)_4$ or $[GunH]_2[Rh(O_2CH)_4Cl_2]$ with thiosalicylic acid in aqueous solution [1]. X-ray photoelectronic data have been recorded for a **dihydrate** ($x=2$), $Rh\,3d_{5/2}=309.5$, $S\,2p_{3/2}=163.5$ eV [2].

$Rh_2(O_2CC_6H_4SH)_4(HO_2CC_6H_4SH)\cdot 7H_2O$ is prepared by treatment of $[NH_4]_3[RhCl_6]$ with thiosalicylic acid in aqueous alcohol at 80°C for 30 min. It is described as a dark red precipitate [1]. X-ray photoelectronic data have been reported, $Rh\,3d_{5/2}=309.5$, $S\,2p_{3/2}=163.6$ eV [2].

$Rh_2(O_2CC_6H_4SH)_4(NH_3)_2\cdot 4H_2O$. This dark red adduct is prepared by dissolving $Rh_2(O_2CC_6H_4SH)_4(H_2O)_2$ in 25% ammonia solution, boiling for 1.5 to 2 h, then evaporating the solution to dryness. The complex is soluble in water but insoluble in alcohol [3].

$Rh_2(O_2CC_6H_4SH)_4en\cdot xH_2O$. This red complex ($x=6$) is prepared either by boiling an aqueous suspension of $Rh_2(O_2CC_6H_4SH)_4\cdot xH_2O$ (x not specified) with ethylenediamine dihydrochloride, or by treatment of $Rh_2(O_2CH)_4(en)\cdot 2H_2O$ with the calculated amount of thiosalicylic acid in aqueous ethanol [3]. X-ray photoelectronic data have been recorded for the **hexahydrate** ($x=6$), $Rh\,3d_{5/2}=309.4$, $N\,1s=401.9$, 400.4, $S\,2p_{3/2}=163.5$ eV, and for the **dihydrate** ($x=2$), $Rh\,3d_{5/2}=309.5$, $N\,1s=401.9$, 400.3, $S\,2p_{3/2}=163.5$ eV [2].

$Rh_2(O_2CC_6H_4SH)_4(HO_2CC_6H_4SH)(en)\cdot 4H_2O$ is prepared by boiling an aqueous mixture of $Rh_2(O_2CC_6H_4SH)_4(HO_2CC_6H_4SH)\cdot 7H_2O$ and ethylenediamine dihydrochloride for 40 min. The product is a red solid, insoluble in water and alcohol. On heating to 100°C two molecules of water are lost [3].

$Rh_2(O_2CC_6H_4SH)_4HCl\cdot 6H_2O$. This dark red complex, which is probably best formulated as $[H_3O][Rh_2(O_2CC_6H_4SH)_4Cl(H_2O)]\cdot 4H_2O$, is prepared by heating a mixture of "$H_3[RhCl_6]$" and thiosalicylic acid in alcohol at 80 to 90°C for 5 min [1].

$Rh_2(O_2CC_6H_4SH)_4HBr\cdot 2H_2O$. This product, which is probably best formulated as $[H_3O][Rh_2(O_2CC_6H_4SH)_4Br(H_2O)]$ is obtained by repeated treatment of $Rh_2(O_2CC_6H_4SH)_4(HO_2CC_6H_4SH)\cdot 7H_2O$ with concentrated HBr [3].

$Rh_2(O_2CC_6H_4SH)_4\cdot 2HBr$. A product of this apparent stoichiometry is obtained when the above synthesis is performed in the presence of a large excess of free Br_2 [3].

$Rh_2(O_2CC_6H_4SH)_4\{(NH_2)_2CS\}_2\cdot xH_2O$. The **trihydrate** ($x=3$) is prepared by boiling an aqueous mixture of thiourea and $Rh_2(O_2CC_6H_4SH)_4(HO_2CC_6H_4SH)\cdot 7H_2O$ for 2 to 2.5 h. On heating to 100°C the solid loses water to form the **anhydrous complex**. A **tetrahydrate** ($x=4$) has been obtained as a red product by boiling a suspension of $Rh_2(O_2CC_6H_4SH)_4\cdot xH_2O$ in saturated aqueous thiourea solution for 2 h [3].

References:

[1] A. G. Maiorova, L. A. Nazarova, G. N. Emel'yanova (Zh. Neorgan. Khim. **18** [1973] 1866/71; Russ. J. Inorg. Chem. **18** [1973] 986/8). – [2] V. I. Nefedov, Ya. V. Salyn', A. G. Maiorova, L. A. Nazarova, I. B. Baranovskii (Zh. Neorgan. Khim. **19** [1974] 1353/7; Russ. J. Inorg. Chem. **19** [1974] 736/9). – [3] A. G. Maiorova, L. A. Nazarova, G. N. Emel'yanova (Zh. Neorgan. Khim. **18** [1973] 1871/5; Russ. J. Inorg. Chem. **18** [1973] 989/90).

1.1.2.9 Complexes with Substituted Benzoic Acids, $Rh_2(O_2CAr)_4$

Preparation and brief details have been reported for complexes with the following acids: 2,6-$(CH_3)_2C_6H_3CO_2H$, 2-, 3- and 4-$CH_3OC_6H_4CO_2H$, 4-$ClC_6H_4CO_2H$, 4-$O_2NC_6H_4CO_2H$ [1]; $C_6F_5CO_2H$ and 2,4-Cl_2-3,5-$(NO_2)_2C_6HCO_2H$ [2].

References:

[1] D. Holland, D. J. Milner (J. Chem. Res. M **1979** 3734/46; J. Chem. Res. S **1979** 317). – [2] A. J. Anciaux, A. Demonceau, A. F. Noeis, A. J. Hubert, R. Warin, P. Teyssié (J. Org. Chem. **46** [1981] 873/6).

1.1.2.10 Mandelates

$Rh_2(O_2CCHOH\cdot C_6H_5)_4$, $Rh_2(O_2CCHOH\cdot C_6H_5)_4L_2$ (L = H_2O, NH_3, N_2H_4, en, en·H_2O, $(NH_2)_2CS$, py, $C_6H_5NH_2$ and $(CH_3)_2SO$), $Cs[Rh_2(O_2CCHOHC_6H_5)_4Cl]\cdot 3H_2O$ and $K_2[Rh_2(O_2CCHOHC_6H_5)_4(NO_2)_2]$. The preparation and characterisation of these complexes has recently been reported.

R. N. Shchelokov, A. G. Maiorova, S. S. Abdullaev, O. N. Evstaf'eva, I. F. Golovaneva, G. N. Emel'yanova (Zh. Neorgan. Khim. **26** [1981] 3308/14; Russ. J. Inorg. Chem. **26** [1981] 1774/8).

1.1.2.11 Pyridine-2,6-dicarboxylates

The following derivatives of pyridine-2,6-dicarboxylic acid are thought to be true rhodium(II) complexes containing $Rh_2(O_2C\text{-})_4$ units linked through the pyridine 2,6-dicarboxylate residues. The nitrogen atom of the pyridine moiety is not thought to be coordinated. These conclusions are based upon electronic spectra (similar to rhodium(II) acetate) and X-ray photoelectronic data.

$Rh(O_2C\cdot C_5H_3N\cdot CO_2)\cdot 3H_2O$. This very insoluble light-green solid is obtained by heating an aqueous solution of rhodium(II) acetate and pyridine-2,6-dicarboxylic acid, $C_5H_3N(CO_2H)_2$, under reflux for 50 to 70 h. The electronic spectrum ($\lambda_{max} \approx 610$ nm) and X-ray photoelectronic data have been recorded, Rh $3d_{3/2} = 312.6$, Rh $3d_{5/2} = 307.8$ eV.

$Rh(O_2C\cdot C_5H_3N\cdot CO_2)(C_5H_5N)(H_2O)$. This brown insoluble product is obtained by treatment of the trihydrate (above) with cold neat pyridine over a period of ~20 h.

$Rh(O_2C\cdot C_5H_3N\cdot CO_2)(\alpha\text{-picoline})(H_2O)$. This brown insoluble product is similarly prepared using neat α-picoline.

R. W. Matthews, A. D. Hamer, D. L. Hoof, D. G. Tisley, R. A. Walton (J. Chem. Soc. Dalton Trans. **1973** 1035/8).

1.1.2.12 Monothioacetates

$Rh_2(OSCCH_3)_4$. The anhydrous complex is prepared by dehydration of $Rh_2(OSCCH_3)_4(H_2O)_2$ at 120°C. The infrared spectrum (4000 to 30 cm^{-1}) is reproduced and assignments are given [1]. A full normal coordinate analysis of all vibrations of the species $Rh_2(OSCCH_3)_4$ and $Rh_2(OSCCD_3)_4$ has been reported [6]. X-ray photoelectronic data have been recorded, Rh $3d_{5/2}$ = 308.9 eV [1].

$Rh_2(OSCCH_3)_4(H_2O)_2$ is prepared by treatment of the corresponding bis-pyridine adduct (see below) with 1:1 aqueous hydrochloric acid, and forms fine dark green crystals which are insoluble in water, sparingly soluble in acetone and soluble in acetonitrile or dimethyl formamide. The infrared spectrum (4000 to 30 cm^{-1}) has been recorded and assignments made; the region 400 to 30 cm^{-1} is reproduced in the paper. X-ray photoelectronic data are recorded, Rh $3d_{5/2}$ = 308.9, S 2p = 162.9 eV [1], Rh $3d_{5/2}$ = 308.8 eV [2].

$Rh_2(OSCCH_3)_4(C_5H_5N)_2$ is prepared by treatment of $Rh_2(OSCCH_3)_4(CH_3COSH)_2$ with pyridine, and is isolated as a fine crystalline orange powder [1, 3]. It is sparingly soluble in acetonitrile and dimethylformamide. The infrared spectrum (4000 to 30 cm^{-1}) has been recorded and assignments given, the region 400 to 30 cm^{-1} is reproduced. A Raman active band at 164 cm^{-1} has been attributed to ν(Rh-Rh) [1]. X-ray photoelectronic data have been reported, Rh $3d_{5/2}$ = 308.4, S 2p = 162.8, N 1s = 399.7 eV [1], Rh $3d_{5/2}$ = 308.4 eV [2].

$Rh_2(OSCCH_3)_4(CH_3CN)_2$ is prepared as fine red crystals by dissolving $Rh_2(OSCCH_3)_4$-$(CH_3COSH)_2$ in neat CH_3CN. The infrared spectrum (4000 to 30 cm^{-1}) has been recorded and assignments made, the region 400 to 30 cm^{-1} is reproduced. A Raman active band at 154 cm^{-1} has been attributed to ν(Rh-Rh) [1]. X-ray photoelectronic data have been recorded, Rh $3d_{5/2}$ = 308.7, S 2p = 162.8, N 1s = 399.9 eV [1]; Rh $3d_{5/2}$ = 308.7 eV [2].

$K_2[Rh_2(OSCCH_3)_4(NCS)_2]\cdot 2H_2O$. This salt is prepared by treatment of $Rh_2(OSCCH_3)_4(H_2O)_2$ with excess saturated aqueous KSCN, and is isolated as fine red crystals. The water of crystallisation is lost over P_2O_5 in vacuo to afford the **anhydrous salt**. The infrared spectrum has been recorded (4000 to 30 cm^{-1}) and assignments made, ν(CN) = 2085 cm^{-1}, the region 400 to 30 cm^{-1} is reproduced in the paper. X-ray photoelectronic data have been recorded, Rh $3d_{5/2}$ = 308.7, S 2p = 163.0, N 1s = 398.5 eV [1].

$Rh_2(OSCCH_3)_4(NH_2NH_3Cl)_2$. This complex is obtained by treatment of $Rh_2(OSCCH_3)_4$-$(C_5H_5N)_2$ with saturated aqueous hydrazine hydrochloride solution [1, 3] and forms dark cherry-red water-insoluble crystals [3]. The infrared spectrum (1700 to 30 cm^{-1}) has been recorded and assignments made, the region 400 to 30 cm^{-1} is reproduced. X-ray photoelectronic data have been recorded, Rh $3d_{5/2}$ = 308.8, S 2p = 162.8, N 1s = 400.6 and 401.9, Cl 2p = 198.2 eV [1].

$Cs_2[Rh(OSCCH_3)_4Cl_2]\cdot H_2O$. This dark blue salt is prepared by treatment of $Rh_2(OSCCH_3)_4$-$(H_2O)_2$ with excess saturated warm aqueous CsCl solution. The infrared spectrum has been recorded (4000 to 300 cm^{-1}) and assignments given, the region 400 to 30 cm^{-1} is reproduced [1]. X-ray photoelectronic data have been reported, Rh $3d_{5/2}$ = 308.7, S 2p = 163.1, Cl 2p = 197.9 eV [1]; Rh $3d_{5/2}$ = 308.7 eV [2].

$Rh_2(OSCCH_3)_4\{(CH_3)_2SO\}_2$. This adduct, which probably contains S-coordinated $(CH_3)_2SO$ ligands, is prepared by treatment of $Rh_2(OSCCH_3)_4(CH_3COSH)_2$ (see below) with dimethylsulphoxide, and forms fine yellow prismatic crystals. The infrared spectrum (1600 to 30 cm^{-1}) has been recorded and assignments made, the region 400 to 30 cm^{-1} is reproduced [1]. X-ray photoelectronic data have been reported, Rh $3d_{5/2}$ = 308.7, S 2p = 162.7 and 166.1 eV [1]; Rh $3d_{5/2}$ = 308.7 eV [2].

$Rh_2(OSCCH_3)_4\{(NH_2)_2CS\}_2$ is obtained as a cherry-red solid by repeated treatment of $Rh_2(OSCCH_3)_4(C_5H_5N)_2$ with saturated aqueous thiourea solution. The infrared spectrum (4000 to 30 cm^{-1}) has been recorded and assignments given, the region 400 to 30 cm^{-1} is reproduced [1]. X-ray photoelectronic data have been recorded, Rh $3d_{5/2}$ = 308.5, S 2p = 162.7, N 1s = 399.9 eV [1]; Rh $3d_{5/2}$ = 308.5 eV [2].

$Rh_2(OSCCH_3)_4(CH_3COSH)_2$ is prepared by heating a mixture of hydrated rhodium(II) formate and excess thioacetic acid on a water bath for 1 h. It forms fine red crystals, insoluble in water and most organic solvents [1, 3].

The crystals are triclinic, space group $P\bar{1}$-C_i^1, a = 8.276(3), b = 8.849(4), c = 9.170(4) Å, α = 101.74(3)°, β = 100.78(3)°, γ = 117.10(3)°, Z = 1; D_{exp} = 2.08, D_{calc} = 1.97 g/cm^3. The molecule has the familiar "lantern" structure shown in **Fig. 7** with axial S-coordinated CH_3COSH ligands. Bond lengths are Rh-Rh = 2.550 Å, Rh-S(axial) = 2.521 Å, Rh-S(equatorial) = 2.261 and 2.257 Å, Rh-O(equatorial) = 2.116 and 2.100 Å [4, 5].

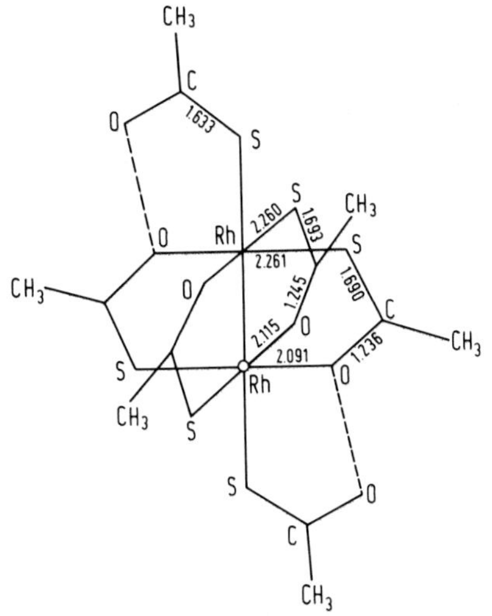

Fig. 7
Molecular structure
of $Rh_2(OSCH_3)_4(CH_3COSH)_2$.

The infrared spectrum (4000 to 30 cm^{-1}) is reproduced and assignments given [1]. X-ray photoelectronic data have been recorded, $Rh\,3d_{5/2}=308.8$, $S\,2p=163.0$ eV [1]; $Rh\,3d_{5/2}=308.8$ eV [2].

References:

[1] I. B. Baranovskii, M. A. Golubnichaya, G. Ya. Mazo, V. I. Nefedov, Ya. V. Salyn', R. N. Shchelokov (Zh. Neorgan. Khim. **21** [1976] 1085/94; Russ. J. Inorg. Chem. **21** [1976] 591/6). – [2] V. I. Nefedov, Ya. V. Salyn', I. B. Baranovskii, A. G. Maiorova (Zh. Neorgan. Khim. **25** [1980] 216/25; Russ. J. Inorg. Chem. **25** [1980] 116/22). – [3] I. B. Baranovskii, M. A. Golubnichaya, G. Ya. Mazo, R. N. Shchelokov (Zh. Neorgan. Khim. **20** [1975] 844; Russ. J. Inorg. Chem. **20** [1975] 475). – [4] L. M. Dikareva, M. A. Porai-Koshits, G. G. Sadikov, I. B. Baranovskii, M. A. Golubnichaya, R. N. Shchelokov (Zh. Neorgan. Khim. **23** [1978] 1044/51; Russ. J. Inorg. Chem. **23** [1978] 578/81). – [5] L. M. Dikareva, G. G. Sadikov, M. A. Porai-Koshits, M. A. Golubnichaya, I. B. Baranovskii, R. N. Shchelokov (Zh. Neorgan. Khim. **22** [1977] 2013/4; Russ. J. Inorg. Chem. **22** [1977] 1093/4).

[6] I. K. Kireeva, G. Ya. Mazo, R. N. Shchelokov (Zh. Neorgan. Khim. **24** [1979] 396/407; Russ. J. Inorg. Chem. **24** [1979] 220/5).

1.1.2.13 Monothiobenzoates

$Rh_2(OSCC_6H_5)_4$ is prepared by treatment of $Rh_2(OSCC_6H_5)_4(C_5H_5N)_2$ (see below) with 1:1 HCl in boiling aqueous suspension [1, 2]. The complex is green and is insoluble in water but soluble in acetonitrile and dimethylformamide. The infrared spectrum is reproduced (4000 to 30 cm^{-1}) and assignments are given [2]. X-ray photoelectronic data have been reported, $Rh\,3d_{5/2}=308.8$, $S\,2p=162.9$ eV [2]; $Rh\,3d_{5/2}=308.8$ eV [3].

$Rh_2(OSCC_6H_5)_4\{HCON(CH_3)_2\}_2$ is prepared as a brown powder by treatment of $Rh_2(OSCC_6H_5)_4$ with neat dimethylformamide. The infrared spectrum (4000 to 30 cm^{-1}) has been recorded and assignments given [2]. X-ray photoelectronic data have been recorded, $Rh\,3d_{5/2}=308.7$, $N\,1s=400.1$, $S\,2p=163.0$ eV [2]; $Rh\,3d_{5/2}=308.7$ eV [3].

$Rh_2(OSCC_6H_5)_4(C_5H_5N)_2$ is prepared as a fine orange precipitate by treatment of finely divided $Rh_2(OSCC_6H_5)_4(C_6H_5COSH)_2$ with excess pyridine [1, 2]. The infrared spectrum (4000 to

30 cm^{-1}) has been recorded and assignments given, the region 400 to 30 cm^{-1} is reproduced [2]. X-ray photoelectronic data have been recorded, Rh $3d_{5/2}$ = 308.7 eV [3]. Thermal decomposition data have been reported [2].

$Rh_2(OSCC_6H_5)_4(CH_3CN)_2$. X-ray photoelectronic data have been recorded, Rh $3d_{5/2}$ = 308.7 eV but no preparative details given [3].

$Rh_2(OSCC_6H_5)_4\{(CH_3)_2SO\}_2$ is obtained as an orange precipitate on treatment of $Rh_2(OSCC_6H_5)_4$ with neat dimethylsulphoxide. The infrared spectrum (4000 to 30 cm^{-1}) has been recorded and assignments given [2]. X-ray photoelectronic data have been recorded, Rh $3d_{5/2}$ = 308.8, S 2p = 162.9 and 166.4 eV [2]; Rh $3d_{5/2}$ = 308.8 eV [3].

$Rh_2(OSCC_6H_5)_4\{(NH_2)_2CS\}_2$ is prepared as a dark red solid by warming finely powdered $Rh_2(OSCC_6H_5)_4$ with a saturated aqueous solution of thiourea. The infrared spectrum (2000 to 30 cm^{-1}) has been recorded and assignments given [2]. X-ray photoelectronic data have been reported, Rh $3d_{5/2}$ = 308.6, N 1s = 400.1, S 2p = 162.7 eV [2]; Rh $3d_{5/2}$ = 308.6 eV [3].

$Rh_2(OSCC_6H_5)_4(C_6H_5COSH)_2$ is prepared by warming finely divided rhodium(II) formate with excess thiobenzoic acid on a water bath for 1 to 1.5 h. It forms dark red crystals [1, 2]. The infrared spectrum has been recorded (4000 to 30 cm^{-1}) and assignments given [2]. X-ray photoelectronic data have been reported, Rh $3d_{5/2}$ = 308.7, S 2p = 162.9 eV [2]; Rh $3d_{5/2}$ = 308.8 eV [3].

References:

[1] I. B. Baranovskii, M. A. Golubnichaya, G. Ya. Mazo, R. N. Shchelokov (Koord. Khim. **1** [1975] 1573; Soviet J. Coord. Chem. **1** [1975] 1299). – [2] I. B. Baranovskii, M. A. Golubnichaya, G. Ya. Mazo, V. I. Nefedov, Ya. V. Salyn', R. N. Shchelokov (Koord. Khim. **3** [1977] 743/9; Soviet J. Coord. Chem. **3** [1977] 576/82). – [3] V. I. Nefedov, Ya. V. Salyn', I. B. Baranovskii, A. G. Maiorova (Zh. Neorgan. Khim. **25** [1980] 216/25; Russ. J. Inorg. Chem. **25** [1980] 116/22).

1.1.3 Rhodium(II/III) Carboxylates

These complexes, which are prepared by one-electron oxidation of the corresponding rhodium(II) species, retain the "lantern" structure of the parent molecule. They do not contain discrete Rh(II) and Rh(III) centres.

1.1.3.1 Formates

The bonding, magnetic properties, electronic spectra and electron spin resonance (ESR) spectra of the cationic species **$[Rh_2(O_2CH)_4]^+$** and **$[Rh_2(O_2CH)_4(H_2O)_2]^+$** have been discussed in terms of molecular orbital calculations [1, 2].

References:

[1] J. G. Norman, H. J. Kolari (J. Am. Chem. Soc. **100** [1978] 791/9). – [2] J. G. Norman, G. E. Renzoni, D. A. Case (J. Am. Chem. Soc. **101** [1979] 5256/67).

1.1.3.2 Acetates

$[Rh_2(O_2CCH_3)_4]^+$. This binuclear cation has been prepared by chemical (Cl_2, PbO_2 or Ce^{4+}) [1, 2] or electrochemical [3, 4, 5, 8] oxidation of $Rh_2(O_2CCH_3)_4$. The cation is orange or violet in colour depending upon the medium, and slowly disproportionates into rhodium(II) acetate and rhodium(III) species. It is readily reduced to rhodium(II) acetate by mossy zinc amalgam. The electronic spectrum has been recorded, λ_{max} = 758, 513 nm in 1M H_2SO_4 solution; λ_{max} = 805, 526 nm in 0.05 M NaCl; λ_{max} = 758, 515, 217 nm in 1M CF_3SO_3H [1]. The following regions have been reproduced: 800 to 400 nm, 1M H_2SO_4 [1], 1000 to 400 nm, 1M $HClO_4$ and 1000 to 400 nm, 3.5 M ammonia solution [3]. The magnetic moment has been measured in solution using Evans NMR method, μ_{eff} = 2.6 ± 0.3 and 2.1 ± 0.4 B.M. [1]. Cyclic voltammetry and polarography data are reported [5]. Kinetics of substitution and redox reactions of the $[Rh_2(O_2CCH_3)_4]^+$ complex with chloride and bromide anions have been reported [8].

$[Rh_2(O_2CCH_3)_4(H_2O)_2][ClO_4]$. This perchlorate salt is prepared by mixing stoichiometric amounts of $Rh_2(O_2CCH_3)_4$ and $Ce(SO_4)_2$ in sulphuric acid solution, and eluting the mixture down a cation exchange column using 2M $HClO_4$. It is a brown-red crystalline solid [2].

The X-ray crystal structure has been determined [2, 6]. The crystals are triclinic, space group $P\bar{1}$-C_i^1, a = 8.187(2), b = 8.114(2), c = 16.120(3) Å, α = 91.85(3)°, β = 101.84(3)°, γ = 115.17(3)°, Z = 2; D_{exp} = 2.09, D_{calc} = 2.10 g/cm^3. The complex cation retains the familiar "lantern" structure, the Rh-Rh distance is 2.3165 Å [6].

The infrared spectrum (1600 to 300 cm^{-1}) has been recorded and assignments given [2]. The electronic spectrum has been recorded, λ_{max} = 764, 519, 330, 250, 219 nm in aqueous solution; λ_{max} = 781, 530, 312, 245 nm in $CHCl_3$ solution; λ_{max} = 788, 537 nm in acetic acid solution. Magnetic properties have been recorded over the temperature range 77 to 294.5 K, μ_{eff} = 2.09 to 2.26 B.M. [2]. X-ray photoelectronic data have been reported, Rh $3d_{5/2}$ = 310.1, Rh $3p_{1/2}$ = 523.8 eV [7].

$[Rh_2(O_2CCH_3)_4][O_2CCH_3]$ is obtained by electrolytic oxidation of $Rh_2(O_2CCH_3)_4$ and isolated as a solid. No other details are given [3].

$Rh_2(O_2CCH_3)_4Cl$ is prepared by electrolytic methods but no details are recorded. The infrared spectrum (440 to 200 cm^{-1}) is reproduced [3].

$Rh_2(O_2CCH_3)_4Br$ has been detected as a reaction intermediate [8].

References:

[1] C. R. Wilson, H. Taube (Inorg. Chem. **14** [1975] 2276/9). – [2] M. Moszner, J. J. Ziolkowski (Bull. Acad. Polon. Sci. Ser. Sci. Chim. **24** [1976] 433/7). – [3] R. D. Cannon, D. B. Powell, K. Sarawek, J. S. Stillman (Chem. Commun. **1976** 31/2). – [4] K. Das, K. M. Kadish, J. L. Bear (Inorg. Chem. **17** [1978] 930/4). – [5] K. M. Kadish, K. Das, R. Howard, A. Dennis, J. L. Bear (Bioelectrochem. Bioenergetics **5** [1978] 741/53).

[6] J. J. Ziolkowski, M. Moszner, T. Glowiak (Chem. Commun. **1977** 760/1). – [7] A. M. Dennis, R. A. Howard, K. M. Kadish, J. L. Bear, J. Brace, N. Winograd (Inorg. Chim. Acta **44** [1980] L139/L141). – [8] R. D. Cannon, D. B. Powell, K. Sarawek (Inorg. Chem. **20** [1981] 1470/4).

1.1.3.3 Methoxyacetates, $[Rh_2(O_2CCH_2OCH_3)_4]^+$

Electrochemical oxidation of $Rh_2(O_2CCH_2OCH_3)_4$ to afford the corresponding monocation has been investigated by polarography, cyclic voltammetry and controlled potential electrolysis [1, 2].

References:

[1] K. Das, K. M. Kadish, J. L. Bear (Inorg. Chem. **17** [1978] 930/4). – [2] K. M. Kadish, K. Das, R. Howard, A. Dennis, J. L. Bear (Bioelectrochem. Bioenergetics **5** [1978] 741/53).

1.1.3.4 Phenoxyacetates, $[Rh_2(O_2CCH_2OC_6H_5)_4]^+$

Electrochemical oxidation of $Rh_2(O_2CCH_2OC_6H_5)_4$ to afford the corresponding monocation has been investigated by polarography, cyclic voltammetry and controlled potential electrolysis [1, 2].

References:

[1] K. Das, K. M. Kadish, J. L. Bear (Inorg. Chem. **17** [1978] 930/4). – [2] K. M. Kadish, K. Das, R. Howard, A. Dennis, J. L. Bear (Bioelectrochem. Bioenergetics **5** [1978] 741/53).

1.1.3.5 Trifluoroacetates

$[Rh_2(O_2CCF_3)_4]^+$. The electrochemical oxidation of $Rh_2(O_2CCF_3)_4$ to afford the corresponding monocation has been investigated by polarography, cyclic voltammetry and controlled-potential electrolysis [1].

$[Rh_2(O_2CCF_3)_4\{P(C_6H_5)_3\}_2]^+$. This species has been generated electrochemically, and its electron spin resonance (ESR) spectrum has been recorded, $g_{\parallel}=2.003$, $g_{\perp}=2.088$ [2].

$[Rh_2(O_2CCF_3)_4\{P(OC_6H_5)_3\}_2]^+$. This species has been generated electrochemically, and its ESR spectrum has been recorded, $g_{\parallel}=2.002$, $g_{\perp}=2.127$ [2].

$[Rh_2(O_2CCF_3)_4\{P(OCH_2)_3CC_2H_5\}_2]^+$. This species has been generated electrochemically, and its ESR spectrum has been recorded, $g_{\parallel}=2.003$, $g_{\perp}=2.154$ [2].

References:

[1] K. Das, K. M. Kadish, J. L. Bear (Inorg. Chem. **17** [1978] 930/4). – [2] T. Kawamura, K. Fukamachi, T. Sowa, S. Hayashida, T. Yonezawa (J. Am. Chem. Soc. **103** [1981] 364/9).

1.1.3.6 Propionates

$[Rh_2(O_2CC_2H_5)_4]^+$. This species has been prepared in solution by electrochemical oxidation of $Rh_2(O_2CC_2H_5)_4$; polarography, cyclic voltammetry and controlled-potential electrolysis data have been reported [1, 2].

$[Rh_2(O_2CC_2H_5)_4\{P(C_6H_5)_3\}_2]^+$. This reddish-purple species has been generated by electrochemical oxidation and its electronic spectrum ($\lambda_{max}=525$ nm) has been recorded [3, 4]. The ESR spectrum of a frozen solution has been recorded at −170°C [3, 4]; $g_{\perp}=2.157$, triplet 15.1 mT; $g_{\parallel}=1.994$, triplet of triplets 21.7 and 1.47 mT. The spectrum has been reproduced [3].

$[Rh_2(O_2CC_2H_5)_4\{P(OC_6H_5)_3\}_2]^+$. This species has been generated by electrochemical oxidation and its ESR spectrum has been recorded ($g_{\perp}=2.174$, $g_{\parallel}=2.001$) and is reproduced [4].

$[Rh_2(O_2CC_2H_5)_4\{P(OCH_2)_3CC_2H_5\}_2]^+$. This species has been generated electrochemically and its ESR spectrum has been recorded, $g_{\perp}=2.203$, $g_{\parallel}=1.995$ [4].

References:

[1] K. Das, K. M. Kadish, J. L. Bear (Inorg. Chem. **17** [1978] 930/4). – [2] K. M. Kadish, K. Das, R. Howard, A. Dennis, J. L. Bear (Bioelectrochem. Bioenergetics **5** [1978] 741/53). – [3] T. Kawamura, K. Fukamachi, S. Hayashida (Chem. Commun. **1979** 945/6). – [4] T. Kawamura, K. Fukamachi, T. Sowa, S. Hayashida, T. Yonezawa (J. Am. Chem. Soc. **103** [1981] 364/9).

1.1.3.7 n-Butyrates

$[Rh_2(O_2CC_3H_7)_4]^+$, $[Rh_2(O_2CC_3H_7)_4L]^+$, $[Rh_2(O_2CC_3H_7)_4L_2]^+$ (L = CH_3CN, pyridine, N-methylimidazole, piperidine, $(C_2H_5)_2S$, 4-picoline-N-oxide, $(CH_3)_2SO$ and a caged phosphite). Electrochemical studies on these systems have been reported [1]. Reduction potentials for the species $[Rh_2(O_2CC_3H_7)_4L]^+$ and $[Rh_2(O_2CC_3H_7)_4L_2]^+$ are tabulated [1] together with ΔG values for the reactions

$[Rh_2(O_2CC_3H_7)_4L] + [Rh_2(O_2CC_3H_7)_4]^+ \rightarrow [Rh_2(O_2CC_3H_7)_4] + [Rh_2(O_2CC_3H_7)_4L]^+$ and for
$[Rh_2(O_2CC_3H_7)_4L_2] + [Rh_2(O_2CC_3H_7)_4L]^+ \rightarrow [Rh_2(O_2CC_3H_7)_4L] + [Rh_2(O_2CC_3H_7)_4L_2]^+$.

The electrochemical oxidation of $Rh_2(O_2CC_3H_7)_4$ to $[Rh_2(O_2CC_3H_7)_4]^+$ has been investigated by polarography, cyclic voltammetry and controlled-potential electrolysis [2].

References:

[1] R. S. Drago, S. P. Tanner, R. M. Richman, J. R. Long (J. Am. Chem. Soc. **101** [1979] 2897/903). – [2] K. Das, K. M. Kadish, J. L. Bear (Inorg. Chem. **17** [1978] 930/4).

1.1.4 Rhodium(III) Carboxylates

1.1.4.1 Acetates (see "Rhodium" 1938, p. 70)

For $[Rh_2(NH_3)_5(O_2CCH_3)]^{2+}$ salts and $[Rh(O_2CH)(NH_3)_5]^+$, see p. 133.

$\{Rh(O_2CCH_3)(HOH\cdots O_2CCH_3)(OH)(NH_3)\}_2$. This binuclear complex is prepared by treatment of $Rh(OH)_3(H_2O)_2(NH_3)$ with hot acetic acid, and is isolated as an amorphous yellow-orange powder. The infrared spectrum (4000 to 600 cm^{-1}) is reproduced and assignments given. The electronic spectrum has been reported (λ_{max} = 410, 310, 250 and 195 nm) and assignments proposed. The proton NMR spectrum and thermogravimetric analysis curve are reproduced. A binuclear structure involving a pair of bridging carboxylate ligands is proposed [1].

$\{Rh(O_2CCH_3)(HOH\cdots O_2CCH_3)(OH)(NH_3)\}_2CH_3CO_2H$. The complex described above is soluble in water, alcohols, dimethylsulphoxide, methylcyanide and acetic acid. From the solution in acetic acid the adduct has been isolated [1].

$[Rh_3O(O_2CCH_3)_6(H_2O)_3](ClO_4)\cdot 2H_2O$ is prepared by treatment of $RhCl_3\cdot 3H_2O$ with the calculated amount of silver acetate in hot aqueous acetic acid solution, followed by removal of silver chloride, evaporation of the filtrate on a water bath, and reprecipitation of the residue in the presence of perchloric acid [3]. It is also obtained by ozonising a solution of rhodium(II) acetate in boiling acetic acid over a period of 2 h [4, 5]. The complex, which loses two molecules of water of crystallisation at 100°C [3] has been described as a finely crystalline brown solid [3, 5] and as a yellow-orange solid [4]. It is very soluble in water, alcohol, and acetone, but insoluble in chloroform and benzene [3].

X-ray powder photographs (data recorded in paper) establish that the rhodium complex $[Rh_3O(O_2CCH_3)_6(H_2O)_3][ClO_4]\cdot 2H_2O$ is isostructural with the analogous chromium(III) complex [3]. The dihydrate is monoclinic, space group $P2_1/c\text{-}C^5_{2h}$, a = 11.737(2), b = 14.962(4), c = 15.200(4) Å, β = 92.11(3)°, Z = 4; D_{exp} = 2.17, D_{calc} = 2.16 g/cm^3. The trinuclear cation contains three rhodium atoms at the vertices of an equilateral triangle with a bridging oxygen at the centre, and each pair of rhodium atoms is bridged by two carboxylate ligands, see **Fig. 8**, p. 60. Bond distances are Rh-O(centre) = 1.941, 1.906 and 1.924 Å, Rh-Rh (non-bonding) distances are 3.322, 3.348 and 3.325 Å [6].

$[Rh_3O(O_2CCH_3)_6(H_2O)_3](ClO_4)\cdot H_2O$. The monohydrate is monoclinic, space group $P2_1\text{-}C^2_2$, a = 8.105(3), b = 13.959(5), c = 11.515(5) Å; β = 93.95(3)°, Z = 2; D_{exp} = 2.16 g/cm^3, D_{calc} = 2.17 g/cm^3. The trinuclear cation has the same oxygen centred structure shown in Fig. 8, p. 60. Bond distances are Rh-O(centre) = 1.93(2), 1.94(2), 1.90(2) Å, Rh-Rh (non-bonding) distances are 3.329(3), 3.333(3), 3.325(3) Å [6].

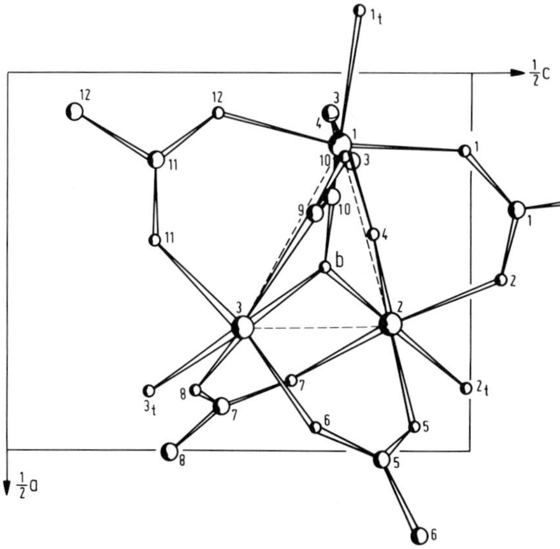

Fig. 8
View of the trinuclear cation along b-axis.

The infrared spectrum (3600 to 150 cm^{-1}) and electronic spectrum have been recorded, λ_{max} = 532 and 357 nm in H_2O solution; λ_{max} = 526, 360 nm, CH_3OH solution; λ_{max} = 540 (shoulder) and 357 nm in CH_3OH solution [4]. The complex is diamagnetic, μ_{eff} = 0.48 B.M. [4]; μ_{eff} = 0.00 B.M. [5]. Vibrational spectra have been tabulated and assigned [8].

$[Rh_3O(O_2CCH_3)_6(C_5H_5N)_3]ClO_4$ is prepared by treatment of $[Rh_3O(O_2CCH_3)_6(H_2O)_3]ClO_4 \cdot 2H_2O$ with pyridine and sodium perchlorate in acetone [3] or warm methanol [4, 5] and is isolated as a pale yellow solid [4, 5]. The salt is soluble in methylcyanide, dimethylsulphoxide, dimethylformamide, chloroform and dichloromethane, slightly soluble in acetone and alcohols, and insoluble in ether, light petroleum, benzene, carbon tetrachloride, acetic acid and water [4]. Vibrational spectra have been tabulated and assigned [8].

X-ray powder diffraction data (tabulated in the paper) establish that the rhodium complex is isomorphous with analogous complexes of Cr^{III}, Mn^{III}, Fe^{III}, Ru^{III} and Ir^{III} [4].

The infrared spectrum (3600 to 150 cm^{-1}) and electronic spectrum have been reported, λ_{max} = 307 nm in H_2O solution [5]; λ_{max} = 312 and 250 nm in CH_3OH solution [4]. The complex is diamagnetic, μ_{eff} = 0.45 B.M. [4].

$[Rh_3O(O_2CCH_3)_6(\beta\text{-picoline})_3]ClO_4$ is prepared by adding β-picoline to an aqueous solution of $[Rh_3O(O_2CCH_3)_6(H_2O)_3](ClO_4) \cdot 2H_2O$, and is isolated as a pale yellow solid. Its solubility properties are similar to those of the pyridine adduct (see above). The electronic spectrum (λ_{max} = 312, 250 nm, CH_3OH solution) has been recorded and a magnetic moment reported, μ_{eff} = 0.61 B.M. [4].

$[Rh_3O(O_2CCH_3)_6\{P(C_6H_5)_3\}_3]ClO_4$ is prepared by heating a mixture of $[Rh_3O(O_2CCH_3)_6(H_2O)_3](ClO_4) \cdot 2H_2O$ and triphenylphosphine in methanol at 60°C for 5 h, and is isolated on evaporation of the solution, as a red-orange solid. The solubility pattern of this complex is similar to that of the pyridine adduct (see above). The electronic spectrum has been recorded, λ_{max} = 485(shoulder) and 350 nm in CH_3OH solution [4].

Miscellaneous Acetates. Treatment of $K_3[RhCl_6]$ with hydrazine acetate, $N_2H_5O_2CCH_3$, affords a series of polymeric acetate hydrazide complexes. Because of their polymeric nature these species are difficult to characterise properly. They decompose at 80°C losing N_2H_4 and H_2O. Proposed formulae with susceptibilities ($\chi_{mol} \times 10^6$) are listed below [7]:

Formula	$\chi \times 10^6$
$Rh_4Cl_{7.5}(O_2CCH_3)_5(N_2H_4)_{11.5}(N_2H_5)_{0.5}$	−320
$Rh_4Cl_7(O_2CCH_3)_{5.5}(N_2H_4)_{9.5}(N_2H_5)_{0.5}$	−450
$Rh_4Cl_7(O_2CCH_3)_{4.5}(N_2H_3)_{0.5}(N_2H_4)_{9.5}$	−560
$Rh_4Cl_7(O_2CCH_3)_{3.5}(N_2H_3)_{1.5}(N_2H_4)_{9.5}$	−390
$Rh_4Cl_7(O_2CCH_3)_{2.5}(N_2H_3)_{2.5}(N_2H_4)_{5.5}$	−410
$Rh_4Cl_{6.5}(O_2CCH_3)_{4.5}(N_2H_3)(N_2H_4)_{9.5}$	−410
$Rh_4Cl_6(O_2CCH_3)_{3.5}(N_2H_3)_{2.5}(N_2H_4)_{6.5}$	−360
$Rh_4Cl_{5.5}(O_2CCH_3)_4(N_2H_3)_{2.5}(N_2H_4)_8$	−280
$Rh_4(O_2CCH_3)_9(N_2H_3)_3(N_2H_4)_2$	−195
$Rh_4(O_2CCH_3)_7(N_2H_3)_5 \cdot 2H_2O$	−180
$Rh_4(O_2CCH_3)_7(N_2H_3)_{3.5}(OH)_{0.5} \cdot 4H_2O$	−160
$Rh_4(O_2CCH_3)_7(N_2H_3)_3(OH)_2$	−210

$[(C_6H_{15}N_3)Rh(OH)_2(O_2CCH_3)Rh(C_6H_{15}N_3)](ClO_4)_3 \cdot H_2O$ ($C_6H_{15}N_3$ = 1,4,7-triazacyclononane). This complex is prepared by treatment of $[(C_6H_{15}N_3)(H_2O)Rh(OH)_2Rh(H_2O)(C_6H_{15}N_3)](ClO_4)_4 \cdot 4H_2O$ with acetic acid in aqueous solution at 100°C, and is isolated as a yellow crystalline solid by addition of sodium perchlorate to the cold concentrated solution. The electronic spectrum is reported, λ_{max} = 339 nm [2].

References:

[1] G. Pannetier, J. Segall (J. Less-Common Metals **22** [1970] 293/303). – [2] K. Wieghardt, W. Schmidt, B. Nuber, B. Prikner, J. Weiss (Chem. Ber. **113** [1980] 36/41). – [3] I. B. Baranovskii, G. Ya. Mazo, L. M. Dikareva (Zh. Neorgan. Khim. **16** [1970] 2602/3; Russ. J. Inorg. Chem. **16** [1970] 1388/9). – [4] S. Uemura, A. Spencer, G. Wilkinson (J. Chem. Soc. Dalton Trans. **1973** 2565/71). – [5] T. Szymańska-Buzar, J. J. Ziolkowski (Koord. Khim. **2** [1976] 1172/91; Soviet J. Coord. Chem. **2** [1976] 897/912).

[6] T. Glowiak, M. Kubiak, T. Szymańska-Buzar (Acta Cryst. B **33** [1977] 1732/7). – [7] L. Cambi, E. Paglia (Atti. Accad. Nazl. Lincei Rend. Classe Sci. Fis. Mat. Nat. **30** [1961] 429/36; C.A. **56** [1962] 6885). – [8] M. K. Johnson, D. B. Powell, R. D. Cannon (Spectrochim. Acta A **37** [1981] 995/1006).

For **$[Rh(NH_3)_5(O_2CR)]^{2+}$** with R = H, CH_3, CH_2F, CHF_2, CF_3 and CCl_3 see p. 133.

1.1.4.2 Chloroacetates

$Rh(O_2CCH_2Cl)_2(OH)(H_2O)_2(NH_3)$. This is prepared by treatment of $Rh(OH)_3(H_2O)_2(NH_3)$ with monochloroacetic acid and is isolated as an amorphous yellow powder. The infrared spectrum (4000 to 400 cm^{-1}) is reproduced and selected assignments given. Thermogravimetric analysis curves for decomposition under hydrogen and nitrogen are reproduced.

$\{Rh(O_2CCHCl_2)(OH)_2(NH_3)\}_2$. This compound is prepared by treatment of $Rh(OH)_3(H_2O)_2(NH_3)$ with dichloroacetic acid, and is isolated as an amorphous yellow powder. The infrared spectrum (4000 to 400 cm^{-1}) is reproduced and assignments given. Thermogravimetric analysis curves (decomposition under hydrogen and nitrogen) are reproduced.

$\{Rh(O_2CCCl_3)(OH)_2(NH_3)\}_2$. This is prepared by treatment of $Rh(OH)_3(H_2O)_2(NH_3)$ with trichloroacetic acid and isolated as an amorphous yellow powder. The infrared spectrum (4000 to

400 cm^{-1}) is reproduced and assignments given. Thermogravimetric analysis curves for decomposition under hydrogen and nitrogen are reproduced.

G. Pannetier, J. Pagniet, C. Laurent (J. Less-Common Metals **25** [1971] 385/95).

1.1.4.3 Pyridine-2,6-dicarboxylates

$Na[Rh\{(O_2C)_2C_5H_3N\}_2]\cdot2H_2O$. This salt is prepared by heating an aqueous mixture of rhodium(III) chloride and the disodium salt of pyridine-2,6-dicarboxylic acid under reflux for 5 h, then evaporating to dryness and crystallising the residue from methanol. It forms orange-brown crystals or yellow powder, and is extremely water-soluble. Selected infrared frequencies are recorded, ν(OH) ≈3450, ν(OCO) = 1670 to 1600 cm^{-1}. The electronic spectrum has been reported (λ_{max} = 380 nm) and ESCA data have been given.

$[NPh_4][Rh\{(O_2C)_2C_5H_3N\}_2]\cdot3H_2O$. This salt is obtained as an insoluble yellow precipitate by addition of $[NPh_4]Cl$ to an aqueous solution of the corresponding sodium salt. X-ray photoelectronic data have been recorded, $Rh\,3d_{3/2}$ = 314.3, $Rh\,3d_{5/2}$ = 309.4 eV.

R. W. Matthews, A. D. Hamer, D. L. Hoof, D. G. Tisley, R. A. Walton (J. Chem. Soc. Dalton Trans. **1973** 1035/8).

1.1.4.4 Miscellaneous Carboxylates

A series of polynuclear rhodium(III) carboxylato-hydrazido complexes have been reported. They are prepared by treatment of $K_3[RhCl_6]$ with excess of a neutral or weakly acid solution of the hydrazine salts of carboxylic acids at controlled pH. The complexes prepared, which are generally polymeric and insoluble in water or organic solvents, are listed below together with the pH value maintained in the preparation and the magnetic susceptibility ($\chi_{mol}\times10^{-6}$).

Compound	pH	$\chi_{mol}\times10^{-6}$
$Rh_4Cl_{4.5}(O_2CN_2H_3)_6(N_2H_3)_{1.5}(N_2H_4)_{3.5}$	7	−150
$Rh_4Cl_6(O_2CC_2H_5)_3(N_2H_3)_3(N_2H_4)_7\cdot3H_2O$	4.5	−205
$Rh_4Cl_{1.5}(O_2CC_2H_5)_7(N_2H_3)_{3.5}(N_2H_4)_{2.5}\cdot3H_2O$	—	−105
$Rh_4Cl_{3.5}(O_2CCH_2OH)_{5.5}(N_2H_3)_3(N_2H_4)_{6.5}$	4	−129
$[Rh_4Cl_5(O_2CCH_2NH_2)_3(N_2H_3)_4(N_2H_4)_{4.5}]$	—	−10
$[Rh_4Cl_5(O_2CCH_2CN)_{4.5}(N_2H_3)_{2.5}(N_2H_4)_8]\cdot H_2O$	6	—
$[Rh_4Cl_4(O_2CCH_2CN)_5(N_2H_3)_3(N_2H_4)_9]\cdot3H_2O$	—	−108
$[Rh_4Cl_3(O_2CCH_2C_6H_5)_6(N_2H_3)_3(N_2H_4)_9]\cdot4H_2O$	8	−195
$[Rh_4Cl_{2.5}(O_2CC_6H_5)_8(N_2H_3)_{1.5}(N_2H_4)_{9.5}]$	7	—
$[Rh_4(O_2C\cdot C_6H_4\cdot OH)_7(N_2H_3)_5(N_2H_4)_{4.5}]$	6	—
$[Rh_4Cl_{3.5}(O_2C\cdot CH_2\cdot CO_2)_4(N_2H_3)_{0.5}(N_2H_4)_{9.5}]\cdot4H_2O$	—	−160
$[Rh_4Cl_{5.5}(O_2CCH_2CH_2CO_2)_3(N_2H_3)_{0.5}(N_2H_4)_8]\cdot6H_2O$	4	−205
$[Rh_4Cl_{4.5}(O_2CCH_2CH_2CO_2)_4(N_2H_4)_9(N_2H_5)_{0.5}]\cdot H_2O$	—	−190
$[Rh_4(O_2CCH_2CH_2CO_2)_{4.5}(N_2H_3)_3(N_2H_4)_4]\cdot5H_2O$	—	—

L. Cambi, E. D. Paglia, G. Bargigia (Atti. Accad. Nazl. Lincei Rend. Classe Sci. Fis. Mat. Nat. **30** [1961] 636/43).

For rhodium(III) nitrosyl carboxylato complexes see p. 110; for binuclear rhodium(III) ammines with acetate, malonate and succinate ions see pp. 154/5; for carboxylato-pentammine complexes see pp. 133/4.

1.1.4.5 Monothioacetate, $[Rh(OSCCH_3)(NH_3)_5]I_2$

The compound is prepared by addition of thioacetic acid to the residue obtained by treatment of $[Rh(H_2O)(NH_3)_5]Br_3$ with the stoichiometric amount of Ag_2O followed by removal of AgBr, and is crystallised from solution by addition of methanolic KI. The infrared spectrum (400 to 30 cm^{-1}) is reproduced.

I. B. Baranovskii, M. A. Golubnichaya, G. Ya. Mazo, V. I. Nefedov, Ya. V. Salyn', R. N. Shchelokov (Zh. Neorgan. Khim. **21** [1976] 1085/94; Russ. J. Inorg. Chem. **21** [1976] 591/6).

1.1.5 Aminocarboxylates

1.1.5.1 Rhodium(I) Leucinates

$(CH_3)_2CHCH_2CH(NH_2)CO_2H$ = leucine = leu H

Rh(leu)($NH_2C_6H_{11}$)$_2$ ($NH_2C_6H_{11}$ = cyclohexylamine) is prepared by treatment of the corresponding dicarbonyl complex, $Rh(leu)(CO)_2$, with excess $NH_2C_6H_{11}$ in acetone solution and is isolated as a yellow solid by evaporation. The electronic spectrum (λ_{max} = 282 nm) and infrared spectrum have been reported, ν(NH) = 3350, 3200; $\nu(CO_2)_{as}$ = 1625 to 1575 cm^{-1}.

Rh(leu){$NH_2CH_2CH(CH_3)_2$}$_2$ is prepared by treatment of $Rh(leu)(CO)_2$ with isobutylamine and isolated as a yellow solid. The electronic spectrum (λ_{max} = 308 nm) and infrared spectrum have been reported, ν(NH) = 3300 to 3200, $\nu(CO_2)_{as}$ = 1600 cm^{-1}.

Rh(leu)($C_{12}H_8N_2$) ($C_{12}H_8N_2$ = o-phenanthroline) is prepared by adding o-phenanthroline to an acetone solution of $Rh(leu)(CO)_2$ and separates as a dark green hygroscopic solid.

Rh(leu)($C_{10}H_8N_2$) ($C_{10}H_8N_2$ = 2,2′-dipyridyl) is prepared by adding 2,2′-dipyridyl to an acetone solution of $Rh(leu)(CO)_2$ and separates as a brown hygroscopic solid.

D. Doweran, M. M. Singh (J. Indian Chem. Soc. **57** [1980] 368/71).

1.1.5.2 Rhodium(II) Aminocarboxylates

For **aminoacetates** (glycinates) see p. 37; for **β-aminopropionates** (β-alaninates) see p. 46; for **t-butyloxycarbonyl-β-aminopropionates** (t-butyloxy-carbonyl-β-alaninates) see p. 47.

γ-Aminobutyrates

$NH_2CH_2CH_2CH_2CO_2H$ = γ-aminobutyric acid = γ-NH_2butH

The blue or green diamagnetic complexes described in this section are thought to contain γ-aminobutyrate ligands bound to metal-metal bonded Rh^{II}-Rh^{II} groups through oxygen and nitrogen donor atoms.

Rh(γ-NH_2but)Cl·2H_2O is prepared by treatment of an aqueous solution of hexachlororhodic acid (from $RhCl_3 \cdot nH_2O$/HCl) with γ-aminobutyric acid under reflux for 2 h and is isolated in 6% yield by evaporation of the green filtrate to small volume. The complex is a powdery green solid insoluble in water, acetone, ether, benzene and acetonitrile, and only slightly soluble in acetic acid.

The infrared spectrum has been recorded, $\nu(NH_2)$ = 3230, 3135; ν(OCO) = 1570, 1400; ν(RhO) = 500 to 400; ν(RhCl) = 324 cm^{-1}. It has been taken to indicate the presence of N,O-bonded γ-aminobutyrate ligands. ESCA data confirm the presence of rhodium(II); Rh $3d_{5/2}$ = 309.3, Cl 2p = 198.6 eV. The compound is diamagnetic.

Rh(γ-NH_2but)Cl is prepared from the dihydrate Rh(γ-NH_2but)Cl·2H_2O by heating to 120 to 130°C. It undergoes stepwise exothermic decomposition at 250 and 390 to 410°C. ESCA data confirm retention of the rhodium(II) oxidation state in the dehydrated complex; Rh $3d_{5/2}$ = 309.2, Cl 2p = 198.8 eV.

Rh(γ-NH_2but)Cl·2H_2O(HCl)·0.5(γ-NH_2butH). A light green solid of this apparent stoichiometry is obtained by addition of acetone to the green mother liquor left after isolation of Rh(γ-NH_2but)Cl·2H_2O (preceding complex). The infrared spectrum displays a band at 1700 cm^{-1} attributed to a nonionised CO_2H group. The electronic spectrum (aqueous solution 700 to 350 nm) is reproduced, λ_{max} values have been recorded as 600 and 430 nm (shoulder). The compound is diamagnetic.

Rh(γ-NH_2but)(O_2CH)H_2O·HCl·0.5(γ-NH_2butH). A solid of this apparent stoichiometry is obtained by addition of formic acid to the mother liquor remaining after isolation of Rh(γ-NH_2but)Cl·2H_2O (see above) and is precipitated from solution by addition of acetone. The complex is quite soluble in water. An infrared band at 1700 cm^{-1} has been reported. The electronic spectrum (700 to 350 nm) is reproduced. ESCA data have been recorded, Rh $3d_{5/2}$ = 309.3 eV.

A. E. Bukanova, I. V. Prokof'eva, Ya. V. Salyn', L. K. Shubochkin (Koord. Khim. **3** [1977] 1739/42; Soviet J. Coord. Chem. **3** [1977] 1365/8).

1.1.5.3 Rhodium(III) Aminocarboxylates

Aminoacetates

$NH_2CH_2CO_2H$ = glycine = glyH

Many of the complexes described in this section are known to exist in several isomeric forms, some of which involve glycine ligands coordinated in the betaine form ($\overset{+}{N}H_3CH_2CO_2^-$). There is considerable uncertainty concerning the stereochemistry of several products.

***fac*-Rh(gly)$_3$·H_2O.** To obtain this complex rhodium trichloride is first dissolved in hydrochloric acid (translation says water) and evaporated; then twice treated with water and evaporated on a steambath before final dissolution to give an aqueous solution of chlororhodic acid (pH = 2 to 3). This solution is treated with glycine (12 molar excess) and ethanol (15% total volume), then heated under reflux for 2 to 3 h. On cooling large green-yellow platelets deposit. The same product is obtained by boiling a solution of $Na_3[RhCl_6]$ (pH = 3 to 4) with excess glycine for 1 h, then keeping the solution for several months at 18 to 20°C. The refractive indices of the crystals are n_g = 1.702, n_p = 1.584, n_m = 1.665 [1].

The X-ray crystal structure has been determined. The crystals are monoclinic, space group $P2_1/a$-C_{2h}^5. Lattice parameters are a = 11.98 ± 0.02, b = 14.72 ± 0.02, c = 6.18 ± 0.02 Å, β = 98° ± 0.5°, Z = 4. The Rh-O and Rh-N distances are in the range 1.9 to 2.1 Å [1].

The electronic spectrum has been reported, λ_{max} 340 and 285 nm [1]. The infrared spectrum (4000 to 400 cm^{-1}) is reproduced [1] and ESCA data have been recorded, Rh $3d_{5/2}$ = 310.5, N 1s = 400.4 eV [2]. The complex is sparingly soluble in water (4.21×10^{-4} mol/l at 25°C) and is a nonelectrolyte, conductance 5.7 $cm^2 \cdot \Omega^{-1} \cdot mol^{-1}$ for a saturated solution at 25°C. A thermogravimetric analysis curve is reproduced, the molecule of water is lost at 115 to 140°C and the **anhydrous compound** decomposes at 300 to 320°C with a marked exothermic effect [1]. Polarographic data have been reported, $E_{1/2}$ = −1.02 V [3]. The reactions of this complex with mineral acids (HCl, HNO_3) are described below.

fac-Rh(gly)$_3$ has also been obtained in 12% yield as the final product from the fusion of $RhCl_3 \cdot 4H_2O$ and glycine. Infrared spectra of various melts are reproduced [4].

References:

[1] A. E. Bukanova, I. V. Prokof'eva, M. A. Porai-Koshits, A. S. Antsyshkina, L. M. Dikareva (Zh. Neorgan. Khim. **17** [1972] 757/61; Russ. J. Inorg. Chem. **17** [1972] 396/8). – [2] V. I. Nefedov, I. V. Prokof'eva, A. E. Bukanova, L. K. Shubochkin, Ya. V. Salyn', V. L. Pershin (Zh. Neorgan. Khim. **19** [1974] 1578/80; Russ. J. Inorg. Chem. **19** [1974] 859/60). – [3] A. E. Bukanova, N. A. Ezerskaya, I. V. Prokof'eva (Zh. Anilit. Khim. **28** [1973] 1149/53; J. Anal. Chem. [USSR] **28** [1973] 1016/9). – [4] V. S. Bondarenko, G. S. Voronina, V. I. Kazbanov (Zh. Neorgan. Khim. **27** [1982] 2037/41; Russ. J. Inorg. Chem. **27** [1982] 1149/52).

***mer*-Rh(gly)$_3$.** This isomer is prepared by treating boiling aqueous Rh(gly)$_2$Cl(glyH) (isomer prepared from $Na_3[RhCl_6]$ and glycine) with one equivalent of sodium hydroxide in aqueous solution, then evaporating the solution on a water bath until precipitation occurs [1]. The complex forms fine yellow needles, refractive indices $n_g = 1.754$, $n_p = 1.609$, $n_m = 1.694$. The compound is sparingly soluble in water ($\sim 3.36 \times 10^{-3}$ mol/l at 25°C) and insoluble in ammonia [1]. A freshly prepared aqueous 10^{-3} M solution has a pH of 5.9 at 25°C and a molar conductance of 3.79 $cm^2 \cdot \Omega^{-1} \cdot mol^{-1}$ [1]. The infrared spectrum (4000 to 400 cm^{-1}) is reproduced [2]. The electronic spectrum has been reported, $\lambda_{max} = 360$, 330 and 285 nm [3]. Polarographic data have been reported, $E_{1/2} = -1.07$ V [4]. A thermogravimetric analysis curve is reproduced, decomposition of the complex commences at 325°C [1].

References:

[1] I. V. Prokof'eva, A. E. Bukanova, O. E. Zvyagintsev (Zh. Neorgan. Khim. **14** [1969] 2802/5; Russ. J. Inorg. Chem. **14** [1969] 1476/8). – [2] A. E. Bukanova, I. V. Prokof'eva, M. A. Porai-Koshits, A. S. Antsyshkina, L. M. Dikareva (Zh. Neorgan. Khim. **17** [1972] 757/61; Russ. J. Inorg. Chem. **17** [1972] 396/8). – [3] A. E. Bukanova, N. A. Ezerskaya, I. V. Prokof'eva (Zh. Analit. Khim. **28** [1973] 1149/53; J. Anal. Chem. [USSR] **28** [1973] 1016/9).

Rh(gly)$_2$Cl(glyH) (isomer A) is obtained indirectly from *fac*-Rh(gly)$_3 \cdot H_2O$. Treatment of *fac*-Rh(gly)$_3 \cdot H_2O$ with excess HCl affords a product of stoichiometry Rh(gly)Cl$_2$(glyH)$_2 \cdot H_2O$ which, on dissolution with two equivalents of CsCl in aqueous solution and subsequent slow evaporation, deposits fine yellow crystals of the required product. The infrared spectrum (4000 to 400 cm^{-1}) is reproduced; a strong band at ~ 1700 cm^{-1}, $\nu(CO_2H)$, distinguishes this product from isomer B [1]. The electronic spectrum has been recorded, $\lambda_{max} = 350$, 285 nm. ESCA data have been reported (Rh $3d_{5/2} = 310.5$, N 1s = 400.6, Cl 2p = 198.7 eV) [2].

Rh(gly)$_2$Cl(glyH) (isomer B) is prepared by treatment of $Na_3[RhCl_6]$ with glycine (12 mol per mol of rhodium salt) in boiling aqueous solution for 6 h and deposits from solution on cooling as very fine needles [3]. The infrared spectrum (2000 to 400 cm^{-1}) is reproduced [4], the absence of a band at ~ 1700 cm^{-1} distinguishes this product from isomer A. The electronic spectrum (560 to 360 nm) is reproduced [3]; maxima at 340 and 300 nm are recorded [2]. ESCA data have been recorded (Rh $3d_{5/2} = 310.5$; N 1s = 400.4 and 402.1, Cl 2p = 198.8 eV); the N 1s values have been attributed to the presence of the glycine ligand in the betaine ($\overset{+}{N}H_3CH_2CO_2^-$) form [2, 5]. Polarographic data have been recorded, $E_{1/2} = -0.53$, -1.03 V [6].

[Rh(gly)$_2$(NH$_3$)(glyNH$_4$)]Cl is prepared from Rh(gly)$_2$Cl(glyH) (isomer B) by treatment with aqueous ammonia, and separates as a yellow precipitate, readily soluble in water [3].

References:

[1] A. E. Bukanova, I. V. Prokof'eva, T. P. Sidorova (Zh. Neorgan. Khim. **19** [1974] 3331/4; Russ. J. Inorg. Chem. **19** [1974] 1825/7). – [2] I. V. Prokof'eva, A. E. Bukanova, V. I. Nefedov, L. K. Shubochkin, Ya. V. Salyn', T. P. Sidorova (Koord. Khim. **1** [1975] 536/8; Soviet J. Coord. Chem. **1** [1975] 437/9). – [3] O. E. Zvyagintsev, I. V. Prokof'eva, A. E. Bukanova (Zh. Neorgan. Khim. **11** [1966] 2070/3; Russ. J. Inorg. Chem. **11** [1966] 1107/10). – [4] I. V. Prokof'eva, A. E. Bukanova, O. E. Zvyagintsev (Zh. Neorgan. Khim. **15** [1970] 1037/9; Russ. J. Inorg. Chem. **15** [1970] 528/30). – [5] V. I. Nefedov, I. V. Prokof'eva, A. E. Bukanova, L. K. Shubochkin, Ya. V. Salyn', V. L. Pershin (Zh. Neorgan. Khim. **19** [1974] 1578/80; Russ. J. Inorg. Chem. **19** [1974] 859/60).

[6] A. E. Bukanova, N. A. Ezerskaya, I. V. Prokof'eva (Zh. Analit. Khim. **28** [1973] 1149/53; J. Anal. Chem. [USSR] **28** [1973] 1016/9).

[Rh(gly)$_2$(glyH)(H$_2$O)]NO$_3$ (isomer A) is prepared by treatment of *fac*-Rh(gly)$_3$·H$_2$O with excess 1:1 HNO$_3$ on a water bath, followed by careful evaporation to dryness and precipitation of the product from concentrated aqueous solution using acetone. It is a pale yellow air-stable powder readily soluble in water [1]. ESCA data have been reported, Rh 3d$_{5/2}$ = 310.7, N 1s = 406.9 and 400.5 eV [2].

[Rh(gly)$_2$(glyH)(H$_2$O)]NO$_3$ (isomer B) is similarly prepared from *mer*-Rh(gly)$_3$ as a yellow hygroscopic powder readily soluble in water [1]. ESCA data have been reported, Rh 3d$_{5/2}$ = 310.8, N 1s = 406.9 and 400.5 eV [2].

References:

[1] A. E. Bukanova, I. V. Prokof'eva, T. P. Sidorova (Zh. Neorgan. Khim. **19** [1974] 3331/4; Russ. J. Inorg. Chem. **19** [1974] 1825/7). – [2] V. I. Nefedov, I. V. Prokof'eva, A. E. Bukanova, L. K. Shubochkin, Ya. V. Salyn', V. L. Pershin (Zh. Neorgan. Khim. **19** [1974] 1578/80; Russ. J. Inorg. Chem. **19** [1974] 859/60).

Rh(gly)Cl$_2$(glyH)$_2$ (isomer A) is prepared by treating *fac*-Rh(gly)$_3$·H$_2$O with excess HCl and warming the mixture. Evaporation of the resultant yellow solution produces short yellow needles. A titration curve (NaOH solution) is reproduced and shows simultaneous neutralisation of the two glycine ligands [1]. The electronic spectrum has been reported, λ_{max} = 385, 320 nm [2]. ESCA data have been recorded, Rh 3d$_{5/2}$ = 310.5, N 1s = 400.5, Cl 2p = 198.9 eV [1, 2, 3]. The infrared spectrum (3800 to 700 cm^{-1}) is reproduced [1].

Rh(gly)Cl$_2$(glyH)$_2$ (isomer B) is prepared by treating Rh(gly)$_2$Cl(glyH) (isomer prepared from Na$_3$[RhCl$_6$] and glycine) with one equivalent of HCl. A titration curve (NaOH solution) is reproduced and shows consecutive neutralisation of the two glycine ligands [1]. The electronic spectrum has been reported, λ_{max} = 360 to 350 nm [2]. ESCA data are taken to indicate the presence of one glycine ligand in the betaine form ($\overset{+}{N}H_3CH_2CO_2^-$); Rh 3d$_{5/2}$ = 310.4, N 1s = 400.5 and 401.9, Cl 2p = 198.9 eV [2].

References:

[1] A. E. Bukanova, I. V. Prokof'eva, T. P. Sidorova (Zh. Neorgan. Khim. **19** [1974] 3331/4; Russ. J. Inorg. Chem. **19** [1974] 1825/7). – [2] I. V. Prokof'eva, A. E. Bukanova, V. I. Nefedov, L. K. Shubochkin, Ya. V. Salyn', T. P. Sidorova (Koord. Khim. **1** [1975] 536/8; Soviet J. Coord. Chem. **1** [1975] 437/9). – [3] V. I. Nefedov, I. V. Prokof'eva, A. E. Bukanova, L. K. Shubochkin, Ya. V. Salyn', V. L. Pershin (Zh. Neorgan. Khim. **19** [1974] 1578/80; Russ. J. Inorg. Chem. **19** [1974] 859/60).

$RhCl_3(glyH)_3$ (isomer A) is prepared by treatment of $Rh(gly)_2Cl(glyH)$ (isomer prepared from $Na_3[RhCl_6]$ and glycine) with 1 M hydrochloric acid, and is crystallised from acetone. It behaves as an acid in aqueous solution [1]. The electronic spectrum has been recorded, λ_{max} = 500 to 490 and 430 nm [2], and is reproduced (560 to 360 nm) [1]. ESCA data have been taken to indicate that one glycine ligand is coordinated in the betaine ($\overset{+}{N}H_3CH_2CO_2^-$) form, Rh 3d$_{5/2}$ = 310.1, N1s = 400.5 and 402.3, Cl2p = 198.6 eV [2].

$RhCl_3(glyH)_3$ (isomer B) is prepared by treatment of *mer*-$Rh(gly)_3$ with hot 6 M HCl followed by evaporation of the solution and crystallisation of the product from acetone. The electronic spectrum (λ_{max} = 400 nm) has been recorded. ESCA data have been reported, Rh 3d$_{5/2}$ = 310.4, N1s = 400.5, Cl2p = 198.8 eV [2].

For $[Rh(glyH)(NH_3)_5](ClO_4)_3$ see p. 134; see also p. 71 for thermodynamic data on rhodium(III) glycinates.

References:

[1] O. E. Zvyagintsev, I. V. Prokof'eva, A. E. Bukanova (Zh. Neorgan. Khim. **11** [1966] 2070/3; Russ. J. Inorg. Chem. **11** [1966] 1107/10). – [2] I. V. Prokof'eva, A. E. Bukanova, V. I. Nefedov, L. K. Shubochkin, Ya. V. Salyn', T. P. Sidorova (Koord. Khim. **1** [1975] 536/8; Soviet J. Coord. Chem. **1** [1975] 437/9).

α-Aminopropionates, Rh(α-ala)$_3$

$NH_2CH(CH_3)CO_2H$ = α-alanine = α-alaH

Racemic Rh(α-ala)$_3$ is prepared by treatment of chlororhodic acid from $RhCl_3 \cdot nH_2O$ and HCl with racemic α-alanine (3 mol per mol of chlororhodic acid). This mixture is warmed, neutralised with saturated aqueous $NaHCO_3$ and then boiled for several hours. Cooling and repeated boiling gives fractions consisting first of white and then yellow acicular crystals (*fac* and *mer* isomers, respectively) [1].

***fac*-Rh(α-ala)$_3$**. The electronic spectrum has been recorded, λ_{max} = 340 to 335 and 285 nm; ESCA data have been reported, Rh 3d$_{5/2}$ = 310.0, N1s = 400.3 eV. The solubility in water is 3.62×10^{-4} mol/l at 25°C and the molar conductance 6.79 $cm^2 \cdot \Omega^{-1} \cdot mol^{-1}$ [1].

***mer*-Rh(α-ala)$_3$**. The electronic spectrum has been recorded, λ_{max} = 330 to 320 and 290 to 285 nm; ESCA data have been reported, Rh 3d$_{5/2}$ = 310.4, N1s = 400.3 eV. The solubility in water is 3.18×10^{-3} mol/l at 25°C and the molar conductance 1.17 $cm^2 \cdot \Omega^{-1} \cdot mol^{-1}$ [1].

***fac*-Rh(L-α-ala)$_3$** is obtained by heating rhodium(III) hydroxide (prepared from $RhCl_3 \cdot 3H_2O$ and $NaHCO_3$ or RhI_3 and Na_2CO_3) with L-α-alanine in aqueous solution under reflux for 2 h. It separates as a white crystalline solid [2].

Separation by fractional crystallisation affords the following optical isomers.

***fac*(+)-Rh(L-α-ala)$_3$** as an off-white solid sparingly soluble in water, λ_{max} = 339 and 284 nm, $[\alpha]_D$ = +256°, $[\alpha]_{546}$ = +223°, $[\alpha]_{300}$ = +2230° [2]. Circular dichroism data have been recorded [3, 4].

***fac*(−)-Rh(L-α-ala)$_3$** as an off-white crystalline solid, completely insoluble in water but soluble in 50% H_2SO_4, λ_{max} = 340 and 283 nm, $[\alpha]_{500}$ = −92°, $[\alpha]_{300}$ = −795°. Infrared data (3200 to 740 cm^{-1}) have been recorded [2].

***mer*(±)-Rh(L-α-ala)$_3$** is obtained by fractional crystallisation of the water-soluble residue left after removal of the insoluble *fac* isomers. Infrared data (3450 to 735 cm^{-1}) have been recorded [2].

***mer*(+)-Rh(L-α-ala)$_3$** is described as a pale yellow solid, readily soluble in water. The electronic spectrum has been recorded, λ_{max} = 335 and 287 nm [2].

References:

[1] I. V. Prokof'eva, A. E. Bukanova, L. K. Shubochkin, V. I. Nefedov, T. P. Sidorova (Zh. Neorgan. Khim. **21** [1976] 1081/4; Russ. J. Inorg. Chem. **21** [1976] 589/91). – [2] J. H. Dunlop, R. D. Gillard (J. Chem. Soc. **1965** 6531/41). – [3] R. D. Gillard (Proc. Roy. Soc. [London] A **297** [1967] 134/40, 138). – [4] R. D. Gillard, N. C. Payne (J. Chem. Soc. A **1969** 1197/203, 1198).

β-Aminopropionates

$NH_2CH_2CH_2CO_2H$ = β-alanine = β-alaH

Rh(β-ala)$_3$·2H_2O is obtained as a yellow-green crystalline precipitate by treatment of chlororhodic acid (from $RhCl_3$·4.5H_2O and HCl) with β-aminopropionic acid in aqueous solution under reflux for 2h. The electronic spectrum (350 to 250 nm) and thermogravimetric analysis curve are reproduced. The conductance is 12.2 $cm^2 \cdot \Omega^{-1} \cdot mol^{-1}$.

Rh(β-ala)$_2$Cl(β-alaH)·H_2O is prepared from the previous compound by treatment with one equivalent of 0.1M hydrochloric acid and is isolated as a pale yellow powder. The electronic spectrum has been reported, λ_{max} = 345, 295 nm; conductance 164 $cm^2 \cdot \Omega^{-1} \cdot mol^{-1}$.

Rh(β-ala)Cl_2(β-alaH)$_2$·H_2O is similarly prepared using two equivalents of 0.1M HCl. The electronic spectrum has been reported, λ_{max} = 380, 340, 300 nm; conductance 328 $cm^2 \cdot \Omega^{-1} \cdot mol^{-1}$.

$RhCl_3$(β-alaH)$_3$ is similarly prepared using excess aqueous HCl, and is isolated as a yellow precipitate. The electronic spectrum has been reported, λ_{max} = 383, 320 nm; conductance 334.7 $cm^2 \cdot \Omega^{-1} \cdot mol^{-1}$.

[Rh(β-alaH)$_3$(H_2O)$_3$][NO_3]$_3$ is prepared by dissolving $RhCl_3$(β-alaH)$_3$ in water and boiling the solution for 1 h, then adding silver nitrate, filtering off silver chloride and evaporating the filtrate. It is described as a pale yellow powder. The electronic spectrum has been reported, λ_{max} = 340, 280 nm; conductivity 492 $cm^2 \cdot \Omega^{-1} \cdot mol^{-1}$.

A. E. Bukanova, I. V. Prokof'eva, L. K. Shubochkin (Zh. Neorgan. Khim. **25** [1980] 1575/9; Russ. J. Inorg. Chem. **25** [1980] 875/7).

α-Aminobutyrates

$CH_3CH_2CH(NH_2)CO_2H$ = α-aminobutyric acid = α-NH_2butH

Rh(α-NH_2but)$_3$·H_2O is obtained in two isomeric forms by boiling an aqueous solution of chlororhodic acid (prepared from $RhCl_3$·nH_2O and HCl) with α-aminobutyric acid (3 mol per mol of rhodium) and ethanol for 2h, then neutralising the filtrate with NaOH. The two isomers separate as white fibres and yellow plates, respectively, on concentration of the solution. They show identical thermal behaviour, water of crystallisation is lost at 120 to 130°C, decomposition begins at 270 to 280°C and is complete at ~400°C. These products are referred to as diastereoisomers of *fac*-Rh(α-NH_2but)$_3$·H_2O [1] but are probably *mer* (yellow plates) and *fac* (white fibres) isomers.

***fac*-Rh(α-NH_2but)$_3$·H_2O** is separated as white fibres from the mixture of isomers described above. The electronic spectrum (λ_{max} = 330 and 278 nm) and ESCA data (Rh3$d_{5/2}$ = 310.6 eV) have been recorded. Solubility in water 5.28 × 10^{-4} mol/l at 25°C; conductance in aqueous solution 39.61 $cm^2 \cdot \Omega^{-1} \cdot mol^{-1}$. A thermogravimetric analysis curve is reproduced [1].

***mer*-Rh(α-NH_2but)$_3$·H_2O** is separated as yellow platelets from the mixture of isomers described above. The electronic spectrum (λ_{max} = 350 and 285 nm) has been recorded. Solubility in water 1.46 × 10^{-3} mol/l at 25°C; conductance in aqueous solution 27.96 $cm^2 \cdot \Omega^{-1} \cdot mol^{-1}$. A thermogravimetric analysis curve is reproduced [1].

Rh(α-NH$_2$but)$_2$Cl(α-NH$_2$butH)·H$_2$O is prepared by boiling an aqueous solution of $Na_3[RhCl_6]$ and α-aminobutyric acid (mole ratio 1:4) for a prolonged period, then evaporating the solution on a water bath. It separates as orange acicular crystals; refractive indices are $n_g = 1.595$, $n_p = 1.531$, $n_m = 1.552$. The infrared spectrum (2000 to 400 cm^{-1}) is reproduced [2].

The complex loses water gradually at 105 to 110°C, begins to decompose at ~280°C (endothermic effect) and exhibits a strong exothermic effect at 305 to 335°C. The complex is readily soluble in water, molar conductance and pH value of an aqueous solution both increase on dilution; conductance 202.4 $cm^2 \cdot \Omega^{-1} \cdot mol^{-1}$ and pH = 2.3 for a 5×10^{-3} M solution, 277 $cm^2 \cdot \Omega^{-1} \cdot mol^{-1}$ and pH = 3 for a 1×10^{-3} M solution, both at 25°C. A titration curve (NaOH) is reproduced [2].

References:

[1] I. V. Prokof'eva, A. E. Bukanova, L. K. Shubochkin, V. I. Nefedov, T. P. Sidorova (Zh. Neorgan. Khim. **21** [1976] 1081/4; Russ. J. Inorg. Chem. **21** [1976] 589/91). – [2] I. V. Prokof'eva, A. E. Bukanova, O. E. Zvyagintsev (Zh. Neorgan. Khim. **15** [1970] 1037/40; Russ. J. Inorg. Chem. **15** [1970] 528/30).

Aspartate, Rh(aspH)$_3$·2H$_2$O

$HO_2CCH_2CH(NH_2)CO_2H$ = aspartic acid = aspH$_2$

This is prepared by heating an aqueous mixture of $RhCl_3 \cdot 4H_2O$ and aspartic acid under reflux for 40 h, then evaporating the filtered solution until a brown glassy residue remains. Crystallisation from ethanol/diethyl ether gives air-sensitive yellow crystals. Infrared data are discussed.

H. Frye, C. Luschak, D. Chinn (Z. Naturforsch. **22b** [1967] 268/9).

Glutamates

$HO_2CCH_2CH_2CH(NH_2)CO_2H$ = glutamic acid = gluH$_2$

Rh(gluH)$_3$·2H$_2$O is prepared by heating an aqueous mixture of $RhCl_3 \cdot 4H_2O$ and glutamic acid under reflux for 40 h, then evaporating the filtered solution until a brown glassy resin remains. Crystallisation from *n*-butanol/diethyl ether gives air-sensitive yellow plates. Infrared data are discussed [1].

Rh(glu)Cl(H$_2$O)$_2$ is obtained by heating an aqueous solution of $RhCl_3 \cdot 4H_2O$ and glutamic acid (molar ratio 1:4) on a steam bath for 1 h, and maintaining the pH at 4.8 by careful addition of 6 M NaOH solution. It forms insoluble, air-stable yellow crystals which decompose at 280°C. Infrared data are discussed [2].

References:

[1] H. Frye, C. Luschak, D. Chinn (Z. Naturforsch. **22b** [1967] 268/9). – [2] H. Kalberer, H. Frye (Inorg. Nucl. Chem. Letters **7** [1971] 215/8).

Complexes with Substituted Glutamic Acids

Rh(L)Cl(H$_2$O)$_2$ (H_2L = β-hydroxyglutamic acid, N-benzoylglutamic acid or N-(p-nitrobenzoyl)glutamic acid). These complexes are prepared by heating an aqueous mixture of $RhCl_3 \cdot 4H_2O$ and the appropriate acid (molar ratio 1:4), and adjusting the pH of the solution to ~4.0 using 6 M NaOH. The products separate as insoluble air-stable yellow crystals. Decomposition temperatures are in the range 266 to 272°C. Infrared data are discussed.

H. Kalberer, H. Frye (Inorg. Nucl. Chem. Letters **7** [1971] 215/8).

Prolinates

$\overline{NH \cdot CH_2CH_2CH_2C}HCO_2H$ = proline = proH

Rh(L-pro)$_3$·H$_2$O is obtained by boiling chlororhodic acid (prepared from $RhCl_3 \cdot nH_2O$ and HCl) with L-proline for 2 h, neutralising the solution with NaOH or Na_2CO_3 and heating again. The monohydrate deposits from the filtered solution on concentration as pale yellow filaments. The water of crystallisation is lost at 105 to 110°C to give the anhydrous complex, further decomposition commences at 320 to 330°C. The hydrated complex is soluble in water, a 1.01×10^{-3} M solution has a pH value of 5.5 and a conductance of 14.59 $cm^2 \cdot \Omega^{-1} \cdot mol^{-1}$. The electronic spectra (400 to 250 nm) of the hydrated and anhydrous forms in 1:1 H_2SO_4 are reproduced. Hydrated and anhydrous forms show identical infrared spectra, indicative of coordination through O and N atoms of the prolinate anion; ν(NH) = 3230, 3180 cm^{-1}, ν(OCO) = 1660, 1410 cm^{-1}. ESCA data have been recorded, $Rh3d_{5/2}$ = 310.2, N1s = 400.4 eV (anhydrous form) and $Rh3d_{5/2}$ = 310.3, N1s = 400.3 eV (hydrated form).

A. E. Bukanova, I. V. Prokof'eva, Ya. V. Salyn', L. K. Shubochkin (Zh. Neorgan. Khim. **26** [1981] 156/9; Russ. J. Inorg. Chem. **26** [1981] 82/4).

Cysteinate, {Rh(L-cys)$_2$Cl}$_2$

$HSCH_2CH(NH_2)CO_2H$ = cysteine = cysH

This is prepared by treatment of $RhCl_3 \cdot 3H_2O$ with L-cysteine hydrochloride in aqueous solution and deposits from solution as an orange-brown precipitate on addition of ethanol. The infrared spectrum is interpreted in terms of a binuclear chloride-bridged structure with cysteinate anions, $^-SCH_2CH(NH_2)CO_2H$, coordinated through N and S atoms; ν(NH_2) = 3200, ν(CO_2H) = 1730, ν($RhCl_2Rh$) = 269 cm^{-1}.

W. Levason, C. A. McAuliffe, D. M. Johns (Inorg. Nucl. Chem. Letters **13** [1977] 123/7).

Oxamate, K$_3$[Rh(oxam)$_3$]·3H$_2$O

$H_2N \cdot C(O)CO_2H$ = oxamic acid = oxamH_2

This complex is prepared by heating under reflux for 3 h an aqueous mixture of Rh_2O_3 and free acid (ratio 1:3), then adjusting the pH to 13 using 3M KOH and precipitating the product with cold ethanol. It forms golden-yellow hygroscopic crystals. Electronic spectra and infrared spectra (4000 to 300 cm^{-1}) have been reported and assignments given; λ_{max} = 405, 355(sh), 315, 280 nm in aqueous solution.

J. K. Kouinis, P. T. Veltsistas, J. M. Tsangaris (Monatsh. Chem. **113** [1982] 155/61).

Complexes with Miscellaneous Amino Acids

Stability and stepwise association constant data for some rhodium(III) amino-acid complexes in aqueous media have been determined by a pH-metric titration procedure. Values for the successive equilibrium constants K_1 and K_2, the overall stability constant K_s (for definition see original paper) and $-\Delta F°$ are tabulated below.

Values of successive equilibrium and overall stability constants for various rhodium(III) chloride-amino acid systems at 25°C according to [1]:

System	log K_1	log K_2	log K_s**)	$-\Delta F°$ in kcal/mol*)
DL-α-alanine-$RhCl_3$	7.19	3.09	10.28	14.1 (58.9)
β-alanine-$RhCl_3$	6.63	3.29	9.92	13.6 (56.8)
L-asparagine-$RhCl_3$	6.86	2.87	9.73	13.4 (56.0)
glycine-$RhCl_3$	7.34	3.03	10.37	14.2 (59.3)
DL-isoleucine-$RhCl_3$	7.16	2.76	9.92	13.6 (56.8)
L-leucine-$RhCl_3$	7.13	2.79	10.10	13.9 (58.1)
DL-methionine-$RhCl_3$	6.69	2.69	9.38	12.9 (53.9)
DL-phenylalanine-$RhCl_3$	6.82	3.12	9.94	13.7 (57.2)
L-proline-$RhCl_3$	8.24	3.00	11.24	15.4 (64.4)
DL-serine-$RhCl_3$	6.92	3.03	9.95	13.7 (57.2)
DL-taurine-$RhCl_3$	6.61	2.68	9.29	12.7 (53.1)
DL-threonine-$RhCl_3$	6.86	3.02	9.88	13.6 (56.8)
DL-valine-$RhCl_3$	7.14	2.66	9.80	13.5 (56.4)

*) in parentheses: kJ/mol. – **) K_s in mol^2/l^2.

Several papers discuss hydrogenation catalysts of poorly defined stoichiometry and structure which involve rhodium and amino acids including L-tyrosine [2, 3, 4], N-phenylanthranilic acid [2, 3, 5], alanine, phenylalanine [6] and aspartic acid [4].

References:

[1] O. Farooq, N. Ahmad (J. Electroanal. Chem. Interfacial Electrochem. **57** [1974] 121/4). – [2] O. N. Efimov, M. L. Khidekel, V. A. Avilov, P. S. Chekrii, O. N. Eremenko, A. G. Ovcharenko (Zh. Obshch. Khim. **38** [1968] 2668/77; J. Gen. Chem. [USSR] **38** [1968] 2581/8). – [3] V. A. Avilov, Yu. G. Borod'ko, V. B. Panov, M. L. Khidekel, P. S. Chekrii (Kinetika Kataliz **9** [1968] 698/9; Kinet. Catal. [USSR] **9** [1968] 582). – [4] I. Rajca, A. Borowski, A. Marzec (Neftekhimiya **17** [1977] 672/7 from C.A. **88** [1978] No. 36672). – [5] O. N. Efimov, V. V. Panov (Izv. Akad. Nauk SSSR Ser. Khim. **1970** 491/3; Bull. Acad. Sci. USSR Div. Chem. Sci. **1970** 451/3).

[6] V. K. Latov, V. M. Belikov, T. A. Belyaeva, A. I. Vinogradova, S. I. Soinov (Izv. Akad. Nauk SSSR Ser. Khim. **1977** 2481/7; Bull. Acad. Sci. USSR Div. Chem. Sci. **1977** 2300/5).

1.1.6 Aminopolycarboxylic Acid Complexes

1.1.6.1 Iminodiacetic Acid Complexes

$NH(CH_2CO_2H)_2$ = iminodiacetic acid = H_2ida

***trans-fac*-H[Rh(ida)$_2$]·3H_2O** is prepared by heating the free ligand with an aqueous suspension of freshly precipitated $Rh(OH)_3$ in a sealed tube at 120 to 130°C for several hours. It is isolated as pale yellow crystals on cooling the filtered reaction mixture and is recrystallised from hot water [1]. The electronic spectrum has been recorded and assignments given; λ_{max} = 400, 328 and 284 nm [1]; λ_{max} = 400, 328 and 282 nm [2]. The spectrum is reproduced (600 to 200 nm) in [1]. Selected infrared bands have been reported, $\nu(CO_2Rh)$ = 1622 cm^{-1}. Proton NMR data have been recorded and the spectrum (before and after N-H/D exchange) is reproduced [1]. A polarogram ($E_{1/2}$ = −1.27 V, pH = 6.5) is reproduced [2].

***cis*-H[Rh(ida)$_2$]·2H_2O** is obtained as bright yellow crystals by addition of ethanol to the filtrate left after removal of the *trans* isomer (see above). Separation of *cis* and *trans* isomers by ion exchange techniques is also mentioned. Further purification is achieved by recrystallisation from water/ethanol. The electronic spectrum has been measured (λ_{max} = 362 and 295 nm) and is reproduced (600 to 200 nm). Selected infrared bands have been recorded, ν(CO_2Rh) = 1624 and 1584 cm^{-1} (shoulder). Proton NMR data have been reported and the spectrum (before and after N-H/D exchange) is reproduced [1]. A polarogram ($E_{1/2}$ = −0.95 V, pH = 6.5) is reproduced [2].

***trans*-K[Rh(ida)$_2$]·2H_2O** is prepared by treatment of *trans*-H[Rh(ida)$_2$]·3H_2O with the equivalent amount of KOH [2]. The electronic spectrum has been recorded and is reproduced, (λ_{max} = 404, 330 and 281 nm) [3]. Selected infrared bands have been reported, ν(NH) = 3110, ν(CH_2) = 2980, 2935 and 2900 cm^{-1} [2]; ν(OH) = 3427, ν(NH) = 3192, ν(CO_2Rh) = 1645 and 1364 cm^{-1} [3]; the spectrum from 4000 to 2500 and 1200 to 900 cm^{-1} is reproduced [2].

Na[Rh(ida)(Hida)Cl]·2H_2O is prepared by heating an aqueous mixture of the free ligand, $RhCl_3$·nH_2O and NaOH (pH = 2.9), first to 95°C, then under reflux for 2h. The product separates from the cooled, concentrated reaction solution as pale yellow crystals. The electronic spectrum (500 to 300 nm) is reproduced. The infrared spectrum has been reported and assignments given, ν(NH) = 3195, ν(CH_2) = 2970 and 2830, ν(CO_2H) = 1735, ν(CO_2Rh) = 1620 and 1370, ν(CN) = 1040, ν(RhO) = 890 and 420, ν(RhCl) = 300 cm^{-1} [4]. A differential thermal analysis (DTA) curve and a gas chromatographic analysis (GCA) curve are reproduced, gaseous decomposition products are water (100°C), CO_2 (340°C), and CO (550°C) [5]. Potentiometric and conductometric curves are reproduced [4].

Na[Rh(Hida)$_2$(OH)$_2$]·H_2O is prepared by heating an aqueous mixture of freshly precipitated $Rh(OH)_3$ and free ligand under reflux for 4 h. The product crystallises from the concentrated, cooled reaction solution and is recrystallised from water/ethanol. The electronic spectrum (500 to 300 nm) is reproduced. The infrared spectrum has been recorded and assignments given; ν(NH) = 3200, ν(CH_2) = 2960 and 2860, ν(CO_2H) = 1710, ν(CO_2Rh) = 1620 and 1370, ν(CN) = 1060, ν(RhO) = 870 and 425 cm^{-1} [4]. A DTA curve and a GCA curve are reproduced: gaseous decomposition products are water (100°C), CO_2 (340°C), and CO (460 and 700°C) [5]. Potentiometric and conductometric titration curves are reproduced [4].

Na[Rh(Hida)(H_2ida)Cl_3]·2H_2O is prepared by addition of 2M HCl to the mother liquor from the preparation of Na[Rh(ida)(Hida)Cl]·2H_2O (see above) and is isolated as a powder by desiccation of the solution over P_2O_5. The electronic spectrum (500 to 300 nm) is reproduced. The infrared spectrum has been recorded and assignments given; ν(NH) = 3190, ν(CH_2) = 2970 and 2825, ν(CO_2H) = 1760 to 1735, ν(CO_2Rh) = 1620 and 1390, ν(CN) = 1065, ν(RhO) = 870 and 430, ν(RhCl) = 300 and 240 cm^{-1} [4]. A DTA curve and a GCA curve are reproduced, gaseous decomposition products are H_2O (120°C), CO_2 (270 and 420°C), CO (420 and 550 to 650°C) [5]. Potentiometric and conductometric titration curves are reproduced [4].

Iminodiacetic acid chelating resins have been used to separate rhodium from Pd/Pt/Rh mixtures [6].

References:

[1] B. B. Smith, D. T. Sawyer (Inorg. Chem. **7** [1968] 922/8). – [2] P. Gouzerh (J. Chim. Phys. **68** [1971] 758/67). – [3] K. Sugiura (Nippon Kagaku Zasshi **90** [1969] 691/6; C.A. **71** [1969] No. 108588). – [4] F. Gonzalez-Vilchez, M. Castillo-Martos (Rev. Chim. Minerale **14** [1977] 58/65). – [5] F. Gonzalez-Vilchez, M. Castillo-Martos (Thermochim. Acta **21** [1977] 127/33).

[6] R. Hering, H. Schiefelbein (Abhandl. Akad. Wiss. DDR Abt. Math. Naturwiss. Tech. **1977** 159/68 from C.A. **88** [1978] No. 142152).

1.1.6.2 Methyliminodiacetic Acid Complexes

$CH_3N(CH_2COOH)_2$ = methyliminodiacetic acid = H_2mida

***trans-fac*-H[Rh(mida)$_2$]·2 H_2O** is prepared by heating an aqueous mixture of the free ligand and freshly precipitated $Rh(OH)_3$ in a sealed tube at 120 to 130°C for several hours. The product is isolated as pale yellow crystals by addition of ethanol to the cooled filtered reaction mixture and is further purified by recrystallisation from water/ethanol. The electronic spectrum has been recorded, λ_{max} = 404, 328 and 282 nm, and is reproduced (600 to 200 nm). Selected infrared bands have been recorded, $\nu(CO_2Rh)$ = 1627 cm^{-1}. The proton spectrum is reproduced and assignments are given.

B. B. Smith, D. T. Sawyer (Inorg. Chem. **7** [1968] 922/8).

1.1.6.3 Nitrilotriacetic Acid Complexes

$N(CH_2CO_2H)_3$ = nitrilotriacetic acid = H_3nta

***trans-fac*-H[Rh(H nta)$_2$]·2 H_2O** is prepared by heating an aqueous mixture of free ligand and freshly prepared $Rh(OH)_3$ in a sealed tube at 120 to 130°C for several hours. The product is isolated as pale yellow crystals by passing the filtered reaction mixture down a Dowex-50W-X12 cation-exchange resin (H^+ form), then concentrating and cooling the eluate. The electronic spectrum has been recorded, λ_{max} = 404, 332.5 and 280 nm, and is reproduced (600 to 200 nm). Selected infrared bands have been reported, $\nu(CO_2Rh)$ = 1615, $\nu(CO_2H)$ = 1723 cm^{-1}. The proton spectrum (D_2O, pH = 0.7 and 7.0) is reproduced and assignments are given [1].

Reference is made to a 1:1 Rh/H_3nta complex present in solution [1].

The electromigration of rhodium and other platinum group metals in aqueous H_3nta solution has been investigated as a possible method of separation. Electromigration plots are reproduced [2].

References:

[1] B. B. Smith, D. T. Sawyer (Inorg. Chem. **7** [1968] 922/8). – [2] E. K. Korchemnaya, V. I. Naumova, N. A. Ezerskaya, A. N. Ermakov (Zh. Analit. Khim. **27** [1972] 1150/6; J. Anal. Chem. [USSR] **27** [1972] 1028/33).

1.1.6.4 Diaminopolycarboxylic Acid Complexes

Ethylenediaminetetraacetic Acid Complexes

$(HO_2CCH_2)_2NCH_2CH_2N(CH_2CO_2H)_2$ = ethylenediaminetetraacetic acid = H_4edta

General Remarks

A review covering ethylenediaminetetraacetic acid complexes of the platinum-group metals discusses preparation of rhodium derivatives and their application in polarographic determination, spectrophotometric analysis, complexometric titration and chromatographic separation of rhodium [1].

The first report of complex formation between rhodium(III) and H_4edta described a yellow solid formed when acid solutions of $RhCl_3·3H_2O$ or Rh_2O_3 were treated with the free ligand. Electronic spectra (550 to 300 nm) are reproduced in the paper. Previous unsuccessful attempts to prepare complexes of this type were attributed to the use of "aged" solutions [2]. Separation of platinum-group metals including rhodium has been achieved by complex-formation chromatography [3] and by electromigration [4] using H_4edta complexes. Quantitative determination of rhodium in the presence of iridium has been achieved by polarographic measurements on rhodium H_4edta complexes in solution, polarograms and electronic spectra (390 to 300 nm) are reproduced and evidence for formation of two rhodium complexes in the solutions has been reported [5]. A later study of reactions between $Na_3[RhCl_6]$ and Na_2H_2edta in aqueous solution at various pH values gave polarographic and spectroscopic evidence for the presence of four rhodium complexes, two of which – Na[Rh(Hedta)Cl]·$2H_2O$ and Na[Rh(H_2edta)Cl_2]·$2H_2O$ – were isolated [6]. Rhodium has been determined spectrophotometrically at 360 nm by complexation with H_4edta at pH = 4 to 5, molar absorptivity is 6.02×10^2 cm^2/mol and the log stability constant for a 1:1 Rh/H_4edta complex is 3.62 [7].

References:

[1] N. A. Ezerskaya (Izv. Sibirsk. Otd. Akad. Nauk SSSR Ser. Khim. Nauk **1970** No. 4, pp. 21/9 from C.A. **74** [1971] No. 106737). – [2] W. McNevin, H. D. McBride, E. A. Hakkila (Chem. Ind. [London] **1958** 101). – [3] M. P. Volynets, A. N. Ermakov, L. P. Nikitina, N. A. Ezerskaya (Zh. Analit. Khim. **25** [1970] 759/64; J. Anal. Chem. [USSR] **25** [1970] 653/7). – [4] E. K. Korchemnaya, V. I. Naumova, N. A. Ezerskaya, A. N. Ermakov (Zh. Analit. Khim. **27** [1972] 1150/6; J. Anal. Chem. [USSR] **27** [1972] 1028/33). – [5] N. A. Ezerskaya, V. N. Filimonova (Zh. Analit. Khim. **17** [1962] 972/8; J. Anal. Chem. [USSR] **17** [1962] 946/51).

[6] N. A. Ezerskaya, V. N. Filimonova (Zh. Neorgan. Khim. **8** [1963] 830/8; Russ. J. Inorg. Chem. **8** [1963] 424/9). – [7] Y. M. Issa, F. M. Issa (Z. Anal. Chem. **276** [1975] 72).

Complexes

dl-NH_4[Rh(edta)(H_2O)]·H_2O is prepared as a pale yellow powder by treating conc. aqueous dl-Rh(Hedta)(H_2O) with $NH_4O_2CCH_3$ and precipitating the product with ethanol [1].

Na[Rh(edta)(H_2O)]. This stoichiometry was tentatively assigned to the yellow product formed in solution on boiling an aqueous mixture of $Na_3[RhCl_6]$ and Na_2H_2edta (Trilon B) at pH = 4.5 to 5.0. The electronic spectrum was reported, λ_{max} = 350 and 310 nm (shoulder), and is reproduced (390 to 290 nm) in the paper. A polarogram is reproduced [2].

Na[Rh(edta)]·$2H_2O$ is prepared by treatment of aqueous Na[Rh(Hedta)Cl]·$2H_2O$ with fresh silver oxide, and is isolated from the filtrate by evaporation. The yellow crystals lose two molecules of water at 130°C. The electronic spectrum with λ_{max} = 346 and 293 nm (aqueous solution) and infrared spectrum with $\nu(CO_2Rh)$ = 1618 and 1310 cm^{-1} have been reported; no infrared bands attributable to free carboxylate groups were observed [3].

The complex has been resolved as the strychninium salt [3, 4]. The least and most soluble fractions of the strychninium salt gave sodium salts Na[Rh(edta)]·$2H_2O$ with $[\alpha]_D$ values of −28° (1.50×10^{-2} M solution) and +35.3° (1.33×10^{-2} M solution), respectively [3]. NMR data have been reported for the $[Rh(edta)]^-$ anion [5].

dl-K[Rh(edta)(H_2O)]·H_2O is prepared as pale yellow crystals by twice treating conc. aqueous solutions of Rh(Hedta)(H_2O) with potassium acetate and precipitating the product by addition of ethanol [1].

l-K[Rh(edta)(H_2O)]·H_2O is prepared by treatment of l-[Co en$_2$(NO_2)$_2$] l-[Rh(edta)(H_2O)]·$2H_2O$ with aqueous KI, and is precipitated from the filtered solution by addition of alcohol. A 0.1% solution gave $[\alpha]_D = -140°$ and $[\alpha]_{546.1} = -180°$ [1].

d-K[Rh(edta)(H_2O)]·H_2O is prepared by addition of KI to the filtrate left after removal of l-K[Rh(edta)(H_2O)] from the racemic K[Rh(edta)(H_2O)]·H_2O, and is precipitated as a yellow powder by addition of ethanol; $[\alpha]_{546.1}$ has been given as +180° [1].

l-[Coen$_2$(NO_2)$_2$] l-[Rh(edta)(H_2O)]·2H_2O is prepared by adding d,l-K[Rh(edta)(H_2O)]·H_2O to a solution of l-[Coen$_2$(NO_2)$_2$][O_2CCH_3] and separates on cooling (5°C) as fine golden needles. A 0.1% solution gave $[\alpha]_{546.1} = -130°$ [1].

Na_2[Rh(edta)Cl]. The kinetics of formation of $[Rh(edta)Cl]^{2-}$ from $[RhCl_6]^{3-}$ and H_4edta in NaCl/HCl media (pH ~ 2.6 to 3.5) have been studied spectrophotometrically at 350 nm in the temperature range 70 to 90°C. Dependence of k_{obs} on HCl and H_4edta concentrations supports a mechanism involving rate-determining reactions of the outer sphere associated species $RhCl_6^{3-}\cdots H_2edta^{2-}$ and $RhCl_6^{3-}\cdots H_3edta^{-}$ [6]. The equilibrium between $[Rh(edta)Cl]^{2-}$ and Cl^- anions has been investigated [7].

References:

[1] F. P. Dwyer, F. L. Garvan (J. Am. Chem. Soc. **82** [1960] 4823/6). – [2] N. A. Ezerskaya, V. N. Filimonova (Zh. Neorgan. Khim. **8** [1963] 830/8; Russ. J. Inorg. Chem. **8** [1963] 424/9). – [3] K. Yamasaki, K. Sugiura (Naturwissenschaften **48** [1961] 552/3). – [4] K. Sugiura, K. Yamasaki (Nippon Kagaku Zasshi **88** [1967] 948/52). – [5] B. B. Smith, D. T. Sawyer (Inorg. Chem. **7** [1968] 2020/6).

[6] D. Banerjea, B. Chattopadhyay (Indian J. Chem. **8** [1970] 993/6). – [7] E. V. Shemyakina, A. Ya. Fridman, N. M. Dyatlova (Koord. Khim. **5** [1979] 905/8; Soviet J. Coord. Chem. **5** [1979] 715/8).

Na_3[Rh(edta)(OH)$_2$]. Formation constant data for the $[Rh(edta)(OH)_2]^{3-}$ anion have been measured by spectrophotometric and pH titrimetric methods. The electronic spectrum (330 to 220 nm) is reproduced [1].

Na_3[Rh(edta)Cl(OH)]. This complex is cited as an intermediate formed during the two-stage substitution of chloride ligands in $Na_3[Rh(edta)Cl_2]$ by hydroxide anions [1].

K_n[Rh(edta)X(OH)] (X = Br, I, SCN, NO_2, SO_2(sic), n = 3; X = S_2O_3, n = 4). The salts have been studied by spectrophotometric methods at 25°C and ionic strength 1M (KNO_3). Electronic transitions have been recorded [2].

Na_3[Rh(edta)Cl_2] is obtained in solution by treatment of "$Na_2[RhCl_5]$" with H_4edta and has been investigated by spectrophotometric and pH titrimetric methods. The electronic spectrum (330 to 220 nm) is reproduced and formation constant data are recorded [1].

K_n[Rh(edta)X_2] (X = Br, I, SCN, NO_2, SO_2(sic), n = 3; X = S_2O_3, n = 5). These salts have been studied spectrophotometrically at 25°C and ionic strength 1M (KNO_3). Electronic transitions have been recorded [2].

References:

[1] E. V. Shemyakina, A. Ya. Fridman, N. M. Dyatlova (Koord. Khim. **5** [1979] 905/8; Soviet J. Coord. Chem. **5** [1979] 715/8). – [2] E. V. Shemyakina, N. M. Dyatlova, O. V. Popov (Koord. Khim. **7** [1981] 619/23).

dl-Rh(Hedta)(H_2O) is prepared by heating an aqueous mixture of H_4edta and freshly precipitated $Rh(OH)_3$ in a sealed tube at 145°C for 6 h and is precipitated as yellow needles by addition of alcohol [1]. A preparation from Na[Rh(edta)(H_2O)]·H_2O has also been reported [2].

The X-ray crystal structure is monoclinic, space group $P2_1/c$-C_{2h}^5, lattice parameters a = 8.454(2), b = 8.708(3), c = 17.639(3) Å, β = 100.58(3)°, Z = 4; D_{obs} = 2.12, D_{calc} = 2.116 g/cm^3. Coordinate bond distances are Rh-N = 2.082, 1.988 Å; Rh-O = 2.096, 2.030, 2.001, 2.027 Å. The uncomplexed $-CH_2CO_2H$ group and the coordinated H_2O are involved in intermolecular H-bonding [3].

(−)$_{589}$-Rh(Hedta)(H$_2$O) is prepared by treatment of (−)$_{589}$-[Rh(Hedta)Cl]$^-$ with silver nitrate. A value of −260° has been recorded for [α]$_{589}$. The CD spectrum of [−]$_{589}$-[Rh(Hedta)(H$_2$O)] at various pH values (0.8 to 12.1) and a plot of Δε (315 nm) against pH are reproduced [4]. By comparison with complexes of known absolute configuration (−)$_{546}$-Rh(Hedta)(H$_2$O) has been assigned to the L-series [5]. The infrared spectrum has been recorded; frequencies in cm^{-1}: 3420s, 2723w, 2611w, 2537w, 1322w, 1314m, 1271m, 1225s, 1160m, 1087s, 1063m, 1044m, 1008w, 1000m, 981w, 964w, 949vw, 934s, 915s, 901m, 882s, 821s, 766m and 739w (s = strong, m = medium, (v)w = (very) weak), bands at 2723, 2611 and 2537 are attributed to H-bonding in the crystal [6]. The proton NMR spectrum has been recorded using solutions in D_2O at pH = 0.5 and 6.0; differences have been attributed to conversion of Rh(Hedta)(H$_2$O) to [Rh(edta)]$^-$ at higher pH [7]:

$$\text{Rh(Hedta)(H}_2\text{O)} \rightleftharpoons [\text{Rh(edta)(H}_2\text{O)}]^-$$
$$\swarrow \qquad \swarrow$$
$$[\text{Rh(edta)}]^- \qquad [\text{Rh(edta)(OH)}]^{2-}$$

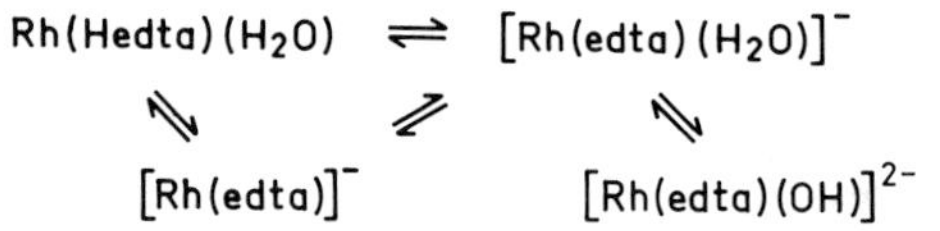

References:

[1] F. P. Dwyer, F. L. Garvan (J. Am. Chem. Soc. **82** [1960] 4823/6). − [2] K. Sugiura, K. Yamasaki (Nippon Kagaku Zasshi **88** [1967] 948/52 from C.A. **68** [1968] No. 83977). − [3] G. H. Y. Lin, J. D. Leggett, R. M. Wing (Acta Cryst. B **29** [1973] 1023/30). − [4] M. Saito, T. Uehiro, Y. Yoshino (Bull. Chem. Soc. Japan **53** [1980] 3531/6). − [5] R. D. Gillard, G. Wilkinson (J. Chem. Soc. **1964** 1368/72, 1370).

[6] R. D. Gillard, G. Wilkinson (J. Chem. Soc. **1963** 4271/2). − [7] B. B. Smith, D. T. Sawyer (Inorg. Chem. **7** [1968] 2020/6).

H[Rh(Hedta)Cl]·2H$_2$O is prepared by passing the sodium salt through Dowex 50W-X1 or 50W-X12 cation exchange resin (H$^+$ form) and evaporating the solution to a small volume [1, 2]. Infrared data have been reported, a pH titration curve is reproduced [1].

Na[Rh(Hedta)Cl]·2H$_2$O is prepared by boiling an aqueous mixture of Na$_3$RhCl$_6$ and Na$_2$H$_2$edta for 1 h at pH = 4.8 to 1.0, then evaporating the solution on a water bath [3], or by heating a mixture of RhCl$_3$·4H$_2$O and Na$_2$H$_2$edta in ethanol [4] or water [1, 5]. It forms yellow needles; refractive indices n_g = 1.631, n_m = 1.627, n_p = 1.610 [3]. Infrared data have been reported [4]: main bands are 1721, 1631 and 1217 cm^{-1} [5]. The complex dehydrates at 130°C. The electronic spectrum (λ_{max} = 373 and 304 nm) has been reported [5].

The complex anion has been resolved by fractional crystallisation of the strychninium salt [1, 5]. The least and most soluble fractions gave sodium salts with rotations [α]$_D$ = −27.3° (5.33 × 10^{-2} M) and +30.0° (5.53 × 10^{-2} M), respectively [5].

Na[Rh(Hedta)Br]·2H$_2$O has been reported as a yellow complex and infrared data given; frequencies in cm^{-1}: 3420, 1731, 1658, 1637, 1355 and 1225. Electronic spectra data have been reported [1].

K[Rh(Hedta)I]·2H$_2$O has been reported as a yellow complex and infrared data given; frequencies in cm^{-1}: 3442, 1733, 1657, 1634, 1350, 1215. Electronic spectra data have been reported [1].

K$_2$[Rh(Hedta)Cl$_2$] is prepared by suspending K[Rh(H$_2$edta)Cl$_2$] in aqueous potassium acetate and gradually adding alcohol [6].

$K_2[Rh(Hedta)Br_2]$ is prepared by dissolving $K[Rh(H_2edta)Br_2]$ in aqueous potassium acetate and precipitating with ethanol. It is orange in colour and more soluble in water than the parent compound [6].

References:

[1] K. Sugiura, K. Yamasaki (Nippon Kagaku Zasshi **88** [1967] 948/52; C.A. **68** [1968] No. 83977). – [2] B. B. Smith, D. T. Sawyer (Inorg. Chem. **7** [1968] 2020/6). – [3] N. A. Ezerskaya, V. N. Filimonova (Zh. Neorgan. Khim. **8** [1963] 830/8; Russ. J. Inorg. Chem. **8** [1963] 424/9). – [4] V. F. Kuznetsov, O. A. Karpeiskaya, A. A. Belyi, M. E. Vol'pin (Izv. Akad. Nauk SSSR Ser. Khim. **1981** 1140/2; Bull. Acad. Sci. USSR Div. Chem. Sci. **30** [1981] 900/1). – [5] K. Yamasaki, K. Sugiura (Naturwissenschaften **48** [1961] 552/3).

[6] F. P. Dwyer, F. L. Garvan (J. Am. Chem. Soc. **82** [1960] 4823/6).

$H[Rh(H_2edta)Cl_2]\cdot H_2O$ is prepared by heating a mixture of $K[Rh(edta)(H_2O)]\cdot H_2O$ and conc. aqueous HCl on a steam bath, and separates on cooling. Proton NMR data have been reported and the spectra for D_2O solutions at pH = 0.5 and 6.0 are reproduced [1].

$NH_4[Rh(H_2edta)Cl_2]$ is prepared by heating an aqueous mixture of $RhCl_3\cdot 3H_2O$ and Na_2H_2edta at 120°C for 5 h, then adding conc. HCl and NH_4Cl to the concentrated filtrate and cooling to induce crystallisation. The complex forms yellowish brown crystals [2].

$Na[Rh(H_2edta)Cl_2]$ is prepared by treatment of $Na[Rh(Hedta)Cl]\cdot 2H_2O$ with boiling 6M HCl and deposits as fine yellow crystals on evaporation of the solution. The electronic spectrum (450 to 300 nm) and polarogram are reproduced in the paper, $E_{1/2} = -0.75$ V [3].

dl-$K[Rh(H_2edta)Cl_2]$ is prepared by heating a paste of dl-$Rh(Hedta)(H_2O)$ and conc. HCl on a steam bath for 7 min, then grinding up the cold orange-yellow solid with cold water followed by ethanol. The solid is redissolved in aqueous potassium acetate and precipitated with conc. HCl. It forms pale orange leaflets [4]. Infrared data have been recorded, $\nu(CO_2H) = 1720$, $\nu(CO_2Rh) = 1598$ cm^{-1} [5].

d-$K[Rh(H_2edta)Cl_2]$ is prepared by heating a slurry of d-$K[Rh(edta)(H_2O)]\cdot H_2O$ and conc. HCl for 1 min, evaporating the mixture slowly to dryness, then dissolving the residue in aqueous potassium acetate and precipitating the product by addition of ethanol. A twice recrystallised sample had $[\alpha]_D = +136°$ and $[\alpha]_{546.1} = +176°$ for a 0.125% solution [4].

$H[Rh(H_2edta)Br_2]\cdot H_2O$ is prepared from $Rh(Hedta)(H_2O)$ and conc. HBr or by dissolving $K[Rh(edta)(H_2O)]H_2O$ in excess conc. HBr and heating the solution on a steam bath. It separates as a bright orange solid on cooling the solution [4]. Proton NMR data have been recorded for D_2O solutions at pH = 0.5 and 6.0 [1].

dl-$K[Rh(H_2edta)Br_2]$ is prepared by heating paste of dl-$K[Rh(edta)(H_2O)]$ and constant boiling HBr on a steam bath for 7 min. The product is triturated with water and ethanol, then reprecipitated with acid from aqueous potassium acetate solution. It forms bright orange leaflets [4]. The infrared spectrum has been recorded, $\nu(CO_2H) = 1717$, $\nu(CO_2Rh) = 1595$ cm^{-1} [5].

l-$K[Rh(H_2edta)Br_2]\cdot 4HO$ is prepared from hydrobromic acid and l-$K[Rh(edta)(H_2O)]\cdot H_2O$. A 0.125% solution of pure material gave $[\alpha]_D = -136°$, $[\alpha]_{546.1} = -164°$ [4].

Retention of configuration has been demonstrated in the interconversion of $[Rh(edta)(H_2O)]^-$, $[Rh(H_2edta)Cl_2]^-$ and $[Rh(H_2edta)Br_2]^-$ anions [4, 6].

References:

[1] B. B. Smith, D. T. Sawyer (Inorg. Chem. **7** [1968] 2020/6). – [2] M. Saito, T. Uehiro, Y. Yoshino (Bull. Chem. Soc. Japan **53** [1980] 3531/6). – [3] N. A. Ezerskaya, V. N. Filimonova (Zh. Neorgan. Khim. **8** [1963] 830/8; Russ. J. Inorg. Chem. **8** [1963] 424/9). – [4] F. P. Dwyer, F. L. Garvan (J. Am. Chem. Soc. **82** [1960] 4823/6). – [5] R. D. Gillard, G. Wilkinson (J. Chem. Soc. **1963** 4271/2).

[6] R. D. Gillard, G. Wilkinson (J. Chem. Soc. **1964** 1368/72).

N-Hydroxyethylethylenediamine-N,N',N'-triacetic Acid Complexes

$$\begin{array}{c}HOCH_2CH_2 \\ HO_2CCH_2\end{array}\!\!NCH_2CH_2N\!\!\begin{array}{c}CH_2CO_2H \\ CH_2CO_2H\end{array} = H_3\text{heedta} =$$

N-hydroxyethylethylenediamine-N,N',N'-triacetic acid

Rh(heedta)(H_2O)·H_2O has been reported. The electronic spectrum (500 to 250 nm) is reproduced, λ_{max} occurs at 346 and 296 nm. Infrared data have been recorded, $\nu(OH) = 3291$, $\nu(CO_2Rh) = 1625, 1325$ cm^{-1} [1].

K[Rh(heedta)Cl]·2H_2O is prepared by evaporating on a water bath an aqueous mixture of $RhCl_3 \cdot nH_2O$, KOH (3 equivalents) and H_3heedta. It forms yellow crystals which lose one mole of H_2O at 100°C and the second at 130°C [2]. The electronic spectrum (500 to 250 nm) is reproduced [1]; λ_{max} values are 373 and 305 nm [1, 2].

The complex has been resolved through fractional crystallisation of the strychninium salt [1, 2]. The least and most soluble fractions gave potassium salts with rotations $[\alpha]_D = +27.0°$ (1.5×10^{-2} M solution) and $+23.0°$ (5.97×10^{-2} M solution), respectively [2].

K[Rh(heedta)Br]·2H_2O has been reported. The electronic spectrum (500 to 250 nm) is reproduced, λ_{max} occurs at 384 and 316 nm. Infrared data have been recorded, $\nu(OH) = 3428$, $\nu(CO_2Rh) = 1634$ and 1350 cm^{-1} [1].

K[Rh(heedta)I]·2H_2O has been reported. The electronic spectrum (500 to 250 nm) is reproduced; λ_{max} occurs at 401 and 294 nm. Infrared data have been recorded, $\nu(OH) = 3448$, $\nu(CO_2Rh) = 1635$ and 1351 cm^{-1} [1].

References:

[1] K. Sugiura (Nippon Kagaku Zasshi **90** [1969] 691/6). – [2] K. Yamasaki, K. Sugiura (Naturwissenschaften **48** [1961] 552/3).

(S,S)-Ethylenediaminedisuccinic Acid Complexes

$$\begin{array}{c}HO_2CCH_2 \\ HO_2C\end{array}\!\!CHNHCH_2CH_2NHCH\!\!\begin{array}{c}CO_2H \\ CH_2CO_2H\end{array} = (S,S)H_4\text{edds} =$$

S,S-ethylenediaminedisuccinic acid

Na[Rh(S,S-edds)]·H_2O is prepared by heating a mixture of $(S,S)H_4$edds·H_2O, 4 M NaOH solution and $RhCl_3 \cdot 3H_2O$ in a sealed tube at 145°C for 6 h, and is precipitated from the filtered reaction mixture by addition of methanol. The yellow complex is hygroscopic and readily soluble in water but insoluble in alcohol, acetone and acetonitrile. The proton NMR spectrum (D_2O solution) is reproduced and the pH dependence of the infrared spectrum (1700 to 1500 cm^{-1}) is illustrated [1]. The purified complex (ion-exchange chromatography, gel filtration) gave $[\alpha]_D = +146°$ and $[\alpha]_{546} = +194°$ for a 0 to 1% aqueous solution. The electronic spectrum (500 to 250 nm) is reproduced and assignments have been reported, λ_{max} = 375, 328 and 285 nm (shoulder). The CD spectrum has been recorded and is reproduced, $\Delta\varepsilon$ = −2.29 and +2.74 [2]. The complex anion has been assigned as the trans(O_5) isomer with the Λ configuration [1, 2].

References:

[1] J. A. Neal, N. J. Rose (Inorg. Chem. **12** [1973] 1226/32). – [2] D. J. Radanovic, K. D. Gailey, M. I. Djuran, B. E. Douglas (J. Coord. Chem. **10** [1980] 115/23).

Ethylenediamine-N,N'-diacetic Acid-N,N'-Dipropionic Acid Complexes

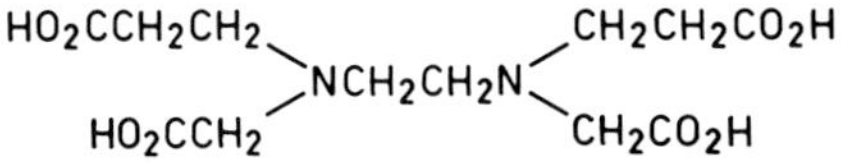

$$\begin{array}{c} HO_2CCH_2CH_2 \\ \diagdown \\ HO_2CCH_2 \end{array} NCH_2CH_2N \begin{array}{c} CH_2CH_2CO_2H \\ \diagup \\ CH_2CO_2H \end{array} = H_4eddadp =$$

ethylenediamine-N,N'-diacetic acid-N,N'-dipropionic acid

***trans*-(O_5)-Na[Rh(eddadp)]·2H$_2$O, *trans*-(O_5,O_6)-Na[Rh(eddadp)]·3H$_2$O.** These are prepared by heating an aqueous mixture of RhCl$_3$·3H$_2$O, H$_4$eddadp and 4 M NaOH in a sealed tube at 145°C for 7h. The yellow solution is filtered, then eluted from a QAE A-25 Sephadex anion exchange resin column in the chloride form using 0.1 M NaCl. The first band affords the *trans* (O_5) isomer, the second yields the *trans*(O_5,O_6) isomer. Both form yellow crystals on addition of ethanol. The electronic spectra (500 to 250 nm) are reproduced and assignments have been given [1]. Carbon-13 NMR spectra are reproduced diagrammatically and ^{13}C and ^{1}H NMR data have been tabulated [2].

Both isomers have been resolved using $(-)_D$-[Co(en)$_2$(C$_2$O$_4$)]Br. Resolution of the *trans*(O_5) isomer gave $(-)_D$-[Co(en)$_2$(C$_2$O$_4$)]$(-)_D$-[Rh(eddadp)]·xH$_2$O (less soluble diastereoisomer) $[\alpha]_D = -401°$ and $(-)_D$-[Co(en)$_2$(C$_2$O$_4$)]$(+)_D$-[Rh(eddadp)]·xH$_2$O (more soluble diastereoisomer) $[\alpha]_D = -345°$. Resolution of the *trans* (O_5,O_6) gave $(-)_D$-[Co(en)$_2$(C$_2$O$_4$)]$(+)_D$-[Rh(eddadp)]·xH$_2$O (less soluble diastereoisomer) $[\alpha]_D = -365°$ and $(-)_D$-[Co(en)$_2$(C$_2$O$_4$)]$(-)_D$-[Rh(eddadp)]·xH$_2$O (more soluble diastereoisomer) $[\alpha]_D = -480°$. The CD spectra of *trans* (O_5)-$(-)_D$-Li[Rh(eddadp)]·5H$_2$O and *trans* (O_5,O_6)-$(-)_D$-Li[Rh(eddadp)]·H$_2$O (500 to 250 nm) are reproduced [1].

References:

[1] D. J. Radanovic, K. D. Gailey, M. I. Djuran, B. E. Douglas (J. Coord. Chem. **10** [1980] 115/23). – [2] K. D. Gailey, D. J. Radanovic, M. I. Djuran, B. E. Douglas (J. Coord. Chem. **8** [1978] 161/7).

1,2-Propylenediaminetetraacetic Acid Complexes

(HO$_2$CCH$_2$)$_2$NCH(CH$_3$)CH$_2$N(CH$_2$CO$_2$H)$_2$ = 1,2-propylenediaminetetraacetic acid = 1,2-H$_4$pdta

Na$_2$[Rh(1,2-pdta)Cl]·H$_2$O is prepared by neutralising an aqueous solution of Rh(1,2-H$_2$pdta)Cl(H$_2$O) with 0.01M NaOH solution, then concentrating the mixture under reduced pressure until crystallisation occurs. The complex is dark yellow. Infrared data have been reported and assignments given; frequencies in cm^{-1}: 3415, 3015, 1580, 1370 and 1090. Thermogravimetric (TGA) and differential thermal analysis (DTA) curves (0 to 600°C) are reproduced together with potentiometric and conductometric titration graphs [1].

L-Ba[Rh(l-1,2-pdta)(H$_2$O)]$_2$·7H$_2$O is prepared by shaking an aqueous mixture of Ba(ClO$_4$)$_2$ and d-[Co(en)$_2$(NO$_2$)$_2$] L-[Rh(l-1,2-pdta)(H$_2$O)], filtering the cooled (4°C) mixture to remove d-[Co(en)$_2$(NO$_2$)$_2$]ClO$_4$ and then precipitating the required barium salt by addition of ethanol. A twice recrystallised sample (water/ethanol) gave $[\alpha]_D = +140°$ and $[\alpha]_{546.1} = -168°$ [2].

d-[Co en$_2$(NO$_2$)$_2$]L-[Rh(l-1,2-pdta)(H$_2$O)] is prepared by extracting L-Rh(l-1,2-Hpdta)(H$_2$O)·H$_2$O with water, treating the extract with Ba(OH)$_2$·8H$_2$O and d-[Co(en)$_2$(NO$_2$)$_2$]SO$_4$, filtering off precipitated barium salts and concentrating the filtrate, then cooling in ice to give the yellow product. A 0.25% solution gave $[\alpha]_D = -82°$, $[\alpha]_{546.1} = -94°$ [2].

L-Rh(l-1,2-Hpdta)(H$_2$O)·H$_2$O is prepared by heating an aqueous mixture of l-H$_4$pdta and Rh(OH)$_3$ at 145°C in a pyrex tube and is fractionally crystallised as a pale yellow solid by addition of ethanol and acetone to the concentrated filtrate. All fractions are levorotatory, $\alpha_{546.1} = -158°$ to $-40°$. Infrared data have been reported, $\nu(CO_2H) = 1732$, $\nu(CO_2Rh) = 1620$ cm^{-1} [2].

This complex, which is reported to exist in solution as [Rh(l-1,2-pdta)]$^-$ at pH≧6.5 [3], undergoes reversible mutarotation in the presence of light (~348 nm) [2]. The mutarotation was originally attributed to aquation but has more recently been investigated by proton NMR (spectra reproduced in paper) and ascribed to reversible formation of a less stable conformer with an axial methyl group. The CD spectrum (500 to 350 nm) before and after photolysis is reproduced [3].

H[Rh(1,2-Hpdta)Cl]·H$_2$O is prepared by treating RhCl$_3$·3H$_2$O with H$_4$pdta and is isolated as yellowish red monoclinic crystals. Infrared frequencies 3420, 3030, 1735, 1600, 1390, 1080 cm^{-1} and electronic spectral data with $\lambda_{max} = 375$, 280 and 245 nm have been reported, $\Delta = 32200$ cm^{-1}, CFSE = 77280 cm^{-1}. Thermogravimetric (TGA) and differential thermal analysis (DTA) curves (0 to 600°C) are reproduced together with potentiometric and conductometric titration graphs [1].

Na[Rh(1,2-Hpdta)Cl]·3H$_2$O is prepared by heating an aqueous mixture of RhCl$_3$·3H$_2$O and Na$_2$H$_2$pdta under pressure at 120°C for 5 h. No further details are given [4].

Rh(1,2-H$_2$pdta)Cl(H$_2$O) is prepared by treatment of RhCl$_3$·xH$_2$O in 10^{-3}M HCl with H$_4$pdta and NaOH, and is isolated as light yellow monoclinic crystals. The infrared spectrum has been recorded, frequencies in cm^{-1}: 3410, 3025, 2980, 1720, 1630, 1380, 1350, 1085, 900, 860. The electronic spectrum has been recorded and assignments made, $\Delta = 32300$ cm^{-1}, CFSE = 77250 cm^{-1}, B = 437 cm^{-1}, $\beta = 0.61$. TGA and DTA curves are reproduced (0 to 600°C) together with potentiometric and conductometric tritration graphs [1].

References:

[1] S. Gonzalez-Garcia, F. Gonzalez-Vilchez (Anales Quim. **66** [1970] 859/74). – [2] F. P. Dwyer, F. L. Garvan (J. Am. Chem. Soc. **83** [1961] 2610/5). – [3] G. L. Blackmer, J. L. Sudmeier, R. N. Thibedeau, R. M. Wing (Inorg. Chem. **11** [1972] 189/91). – [4] M. Saito, T. Uehiro, Y. Yoshino (Bull. Chem. Soc. Japan **53** [1980] 3531/6).

1,3-Propanediaminetetraacetic Acid Complex, Na[Rh(1,3-pdta)]·2H$_2$O

(HO$_2$CCH$_2$)$_2$NCH$_2$CH$_2$CH$_2$N(CH$_2$CO$_2$H)$_2$ = 1,3-propanediaminetetraacetic acid = 1,3-H$_4$pdta

A general reference has been given to the preparation of this complex salt by treatment of RhCl$_3$ with aqueous 1,3-H$_4$pdta in a sealed tube at 145°C. The ^{13}C and ^{1}H NMR spectra are reproduced and assignments given. Infrared data have been reported, $\nu(CO_2Rh) = 1690$, 1640 cm^{-1}; the region 1700 to 1500 cm^{-1} is reproduced [1].

Details of the preparation have been given in a later paper, the complex is described as a yellow crystalline **trihydrate**. The electronic absorption (500 to 250 nm) and CD spectra are reproduced. The $(-)_D$ isomer is assigned to the Λ configuration [2].

References:

[1] K. D. Gailey, D. J. Radanović, M. Djuran, B. E. Douglas (J. Coord. Chem. **8** [1978] 161/7). – [2] D. J. Radanović, M. I. Djuran, K. D. Gailey, B. E. Douglas (J. Coord. Chem. **11** [1982] 247/50).

trans-1,2-Cyclohexanediaminetetraacetic Acid Complexes

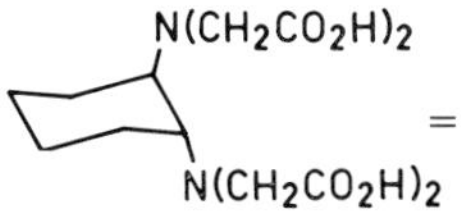

= *trans*-1,2-cyclohexanediaminetetraacetic acid = H_4cdta

K[Rh(Hcdta)Cl]·KCl·2H₂O is prepared by boiling a solution of $H_3[RhCl_6]$ and Na_2H_2cdta for 4h and is precipitated from the concentrated filtered solution by addition of acetone. It is described as a light yellow powdery precipitate. Selected infrared data have been reported, frequencies (in cm^{-1}): ν(OH) = 3400, ν(CO_2H) = 1740, ν(CO_2Rh) = 1650, 1370, ν(CN) = 1160, 1090; ESCA data: Rh $3d_{5/2}$ = 310.3, N1s = 401.0, Cl $2p_{3/2}$ = 198.5 eV. Results of thermogravimetric analysis and products of thermal decomposition have been tabulated. Electronic spectra, polarograms, molar conductance and pH data have been recorded for fresh and aged solutions [1].

H[Rh(H_2cdta)Cl_2]·2H_2O is prepared by boiling an aqueous HCl solution of K[Rh(Hcdta)Cl]KCl·2H_2O for 2h, then evaporating the solution until crystallisation occurs. It forms yellow-orange crystals [1].

An X-ray crystal structure determination has been reported. The crystals are monoclinic, space group Bb-C_s^4. Lattice parameters are a = 13.591(6), b = 14.416(6), c = 10.704(7) Å, γ = 101.29(1)°, Z = 4. The coordination polyhedron is a distorted octahedron: two chlorine atoms, two nitrogen atoms and two oxygen atoms (acetate). Bond distances are Rh-Cl = 2.375(6) and 2.335(6), Rh-N = 2.093(8) and 2.093(8), Rh-O = 2.083(7) and 1.989(7) Å. The cyclohexane ring has a "chair" configuration [2]. The complex is incorrectly formulated in the translation.

Selected infrared frequencies in cm^{-1}: ν(OH) = 3600, 3500; ν(CO_2H) = 1720; ν(CO_2Rh) = 1650 to 1600, 1400; ν(CN) = 1100, 1070 cm^{-1}; ESCA data: Rh $3d_{5/2}$ = 310.3, N1s = 400.9, Cl $2p_{3/2}$ = 198.5 eV. Results of thermogravimetric analysis (TGA) and products of thermal decomposition have been tabulated. Electronic spectra, polarograms, molar conductance and pH data have been recorded for fresh and aged solutions [1].

K[Rh(H_2cdta)Cl_2]·KCl·2H_2O is prepared by adding HCl to a solution of K[Rh(Hcdta)Cl]KCl·2H_2O formed during the reaction between $RhCl_6^{3-}$ and K_2H_2cdta. Infrared frequencies in cm^{-1}: ν(OH) = 3570, 3400, ν(CO) = 1740, $ν_{as}$(COO) = 1670, 1600, $ν_{sym}$(COO) = 1430, 1410, ν(CN) = 1100, 1070 cm^{-1} [1].

Cyclohexanediaminetetraacetic acid has been used to determine rhodium spectrophotometrically at 356 nm and pH = 4 to 5; the molar absorptivity (8.3×10^2 cm^2/mol) and log stability constant (3.82) have been measured [3].

References:

[1] N. A. Ezerskaya, T. P. Solovykh, L. K. Shubochkin, Ya. V. Salyn' (Koord. Khim. **3** [1977] 750/6; Soviet J. Coord. Chem. **3** [1977] 582/8). – [2] T. V. Filippova, T. N. Polynova, A. L. Il'inskii, M. A. Porai-Koshits, N. A. Ezerskaya (Zh. Neorgan. Khim. **26** [1981] 1418/9; Russ. J. Inorg. Chem. **26** [1981] 761/2). – [3] Y. M. Issa, F. M. Issa (Z. Anal. Chem. **276** [1975] 72).

1.1.6.5 Diethylenetriamine-N,N,N',N'',N''-pentaacetic Acid Complexes

(HO$_2$CCH$_2$)$_2$NCH$_2$CH$_2$NCH$_2$CH$_2$N(CH$_2$CO$_2$H)$_2$
  |
  CH$_2$CO$_2$H

= diethylenetriamine-N,N,N',N'',N''-pentaacetic acid = H$_5$detpa

Rh(H$_2$detpa)·3H$_2$O. This complex is incorrectly formulated in C.A. **71** [1969] No. 108588. The preparation is reported, the electronic spectrum has been recorded (λ_{max} = 263 nm) and is reproduced (500 to 250 nm). Selected infrared bands in cm^{-1}: ν(OH) = 3500, ν(CO$_2$H) = 1721, 1233, ν(CO$_2$Rh) = 1576, 1364 [1].

Rh(H$_3$detpa)Cl·H$_2$O is prepared by heating equimolar amounts of H$_3$RhCl$_6$ and free ligand in water on a steam bath for 3 to 4 h and deposits as a pale yellow precipitate. The electronic spectrum (600 to 250 nm) is reproduced, ESCA data are recorded (Rh3d$_{5/2}$ = 310.4, N1s = 400.7, Cl2p$_{3/2}$ = 198.6 eV) and half-wave potentials are reported [2].

Rh(H$_4$detpa)Cl$_2$·H$_2$O deposits as a yellow powder on partial evaporation of the mother liquor from the preceding preparation. The electronic spectrum (600 to 250 nm) is reproduced, ESCA data are recorded (Rh3d$_{5/2}$ = 310.6, N1s = 401.2, Cl2p$_{3/2}$ = 198.7 eV) and half-wave potentials are reported [2].

Rh(H$_5$detpa)Cl$_3$·4H$_2$O deposits as a bright yellow powder on further evaporation of the mother liquor from the preceding preparation. The electronic spectrum (600 to 250 nm) is reproduced, ESCA data are recorded (Rh3d$_{5/2}$ = 310.5, N1s = 401.2, Cl2p$_{3/2}$ = 198.6 eV) and half-wave potentials are reported [2].

Rh(H$_6$detpa)Cl$_4$·4H$_2$O is prepared by boiling Rh(H$_5$detpa)Cl$_3$·4H$_2$O or Rh(H$_4$detpa)Cl$_2$·H$_2$O in 1:1 aqueous HCl for 1 h and deposits as a water-soluble pink powder on concentration and addition of acetone. The electronic spectrum (600 to 250 nm) is reproduced, ESCA data are recorded (Rh3d$_{5/2}$ = 310.4, N1s = 401.7, Cl2p$_{3/2}$ = 198.8 eV) and half-wave potentials are reported [2].

The structures and stereochemistry of the preceding complexes are discussed and infrared spectra are described by [2].

References:

[1] K. Sugiura (Nippon Kagaku Zasshi **90** [1969] 691/6). – [2] N. A. Ezerskaya, T. P. Solovykh, Ya. V. Salyn', O. N. Evstaf'eva, L. K. Shubochkin (Zh. Neorgan. Khim. **27** [1982] 1261/6; Russ. J. Inorg. Chem. **27** [1982] 707/10).

1.2 Rhodium β-Diketonates

General Literature:

R. C. Mehrotra, R. Bohra, D. P. Gaur, Metal β-Diketonates and Allied Derivatives, Academic Press, London 1978.

G. Guiochon, C. Pommier, Gas Chromatography in Inorganics and Organometallics, Ann Arbor Sci., Ann Arbor 1973.

G. Guiochon, C. Pommier, La Chromatographie en Phase Gazeuse en Chimie Inorganique, Gauthier-Villars, Paris 1971.

J. P. Fackler, Metal β-Ketoenolate Complexes, Progr. Inorg. Chem. **7** [1966] 361/425, 407.

R. W. Moshier, R. E. Sievers, Gas Chromatography of Metal Chelates, Pergamon Press, London 1965.

J. P. Collman, Reactions of Metal Acetylacetonates, Angew. Chem. Intern. Ed. Engl. **4** [1965] 132/8.

J. P. Collman, Reactions of Coordinated Ligands and Homogeneous Catalysis, Advan. Chem. Series **37** [1963] 78/98.

1.2.1 Rhodium(I) β-Diketonates

Most rhodium(I) β-diketonato complexes contain carbonyl or other carbon donor ligands and are therefore not eligible for inclusion in this volume.

Rhodium(I) Acetylacetonates (acacH = $CH_3CO \cdot CH_2CO \cdot CH_3$ = 2,4-pentanedione)

Three partially characterised acetylacetonato complexes derived from $[Rh(acac)(CO)_2]$ are thought to contain rhodium(I).

$Na_4[(acac)Rh(O_2)_2Rh(OH)(DMF)]$ (DMF = dimethylformamide, $HCON(CH_3)_2$) is prepared by addition of sodium borohydride to an aerobic solution of $Rh(acac)(CO)_2$ in dimethylformamide and is isolated as a brown precipitate on addition of diethyl ether [1].

$NaH_3[(acac)Rh(O_2)_2Rh(OH)(H_2O)]$ is prepared by dissolving the preceding compound in water and re-precipitating with acetone. It is a brown solid, readily soluble in water and methanol but insoluble in acetone, benzene, dimethylformamide and diethyl ether [1].

$Ag_4[(acac)Rh(O_2)_2Rh(OH)(H_2O)]$ is prepared by addition of an aqueous solution of the preceding compound to a stirred aqueous solution of silver nitrate, and precipitates as an insoluble solid. A fourth product obtained from this system under anaerobic conditions was not characterised but is thought to contain bridging carbonyl ligands [1].

Several of the above products when deposited on solid supports are catalysts for olefin hydrogenation [2]; see also rhodium(I)phosphine-β-diketonato complexes in "Rhodium" Suppl. Vol. B 3.

References:

[1] A. S. Berenblyum, L. I. Lakhman, M. L. Khidekel (Izv. Akad. Nauk SSSR Ser. Khim. **1974** 910/2; Bull. Acad. Sci. USSR Div. Chem. Sci. **1974** 874/6). – [2] A. S. Berenblyum, L. I. Lakhman, I. I. Moiseev, E. D. Radchenko, I. V. Kalechits, M. L. Khidekel' (Izv. Akad. Nauk SSSR Ser. Khim. **1975** 1283/7; Bull. Acad. Sci. USSR Div. Chem. Sci. **1975** 1184/8).

1.2.2 Rhodium(II) β-Diketonates

See rhodium(II) acetato-β-diketonato complexes, pp. 35/6.

1.2.3 Rhodium(III) β-Diketonates

1.2.3.1 Rhodium(III) Acetylacetonate, $Rh(acac)_3$ (acacH = $CH_3CO \cdot CH_2CO \cdot CH_3$)

Preparation. This compound is conveniently prepared by adding acetylacetone to a solution of $Rh(NO_3)_3 \cdot nH_2O$, nitric acid and sodium bicarbonate at pH = 4, and heating the mixture under reflux until the required product deposits as orange-yellow crystals. Recrystallisation from aqueous methanol gives orange-yellow monoclinic plates. $Rh(OH)_3 \cdot nH_2O$ and "sodium rhodate(III)" are reported not to form $Rh(acac)_3$ on treatment with acetylacetone [1]. However, the water-soluble rhodium sulphate obtained by fusion of rhodium sponge with $KHSO_4$ does deposit the required product on treatment with acetylacetone at 80°C [2] as does an

$RhCl_3 \cdot nH_2O$/KOH mixture when heated with acetylacetone and a few drops of water at 120°C for 6 h [3]. The melting point is given as 260°C [1] and 259.7 to 260.6°C [4]; sublimation occurs at 240°C and 1 Torr [1] and at 140 to 150°C (pressure not specified) [3], and decomposition to a rhodium mirror commences at 280°C [1].

Separation of $Rh(acac)_3$ from other metal acetylacetonato complexes has been achieved by fractional sublimation [5, 6] and by zone melting techniques [4]. Liquid-solid chromatography has been investigated as a method for separation of metal acetylacetonates including $Rh(acac)_3$ [7].

$^{105}Rh(acac)_3$ has been detected amongst the products of fission recoil experiments [8].

References:

[1] F. P. Dwyer, A. M. Sargeson (J. Am. Chem. Soc. **75** [1953] 984/5). – [2] J. A. S. Smith, J. D. Thwaites (Discussions Faraday Soc. No. **34** [1962] 143/6). – [3] H. Awano, H. Watarai, N. Suzuki (J. Inorg. Nucl. Chem. **41** [1979] 124/6). – [4] H. Kaneko, H. Kobayashi, K. Ueno (Talanta **14** [1967] 1403/9). – [5] E. W. Berg, F. R. Hartlage (Anal. Chim. Acta **33** [1965] 173/81).

[6] E. W. Berg, F. R. Hartlage (Anal. Chim. Acta **34** [1966] 46/52). – [7] C. A. Tollinche, T. H. Risby (J. Chromatog. Sci. **16** [1978] 448/54). – [8] H. Meinhold, P. Reichold (Radiochim. Acta **11** [1969] 175/81).

Optical Activity. Partial optical resolution of $Rh(acac)_3$ has been achieved by solvent extraction using d-diisoamyl tartrate [1, 2]; by column chromatography on D-(+)-lactose [3, 4, 5]; and by differential partitioning between immiscible chiral solvents (diethyl(+)-tartrate/(−)-α-pinene) [6]. $Rh(acac)_3$ is optically stable over a period of 10 h at 165°C in chlorobenzene [3]. Electronic (500 to 200 nm) and circular dichroism (CD) spectra for $\Lambda(-)$-$[Rh(acac)_3]$ in ethanol are reproduced in [6]. The optical rotatory dispersion (ORD) curve for partially resolved $Rh(acac)_3$ (first eluted fraction from D-(+)-lactose column) has been reproduced [3].

References:

[1] N. S. Bowman, V. G'ceva, G. K. Schweitzer, I. R. Supernaw (Inorg. Nucl. Chem. Letters **2** [1966] 351/4). – [2] G. K. Schweitzer, I. R. Supernaw, N. S. Bowman (J. Inorg. Nucl. Chem. **30** [1968] 1885/90). – [3] R. C. Fay, A. Y. Girgis, U. Klabunde (J. Am. Chem. Soc. **92** [1970] 7056/60). – [4] J. P. Collman, R. P. Blair, A. L. Slade, R. L. Marshall (Chem. Ind. [London] **1962** 141). – [5] J. P. Collman, R. P. Blair, R. L. Marshall, L. Slade (Inorg. Chem. **2** [1963] 576/80).

[6] S. F. Mason, R. D. Peacock, T. Prosperi (J. Chem. Soc. Dalton Trans. **1977** 702/4).

Crystallographic Properties. The X-ray crystal structure of $Rh(acac)_3$ has been determined. The crystals from aqueous ethanol are monoclinic, space group $P2_1/c$-C^5_{2h}, lattice parameters a = 13.925(4), b = 7.483(4), c = 16.392(7) Å, β = 98.63(2)°, Z = 4, D_{obs} = 1.56 g/cm^3, D_{calc} = 1.574 g/cm^3. The central rhodium atom is octahedrally coordinated by six oxygen atoms; mean bond lengths and angles are Rh-O = 1.992, C-O = 1.272, C-C(ring) = 1.383 Å, ∢O-Rh-O = 95.3°, ∢Rh-O-C = 121.8°, ∢O-C-C = 126.4°, ∢C-C-C = 128.0°.

J. C. Morrow, E. B. Parker (Acta Cryst. B **29** [1973] 1145/6).

Spectra. The electronic spectrum of $Rh(acac)_3$ has been measured, λ_{max} = 316 nm [1], and the spectrum reproduced [2]. The phosphorescence spectrum has also been recorded [2, 3] and reproduced [2]; however, data from the two papers are conflicting [3]. Displacement of the bands in the electronic spectrum of $Rh(acac)_3$ to shorter wavelengths relative to those of $Co(acac)_3$ has been attributed to a metal ligand π interaction in the rhodium complex [4].

The vibrational spectrum has been measured over the range 4000 to 60 cm^{-1} and assignments given; the range 1600 to 200 cm^{-1} is reproduced diagrammatically in the paper [5]. The complex is transparent in the near infrared region (15000 to 3500 cm^{-1}) [6].

Proton NMR data ($\tau(CH_3) = 7.90$, $\tau(CH) = 4.70$ in CCl_4 solution) have been recorded [7] and a ^{103}Rh resonance has been measured directly [8]. The effects of substituents in the chelate rings of $Rh(acac)_3$ and other metal acetylacetonates have been studied by ^1H and ^{13}C NMR [9]. Long range anisotropic shielding in unsymmetrically substituted rhodium(III) acetylacetonato complexes has been reported [10]. Electron spin resonance (ESR) spectra for mixtures of $Rh(acac)_3$ and CF_3NO in carbon tetrachloride solution: $g = 2.0060$, $a_N = 14.7 \pm 0.3$, $a_F = 12.5 \pm 0.3$, $a_{Rh} = 4.3 \pm 0.3$. The results have been tentatively interpreted by [11] in terms of a tridentate ligand structure of the form

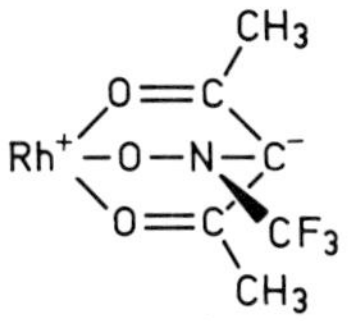

References:

[1] P. D. Hopkins, B. E. Douglas (Inorg. Chem. **3** [1964] 357/60). – [2] M. K. DeArmond, J. E. Hillis (J. Chem. Phys. **49** [1968] 466/7). – [3] G. A. Crosby, R. J. Watts, S. J. Westlake (J. Chem. Phys. **55** [1971] 4663/4). – [4] S. D. Nasirdinov, E. A. Shugam, V. V. Zelentsov, O. A. Osipov (Tr. Vses. Nauchn. Issled. Inst. Khim. Reaktivov Osobo Chist. Khim. Veshchestv No. 31 [1969] 328/36 from C.A. **74** [1971] No. 117865). – [5] M. Mikami, I. Nakagawa, T. Shimanouchi (Spectrochim. Acta A **23** [1967] 1037/53).

[6] R. Dingle (J. Mol. Spectrosc. **18** [1965] 276/87, 278). – [7] J. A. S. Smith, J. D. Thwaites (Discussions Faraday Soc. No. 34 [1962] 143/6, 144). – [8] K.-D. Grüninger, A. Schwenk, B. E. Mann (J. Magn. Resonance **41** [1980] 354/7). – [9] J. C. Hammel, J. A. S. Smith (J. Chem. Soc. A **1970** 1855/9). – [10] J. P. Collman, R. L. Marshall, W. L. Young (Chem. Ind. [London] **1962** 1380/2).

[11] D. Plancherel, D. R. Eaton (Can. J. Chem. **59** [1981] 156/63).

Ionisation. High pressure (0.01 Torr) mass spectra of $Rh(acac)_3$ show the presence of $[Rh(acac)_3]^+$ (70% abundance), $[Rh_2(acac)_5]^+$ (up to 30%), $[Rh_2(acac)_2]^+$ and $[Rh_3(acac)_7]^+$ (each less than 2%) ions [1]. Chemical ionisation mass spectral data for $Rh(acac)_3$ are in good agreement with calculated isotopic distribution for the ion $[Rh(acac)_3H]^+$ [2]. The mass spectrum (m/e = 400 to 40) is reproduced and has been rationalised in terms of ion reactions [3]. The first ionisation energy of $Rh(acac)_3$, measured by electron impact methods, is reported to be 7.7(5) eV [4] and 7.34 ± 0.01 eV [5]. Gas chromatography-mass spectrometry measurements on metal acetylacetonates, including $Rh(acac)_3$, indicate that detection sensitivity in gas chromatography is largely dependent upon thermal stability of the chelate [6, 7].

References:

[1] S. M. Schildcrout (J. Phys. Chem. **80** [1976] 2834/8). – [2] S. R. Prescott, J. E. Campana, P. C. Jurs, T. H. Risby, A. L. Yergey (Anal. Chem. **48** [1976] 829/32). – [3] C. G. Macdonald, J. S. Shannon (Australian J. Chem. **19** [1966] 1545/66, 1552). – [4] F. Bonati, D. Distefano, G. Innorta, G. Minghetti, S. Pignataro (Z. Anorg. Allgem. Chem. **386** [1971] 107/12, 109). – [5] M. M. Bursey, P. F. Rogerson (Inorg. Chem. **10** [1971] 1313/5).

[6] K. Tanikawa (Chem. Pharm. Bull. [Tokyo] **24** [1976] 1954/6). – [7] K. Tanikawa (Bunseki Kagaku **25** [1976] 721/5 from C.A. **87** [1977] No. 47626).

Chemical Reactions

The **thermal gravimetric analysis** (TGA) curve of $Rh(acac)_3$ (0 to 300°C) has been published [1].

Protonation of $Rh(acac)_3$ by trifluoroacetic acid in deuterochloroform leads to formation of $[Rh(acac)_2(acacH)]^+$ [7]. $Rh(acac)_3$ undergoes ligand exchange in neat acetylacetone at 185°C without decomposition, a first order rate constant of $2.4 \times 10^{-5}\ s^{-1}$ is reported [2]. The rate of monochlorination of $Rh(acac)_3$ by N-chlorosuccinimide in chloroform has been measured by stopped-flow pulse Fourier transform ^1H NMR; a second order rate constant of 0.068 ± 0.004 $dm^3 \cdot mol^{-1} \cdot s^{-1}$ is recorded. Bromination by N-bromo-succinimide is slower by a factor of $\sim 10^6$ [3]. Products obtained by formylation, acylation, chlorination, bromination, iodination and nitration of $Rh(acac)_3$ are described in the succeeding section, below.

Many acetylacetonates including $Rh(acac)_3$ when used in conjunction with aluminium alkyls catalyse polymerisation of propylene oxide [4] and olefins [5].

Treatment of $Rh(acac)_3$ with triphenylphosphine and aluminium triethyl in tetrahydrofuran affords a route to $RhH(PPh_3)_4$ [6].

Distribution of $Rh(acac)_3$ between water and various organic solvents (n-C_7H_{16}, C_6H_6, CCl_4 and $CHCl_3$) has been determined as a function of temperature. Thermodynamic data indicate that $Rh(acac)_3$ is less hydrated than acetylacetonates of Al^{III}, Co^{III} and Cr^{III} [8]. Partition coefficients for $Rh(acac)_3$ in heptane-water/DMSO systems at 25°C have been reported [9].

References:

[1] K. J. Eisentraut, R. E. Sievers (J. Inorg. Nucl. Chem. **29** [1967] 1931/6, 1936). – [2] H. Kido (Bull. Chem. Soc. Japan **53** [1980] 82/7). – [3] A. J. Brown, O. W. Howarth, P. Moore (J. Am. Chem. Soc. **100** [1978] 713/8). – [4] G. E. Foll (Soc. Chem. Ind. [London] Monograph No. 26 [1967] 103/130, 105). – [5] J. Krepelka, J. Zachoval, P. Strop (Sb. Vys. Sk. Chem. Technol. Praze Org. Chem. Technol. C **18** [1973] 5/15 from C.A. **80** [1974] No. 3888).

[6] T. Ito, S. Kitazume, A. Yamamoto, S. Ikeda (J. Am. Chem. Soc. **92** [1970] 3011/6). – [7] A. J. C. Nixon, D. R. Eaton (Can. J. Chem. **56** [1978] 1928/32). – [8] P. D. Hopkins, B. E. Douglas (Inorg. Chem. **3** [1964] 357/60). – [9] H. Awano, H. Watarai, N. Suzuki (J. Inorg. Nucl. Chem. **41** [1979] 124/6).

See also rhodium(III)phosphine acetylacetonato complexes in "Rhodium" Suppl. Vol. B 3.

1.2.3.2 Rhodium(III) Complexes of 3-Substituted Acetylacetones

$Rh(3\text{-}CH_3\text{-}acac)_3$ ($3\text{-}CH_3\text{-}acacH = CH_3COCH(CH_3)COCH_3$ = 3-methyl-2,4-pentanedione) is prepared from the corresponding β-diketone but no details are reported [1, 2]. Proton NMR data (CCl_4) have been reported, chemical shifts $\tau(CH_3) = 7.83$ and 8.07 [1] or 7.79 and 8.10 [2].

$Rh(acac)_2(3\text{-}OHC\text{-}acac)$ ($3\text{-}OHC\text{-}acacH = CH_3CO \cdot CH(CHO)CO \cdot CH_3$ = 3-formyl-2,4-pentanedione) is prepared by treatment of $Rh(acac)_3$ with dimethylformamide and $POCl_3$ at room temperature for 105 min followed by hydrolysis, extraction with dichloromethane and chromatography on alumina (benzene/dichloromethane). It forms yellow crystals, melting point 233 to 235°C, from benzene/heptane. The proton NMR spectrum (CCl_4) has been reported, $\tau(CHO) = -0.03$, $\tau(CH) = 4.51$, $\tau(CH_3) = 7.42$ and 7.89 [3].

d-$Rh(acac)_2(3\text{-}CH_3CO \cdot acac)$ ($3\text{-}CH_3CO \cdot acacH = CH_3COCH(COCH_3)COCH_3$ = 3-acetyl-2,4-pentanedione) is obtained in 12% yield by treatment of d-$Rh(acac)_3$ with prepurified acetyl chloride and aluminium trichloride in 1,2-dichloroethane at −20 to −15°C, and is separated from residual d-$Rh(acac)_3$ by chromatography on alumina, $M[\alpha]_{5461} = 56°$ [4].

Rh(acac)$_2$(3-Cl-acac) (3-Cl-acacH = $CH_3COCHCl \cdot COCH_3$ = 3-chloro-2,4-pentanedione) is prepared by treatment of Rh(acac)$_3$ with one equivalent of N-chlorosuccinimide and is separated by preparative TLC (neutral alumina/toluene). Proton NMR data ($CHCl_3$) have been reported, $\tau(CH) = 4.51$, $\tau(CH_3) = 7.59$, 7.84 [2].

Rh(acac)(3-Cl-acac)$_2$ is similarly prepared using two equivalents of N-chlorosuccinimide and is separated by TLC as for the preceding compound. Proton NMR data ($CHCl_3$) have been recorded, $\tau(CH) = 4.50$, $\tau(CH_3) = 7.60$, 7.59 [2].

Rh(3-Cl-acac)$_3$ is prepared by heating Rh(acac)$_3$ under reflux with N-chlorosuccinimide in chloroform solution for 15 min, and is crystallised from hot benzene-heptane as yellow crystals, melting point 252 to 253°C [4]. The electronic spectrum ($CHCl_3$) $\lambda_{max} = 270$, 342 nm [4], vibrational spectrum (KBr pellet) 1550, 1445, 1415, 1335, 1275, 1035, 920 cm^{-1} [4], and proton NMR spectrum (CCl_4) $\tau(CH_3) = 7.57$ [4], 7.61 [1] and 7.59 [5] have all been reported.

Rh(3-Cl-acac)$_2$(3-Br-acac) (3-Br-acacH = $CH_3COCHBrCOCH_3$ = 3-bromo-2,4-pentanedione) has been mentioned but no preparative procedure is reported. Proton NMR data have been recorded, $\tau(CH_3) =$ 7.49 and 7.58 [2].

Rh(3-Br-acac)$_3$ is prepared by treatment of Rh(acac)$_3$ with N-bromosuccinimide in chloroform under reflux for 10 min, and is isolated by extraction with water to remove excess N-bromosuccinimide followed by evaporation of the organic solution and recrystallisation from 1:1 chloroform-ethanol/benzene. It deposits as yellow crystals of a **chloroform solvate**, melting point 245 to 248°C (decomposition). The electronic spectrum ($CHCl_3$) with $\lambda_{max} = 268$, 343 nm, vibrational spectrum (KBr pellet) at 1540, 1410, 1325, 1270, 1212, 1020, 975, 685 and 660 cm^{-1}, and NMR spectrum (CCl_4) with $\tau(CH_3) = 7.49$, $\tau(CHCl_3) = 2.70$ have been reported [4]. The chloroform is removed by heating in vacuo at 65°C for 5 h to afford the **non-solvated complex** as a yellow powder, melting point 230 to 260°C (decomposition). The vibrational spectrum (KBr pellet) at 1540, 1410, 1326, 1270, 975, 920, 683 and 660 cm^{-1} [4], and proton NMR spectrum with $\tau(CH_3) = 7.49$ [4], 7.52 [1] have been reported. A neutron irradiation study has been described [6].

Rh(3-I-acac)$_3$ (3-I-acacH = $CH_3COCHICOCH_3$ = 3-iodo-2,4-pentanedione). Reference has been made to the preparation of this complex by treatment of Rh(acac)$_3$ with iodine chloride in an acetic acid buffer. The proton NMR spectrum (CCl_4) has been reported, $\tau(CH_3) = 7.33$ [1].

Rh(acac)$_2$(3-O$_2$N-acac) (3-O$_2$N-acacH = $CH_3COCH(NO_2)COCH_3$ = 3-nitro-2,4-pentanedione) is prepared by treatment of Rh(acac)$_3$ with one equivalent of copper(II) nitrate in acetic anhydride, and is separated by preparative TLC (neutral alumina/toluene) [2]. The first ionisation potential (electron impact) is given as 7.65 ± 0.02 eV [7].

Rh(acac)(3-O$_2$N-acac)$_2$ is similarly prepared using 2 equivalents of copper(II) nitrate and is separated by TLC [2]. The first ionisation potential (electron impact) is given as 7.97 ± 0.03 eV [7].

Rh(3-O$_2$N-acac)$_3$ is similarly prepared using 3 equivalents of copper(II) nitrate at 0°C and forms yellow crystals, melting point 290 to 300°C (decomposition) [3]. The proton NMR spectrum (C_6H_6) has been reported, $\tau(CH_3) = 7.59$ [3]. The first ionisation potential (electron impact) is given as 8.39 ± 0.04 eV [7].

References:

[1] J. A. S. Smith, J. D. Thwaites (Discussions Faraday Soc. No. 34 [1962] 143/6). – [2] A. J. Brown, O. W. Howarth, P. Moore (J. Am. Chem. Soc. **100** [1978] 713/8). – [3] J. P. Collman, R. L.

Marshall, W. L. Young, S. D. Goldby (Inorg. Chem. **1** [1962] 704/10, 706, 709). – [4] J. P. Collman, R. P. Blair, R. L. Marshall, L. Slade (Inorg. Chem. **2** [1963] 576/80). – [5] J. C. Hammel, J. A. S. Smith (J. Chem. Soc. Dalton Trans. **1970** 1855/9).

[6] B. J. Kavathekar, J. N. Khanvilkar, M. P. Sahakari, A. J. Mukhedkar (Radiochim. Acta **23** [1976] 88/91). – [7] M. M. Bursey, P. F. Rogerson (Inorg. Chem. **10** [1971] 1313/5).

1.2.3.3 Rhodium(III) 1,1,1-Trifluoroacetylacetonates

tfacacH = $CF_3COCH_2COCH_3$ = 1,1,1-trifluoro-2,4-pentanedione

cis/trans-Rh(tfacac)$_3$ is prepared by heating an aqueous mixture of rhodium(III) nitrate, sodium bicarbonate and trifluoroacetylacetone at pH ≈ 4 to 5 for 75 min, and is purified by chromatography [1, 4]. Purification by vacuum sublimation has also been reported [2]. The melting point is recorded as 170 to 172°C [3]. Single crystal and powder X-ray data have been reported. The crystals are orthorhombic, space group Pbca-D_{2h}^{15}, Z = 8, a = 13.501, b = 19.984, c = 14.964 Å; D_{calc} = 1.87 g/cm^3 [4].

cis-Rh(tfacac)$_3$. The more soluble *cis* isomer is extracted from the *cis/trans* mixture with benzene-hexane or ethanol, and purified by chromatography (acid-washed alumina/benzene-hexane) [1]. Separation by gas chromatography (1% XE −60 on chromosorb W) has been achieved [3]. The *cis* isomer forms tiny yellow needles, melting point 148.5 to 149°C. The X-ray powder pattern has been recorded (for d spacings see original paper) [1].

trans-Rh(tfacac)$_3$. The less soluble *trans* isomer concentrated in the residue on extraction of the *cis* isomer is similarly purified by chromatography. The *trans* isomer forms yellow crystals (melting point 189.5 to 190°C). The X-ray powder pattern has been recorded (for d spacings and visually estimated intensities see original paper) [1]. The electronic spectrum of the *trans* isomer has been reproduced over the ranges 370 to 230 nm [5] and 400 to 320 nm [6].

The ^1H and ^{19}F NMR spectra of *cis*- and *trans*-Rh(tfacac)$_3$ have been reproduced [1]; they show that the complex is stable with respect to *cis/trans* isomerisation at 160 to 165°C [7]. Luminescence data (λ_{max} = 467 nm) have been recorded for Rh(tfacac)$_3$ [8]. The photochemical behaviour of *trans*-Rh(tfacac)$_3$ has been investigated [5, 6, 9]; in deaerated cyclohexane irradiation at 254 nm leads to *trans/cis* isomerisation whereas in deaerated 2-propanol only decomposition is observed. Both reactions are quenched by dioxygen [5, 9]. Flash photolysis studies on Rh(tfacac)$_3$ in alcoholic solvents show evidence for the formation of two short-lived intermediates tentatively formulated as *trans*-Rh(tfacac)$_2$(ROH)$_2$ and Rh(tfacac)$_2$(ROH)H [6].

References:

[1] R. C. Fay, T. S. Piper (J. Am. Chem. Soc. **85** [1963] 500/4). – [2] E. W. Berg, F. R. Hartlage (Anal. Chim. Acta **34** [1966] 46/52). – [3] K. Tanikawa, K. Hirano, K. Arakawa (Chem. Pharm. Bull. [Tokyo] **15** [1967] 915/20, 919). – [4] L. Davignon, A. Dereigne, J. M. Manoli, P. Davous (J. Less-Common Metals **21** [1970] 345/51). – [5] C. Kutal, P. A. Grutsch, G. Ferraudi (J. Am. Chem. Soc. **101** [1979] 6884/90).

[6] G. Ferraudi, P. A. Grutsch, C. Kutal (J. Chem. Soc. Chem. Commun. **1979** 15/6). – [7] R. C. Fay, T. S. Piper (Inorg. Chem. **3** [1964] 348/56). – [8] M. K. DeArmond, J. E. Hillis (J. Chem. Phys. **49** [1968] 466/7). – [9] P. A. Grutsch, C. Kutal (J. Am. Chem. Soc. **99** [1977] 7397/9).

Various Data. Thermogravimetric analysis (TGA) has been used to determine the relative volatilities of metal β-diketonate complexes and the TGA curve for Rh(tfacac)$_3$ is reproduced in the paper [1]. Chemical ionisation mass spectroscopy data have been reported for Rh(tfacac)$_3$

and have been shown to agree with calculated values for isotopic distribution in [Rh-$(tfacac)_3H]^+$ [2]. The mass spectrum is reproduced in [3].

Gas chromatography has been applied to the separation [4 to 7] and analysis [8] of trifluoroacetylacetonate complex mixtures containing $Rh(tfacac)_3$. Liquid-solid chromatography has been similarly applied to $Rh(tfacac)_3$ [9].

Plating of rhodium from the vapour state has been achieved using hydrogen reduction of $Rh(tfacac)_3$ [10].

References:

[1] K. J. Eisentraut, R. E. Sievers (J. Inorg. Nucl. Chem. **29** [1967] 1931/6). – [2] S. R. Prescott, J. E. Campana, P. C. Jurs, T. H. Risby, A. L. Yergey (Anal. Chem. **48** [1976] 829/32). – [3] J. E. Campana, T. H. Risby, P. C. Jurs (Anal. Chim. Acta **112** [1979] 321/40). – [4] K. Tanikawa, K. Hirano, K. Arakawa (Chem. Pharm. Bull. [Tokyo] **15** [1967] 915/20). – [5] R. E. Sievers, B. W. Ponder, M. L. Morris, R. W. Moshier (Inorg. Chem. **2** [1963] 693/8).

[6] K. Tanikawa (Chem. Pharm. Bull. [Tokyo] **24** [1976] 1954/6). – [7] K. Tanikawa (Bunseki Kagaku **25** [1976] 721/5 from C.A. **87** [1977] No. 47626). – [8] W. D. Ross, R. E. Sievers, G. Wheeler (Anal. Chem. **37** [1965] 598/600). – [9] C.A. Tollinche, T. H. Risby (J. Chromatog. Sci. **16** [1978] 448/54). – [10] R. L. Van Hemert, L. B. Spendlove, R. E. Sievers (J. Electrochem. Soc. **112** [1965] 1123/6).

1.2.3.4 Rhodium(III) 1,1,1,5,5,5-Hexafluoroacetylacetonates

hfacac H = $CF_3COCH_2COCF_3$ = 1,1,1,5,5,5-hexafluoro-2,4-pentanedione

$Rh(hfacac)_3$. An early report outlines a general preparative method involving addition of an aqueous solution of the neutralised ligand to an aqueous solution of Rh^{3+} ions [1]. More recent papers describe formation of the required product by heating aqueous suspensions of $Rh(OH)_3$ with the free ligand at pH = 5 to 6 over a period of ~1 h [2, 3, 4]. The complex is yellow-orange [2] or yellow [3] and the melting point is given as 114 to 115°C [1, 3] and as 105 to 106.6°C [4].

The crystals exist in two polymorphic forms: 1) hexagonal needles, space group P6/mmm, $P\bar{6}m2$ ($P\bar{6}2m$) or P6mm, a = 18.27 ± 0.01, c = 6.175 ± 0.006 Å, D_{pykn} = 2.04 g/cm³; 2) monoclinic plates, space group $P2_1/n$, a = 19.11 ± 0.01, b = 13.50 ± 0.01, c = 9.15 ± 0.006 Å, β = 93.10° ± 0.06° [2]. The infrared spectrum has been reported (4000 to 700 cm^{-1}) [1, 3, 5] and assignments given [1, 5]. Luminescence data (λ_{max} = 471, 493 nm) have been recorded [6]. Proton NMR data have been reported for samples dissolved in CCl_4 (τ(CH) = 3.42) and $(CH_3)_2CO$ (τ(CH) = 3.07) [5].

The first ionisation potential (electron impact) is given as 10.15 ± 0.12 eV [7].

The thermogravimetric analysis curve has been reproduced in a paper [8].

Vapour pressure measurements and gas-chromatographic studies on the solution thermodynamics of $Rh(hfacac)_3$ have been described [9]. Gas chromatography has been used to separate $Rh(hfacac)_3$ from mixtures of other hexafluoroacetylacetonato complexes [4, 10 to 13].

$[RhCl(hfacac)_2 \cdot 3H_2O]_2$ is prepared by heating a mixture of rhodium(III) chloride trihydrate and hexafluoroacetylacetonate in ethanol under reflux for 4 h, evaporating the solution almost to dryness in a stream of dry air (all ethanol must be removed) and then adding water to precipitate the required product as fine mustard-yellow crystals (melting point 129 to 131°C, darkens at 126 to 129°C). The molecular weight (acetone) is determined as 618 and 620 Daltons (formula weight 607 Daltons). The infrared spectrum (4000 to 700 cm^{-1}) has been recorded and assignments given. The proton NMR spectrum has been reported for samples dissolved in CH_3OH, $\tau(CH) = 3.40$, and $(CH_3)_2CO$, $\tau(CH) = 3.30$ [5].

$[RhCl(hfacac)_2]_2$. Two isomers of this stoichiometry are obtained by vacuum sublimation of $[RhCl(hfacac)_2 \cdot 3H_2O]_2$ (0.1 Torr, 100°C, 21 h) followed by fractional crystallisation and manual separation [5].

Isomer (I) forms small brownish-yellow needles, soluble in carbon tetrachloride, diethyl ether, tetrahydrofuran and acetone but almost insoluble in methanol, ethanol, chloroform, p-dioxan, nitromethane and benzene, and totally insoluble in water [5].

Isomer (II) forms larger ruby-red crystals very soluble in carbon tetrachloride, soluble in acetone, almost insoluble in ethanol and totally insoluble in water [5].

Infrared (4000 to 700 cm^{-1}) and proton NMR data have been reported for both isomers and assignments given, $\tau(CH) = 3.06$ (I) and 3.46 (II). Binuclear chloride-bridged structures differing only in the relative orientation of the β-diketonate ligands have been proposed [5].

References:

[1] M. L. Morris, R. W. Moshier, R. E. Sievers (Inorg. Chem. **2** [1963] 411/2). – [2] L. Davignon, J. M. Manoli, A. Dereigne (J. Less-Common Metals **21** [1970] 341/4). – [3] J. P. Collman, R. L. Marshall, W. L. Young, S. D. Goldby (Inorg. Chem. **1** [1962] 704/10). – [4] K. Arakawa, K. Tanikawa (Bunseki Kagaku **16** [1967] 812/5 from C. A. **68** [1968] No. 92549). – [5] S. C. Chattoraj, R. E. Sievers (Inorg. Chem. **6** [1967] 408/11).

[6] M. K. DeArmond, J. E. Hillis (J. Chem. Phys. **49** [1968] 466/7). – [7] S. M. Schildcrout, R. G. Pearson, F. E. Stafford (J. Am. Chem. Soc. **90** [1968] 4006/10). – [8] K. J. Eisentraut, R. E. Sievers (J. Inorg. Nucl. Chem. **29** [1967] 1931/6, 1936). – [9] W. R. Wolf, R. E. Sievers, G. H. Brown (Inorg. Chem. **11** [1972] 1995/2002). – [10] R. E. Sievers, B. W. Ponder, M. L. Morris, R. W. Moshier (Inorg. Chem. **2** [1963] 693/8).

[11] R. S. Juvet, R. P. Durbin (J. Gas Chromatog. **1** No. 12 [1963] 14/7). – [12] K. Tanikawa (Chem. Pharm. Bull. [Tokyo] **24** [1976] 1954/6). – [13] K. Tanikawa (Bunseki Kagaku **25** [1976] 721/5 from C.A. **87** [1977] No. 47626).

1.2.3.5 Rhodium(III) 2,6-Dimethyl-3,5-heptanedionate, $Rh(dmhd)_3$

dmhd H = $(CH_3)_2CHCO \cdot CH_2COCH(CH_3)_2$ = 2,6-dimethyl-3,5-heptanedione

This compound is prepared by warming together a mixture of the free ligand and an aqueous solution of rhodium(III) ions at $pH \approx 4.0$ and is isolated as a yellow solid (melting point 86 to 92°C) from dichloromethane. The proton NMR spectrum has been reported for a 0.23 M solution in 84% $CF_3 \cdot C_6H_5$/16% CH_2Cl_2, $\tau(CH_3) = 8.93$, 8.90, J = 7.0 Hz; $\tau(C\underline{H})(^iPr) = 7.46$, J = 7.0 Hz; $\tau(C\underline{H}) = 4.54$.

M. Pickering, B. Jurado, C. S. Springer (J. Am. Chem. Soc. **98** [1976] 4503/15, 4504).

1.2.3.6 Rhodium(III) 2,2,6,6-Tetramethyl-3,5-heptanedionate, $Rh(tmhd)_3$

tmhd H = $(CH_3)_3C \cdot CO \cdot CH_2CO \cdot C(CH_3)_3$ = 2,2,6,6-tetramethyl-3,5-heptanedione = dipivaloylmethane

This compound is prepared by treating an aqueous solution of a rhodium(III) salt (unspecified) with aqueous potassium bicarbonate until precipitation of hydroxide is imminent, then adding an ethanolic solution of dipivaloylmethane and stirring the mixture for 5 h. It is isolated as an orange-yellow solid, melting point 185°C. Selected infrared bands and proton NMR resonances have been reported, ν(C⋯C), ν(C⋯O) = 1545 cm^{-1}; τ($C\underline{H}_3$) = 8.87; τ($C\underline{H}$) = 4.40. Fractional sublimation data have been recorded.

E. W. Berg, N. M. Herrera (Anal. Chim. Acta **60** [1972] 117/25).

1.2.3.7 Rhodium(III) (+)-Formylcamphorates

$C_{11}H_{15}O_2$ = (+)-formylcamphorate or (+)-hydroxymethylenecamphorate

1,2,6-[Rh($C_{11}H_{15}O_2$)$_3$] is prepared by heating a mixture of rhodium(III) chloride trihydrate in water and (+)-formylcamphor in benzene under reflux while maintaining the pH at 4 by addition of sodium bicarbonate, and is isolated by evaporation of the benzene layer. The infrared spectrum (1600 to 700 cm^{-1}) has been recorded: 1580(s), 1515(s), 1280(m), 1210(m), 1020(m), 935(m), 790(w), 775(m), 720(w), 710(w) cm^{-1}, all ± 5 cm^{-1}. The electronic spectrum (400 to 250 nm) is reproduced.

The crude product is separated into diastereoisomers by extraction with light petroleum (boiling point 40 to 60°C) followed by evaporation of the extracts to dryness and chromatography of the residue (alumina/light petroleum).

1,2,6-(−)-[Rh($C_{11}H_{15}O_2$)$_3$] is obtained from the faster moving fraction of the first band and is isolated as a yellow crystalline solid after three recrystallisations from ligroin. It is not further separated by chromatography on (+)-lactose. Specific rotations ($CHCl_3$) are $[\alpha]_D = -463°$ and $[\alpha]_{370} = -21800°$. The electronic spectrum ($CHCl_3$) has been measured (λ_{max} = 343 nm) and the optical rotatory dispersion (ORD) curve (1000 to 300 nm) reproduced. Proton NMR data have been recorded and assignments given.

1,2,6-(+)-[Rh($C_{11}H_{15}O_2$)$_3$] is obtained in optically impure form from the slower moving fractions of the first band and is partially separated from remaining amounts of the less soluble laevorotatory isomer by fractional crystallisation from chloroform-ligroin. It is isolated as a yellow crystalline solid, more soluble in organic solvents than the laevorotatory isomer. The electronic spectrum ($CHCl_3$) has been measured (λ_{max} = 343 nm) and the ORD curve (1000 to 300 nm) is reproduced in the paper. Specific rotations ($CHCl_3$) are $[\alpha]_D \approx +1120°$ and $[\alpha]_{370} = +10700°$. Infrared data have been recorded: 1620(s), 1320(ms), 1280(ms), 1215(w), 1190(w), 1155(m), 1110(m), 1075(m), 1050(w), 975(w), 915(w), 795(w), 750(m), 725(m) cm^{-1}, all ± 5 cm^{-1}.

J. H. Dunlop, R. D. Gillard, R. Ugo (J. Chem. Soc. A **1966** 1540/7, 1543, 1546).

1.2.3.8 Rhodium(III) (+)-Acetylcamphorates

$C_{12}H_{17}O_2$ = (+)-acetylcamphorate

Rh($C_{12}H_{17}O_2$)$_3$ is prepared by heating a suspension of rhodium(III) chloride trihydrate and potassium hydroxide in neat (+)-3-acetylcamphor at 80°C for 20 h, and is isolated by vacuum distillation of the excess ligand followed by chromatography of the residue (silica gel/benzene). Thin layer chromatography separates the crude product into four diastereoisomers tentatively identified as Δ-*trans*-Rh($C_{12}H_{17}O_2$)$_3$, Λ-*trans*-Rh($C_{12}H_{17}O_2$)$_3$, Δ-*cis*-Rh($C_{12}H_{17}O_2$)$_3$ and Λ-*cis*-Rh($C_{12}H_{17}O_2$)$_3$. The electronic spectrum of Λ-*trans*-Rh($C_{12}H_{17}O_2$)$_3$ and the circular dichroism spectra of all four diastereoisomers are reproduced in the paper. X-ray powder data (interplanar spacings) are tabulated for all four diastereoisomers. Configurational assignments have been made on the basis of NMR, X-ray and CD data.

G. W. Everett, A. Johnson (Inorg. Chem. **13** [1974] 489/90).

1.2.3.9 Rhodium(III) Benzoylacetonates

bzac H = $PhCO \cdot CH_2COCH_3$ = 1-phenyl-1,3-butanedione

$Rh(bzac)_3$. The crude *cis/trans* isomer mixture is prepared by neutralising a rhodium(III) nitrate solution to pH = 4 with sodium bicarbonate when rhodium(III) hydroxide begins to separate, then adding benzoylacetone and refluxing the mixture at pH = 4 for 45 min. The yellow-orange gummy product is washed with water and purified by chromatography (alumina/benzene).

***cis*-$Rh(bzac)_3$** is separated from the *cis/trans* mixture by extraction with hot ethanol followed by evaporation of the extract and column chromatography (alumina/benzene-hexane) of the residue. It forms yellow hexagonal plates, isomorphous with the corresponding cobalt(III) and chromium(III) compounds. The complex lacked a sharp melting point because of *cis-trans* isomerisation during heating. X-ray powder data (d-spacings and intensities) have been recorded and the proton NMR spectrum reproduced.

***trans*-$Rh(bzac)_3$** is obtained from the residue remaining after ethanol extraction of the *cis* isomer by thrice recrystallising from benzene/pentane. It forms yellow plates (melting point 260.5 to 261.5°C with decomposition) isomorphous with the corresponding cobalt(III) complex. X-ray powder data (d-spacings and intensities) are recorded. The proton NMR spectrum is reproduced.

R. C. Fay, T. S. Piper (J. Am. Chem. Soc. **84** [1962] 2303/8).

1.2.3.10 Rhodium(III) Benzoyltrifluoroacetonate, $Rh(bztfac)_3$

bztfac H = $PhCO \cdot CH_2COCF_3$ = 1-phenyl-4,4,4-trifluoro-1,3-butanedione

A general method of preparation involving addition of free ligand to an aqueous buffered (NaO_2CCH_3) solution of rhodium nitrate is cited. Sublimation data have been recorded (the material condenses to a liquid).

E. W. Berg, F. R. Hartlage (Anal. Chim. Acta **34** [1966] 46/52).

1.2.3.11 Rhodium(III) Dibenzoylmethane Complex, $[Rh(C_{15}H_{11}O_2)_3]$

$C_{15}H_{12}O_2$ = $PhCOCH_2COPh$ = 1,3-diphenyl-1,3-propanedione

The compound is prepared by heating a mixture of $Rh(acac)_3$ and dibenzoylmethane in ethylbenzoate at 160°C, and forms yellow-green crystals (melting point 305 to 308°C).

L. Wolf, E. Butter, H. Weinelt (Z. Anorg. Allgem. Chem. **306** [1960] 87/93).

1.2.3.12 Rhodium(III) Di-p-nitrobenzoylmethane Complex, $[Rh(C_{15}H_9N_2O_6)_3]$

$C_{15}H_{10}N_2O_6$ = $O_2NC_6H_4COCH_2COC_6H_4NO_2$ = 1,3-di(p-nitrophenyl)-1,3-propanedione

This compound is prepared by heating a mixture of $Rh(acac)_3$ and di-p-nitrobenzoylmethane in pyridine. It forms red-brown crystals (melting point 365°C with decomposition).

L. Wolf, E. Butter, H. Weinelt (Z. Anorg. Allgem. Chem. **306** [1960] 87/93).

1.2.3.13 Rhodium(III) Thenoyltrifluoroacetonate, $Rh(ttfac)_3$
ttfacH = $C_4H_3S \cdot COCH_2COCF_3$ = thenoyltrifluoroacetone

The compound has been prepared and the *cis/trans* isomerisation reaction has been studied by ^1H and ^{19}F NMR. Equilibrium constants for the isomerisation in $(CH_3)_2SO$, CH_3COBu^i and cyclohexanone and at a range of temperatures have been measured.

C. Shibata, M. Yamazaki, T. Takeuchi (Nippon Kagaku Kaishi **1976** No. 11, pp. 1725/8 from C.A. **86** [1977] No. 100136).

1.2.3.14 Rhodium(III) 1-Phenyl-3-methyl-4-trifluoroacetylpyrazolonate
= $C_{12}H_9F_3N_2O_2$ = 1-phenyl-3-methyl-4-trifluoroacetylpyrazolone-5

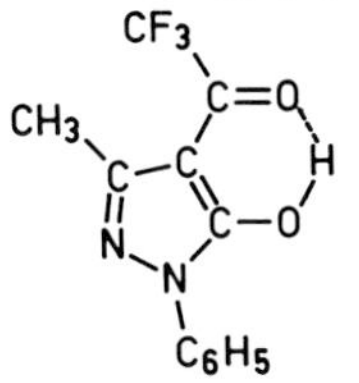

This β-diketone forms a rhodium complex with partially extracts (20% at pH = 7, 35% at pH = 10) from aqueous solution into chloroform, an extraction/pH curve is reproduced.

S. M. Hasany, I. Hanif (J. Radioanal. Chem. **45** [1978] 115/23).

1.2.3.15 Rhodium(III) Thiothenoyltrifluoroacetonate, $Rh(tttfac)_3$
tttfacH = $C_4H_3S \cdot CS \cdot CH_2 \cdot COCF_3$ = 1-thenoyl-4,4,4-trifluorobutane-1-thione-3-one

This compound is prepared by treatment of a buffered rhodium(III) solution with the free ligand ($C_4H_3S \cdot CS \cdot CH_2 \cdot COCF_3$) and has been investigated as an agent for the separation of rhodium by TLC. Good separations are obtained using silica gel-chloroform/hexane.

H. Müller, R. Rother (Anal. Chim. Acta **66** [1973] 49/55).

1.3 Complexes with Miscellaneous Oxygen Donor Ligands

1.3.1 Complex with N,N-Dimethylacetamide, $[AsPh_4]_2[Rh_2Cl_6(dma)_2]$
C_4H_9NO = N,N-dimethylacetamide = dma

This compound is prepared by heating a solution of $[RhCl(CO)_2]_2$ and LiCl in N,N-dimethylacetamide in the presence of oxygen at 75°C for 10 h. After removal of Li_2CO_3 and addition of $[AsPh_4]Cl$ the product is precipitated as an oil by addition of diethyl ether and converted to a green solid by treatment with cold water. The electronic spectrum (λ_{max} = 670 nm, in dma) and selected infrared bands have been reported, ν(RhCl) = 330, 318, 272 cm^{-1}. ESR data have been tabulated: g_1 = 2.103, g_2 = 2.031, g_3 = 1.971, g_{av} = 2.035, g_{iso} = 2.036. The magnetic susceptibility was reported to be $\chi_{mol} = 380 \times 10^{-6}$ cgs units at 22°C. The conductivity (dma, 23°C) was given as 115 $cm^2 \cdot \Omega^{-1} \cdot mol^{-1}$.

B. R. James, G. Rosenberg (Coord. Chem. Rev. **16** [1975] 153/9, 156).

1.3.2 Complexes with 4-Aminoisoxazolidin-3-one (Cycloserine)

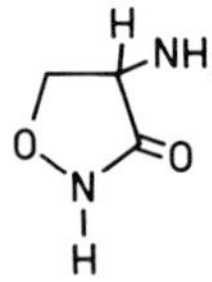
= cycloserine = $C_3H_6N_2O_2$

***mer*-$RhX_3(C_3H_6N_2O_2)_3 \cdot nH_2O$** (X = Cl, n = l; X = Br, n = 2; X = I, n = 0). These complexes are prepared by refluxing stoichiometric amounts of the appropriate rhodium(III) halide and cycloserine in ethanol for 2 h, and deposit from solution as orange or brown diamagnetic solids. Electronic spectra have been recorded and ligand field parameters calculated. Infrared spectra (4000 to 200 cm^{-1}) have been recorded and assignments given. On the basis of far infrared data the complexes are assigned *mer* stereochemistry with O-coordinated cycloserine ligands.

Complex	Colour	Decomp. Point at °C	λ_{max} in nm	ν(Rh-X) in cm^{-1}	ν(Rh-O) in cm^{-1}
$RhCl_3(C_3H_6N_2O_2)_3 \cdot H_2O$	orange-yellow	170 to 175	495, 378	349, 325, 290	472, 422, 370
$RhBr_3(C_3H_6N_2O_2)_3 \cdot 2H_2O$	orange-brown	195 to 200	510, 390	206, 183, 171	480, 418, 376
$RhI_3(C_3H_6N_2O_2)_3$	light brown	205 to 210	529, 397	150, 133, 119	460, 420, 382

C. Preti, G. Tosi (Australian J. Chem. **33** [1980] 57/68).

1.3.3 Complexes with 4,4′-[1,4-Phenylenebis(methylidynenitrilo)]bis(isoxazolidin-3-one)

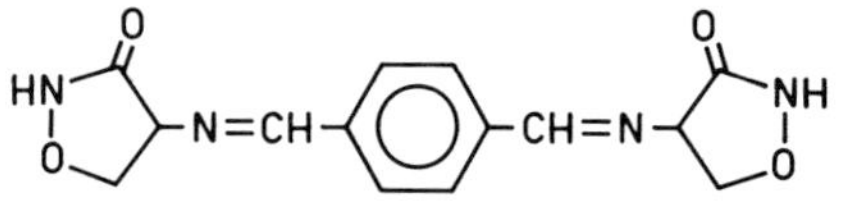
= terizidone = trz

$Rh(trz)_3X_3 \cdot nH_2O$ (X = Cl, n = 1; X = Br, n = 4; X = I, n = 2). These complexes are obtained as orange-yellow to brown solids by heating a mixture of the free ligand and the appropriate rhodium halide under reflux in acetone. The complexes are diamagnetic non-electrolytes with trz ligands coordinated through the imino group of an oxazole ring. Electronic spectra, vibrational spectra (4000 to ~150 cm^{-1}) and thermogravimetric data are reported.

C. Preti, L. Tassi, G. Tosi, P. Zannini, A. F. Zanoli (Australian J. Chem. **35** [1982] 1829/40).

1.3.4 Complex with 3,4,4-Trimethyl-2-carboethoxy-but-2-en-1,4-olide, $[Rh(C_{10}H_{14}O_4)_6]Cl_3 \cdot H_2O$

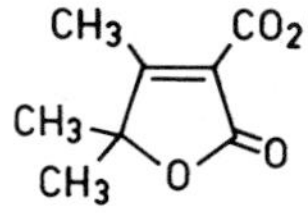
= 3,4,4-trimethyl-2-carboethoxybut-2-en-1,4-olide = $C_{10}H_{14}O_4$

This compound is prepared by treating the ligand with aqueous $RhCl_3 \cdot 3H_2O$ solution in the presence of KCl. Water of crystallisation is lost at 160°C and decomposition commences above 250°C. The organic ligands are coordinated through the oxygen atoms of the carbonyl groups.

S. N. Sarkisyan, A. A. Avetisyan, L. I. Sagradyan (Arm. Khim. Zh. **31** [1978] 539/41 from C.A. **89** [1978] No. 172841).

1.3.5 Complex with 1-Phenyl-3-pyrazolidone (Phenidone)

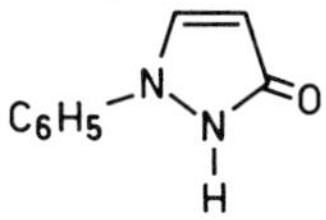
= phenidone = $C_9H_8N_2O$

Complex formation between rhodium(III) and phenidone in aqueous perchloric acid solution (2, 4 and 6 M) has been reported. A change in colour from light pink through red and brown to blue-green has been observed and electronic spectra have been recorded, $\lambda_{max} = 598$ nm. A Rh^{III}/ligand ratio of 1:1 has been determined for the complex by the method of continuous variations. The calculated instability constant was found to be 2.4×10^{-2}.

A. Fabrikanos, G. Pallikaris, V. Havredaki (Chem. Ind. [London] **1975** 281/2).

1.3.6 Complex with Benzoylphenylhydroxylamine, $Rh(bpha)_3$

$C_6H_5C(O)N(OH)C_6H_5$ = benzoylphenylhydroxylamine = H bpha

$Rh(bpha)_3$ is prepared by boiling an aqueous solution of a rhodium(III) salt and the free ligand. The complex is very soluble in chloroform and can be extracted into this medium from aqueous solution.

H. Förster, I. Schoefer (J. Radioanal. Chem. **28** [1975] 153/9).

1.3.7 Complexes with 5-Hydroxy-1,4-naphthoquinone (Juglone)

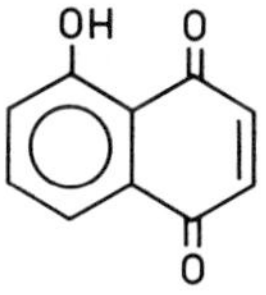
= juglone = $C_{10}H_6O_3$

Stability constants and thermodynamic parameters at 20 and 40°C have been reported for the rhodium(III)-juglone system. Formation of 1:1, 1:2 and 1:3 complexes is indicated.

B. M. L. Bhatia, S. D. Matta, S. S. Sawhney (Thermochim. Acta **47** [1981] 367/9).

1.3.8 Complexes with Catechol and Substituted Catechols

Catecholate, cat = $C_6H_4O_2^{2-}$, en = ethylenediamine, ent = enterobactin^{6-}, $C_{30}H_{28}O_{15}N_3^{6-}$

$K_3[Rh\,cat_3] \cdot H_2O$ is made by ion-exchange from $[Co\,en_3][Rh\,cat_3] \cdot 2H_2O$. It was resolved by using Δ-$[Co\,en_3]I_3 \cdot H_2O$ in a $KHCO_3$-K_2CO_3 buffer and the Δ-salt likewise using Δ-$[Co\,en_3]I_3 \cdot H_2O$.

The electronic absorption spectrum in aqueous basic solution was measured, as was the circular dichroism spectrum (280 to 800 nm, reproduced in paper). The ^1H magnetic resonance spectrum was measured (reproduced in paper).

$[Co\,en_3][Rh\,cat_3] \cdot 2H_2O$ is made from $RhCl_3 \cdot 3H_2O$ in a $KHCO_3$-K_2CO_3 aqueous buffer (pH = 10) with catechol in the absence of air, using $[Co\,en_3]^{3+}$ as precipitant.

$K_3[Rh\,ent]$ is made from $Rh(ClO_4)_3 \cdot 6H_2O$ in ethanol with enterobactin in aqueous phosphate buffer (pH = 7) in the absence of air, followed by ion-exchange. The compound was not isolated pure, but was greenish-gold in colour.

The electronic absorption spectrum was measured and also the circular dichroism spectrum (700 to 200 nm, reproduced in paper). The ^{1}H NMR spectrum was recorded. The Δ-*cis* form is favoured in solution.

J. V. McArdle, S. R. Sofen, S. R. Cooper, K. N. Raymond (Inorg. Chem. **17** [1978] 3075/8).

1.3.9 Complex with Tropolone, $Rh(trop)_3$

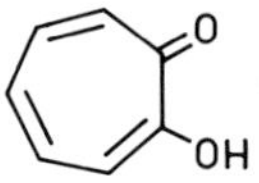
=tropolone=$Htrop_4$

$Rh(trop)_3$ is prepared by heating a mixture of tropolone, $NaO_2CCH_3 \cdot 3H_2O$ and $RhCl_3 \cdot 3H_2O$ in water on a steam bath for 2 h. It is recrystallised from hot chloroform as an orange solid decomposing at 372 to 375°C.

E. L. Muetterties, C. M. Wright (J. Am. Chem. Soc. **87** [1965] 4706/17, 4715).

1.3.10 Complex with 2-Hydroxy-3(3-methyl-2-butenyl)1,4-naphthoquinone (Lapachol)

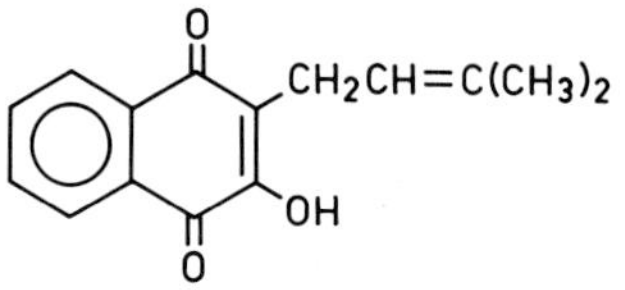
=lapachol=$C_{15}H_{14}O_3$

Stability constants and thermodynamic parameters for 1:1, 1:2 and 1:3 complexes of lapachol with rhodium(III) have been reported.

S. S. Sawhney, B. M. L. Bhatia (Thermochim. Acta **42** [1980] 105/7).

1.3.11 Complexes with Oximidobenzotetronic Acid

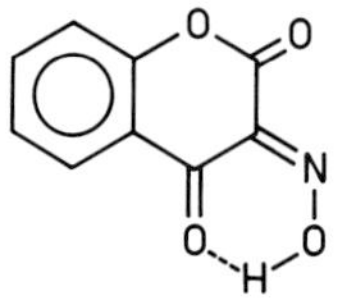
=oximidobenzotetronic acid=Hobta

Rhodium(III) forms a yellow-brown complex $[RhCl_2(obta)_2]^-$ (charge omitted in original paper) at pH=2 to 8 and a red-brown complex with the same Rh/obta ratio at pH=11.5. Electronic spectra (600 to 360 nm, pH range 5 to 12.5) and Job's curves are reproduced. A recommended procedure for the determination of rhodium(III) is given. The obta anion is formulated as an O,O' chelate ligand.

G. S. Manku, A. N. Bhat, B. D. Jain (Talanta **14** [1967] 1229/36).

1.3.12 Complexes with 8-Hydroxyquinoline N-Oxide

$C_9H_7NO_2$=8-hydroxyquinoline N-oxide

Yellow-brown complexes having molar compositions 1:1 and 1:2 (metal:ligand) are obtained by heating $Na_3[RhCl_6]$ with excess free ligand at 100°C for 30 to 60 min. Electronic spectra (560 to 380 nm) measured at pH values in the range 3.1 to 8.8 are reproduced; stability constants (log $K_1 \approx 10.2$, log $K_2 \approx 6.05$) and solvent extraction data have been reported.

R. D. Gupta, G. S. Manku, A. N. Bhat, B. D. Jain (Talanta **17** [1970] 772/81).

1.3.13 Complexes with 1,2,3-Triazole-4,5-dicarboxylic Acid

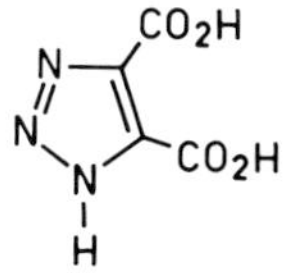

= 1,2,3-triazole-4,5-dicarboxylic acid = $C_4H_3N_3O_4$

Formation of a yellowish precipitate with rhodium(III) at pH = 3 to 6 has been mentioned but no further details were recorded. The mode of coordination of the organic ligand and the stoichiometry of the complex formed were not determined.

F. Capitán, F. Salinas, F. J. Alonso (Anales Quim. **69** [1973] 1133/9).

1.3.14 Complex with o-Hydroxycinnamic Acid (o-Coumaric Acid)

o-HO·C_6H_4·CH=CH·CO_2H = o-coumaric acid = $C_9H_8O_3$

The free energy and the formation constant (at 5 different ionic strengths) of a 1:1 rhodium(III) o-coumaric acid complex have been determined by a titration method. Formation curves are reproduced.

S. C. Tripathi, S. Paul (J. Inorg. Nucl. Chem. **35** [1973] 2465/70).

1.3.15 Complexes with Trimetaphosphimate Anion

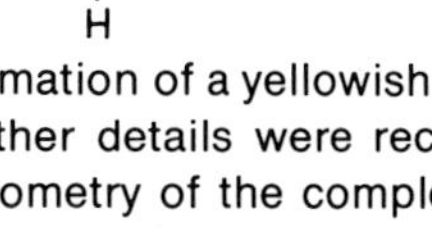

= trimetaphosphimate anion = $[H_3N_3O_6P_3]^{3-}$

$Rh(H_3N_3O_6P_3)(NH_3)_3 \cdot 6H_2O$ is prepared by mixing aqueous solutions of $Rh(NO_3)_3(NH_3)_3$ and the trisodium salt of the trimetaphosphimate anion. It separates as a microcrystalline yellow precipitate [1].

$Rh(H_3N_3O_6P_3)(NH_3)_3 \cdot 3H_2O$ forms on heating the hexahydrate at 180°C [1].

$Rh(H_3N_3O_6P_3)(NH_3)_3 \cdot H_2O$ separates from the mother liquor as well-formed golden-yellow prisms after crystallisation of the hexahydrate (see above). The crystals are orthorhombic, space group $P2_1cn$-C_{2v}^9, lattice parameters a = 9.924(1), b = 9.978(1), c = 12.674(2) Å, Z = 4; D_{exp} = 2.30, D_{calc} = 2.31 g/cm³. The structure contains octahedrally coordinated rhodium bound to the nitrogen atoms of three NH_3 ligands (Rh-N = 2.087, 2.080 and 1.982 Å) and three oxygen atoms of the trimetaphosphimate anion (Rh-O = 2.079, 2.071 and 2.072 Å) [1].

The vibrational spectrum of the monohydrate has been studied (4000 to 50 cm^{-1}) and the theoretical spectrum has been calculated assuming valence force field. The infrared spectrum (3900 to 700 and 350 to 50 cm^{-1}) and Raman spectrum (3300 to 2900 and 1400 to 100 cm^{-1}) are reproduced. Assignments are given [2].

References:

[1] L. K. Shubochkin, O. V. Popov, E. F. Shubochkina, V. I. Sokol, I. A. Rozanov, L. A. Butman (Koord. Khim. **3** [1977] 902/6; Soviet J. Coord. Chem. **3** [1977] 700/4). – [2] L. M. Sukova, O. I. Kondratov, K. I. Petrov, I. A. Rozanov, G. G. Novitskii (Zh. Neorgan. Khim. **24** [1979] 2718/22; Russ. J. Inorg. Chem. **24** [1979] 1509/12).

2 Complexes with Ligands Containing Nitrogen

2.1 Monodentate Nitrogen Donors

2.1.1 Nitrosyl Complexes

General Literature:

J. A. McCleverty, Reactions of Nitric Oxide Coordinated to Transition Metals, Chem. Rev. **79** [1979] 53/76.

R. Eisenberg, C. D. Meyer, The Coordination Chemistry of Nitric Oxide, Accounts Chem. Res. **8** [1975] 26/34.

K. G. Caulton, Synthetic Methods in Transition Metal Nitrosyl Chemistry, Coord. Chem. Rev. **14** [1975] 317/55.

2.1.1.1 Nitrosyl-nitrile Complexes, $[Rh(NO)(NCR)_4]X_2$

$[Rh(NO)(NCMe)_4][BF_4]_2$ (Me = methyl). This complex is prepared by reaction of either $[Rh(C_8H_{12})(NCMe)_2](BF_4)$ or $[Rh(C_8H_{12})_2](BF_4)$ with $NOBF_4$ in acetonitrile. The compound is formed as emerald-green crystals, and ^1H and ^{15}N NMR spectral studies have been made; exchange occurs between coordinated CH_3CN and CD_3CN in solution. Infrared spectral studies reveal ν_{NO} at 1758 cm^{-1} and ν_{CN} at 2345 and 2325 cm^{-1}. The related $[PF_6]^-$ complex has also been prepared [1, 2].

$[Rh(NO)(NCMe)_4](BF_4)_2$ reacts with CO giving $[Rh(CO)_2(NCMe)_2](BF_4)$ [1, 2], and can catalyse the isomerisation of terminal to internal olefines, the oligomerisation of branched olefines and the stereospecific polymerisation of butadiene to trans-1,4-polybutadiene [3].

$[Rh(NO)(CNBu^t)_4](PF_6)_2$ (Bu^t = t-butyl). This is prepared by a route similar to that of its BF_4^- analogue, using $NOPF_6$. This compound is also green with ν_{NO} at 1765 and ν_{CN} = 2310 cm^{-1} [1, 2].

$[Rh(NO)(NCMe)_3Cl](PF_6)$. This is obtained either by reaction of $NOPF_6$ in acetonitrile with $[Rh(C_8H_{12})Cl]_2$ or $[Rh(CO)_2Cl]_2$ or by treatment of $[Rh(NO)(NCMe)_4](PF_6)_2$ with the stoichiometric amount of Cl^-. The complex is green-brown and exhibits ν_{NO} at 1700 cm^{-1}, ν_{CN} at 2335 and 2310 cm^{-1}. It is thought that these complexes have square pyramidal geometry with a bent Rh-N-O bond [1, 2].

References:

[1] N. G. Connelly, P. T. Draggett, M. Green, T. A. Kuc (J. Chem. Soc. Dalton Trans. **1977** 70/3). – [2] N. G. Connelly, M. Green, T. A. Kuc (Chem. Commun. **1974** 542/3). – [3] N. G. Connelly, P. T. Draggett, M. Green (J. Organometal. Chem. **140** [1977] C10/C11).

2.1.1.2 Nitrosyl Halides

Nitrosyl Dihalides, $[Rh(NO)X_2]_n$. These compounds are made by reaction of NOX (X = Cl or Br) with $[Rh(C_8H_{12})Cl]_2$ (C_8H_{12} = cycloocta-1,5-diene) in dichloromethane or chloroform. The **chloro** complex is formed as an air-sensitive red-brown solid for which no reliable elemental analytical data could be obtained, and from which NO is easily lost. The **bromo** complex is isolated in a pure form as a dark red precipitate, is involatile and may be polymeric with bromine bridges. The infrared spectrum of $[Rh(NO)Br_2]_n$ shows ν_{NO} = 1710 and 1668 cm^{-1} (in Nujol). These compounds react with PPh_3 and $AsPh_3$(L) giving $Rh(NO)L_2X_2$ [1].

Nitrosyl Trichloride, $[Rh(NO)Cl_3]_n$. This compound is thought to be formed when NO is passed through an ethanol solution of $RhCl_3 \cdot nH_2O$. The electron spin resonance (ESR) spectrum in ethanol revealed g = 2.17, and treatment of the solution with PPh_3 afforded $Rh(NO)(PPh_3)_2Cl_3$ and $Rh(NO)(PPh_3)_2Cl_2$ [3].

Dinitrosyl Halides, $[Rh(NO)_2X]_n$. These compounds are prepared by passing NO over $[Rh(CO)_2Cl]_2$ at 60°C for 2 d [3, 4] or through petroleum ether or CCl_4 solutions of $[Rh(CO)_2X]_2$ (X = Cl, Br or I) [5, 6]. They form brown-black microcrystalline powders which are air-sensitive, involatile and have no definite melting point.

The compounds appear to be oligo- or polymeric and their infrared spectra contain two bands due to ν_{NO} = 1721, 1575 cm^{-1} (X = Cl); 1718, 1568 cm^{-1} (X = Br); and 1710, 1559 cm^{-1} (X = I) [4, 5]. X-ray photoelectron spectroscopic data obtained from $[Rh(NO)_2Cl]_n$ have been interpreted in terms of substantial delocalisation of charge on to the NO ligands [7].

Reaction of $[Rh(NO)_2X]_n$ with ammonia and aliphatic amines affords green-grey insoluble and air-sensitive compounds, probably $[Rh(NO)_2L_2]X$ (L = NH_3 or amine). Similar reactions apparently occur with pyridine, morpholine and o-phenanthroline but the products have not been characterised [5]. The dinitrosyl halides also react with PPh_3, $AsPh_3$ and $SbPh_3$ (L) forming $Rh(NO)L_3$ and $Rh(NO)L_2X_2$ [8].

References:

[1] G. R. Crooks, B. F. G. Johnson (J. Chem. Soc. Dalton Trans. **1970** 1662/5). – [2] M. C. Baird (Inorg. Chim. Acta **5** [1971] 46/8). – [3] J. Manchot, J. König (Chem. Ber. **60** [1927] 2130/3). – [4] W. P. Griffith, J. Lewis, G. Wilkinson (J. Chem. Soc. **1959** 1775/9). – [5] W. Hieber, K. Heinicke (Z. Naturforsch. **14b** [1959] 819/20).

[6] G. Dolcetti, O. Gandolfi, M. Ghedini, N. W. Hoffman (Inorg. Syn. **16** [1976] 32/5). – [7] V. I. Nefedov, N. M. Sinitsyn, Ya. V. Salyn', L. Baier (Koord. Khim. **1** [1975] 1618/24; Soviet J. Coord. Chem. **1** [1975] 1332/7). – [8] W. Hieber, K. Heinicke (Z. Anorg. Allgem. Chem. **316** [1962] 321/6).

2.1.1.3 Nitrosyl-dithiocarbamato Complexes, $[Rh(NO)(S_2CNR_2)_3]X$

$[Rh(NO)(S_2CNMe_2)_3]X$. The **tetrafluoroborate** and **hexafluorophosphate** are prepared by treating $[Rh(NO)(NCMe)_4]X_2$ with NaS_2CNMe_2 in dichloromethane. The complex is isolated as brown crystals, ν_{NO} = 1545 cm^{-1} (X = BF_4). It is believed that this complex contains a bent Rh-N-O bond and may be either six-coordinate (one monodentate S_2CNMe_2) or seven-coordinate.

N. G. Connelly, P. T. Draggett, M. Green, T. A. Kuc (J. Chem. Soc. Dalton Trans. **1977** 70/3).

2.1.1.4 Nitrosyl-porphyrin Complexes

Rh(NO)(OEP) (OEP = octaethylporphyrin). This complex was obtained by reaction of $[Rh(OEP)]_2$, Rh(OEP)H or Rh(OEP)Cl with NO in toluene. It is diamagnetic and exhibited ν(NO) at 1630 cm^{-1} (Nujol mull). The electronic spectrum (400 to 700 nm) was shown in the paper. It underwent two one-electron oxidations voltammetrically in CH_2Cl_2, forming a paramagnetic mono- and diamagnetic di-cation, with $E_{1/2}$ = 0.77 and 1.27 V, respectively, versus the standard calomel electrode. The compound reacted with an excess of NO giving the unstable $Rh(NO)_2(OEP)$ which is paramagnetic (ESR and electronic spectrum 400 to 700 nm shown in paper).

$[Rh(NO)(OEP)Cl]_n$ is formed in toluene solution as an intermediate in the reaction between Rh(OEP)Cl and NO. It is paramagnetic and its electronic (400 to 700 nm) and ESR spectrum is shown in the paper. There is ESR spectroscopic evidence that this species is associated in solution.

Rh(NO)TPP (TPP = tetraphenylporphyrinate) is prepared by reaction of Rh(TPP)Cl with NO in toluene. The compound is diamagnetic but underwent two one-electron oxidations voltammetrically in dichloromethane, with $E_{1/2} = 0.94$ and 1.36 V versus the standard calomel electrode. Rh(NO)(TPP) reacted with NO giving the unstable paramagnetic adduct $Rh(NO)_2(TPP)$.

Rh(NO)(TPP)Cl was formed during the reaction between Rh(TPP)Cl and NO, but does not associate in solution. It is paramagnetic.

B. B. Wayland, A. R. Newman (Inorg. Chem. **20** [1981] 3093/7).

2.1.1.5 Nitrosyl-phosphine Complexes

2.1.1.5.1 Complexes with PF_3

$Rh(NO)(PF_3)_3$. This compound is prepared by reacting $[Rh(PF_3)_4]^-$ with NO_2^- and CO_2 in aqueous solution [1], by reducing $[Rh(PF_3)_2Cl]_2$ with Cu, PF_3 and NO at room temperature [2], or by treatment of $Rh(\pi\text{-}C_3H_5)(PF_3)_3$ with NO or NOCl [3]. It is a dark yellow volatile liquid.

Its structure has been determined by electron diffraction. The molecule has pseudotetrahedral geometry with Rh-N = 1.858(18), Rh-P = 2.245(5) and N-O = 1.149(19) Å; the P-Rh-P bond angle is 110.4(5)° and the Rh-N-O bond is linear [2]. The ^{19}F NMR spectrum of the compound has been analysed [4].

References:

[1] T. Kruck (Angew. Chem. Intern. Ed. Engl. **6** [1969] 53/67). – [2] D. M. Bridges, D. W. H. Rankin, D. A. Clement, J. F. Nixon (Acta Cryst. B **28** [1972] 1130/6). – [3] J. F. Nixon, B. Wilkins (J. Organometal. Chem. **80** [1974] 129/37, 134). – [4] J. F. Nixon (J. Fluorine Chem. **3** [1973] 179/85).

2.1.1.5.2 Cationic Nitrosyl Complexes, $[Rh(NO)_2(PPh_3)_2]^{n+}$

Bu^n = n-butyl, Et = ethyl, Ph = phenyl

$[Rh(NO)_2(PPh_3)_2](PF_6)$. The **hexafluorophosphate** ($X = PF_6$) is prepared by passing NO through a dichloromethane solution of $[Rh(C_8H_{12})(PPh_3)_2](PF_6)$ (C_8H_{12} = cycloocta-1,5-diene) at atmospheric pressure. The related $[Rh(^{15}NO)_2(PPh_3)_2](PF_6)$ is obtained using ^{15}NO. The compound forms black crystals [1], and the infrared spectrum contains two ν_{NO} bands (1978, 1714 cm^{-1} in Nujol; 1742 cm^{-1} in CH_2Cl_2). The **tetrafluoroborate** ($X = BF_4$) is formed by treatment of $Rh(CO)(PPh_3)_2Cl$ with $NOBF_4$ in ethanol, and forms black crystals [2]. The infrared spectrum contains two bands due to $\nu_{NO} = 1765$, 1716 cm^{-1} (in Nujol). The **perchlorate** ($X = ClO_4$) is obtained either by reaction of $Rh(NO)(PPh_3)_3$ with t-butyl nitrite and $HClO_4$ in benzene [3] or, in low yield, by treatment of $Rh(NO)(PPh_3)_3$ with $Ag(ClO_4)$ [4].

The related complex $[Rh(NO)_2(PMePh_2)_2]^+$ has been reported briefly, but no synthetic details were given [1].

The structure of $[Rh(NO)_2(PPh_3)_2](ClO_4)$ has been determined crystallographically [3]. The compound forms monoclinic crystals, space group C2/c-C_{2h}^6 (Z = 4), a = 17.134(4), b = 12.327(3), c = 17.166(4) Å, β = 108.17(2)°. The coordination geometry of the cation is best described as intermediate between tetrahedral and square planar, N-Rh-N = 157.5(3)°, P-Rh-P = 115.88(5)°, dihedral angle between N-Rh-N and P-Rh-P planes 86.0(1)°. The Rh-N and Rh-P distances are 1.818(4) and 2.355(1) Å, respectively, and the Rh-N-O bond angles are 158.9(4)°. X-ray photoelectron spectroscopic studies of $[Rh(NO)_2(PPh_3)_2][PF_6]$ gave a N1s binding energy of 401.1 eV [5].

The cation $[Rh(NO)_2(PPh_3)_2]^+$ reacts generally with CO giving $[Rh(CO)_3(PPh_3)_2]^+$ and a mixture of CO_2 and N_2O [1, 3, 6]. The mechanism of this reaction has been investigated spectroscopically (IR, NMR) using mixtures of $[Rh(^{14}NO)_2(PPh_3)_2]^+$ and $[Rh(^{15}NO)_2(PPh_3)_2]^+$ [7, 8]. Reaction of $[Rh(NO)_2(PPh_3)_2]^+$ with $L^1 = Ph_2PCH_2CH_2PPh_2$, PBu^n_3 or PEt_3 was rapid, giving $[RhL^1_4]^+$, PPh_3, Ph_3PO and N_2O. ^{31}P NMR spectral studies of the dinitrosyl cation indicated that there was little or no dissociation of PPh_3 in solution [1]. Reaction of $[Rh(NO)_2(PPh_3)_2](BF_4)$ with NaN_3 gave $Rh(NO)_2(PPh_3)_2N_3$ [2], and of $[Rh(NO)_2(PMePh_2)_2]^+$ with NO, $[Rh(NO)(PMePh_2)_2(NO_2)]^+$ [7].

References:

[1] S. Bhaduri, K. Grundy, B. F. G. Johnson (J. Chem. Soc. Dalton Trans. **1977** 2085/91). – [2] W. Beck, K. v. Werner (Chem. Ber. **106** [1977] 868/73). – [3] J. A. Kaduk, J. A. Ibers (Inorg. Chem. **14** [1975] 3070/3). – [4] G. Dolcetti, N. W. Hoffman, J. P. Collman (Inorg. Chim. Acta **6** [1972] 531/42). – [5] P. Finn, W. L. Jolly (Inorg. Chem. **11** [1972] 893/5).

[6] B. L. Haymore, J. A. Ibers (J. Am. Chem. Soc. **96** [1974] 3325/7). – [7] S. Bhaduri, B. F. G. Johnson, C. J. Savory, J. A. Segal, R. H. Walter (Chem. Commun. **1974** 809/10). – [8] S. Bhaduri, B. F. G. Johnson (Transition Metal Chem. [Weinheim] **3** [1978] 156/9).

2.1.1.5.3 Complexes $Rh(NO)(PR_3)_3$

Ph = phenyl; MNTS = N-methyl-N-nitroso-p-tolylsulphonamide, p-$MeC_6H_4SO_2N(NO)Me$; $DMGH_2$ = dimethylglyoxime, THF = tetrahydrofuran, Me = methyl

$Rh(NO)(PPh_3)_3$ can be prepared by a variety of methods. The highest yield syntheses involve treatment of $RhCl_3 \cdot nH_2O$ with PPh_3, NO and zinc dust in THF [1, 2], reduction of either $[Rh(NO)_2Cl]_n$ [1, 3, 16] or $Rh(NO)(PPh_3)_2Cl_2$ [1] with sodium amalgam in the presence of PPh_3, reduction of $RhCl_3 \cdot nH_2O$ by $NaBH_4$ in the presence of PPh_3 [1] (which gives the intermediate $Rh(PPh_3)_4H$), and treatment of this intermediate with MNTS, or reaction of $RhCl_3 \cdot nH_2O$ with PPh_3 and MNTS in boiling ethanol with, or without, KOH [4, 5]. The complex can also be obtained by reaction of $Rh(PPh_3)_3Cl$ with NO in benzene followed by reduction of the product, possibly $Rh(NO)(PPh_3)_2Cl(NO_2)$, in the presence of PPh_3 by sodium amalgam in THF; reaction of $Rh(NO)(PPh_3)_2Cl_2$ with H_2 and K_2CO_3 in THF, also in the presence of PPh_3 [1]; by reaction of $Rh(PPh_3)_3Cl$ or $Rh(CO)(PPh_3)_3H$ with $Co(NO)(DMGH)_2(MeOH)$ [6], and by nitrosylation of $Rh(CO)(PPh_3)_3H$ with MNTS [6]. The complex forms deep red crystals, melting point 205 to 206°C under N_2 [5], and exhibits one NO stretching frequency at 1610 cm^{-1} [1].

A single crystal X-ray structural analysis shows that crystals of $Rh(NO)(PPh_3)_3$ are trigonal; space group P3-C_3^1 with Z = 3; a = 19.057(4), c = 10.799(2) Å [7]. The molecule has pseudo-tetrahedral geometry with Rh-N = 1.759(13) and Rh-P = 2.350(3) Å; N-O = 1.27(2) Å. The P-Rh-P, N-Rh-P and Rh-N-O bond angles are 102.2(6)°, 116.0(5)° and 156.7(26)°, respectively.

^{31}P NMR spectral studies of $Rh(NO)(PPh_3)_3$ indicated that there is slow phosphine exchange, via a dissociative process, with free PPh_3 in solution [8]. The heat of dissociation of PPh_3 from $Rh(NO)(PPh_3)_3$, determined spectrophotometrically in benzene, is 33.6 kJ/mol [9]. The binding energy of the N 1s electron is 400.8 eV [10], and the X-ray photoelectron spectroscopic data have been interpreted in terms of delocalisation of negative change into the NO ligand [11].

$Rh(NO)(PPh_3)_3$ reacts with CO giving $Rh(NO)(CO)(PPh_3)_2$, and with dry HCl in benzene affording $Rh(NO)(PPh_3)_2Cl_2$ [1]. With an excess of dry HCl in dichloromethane, $Rh(NH_2OH)(PPh_3)_2Cl_3$ is formed [1], and not a mixture of $Rh(NHO)(PPh_3)_2Cl_3$ and $Rh(NOH)(PPh_3)_2Cl_3$ as previously suggested by [12]. With PhCOCl, the nitrosyl complex gives $Rh(CO)(PPh_3)_2Cl$, PPh_3 and, presumably, PhNO, and with p-$MeC_6H_4SO_2Cl$, $Rh(NO)(PPh_3)_2Cl_2$ and [p-$MeC_6H_4SO_2]_2$ is formed. With HgI_2 or I_2 itself, $Rh(NO)(PPh_3)_2I_2$ is produced, and with $NOPF_6$, $[Rh(NO)_2(PPh_3)_2](PF_6)$ is formed [1]. The nitrosyl complex engages in NO transfer reactions with triphenylphosphine chloro complexes of Ni, Co and Fe, usually affording an appropriate metal nitrosyl and $Rh(PPh_3)_3Cl$ [13]. Reaction of $Rh(NO)(PPh_3)_3$ with $[Ph_3C][ClO_4]$ in acetone gives $[Rh(PPh_3)_3(OCMe_2)](ClO_4)$ [1].

$Rh(NO)(PPh_3)_3$ catalyses both the hydrogenation of terminal and cyclic olefines in dichloromethane, and the formation of aromatic ketones by reaction of benzyl alcohols with conjugated dienes [1, 12, 14, 15].

Complexes with R = p-C_6H_4X, X = F, Cl, Me; PR_3 = $PMePh_2$. These are summarised in the following table. They are obtained by reacting $[Rh(NO)_2Cl]_n$ with sodium amalgam and $PMePh_2$ in THF [1], or by reaction of $RhCl_3 \cdot nH_2O$ in boiling ethanol with MNTS and an excess of the appropriate triphenylphosphines [5].

Complex	Colour	Melting Point	ν_{NO} in cm^{-1}	Ref.
$Rh(NO)(PPh_3)_3$	red	205 to 206°C (under N_2)	1610	[4]
			1612	[1]
$Rh(NO)(PMePh_2)_3$	red-orange		1570	[1]
$Rh(NO)\{P(p\text{-}MeC_6H_4)_3\}_3$	red	208 to 209°C (under N_2)	1600	[1]
		140 to 143°C (in air)	1602	[5]
$Rh(NO)\{P(p\text{-}FC_6H_4)_3\}_3$	red		1630	[1]
$Rh(NO)\{P(p\text{-}ClC_6H_4)_3\}_3$	red-brown	100 to 106°C (in air)	1598	[5]

References:

[1] G. Dolcetti, N. W. Hoffman, J. P. Collman (Inorg. Chim. Acta **6** [1972] 531/42). – [2] G. Dolcetti, O. Gandolfi, M. Gherdini, N. W. Hoffman (Inorg. Syn. **16** [1976] 32/5). – [3] W. Hieber, K. Heinicke (Z. Naturforsch. **16b** [1961] 554/5). – [4] N. Ahmad, J. J. Levison, S. D. Robinson, M. F. Uttley (Inorg. Syn. **15** [1974] 45/64). – [5] N. Ahmad, S. D. Robinson, M. F. Uttley (J. Chem. Soc. Dalton Trans. **1972** 843/7).

[6] C. B. Ungermann, K. G. Caulton (J. Am. Chem. Soc. **98** [1976] 3862/8). – [7] J. A. Kaduk, J. A. Ibers (Israel J. Chem. **15** [1977] 143/8). – [8] K. G. Caulton (Inorg. Chem. **13** [1974] 1774/6). – [9] G. Reichenback, S. Santini, G. Dolcetti (J. Inorg. Nucl. Chem. **38** [1976] 1572/3). – [10] P. Finn, W. L. Jolly (Inorg. Chem. **11** [1972] 893/5).

[11] V. I. Nefedov, N. M. Sinitsyn, Ya. V. Salyn', L. Baier (Koord. Khim. **1** [1975] 1618/24; Soviet J. Coord. Chem. **1** [1975] 1332/7). – [12] J. P. Collman, N. W. Hoffman, D. E. Morris (J. Am. Chem. Soc. **91** [1969] 5659/60). – [13] A. Sacco, G. Vasapollo, P. Gianoccaro (Inorg. Chim. Acta **32** [1979] 171/4). – [14] W. Strohmeier, R. Enders (Z. Naturforsch. **27b** [1972] 1415/6). – [15] K. Saeki, Mitsui Petrochemical Industries Ltd. (Japan. Kokai 74-20139 [1974]; C.A. **81** [1974] No. 25374).

[16] W. Hieber, K. Heinicke (Z. Anorg. Allgem. Chem. **316** [1962] 321/6).

2.1.1.5.4 Nitrosyl-phosphine-halide Complexes, $Rh(NO)(PR_3)_2X_n$

Monohalide Complexes

$Rh(NO)(PPh_3)_2X$ (X = Cl, Br, I) can apparently be formed by passing NO through chloroform solutions containing $Rh(PPh_3)_3X$, or by reaction of NO with $Rh(CO)(PPh_3)_2X$ or $Rh(CO)(PPh_3)_2X_3$ [1]. They also seem to be formed when dimethylformamide solutions of $Rh(NO)_2(PPh_3)_2X$ are allowed to cool [2]. The compounds are obtained as brown crystals which decompose at >160°C, and their infrared spectra contain one NO absorption at 1628 (X = Cl) or 1626 cm^{-1} (X = Br). The compounds are oxidised to $Rh(NO)(PPh_3)_2X_3$ (X = Cl or Br) on treatment with Cl_2 or Br_2 in CCl_4 solution.

$[Rh(NO)_2(PPh_3)_2Cl](BF_4)$ is prepared by adding $NOBF_4$ in methanol to a solution of $Rh(PPh_3)_3Cl$ in a mixture of benzene and methanol containing an excess of PPh_3. The compound forms red-violet crystals, ν_{NO} = 1720 cm^{-1} (Nujol), and may contain a non-linear Rh-N-O bond [3].

$Rh(NO)_2(PPh_3)_2Cl$ is formed by reaction of $RhCl_3 \cdot 3H_2O$ with PPh_3 and NOCl in ethanol. The compound forms brown crystals, melting point 238 to 240°C and has a broad NO absorption in the range 1600 to 1650 cm^{-1}. It reacts with hydrazine hydrate or with $NaBH_4$ in the presence of PPh_3 giving $Rh(NO)(PPh_3)_3$ [4].

References:

[1] Yu. N. Kukushkin, M. M. Singh (Zh. Neorgan. Khim. **15** [1970] 2741/5; Russ. J. Inorg. Chem. **15** [1970] 1425/7). – [2] Yu. N. Kukushkin, M. M. Singh (Zh. Neorgan. Khim. **14** [1969] 3167/9; Russ. J. Inorg. Chem. **14** [1969] 1670/2). – [3] L. Busetto, A. Palazzi, R. Ros, M. Graziani (Gazz. Chim. Ital. **100** [1970] 849/50). – [4] K. K. Pandey, U. C. Agarwala (J. Inorg. Nucl. Chem. **42** [1980] 293/4).

Dihalide Complexes

$Rh(NO)(PR_3)_2X_2$ (R = Ph, p-XC_6H_4 where X = Me, MeO or Cl, Et, Pr^i, Bu^n; PR_3 = $PMePh_2$). These complexes are prepared in good yield by reaction of $RhX_3 \cdot nH_2O$ (X = Cl, Br) with PPh_3 and either MNTS or pentylnitrite in ethanol [1, 2]. The corresponding iodides are obtained using a mixture of $RhCl_3 \cdot nH_2O$ with LiI or KI. The compounds are listed in the following table [1].

NO stretching frequencies ν_{NO} and melting points M.P. of $Rh(NO)(PR_3)_2X_2$:

X	R	M.P. in °C	ν_{NO} in cm^{-1} (Nujol)
Cl	Ph	207 to 210	1630
	p-MeC_6H_4	179 to 182	1629
	p-$MeOC_6H_4$	173 to 182	1618, 1638 m
	p-ClC_6H_4	169 to 175	1671, 1644 m
	Et	120 to 122	1642
	Pr^i	175 to 185	1655
	Bu^n	73 to 75	1641
Br	Ph	227 to 230	1634
	p-$MeOC_6H_4$	180 to 188	1616, 1639
	p-ClC_6H_4	183 to 195	1669
	Et	120 to 122	1642
	Pr^i	188	1657
	Bu^n	64 to 68	1641

X	R	m.p. in °C	ν_{NO} in cm^{-1} (Nujol)
I	Ph	200 to 205	1627
	p-MeC_6H_4	198 to 204	1629
	p-ClC_6H_4	161 to 163	1641
	Et	92 to 95	1639
	Bu^n	67 to 69	1641
$Rh(NO)(PMePh_2)_2Cl_2$		145 to 155	1630
$Rh(NO)(PMePh_2)_2Br_2$		167 to 171	1630

All bands are strong unless otherwise indicated (m = medium).

$Rh(NO)(PPh_3)_2Cl_2$ and $Rh(NO)\{P(C_6H_{11})_3\}_2Cl_2$ could also be prepared by reaction of $[Rh(NO)_2Cl]_n$ with PPh_3 or $P(C_6H_{11})_3$ in THF [3], and $Rh(NO)(PPh_3)_2X_2$ (X = Cl or Br) can be made by addition of PPh_3 to $[Rh(NO)X_2]_n$ [4]. Reaction of $RhL(PPh_3)_2Cl$ (L = CO or PPh_3) with KNO_2 in the presence of acid (HCl, $HClO_4$, CH_3CO_2H) in chloroform/methanol mixtures [5], of $Rh(CO)(PPh_3)_2NO_2$ with HCl or KCl [6], or of either $Rh(PPh_3)_3Cl$ with $Co(NO)(DMGH)_2(MeOH)$ or $Rh(NO)(PPh_3)_3$ with $CoCl(DMGH)_2(PPh_3)$ ($DMGH_2$ = dimethylglyoxime) [7], also gives $Rh(NO)(PPh_3)_2Cl_2$. Treatment of $[Rh(NO)(NCMe)_4]X_2$ (X = BF_4 or PF_6) with an excess of $[N(PPh_3)_2]Cl$ or $[NMe_4]I$ affords $Rh(NO)(PPh_3)_2Y_2$ (Y = Cl or I) [8]. Reaction of $RhCl_3 \cdot 3H_2O$, PPh_3 and N_2O_3 in ethanol also affords $Rh(NO)(PPh_3)_2Cl_2$ [20].

The structure of $Rh(NO)(PPh_3)_2Cl_2$ has been determined crystallographically. The compound forms monoclinic crystals, space group C2/c-C^6_{2h}, Z = 4, a = 22.019(4), b = 9.604(2), c = 15.854(2) Å, β = 104.57(1)° [9]. The molecular geometry is that of a distorted square pyramid with a bent NO group in the apical position, see **Fig. 9**.

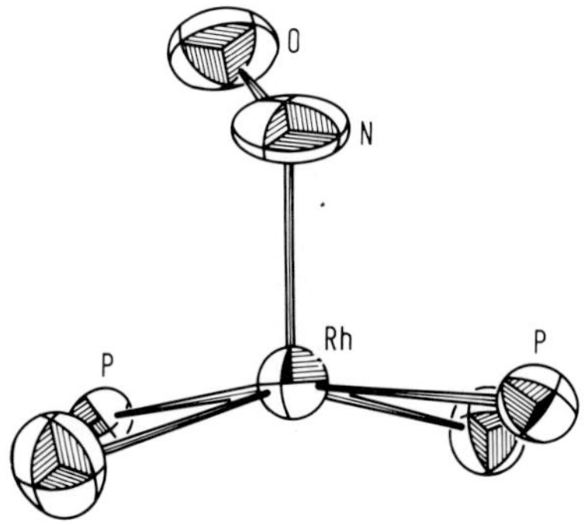

Fig. 9
Structure of
$Rh(NO)(PPh_3)_2Cl_2$.

The important bond distances are Rh-N = 1.912(10), Rh-Cl = 2.3439(14) and Rh-P = 2.3672(13) Å, and the angle Rh-N-O is 124.8(16)°. The direction of the bend in the Rh-N-O group is such that the planes defined by Rh, N and O and by Rh and the two PPh_3 groups form a dihedral angle of 10.3(29)° [9].

The infrared spectra of $Rh(NO)(PR_3)_2X_2$ exhibit one NO stretching frequency (see table on pp. 104/5). X-ray photoelectron spectral studies of $Rh(NO)(PR_3)_2X_2$ have been made: the N 1s binding energy for R = Ph, X = Cl is 401.9 [10, 11] or 401.5 eV [11] and for PR_3 = PMe_2Ph, X = I, 400.3 eV [12]. On the basis of X-ray photoelectron spectroscopy it has been suggested that there is significant transfer of charge from Rh to NO [13]. ^{31}P NMR spectral studies of $Rh(NO)(PPh_3)_2Cl_2$ suggest that there is very slow or no exchange between complexed and free PPh_3 in solution [14].

The nature of the bonding in $Rh(NO)(PPh_3)_2X_2$ (X = Cl or Br) has been probed using optical techniques and dipole moment data [15].

$Rh(NO)(PPh_3)_2Cl_2 \cdot NOBr$ is obtained when $RhCl_3 \cdot nH_2O$ reacts with PPh_3 and $NOBr_3$ in ethanol [16]. The compound forms brown crystals melting at 232°C, exhibits one ν_{NO} at 1630 cm^{-1} (Nujol) and reacts further with $NOBr_3$ in ethanol containing PPh_3 giving $[Rh(PPh_3)_2Br_2]N_2O_2 \cdot CH_2Cl_2$, after recrystallisation from dichloromethane [16].

Reaction of $Rh(NO)(PPh_3)_2Cl_2$ with CO affords $Rh(NO)(CO)(PPh_3)_2Cl_2$ whose infrared spectrum contains ν_{NO} at 1630 cm^{-1} and ν_{CO} at 2080 cm^{-1} (Nujol) [17]. With N_2O_4 or O_2 (under UV light), $Rh(NO)(PPh_3)_2Cl_2$ is converted into $Rh(PPh_3)_2Cl_2(NO_3)$ [18]. Reduction of the nitrosyl dichloride by $AlHBu^i_2$ in the presence of PPh_3 affords $Rh(PPh_3)_3Cl$ [19].

References:

[1] S. D. Robinson, M. F. Uttley (J. Chem. Soc. A **1971** 1254/7). – [2] N. Ahmad, J. J. Levison, S. D. Robinson, M. F. Uttley (Inorg. Syn. **15** [1974] 45/64, 60). – [3] W. Hieber, K. Heinicke (Z. Anorg. Allgem. Chem. **316** [1962] 321/6). – [4] G. R. Crooks, B. F. Johnson (J. Chem. Soc. Dalton Trans. **1970** 1662/5). – [5] Yu. N. Kukushkin, L. I. Danilina (Zh. Neorgan. Khim. **17** [1972] 3355/7; Russ. J. Inorg. Chem. **17** [1972] 1762/3).

[6] Yu. N. Kukushkin, L. I. Danilina (Zh. Neorgan. Khim. **19** [1974] 1349/52; Russ. J. Inorg. Chem. **19** [1974] 734/6). – [7] C. B. Ungermann, K. G. Caulton (J. Am. Chem. Soc. **98** [1976] 3862/8). – [8] N. G. Connelly, P. T. Draggett, M. Green, T. A. Kuc (J. Chem. Soc. Dalton Trans. **1977** 70/3). – [9] S. Z. Goldberg, C. Kubiak, C. D. Meyer, R. Eisenberg (Inorg. Chem. **14** [1975] 1650/4). – [10] D. T. Clark, I. S. Wooley, S. D. Robinson, K. R. Laing, J. N. Wingfield (Inorg. Chem. **16** [1977] 1201/6).

[11] C.-C. Cheng, J. W. Faller (J. Organometal. Chem. **84** [1975] 53/64). – [12] P. Finn, W. L. Jolly (Inorg. Chem. **11** [1972] 893/5). – [13] V. I. Nefedov, N. M. Sinitsyn, Ya. V. Salyn', L. Baier (Koord. Khim. **1** [1975] 1618/24; Soviet J. Coord. Chem. **1** [1975] 1332/7). – [14] K. G. Caulton (Inorg. Chem. **13** [1974] 1774/6). – [15] G. Avetikyan, Yu. N. Kukushkin, K. A. Khokhryakov, L. I. Danilina, E. F. Strizhev (Izv. Vysshikh Uchebn. Zavedenii Khim. Khim. Tekhnol. **21** No. 1 [1978] 7/10; C.A. **88** [1978] No. 181380).

[16] K. K. Pandey, S. Datta, U. C. Agarwala (Transition Metal Chem. [Weinheim] **4** [1979] 337/9). – [17] J. P. Collman, P. Farnham, G. Dolcetti (J. Am. Chem. Soc. **93** [1971] 1788/90). – [18] M. Kubota, C. A. Keorntgen, G. W. McDonald (Inorg. Chim. Acta **30** [1978] 119/26). – [19] G. Dolcetti, N. W. Hoffman, J. P. Collman (Inorg. Chim. Acta **6** [1972] 531/42). – [20] K. K. Pandey, U. C. Agarwala (Z. Anorg. Allgem. Chem. **457** [1979] 235/7).

Trihalide Complexes, $Rh(NO)(PPh_3)_2X_3$

$Rh(NO)(PPh_3)_2Cl_3$. This complex is thought to be formed, together with $Rh(NO)(PPh_3)_2Cl_2$, when $Rh(NO)Cl_3$ in ethanol is treated with PPh_3 [1]. The orange product is paramagnetic, the electron spin resonance spectrum showing $g_{\parallel} = 2.03$, $g_{\perp} = 2.01$. The infrared spectrum shows two ν_{NO} at 1630 cm^{-1} (probably due to $Rh(NO)(PPh_3)_2Cl_2$), and a weak band at 1660 cm^{-1}. It is claimed that $Rh(NO)(PPh_3)_2X_3$ is formed when a mixture of $Rh(PPh_3)_3Cl$ and NO is treated with X_2 (X = Cl, Br or I), and when $Rh(CO)(PPh_3)_2X_3$ reacts with NO [2]. These brown species exhibit one NO stretching frequency: 1627 (X = Cl), 1628 (X = Br), 1623 (X = I), 1625 cm^{-1} ($X_3 = I_2Cl$). These values of ν_{NO} are very similar to those obtained from $Rh(NO)(PPh_3)_2X_2$. When $Rh(PPh_3)_3Cl$ is treated with NOCl, a mixture of $Rh(NO)(PPh_3)_2Cl_3$ and $Rh(NO)(PPh_3)_2Cl_2$ is apparently formed. Similar mixtures were formed using NOBr [3].

References:

[1] M. C. Baird (Inorg. Chim. Acta **5** [1977] 46/8). – [2] Yu. N. Kukushkin, M. M. Singh (Zh. Neorgan. Khim. **15** [1970] 2741/5; Russ. J. Inorg. Chem. **15** [1970] 1425/7). – [3] K. K. Pandey, U. C. Agarwala (Indian J. Chem. A **19** [1980] 805/7).

2.1.1.5.5 Miscellaneous Phosphine Complexes

$[Rh(NO)(PPh_3)_2(NCMe)_n]X_2$. The **tetrafluoroborate** and **hexafluorophosphate** (n = 2) are prepared either by reaction of $[Rh(NO)(NCMe)_4]X_2$ with PPh_3 in dichloromethane [1], or by treatment of $Rh(NO)(PPh_3)_2I_2$ with $AgPF_6$ in acetonitrile [2]. The salts are olive-green, and exhibit one NO stretching frequency: X = BF_4, 1734 (Nujol) or 1703 cm^{-1} (CH_2Cl_2) [1]; X = PF_6, 1720 (Nujol) [1] or 1730 cm^{-1} (KBr) [2]. The ^{15}N labelled species $[Rh(^{15}NO)(PPh_3)_2(NCMe)_n](BF_4)$ has also been prepared.

The number of acetonitrile molecules in the cation is in some doubt because an X-ray structure determination of the **hexafluorophosphate** reveals it to be $[Rh(NO)(PPh_3)_2(NCMe)_3](PF_6)_2$, see **Fig. 10** [3]. The compound crystallises within the monoclinic space group $P2_1/n-C^5_{2h}$ (Z = 4), a = 14.053(8), b = 27.512(15), c = 11.914(8) Å, β = 97.29(5)°. The molecule has pseudo-octahedral geometry, with Rh-N(O) = 2.026(8) Å, while Rh-N(CMe) distances are 2.030(7), 2.104(7) and 2.308(8) Å (*trans* to NO). The Rh-P bond length is 2.405(7) Å and the Rh-N-O angle 118.4(6)°.

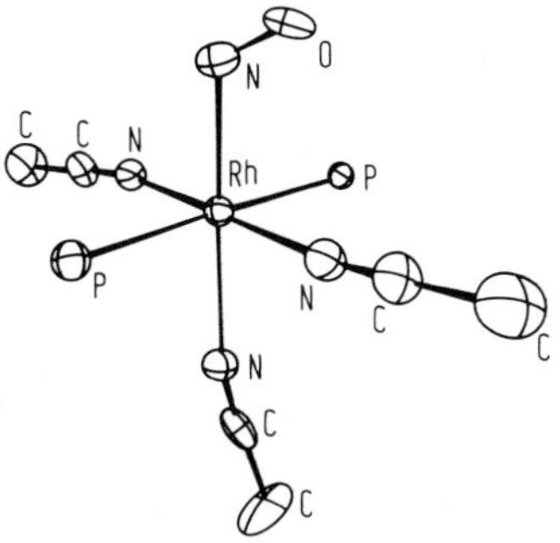

Fig. 10
Structure of
$[Rh(NO)(PPh_3)_2(NCMe)_3](PF_6)_2$.

Treatment of $[Rh(NO)(PPh_3)_2(NCMe)_n]X_2$ with NaS_2CNR_2 (R = Me or Et) gives $[Rh(PPh_3)_2(S_2CNR_2)_2]X$ [1] and not $[Rh(NO)(PPh_3)_2(S_2CNMe_2)](PF_6)$ as suggested earlier [4]. Reduction of $[Rh(NO)(PPh_3)_2(NCMe)_n]X_2$ with hydrazine in the presence of PPh_3 gives $Rh(NO)(PPh_3)_3$ [1].

Addition to $[Rh(NO)(PPh_3)_2(NCMe)_2](PF_6)_2$ of bipyridyl and a series of o-phenanthrolines (L-L) in ethanol affords the species $[Rh(NO)(PPh_3)_2(L\text{-}L)](PF_6)_2$ [2, 5]. These are light brown, and their infrared spectra exhibit one NO stretching frequency (see table).

NO stretching frequencies in $[Rh(NO)(PPh_3)_2(L\text{-}L)]X_2$ [5]:

Ligand L-L	ν_{NO} in cm^{-1} (Nujol)
bipyridyl	1820
4,4′-dimethylbipyridyl	1797
o-phenanthroline	1830
2,9-dimethylphenanthroline	1829
3,4,7,8-tetramethylphenanthroline	1814
5,6-dimethylphenanthroline	1810

The ^{1}H and ^{31}P NMR spectra of the complexes have been obtained and it is thought that the compounds may have the structures shown in **Fig. 11**. The occurrence of H_2O in these species was suggested on the basis of IR spectral measurements [5].

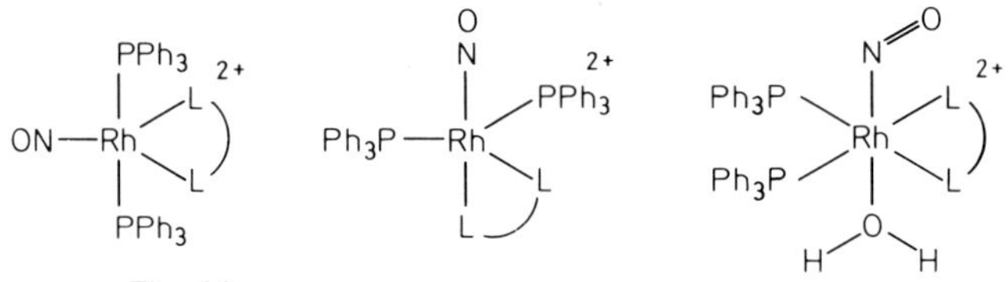

Fig. 11
Possible structures of $[Rh(NO)(PPh_3)_2(L\text{-}L)](PF_6)_2$.

$[Rh(NO)(PPh_3)_2(NCMe)Cl](PF_6)$ has been made by reaction of $[Rh(NO)(NCMe)_3Cl](PF_6)$ with PPh_3 or by addition of the stoichiometric amount of $[N(PPh_3)_2]Cl$ to $[Rh(NO)(PPh_3)_2(NCMe)_n](PF_6)_2$. It is a yellow solid; $\nu_{NO} = 1698\ cm^{-1}$ ($CHCl_3$) [1].

References:

[1] N. G. Connelly, P. T. Draggett, M. Green, T. A. Kuc (J. Chem. Soc. Dalton Trans. **1977** 70/3). – [2] M. Ghedini, G. Dolcetti, O. Gandolfi, B. Giovannitti (Inorg. Chem. **15** [1976] 2385/8). – [3] B. A. Kelly, A. J. Welch, P. Woodward (J. Chem. Soc. Dalton Trans. **1977** 2237/42). – [4] N. G. Connelly, M. Green, T. A. Kuc (Chem. Commun. **1974** 542). – [5] M. Ghedini, G. Dolcetti, G. Denti (Transition Metal Chem. [Weinheim] **3** [1978] 177/81).

$Rh(NO)(SO_2)(PPh_3)_2$. This complex is prepared by reaction of $Rh(NO)(PPh_3)_3$ with an excess of SO_2 in benzene. The infrared spectrum shows ν_{NO} at 1600 cm^{-1} and ν_{SO_2} at 1138 and 948 cm^{-1}. The structure has been determined crystallographically, see **Fig. 12**, and the molecule crystallises with the orthorhombic space group, Pbca-D_{2h}^{15}, Z = 8, a = 10.338(2), b = 18.500(4), c = 33.933(7) Å [2].

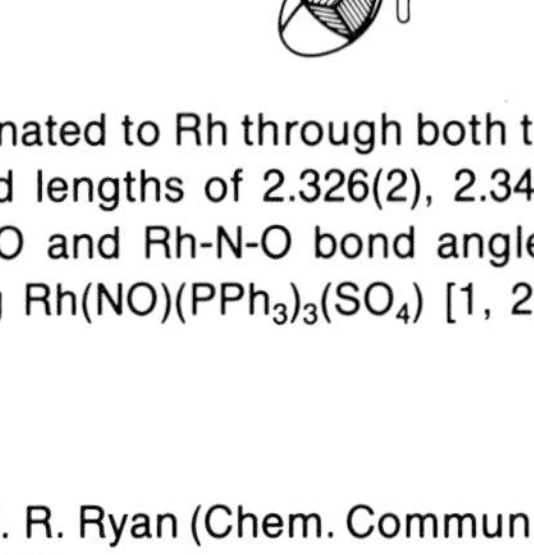

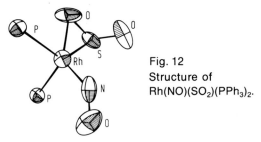

Fig. 12
Structure of
$Rh(NO)(SO_2)(PPh_3)_2$.

The SO_2 is coordinated to Rh through both the S and one of the O atoms of SO_2, with Rh–S, Rh–O, and S=O bond lengths of 2.326(2), 2.342(5) and 1.493(5) Å, respectively, while S-O is 1.430(5) Å. The O-S-O and Rh-N-O bond angles are 115.1(4)° and 140.4(6)°. The compound reacts with O_2 giving $Rh(NO)(PPh_3)_3(SO_4)$ [1, 2].

References:

[1] D. C. Moody, R. R. Ryan (Chem. Commun. **1976** 503). – [2] D. C. Moody, R. R. Ryan (Inorg. Chem. **16** [1977] 2473/8).

$Rh(NO)(PPh_3)_2XY$ Complexes

$Rh(NO)(PPh_3)_2Cl(NO_2)$. Reaction of $Rh(CO)(PPh_3)_2Cl$ or $Rh(PPh_3)_3Cl$ with NO in benzene, chlorobenzene or chloroform gives this complex. It is also formed when $Rh(CO)(PPh_3)_2NO_2$ is treated with NOCl [1, 2, 3]. The compound is green, melting point 154 to 156°C with decomposition, whose infrared spectrum exhibits ν_{NO} at 1660(s) and 1640(w) cm^{-1} [1], 1660 cm^{-1} [2] or 1659(vs) and 1632(w) cm^{-1} (Nujol). The corresponding bromo complex $Rh(NO)(PPh_3)_2Br(NO_2)$ has also been prepared. Its infrared spectrum shows ν_{NO} at 1653(vs) and 1625(w) cm^{-1} [4]. $Rh(NO)(PPh_3)_2Cl(NO_2)\cdot nCH_2Cl_2$ is obtained by reaction of $Rh(CO)(PPh_3)_2Cl$ (n = 1) or $Rh(PPh_3)_3Cl$ (n = 0.8) with N_2O_3 in dichloromethane. It is obtained as a red-yellow solid, melting point 197 to 200°C [10].

X-ray photoelectron spectroscopic studies have been made of $Rh(NO)(PPh_3)_2Cl(NO_2)$, with particular reference to the C binding energies of coordinated PPh_3 [5].

Treatment of $Rh(NO)(PPh_3)_2Cl(NO_2)$ with CO and with NO gives $Rh(CO)(PPh_3)_2Cl$ [2] and $Rh(PPh_3)_2Cl(NO_2)_2$ [4], respectively.

$Rh(CO)(NO)(PPh_3)_2XCl$. The dichloro complex (X = Cl) is prepared by reaction of $Rh(CO)(PPh_3)_2Cl$ with $NaNO_2$ in chloroform in the presence of HCl [6]. The related species with X = Br or I are made similarly from $Rh(CO)(PPh_3)_2X$. All are isolated as brown crystals, and their infrared spectra reveal ν_{NO} at 1630 cm^{-1} (all X) and ν_{CO} at 2110 cm^{-1} (X = Cl), 2100 cm^{-1} (X = Br) and 2085 cm^{-1} (X = I).

$[Rh(NO)(PPh_3)_2(CO_2R)(ROH)](BF_4)$. Reaction of $Rh(CO)(PPh_3)_2NCO$ with $NOBF_4$ at 5°C in methanol or ethanol affords $[Rh(NO)(CO)(PPh_3)_2(NCO)](BF_4)$ which reacts further with the solvent alcohol giving $[Rh(NO)(PPh_3)_2(CO_2R)(ROH)](BF_4)$ (R = Me or Et). This salt can also be obtained directly from $Rh(CO)(PPh_3)_2N_3$ by the same method. If the reaction is carried out at 40°C, then only $[Rh(NO)_2(PPh_3)_2](BF_4)$ is formed [7].

$[Rh(NO)(PPh_3)_2(CO_2R)(ROH)](BF_4)$ is dark green, melting point 170 to 172°C with decomposition (R = Me) and 175°C with decomposition (R = Et). In their infrared spectra, ν_{NO} and ν_{CO} occur at 1718 and 1663 cm^{-1} (R = Me), and 1713 and 1658 cm^{-1} (R = Et) [7].

$[Rh(NO)(PPh_3)_2(CO_2Me)(OCMe_2)](BF_4)$. Reaction of $[Rh(NO)(PPh_3)_2(CO_2Me)(MeOH)](BF_4)$ with acetone affords this compound which is green, melting point 140 to 142°C with decomposition, and its infrared spectrum contains ν_{NO} at 1679 cm^{-1} and ν_{CO} at 1664 and 1650 cm^{-1}, respectively. Treatment of $[Rh(NO)(PPh_3)_2(CO_2Me)(MeOH)](BF_4)$ with NaN_3 gives **$Rh(NO)(PPh_3)_2(CO_2Me)N_3$** which is a yellow-brown compound, melting point 112 to 118°C with decomposition, having ν_{NO}, ν_{CO} and ν_{asN_3} at 1615, 1644 and 2040 cm^{-1}, respectively [7].

$[Rh(NO)(PPh_3)_2](NCS)_2$. This complex is prepared by reaction of $[Rh(NO)(PPh_3)_2](ClO_4)$ with KNCS in acetone or in chloroform containing a small amount of methanol and acetic acid [8, 9]. It is brown, and its infrared spectrum exhibits ν_{NO} at 1690 cm^{-1} and ν_{NCS} at 2080 cm^{-1}. The mode of bonding of the thiocyanato ligand is not established with certainty.

References:

[1] W. B. Hughes (Chem. Commun. **1969** 1126). – [2] J. Kiji, S. Yoshikawa, J. Furukawa (Bull. Chem. Soc. Japan **43** [1970] 2614/5). – [3] Yu. N. Kukushkin, L. I. Danilina, M. M. Singh (Zh. Neorgan. Khim. **16** [1971] 2718; Russ. J. Inorg. Chem. **16** [1971] 1449/50). – [4] E. Miki, K. Muzumachi, T. Ishimori (Bull. Chem. Soc. Japan **48** [1975] 2975/6). – [5] R. Larsson, B. Folkesson (Chem. Scr. **9** [1976] 148/50).

[6] A. Dowerah, M. M. Singh (Transition Metal Chem. [Weinheim] **2** [1974] 74/5). – [7] W. Beck, K. v. Werner (Chem. Ber. **106** [1973] 868/73). – [8] Yu. N. Kukushkin, L. I. Danilina (Zh.

Neorgan. Khim. **19** [1974] 1349/52; Russ. J. Inorg. Chem. **19** [1974] 734/6). – [9] Yu. N. Kukushkin, L. I. Danilina (Zh. Neorgan. Khim. **20** [1975] 2836/8; Russ. J. Inorg. Chem. **20** [1975] 1569/70). – [10] K. K. Pandey, U. C. Agarwala (Z. Anorg. Allgem. Chem. **457** [1979] 235/7).

Tetraazadiene and Related Complexes

$Rh(NO)(PPh_3)\{N_4(p\text{-}MeC_6H_4SO_2)_2\}$. This compound is obtained by reaction of $Rh(NO)(PPh_3)_3$ with p-toluenesulphonylazide. The complex is violet, melting point 148°C, and its infrared (ν_{NO} = 1890 cm^{-1}) and ^{1}H NMR spectra have been reported [1]. The compound reacts with HCl giving $Rh(NO)(PPh_3)_2Cl_2$, and with PPh_3 giving the unstable $Rh(NO)(PPh_3)_2(N_4R_2)$ (R = p-$MeC_6H_4SO_2$) which is deep brown, melting point 132°C, and has ν_{NO} at 1765 cm^{-1} (Nujol). The monophosphine complex also reacts with CO giving the more stable $Rh(NO)(CO)(PPh_3)(N_4R_2)$ which is light brown, melting point 155°C, and whose infrared spectrum (Nujol) contains ν_{NO} and ν_{CO} at 1700 and 2090 cm^{-1}. The monophosphine tetraazadiene complex also reacts with pyridine (py), apparently giving a mixture of $Rh(NO)(PPh_3)(py)(N_4R_2)$ and $Rh(NO)(py)_2(N_4R_2)$, but this could not be confirmed [1, 2].

From ^{1}H NMR studies, it seems possible that the N_4R_2 ring system may exist in the tautomeric forms

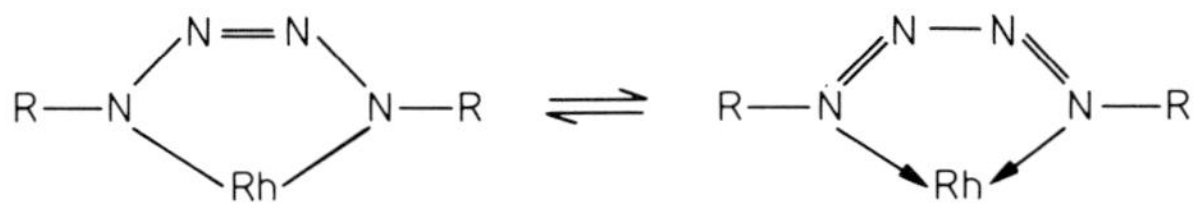

and that, in certain circumstances, the chelate ring can open, forming $Rh(NO)(PPh_3)(NR)(N_3R)$ [1, 2].

$Rh(NO)(PPh_3)_2(p\text{-}MeC_6H_4SO_2NCONSO_2C_6H_4Me)$. This compound is prepared by reacting $Rh(NO)(CO)(PPh_3)_2$ with p-toluenesulphonylazide in benzene, or $Rh(NO)(PPh_3)_3$ with p-MeC_6H_4-SO_2NCO [3]. It is yellow-orange, melting point 209 to 210°C, and its infrared spectrum (Nujol) contains bands at 1792 (ν_{NO}) and 1667 cm^{-1} (ν_{CO}). The complex reacts with HCl giving $Rh(NO)(PPh_3)_2Cl_2$ [3].

References:

[1] G. La Monica, P. Sandrini, F. Zingales, S. Cenini (J. Organometal. Chem. **50** [1973] 287/96). – [2] S. Cenini, P. Fantucci, M. Pizzotti, G. La Monica (Inorg. Chim. Acta **13** [1975] 243/5). – [3] W. Beck, W. Rieber, S. Cenini, F. Porta, G. La Monica (J. Chem. Soc. Dalton Trans. **1974** 298/304).

Carboxylato Complexes, $Rh(NO)(PPh_3)_2(OCOR)_2$

These complexes are formed by reaction of $Rh(NO)(PPh_3)_3$ with acetic acid or RCO_2H (R = CF_3, C_2F_5, C_6F_5) in acetone. The **acetate** is green, melting point 210 to 211°C, and its infrared spectrum exhibits ν_{NO} at 1614 cm^{-1}. The **trifluoromethyl-, pentafluoroethyl-** and **pentafluorophenyl-carboxylates** exhibit ν_{NO} at ~1665 cm^{-1}. These compounds are thought to be structurally analogous to $Rh(NO)(PPh_3)_2Cl_2$ [1, 2, 3].

References:

[1] A. Dobson, S. D. Robinson, M. F. Uttley (Inorg. Syn. **17** [1977] 124/32). – [2] S. D. Robinson, M. F. Uttley (Chem. Commun. **1972** 1047/8). – [3] A. Dobson, S. D. Robinson (J. Organometal. Chem. **99** [1975] C63/C64).

Complexes Derived from Quinones, $Rh(NO)(PPh_3)_2Q$ (Q = o- or p-quinone)

Reaction of $[Rh(NO)(PPh_3)_2(NCMe)_2](PF_6)_2$ with $C_6X_4O_2$ (X = Cl or Br) or of $Rh(NO)(PPh_3)_3$ with a series of quinones gives $Rh(NO)(PPh_3)_2Q$ (see following table) [1, 2].

Quinone complexes of rhodium nitrosyls, $Rh(NO)(PPh_3)_2Q$:

Quinone	Colour	Melting Point in °C	ν_{NO} in cm^{-1}	ν_{CO} in cm^{-1}	Ref.
1,2-$C_6Cl_4O_2$	brown	165	1632[a]		[1]
			1648[b]	1423	[2]
1,2-$C_6Br_4O_2$	brown		1655[a]		[1]
1,4-$C_6H_4O_2$	brown	205	1645[b]	1615 1582	[2]
1,4-naphthoquinone	red-brown	90d	1667[b]	1635 1597	[2]
1,4-$C_6Cl_4O_2$	brown	196	1690[b]	1665 1630	[2]
1,2-naphthoquinone	brown	115d	1623[b]	1433	[2]

[a] KBr, [b] Nujol, d = decomposition.

It is thought that those complexes containing o-quinones have a bent Rh-N-O bond and that the quinone is attached via both O atoms. In the p-quinone complexes, it seems probably that the ligand is attached via a C=C bond, and ^1H NMR and infrared spectral measurements have been made which seem to support these views.

References:

[1] M. Ghedini, G. Dolcetti, B. Giovannitti, G. Denti (Inorg. Chem. **16** [1977] 1725/9). – [2] G. La Monica, G. Navazio, P. Sandrini, S. Cenini (J. Organometal. Chem. **31** [1971] 89/101, 90).

Nitrosyl Nitrato, Nitrito, Sulphato and Azido Complexes

$Rh(NO)(PPh_3)_2(NO_3)_2$. This complex is obtained by reaction of $Rh(NO)(CO)(PPh_3)_2XCl$ (X = Cl, Br or I) with $AgNO_3$ in chloroform. It forms green crystals and exhibits ν_{NO} at 1655 cm^{-1} [1].

$[Rh(NO)(PPh_3)_2(NO_3)]_n$ is prepared either by treatment of $Rh(CO)(PPh_3)_2NO_2$ with HNO_3 or KNO_3 in chloroform containing some ethanol, or of $[Rh(NO)_2(PPh_3)_2](ClO_4)$ or $Rh(NO)(PPh_3)_3$ with HNO_3 [2, 3]. It is a green compound, whose infrared spectrum exhibits ν_{NO} at 1645 cm^{-1} and bands due to coordinated NO_3. The compound is formulated either with n = 1 or 2, but this has not been clarified. It reacts with HCl, HCl + NO, KNO_2 and PPh_3, respectively, giving $Rh(NO)(PPh_3)_2Cl_2$, $[Rh(NO)_2(PPh_3)_2](ClO_4)$, $Rh(NO)_2(PPh_3)_2NO_2$ and $Rh(NO)(PPh_3)_3$ [2].

$[Rh(NO)(PPh_3)_2(NO_2)]_n$. This complex is obtained by reaction of $Rh(PPh_3)_3Cl$ with KNO_2 in acetone or chloroform containing some $HClO_4$ and methyl acetate [4, 5]. It is a green solid whose infrared spectrum shows ν_{NO} at 1630 cm^{-1}. It is formulated with n = 1 or 2, but this has not been clarified. It reacts with HCl, $HClO_4$, HNO_3 and PPh_3, respectively, giving $Rh(NO)(PPh_3)_2Cl_2$, $[Rh(NO)_2(PPh_3)_2](ClO_4)$, $[Rh(NO)(PPh_3)_2(NO_3)]_n$ and $Rh(NO)(PPh_3)_2$ [2].

$Rh(NO)_2(PPh_3)_2NO_2$ can be prepared either by reaction of $[Rh(NO)(PPh_3)_2(NO_3)]_n$, $[Rh(NO)_2(PPh_3)_2](ClO_4)$ or $Rh(NO)(PPh_3)_3$ with KNO_2 in chloroform or by treatment of

$Rh(NO)(PPh_3)_2Cl$ with KNO_2 in acetone [2, 3]. It is a green complex whose infrared spectrum contains ν_{NO} at 1665 and 1645 cm^{-1}. It reacts with HCl, $HClO_4$, HNO_3 and PPh_3 giving $Rh(NO)(PPh_3)_2Cl_2$, $[Rh(NO)_2(PPh_3)_2](ClO_4)$, $[Rh(NO)(PPh_3)_2(NO_2)]_n$ and $Rh(NO)(PPh_3)_3$, respectively [2, 6]. It is reported that $Rh(CO)(PPh_3)_3H$ reacts with N_2O_3 in CH_2Cl_2 giving pale yellow $Rh(NO)_2(PPh_3)_2(NO_2) \cdot 0.5CH_2Cl_2$, with a melting point of 195 to 197°C and ν_{NO} = 1560 and 1530 cm^{-1} [10].

$Rh(NO)(PPh_3)_2(NO_2)_2$. Reaction of $Rh(CO)(PPh_3)_2(OCOCH_3)$ with N_2O_3 in CH_2Cl_2 affords this compound as a red-yellow solid, melting point 215°C, ν_{NO} = 1560 cm^{-1} [10].

$Rh(NO)(PPh_3)_2(SO_4)$ is obtained by reaction of $Rh(NO)(PPh_3)_3$ with an excess of SO_2 in the presence of O_2 in benzene [7, 8]. It is a green complex whose infrared spectrum contains ν_{NO} at 1665 cm^{-1} [7].

The structure of this complex has been determined crystallographically: monoclinic, space group $P2_1/c\text{-}C^5_{2h}$, Z = 4, a = 17.489(4), b = 10.307(2), c = 19.190(7) Å, β = 109.68(2)°; D_{cryst} = 1.54 g/cm^3. The complex is square pyramidal with an apical NO group and bidentate sulphate ligand. The Rh-O distances are 2.079(4) and 2.091(8) Å, the Rh-P distances 2.285(4) and 2.324(3) Å, while the Rh-N and N-O bond lengths are 1.91(1) and 1.11(1) Å, respectively. The Rh-N-O bond angle is 122° [8].

$Rh(NO)(PPh_3)_2N_3$. This complex is obtained by reaction of $[Rh(NO)(PPh_3)_2(CO_2R)(ROH)]$-$(BF_4)$ with NaN_3. It is a red-brown solid, melting point 100 to 105°C with decomposition, its infrared spectrum (Nujol) exhibiting ν_{NO} at 1620 and 1540 cm^{-1}, and ν_{asN_3} at 2036 cm^{-1} [9].

References:

[1] A. Dowerah, M. M. Singh (Transition Metal Chem. [Weinheim] **2** [1974] 74/5). – [2] Yu. N. Kukushkin, L. I. Danilina, A. I. Osokin (Koord. Khim. **4** [1978] 431/5). – [3] Yu. N. Kukushkin, L. I. Danilina (Zh. Neorgan. Khim. **19** [1974] 1349/52; Russ. J. Inorg. Chem. **19** [1974] 734/6). – [4] Yu. N. Kukushkin, L. I. Danilina (Zh. Neorgan. Khim. **17** [1972] 1182/4; Russ. J. Inorg. Chem. **17** [1972] 617/8). – [5] Yu. N. Kukushkin, L. I. Danilina (Zh. Neorgan. Khim. **17** [1972] 3355/7; Russ. J. Inorg. Chem. **17** [1972] 1762/3).

[6] Yu. N. Kukushkin, L. I. Danilina (Zh. Neorgan. Khim. **20** [1975] 2836/8; Russ. J. Inorg. Chem. **20** [1975] 1569/70). – [7] J. Valentine, D. Valentine, J. P. Collman (Inorg. Chem. **10** [1971] 219/25). – [8] B. C. Lucas, D. C. Moody, R. R. Ryan (Cryst. Struct. Commun. **6** [1977] 57/60). – [9] W. Beck, K. v. Werner (Chem. Ber. **106** [1973] 868/73). – [10] K. K. Pandey, U. C. Agarwala (Z. Anorg. Allgem. Chem. **457** [1979] 235/7).

2.1.1.5.6 Nitrosyl Complexes with Polydentate Phosphines

$Rh(NO)[PPh\{(CH_2)_3PPh_2\}_2]$. This is obtained by exchange between the polydentate phosphine and PPh_3 in $Rh(NO)(PPh_3)_3$ in warm or refluxing benzene. The compound is red-brown and exhibits, in its IR spectrum, ν_{NO} at 1610 cm^{-1}. The existence of similar complexes derived from the ligands $Ph_2P(CH_2)_nPPh_2$ (n = 2, 4), $Ph_2P(CH_2)_2PMe_2$, $Ph_2P(CH_2)_3PHPh$, Ph_2P-$(CH_2)_2PPh(neo\text{-}C_{10}H_{19})$, $Ph_2P(CH_2)_3PPh(CH_2)_3NMe_2$, $PhP(CH_2CH_2PPh_2)_2$, $Ph_2P(CH_2)_2PPh(CH_2)_3$-PPh_2, $PhP(CH_2CH_2PMe_2)_2$, $PhP\{CH_2CH_2P(C_6H_{11})_2\}_3$, $P(CH_2CH_2PPh_2)_3$ and $[Ph_2PCH_2CH_2$-$PPhCH_2]_2$ has been established by ^{31}P NMR spectroscopy, but the complexes have not been isolated [1].

Rh(NO)(PPh$_3$)[Ph$_2$P(CH$_2$)$_3$PPh$_2$] has been prepared by the exchange of phosphines described above, but no spectroscopic details were given [1].

[Rh(NO)(Ph$_2$PCH$_2$CH$_2$PPh$_2$)$_2$](BF$_4$)$_2$ is obtained by addition of Ph$_2$PCH$_2$CH$_2$PPh$_2$ to [Rh(NO)(NCMe)$_4$](BF$_4$)$_2$. It is a green solid and its infrared spectrum contains ν_{NO} at 1730 cm^{-1} (Nujol) or 1718 cm^{-1} (CH$_2$Cl$_2$ solution) [2].

[Rh(NO){PPh(CH$_2$CH$_2$CH$_2$PPh$_2$)$_2$}Cl]X is prepared by addition of NOX (X = BF$_4$ or PF$_6$) to Rh{PPh(CH$_2$CH$_2$CH$_2$PPh$_2$)$_2$}Cl in methanol/benzene mixtures [3]. The **tetrafluoroborate** was isolated as green crystals (with 1.5 C$_6$H$_6$ of crystallisation) and both salts exhibit ν_{NO} = 1692 (X = PF$_6$), 1699 cm^{-1} (X = BF$_4$, Nujol). The structure of the **hexafluorophosphate** has been determined crystallographically: space group Pnma-D_{2h}^{16} or P2$_1$ma-C_{2v}^{2}, Z = 4, a = 24.350(5), b = 15.196(2), c = 9.914(2) Å. The molecule has a five-coordinate pyramidal geometry, with NO occupying the unhindered apical site. The metal lies 0.12 Å above the P$_3$Cl basal plane, and the Rh-P bond lengths are 2.374(3) and 2.282 Å (trans to Cl), while Rh-Cl and Rh-N are 2.408(4) and 2.909(15) Å, respectively. The Rh-N-O bond angle is 131(1)° [3].

Rh(NO)(Ph$_2$PCH$_2$CH$_2$PPh$_2$)Cl(NO$_2$) is prepared by reaction of Rh(NO)(Ph$_2$PCH$_2$CH$_2$PPh$_2$)Cl with NO in benzene. It is yellow-green, and its infrared spectrum shows ν_{NO} at 1665 cm^{-1} (Nujol) [4].

References:

[1] T. J. Mazanec, K. D. Tau, D. W. Meek (Inorg. Chem. **19** [1980] 85/91). – [2] N. G. Connelly, P. T. Draggett, M. Green, T. A. Kuc (J. Chem. Soc. Dalton Trans. **1977** 70/3). – [3] T. E. Nappier, D. W. Meek, R. M. Kirchner, J. A. Ibers (J. Am. Chem. Soc. **95** [1973] 4194/210, 4198). – [4] J. Kiji, S. Yoshikawa, J. Furukawa (Bull. Chem. Soc. Japan **43** [1970] 2614/5).

2.1.1.6 Nitrosyl-phosphite Complexes

Rh(NO)[P(OPh)$_3$]$_2$Br$_2$. Although details of the preparation of this complex have not been given, it appears to be formed by reaction of [Rh(NO)Br$_2$]$_n$ with an excess of P(OPh)$_3$ in ethanol-acetone mixtures [1, 2]. It is formed as green crystals, whose IR spectrum exhibits ν_{NO} at 1750 cm^{-1} [2].

The structure of the compound has been determined crystallographically [2]. The complex forms monoclinic crystals, space group P2$_1$/c-C_{2h}^{5}, Z = 4, a = 14.78(2), b = 13.50(2), c = 20.35(2) Å, β = 109.6(2)°. The structure of this compound is very similar to that of Rh(NO)(PPh$_3$)$_2$Cl$_2$; important bond lengths are Rh-N = 2.04(4), Rh-Br = 2.54(1), Rh-P = 2.33(2) Å, and the Rh-N-O angle is 109(5)°.

References:

[1] G. R. Crooks, B. F. G. Johnson (J. Am. Chem. Soc. A **1970** 1662/5). – [2] R. B. English, L. R. Nassimbeni, R. J. Haines (Acta Cryst. B **32** [1976] 3299/301).

2.1.1.7 Nitrosyl-tertiary-arsine Complexes

Rh(NO)(AsPh$_3$)$_3$ is prepared, together with Rh(NO)(AsPh$_3$)$_2$Cl$_2$, by reaction of [Rh(NO)$_2$Cl$_2$]$_n$ with AsPh$_3$ in THF [1]. It forms a dark red oil which solidifies at −17°C and, when pure, the compound melts at 137°C with decomposition. The infrared spectrum exhibits ν_{NO} at ~1650 cm^{-1} [1].

$[Rh(NO)(AsPh_3)_2(NCMe)_2](BF_4)_2$ is formed by addition of $AsPh_3$ to $[Rh(NO)(NCMe)_4](BF_4)_2$ in dichloromethane [2]. It is an olive-green solid, whose infrared spectrum (Nujol) exhibits ν_{NO} at 1720 cm^{-1} and ν_{CN} at 2320 and 2360 cm^{-1} [2].

References:

[1] W. Hieber, K. Heinicke (Z. Anorg. Allgem. Chem. **316** [1962] 321/6). – [2] N. G. Connelly, P. T. Draggett, M. Green, T. A. Kuc (J. Chem. Soc. Dalton Trans. **1977** 70/3).

2.1.1.8 Nitrosyl-arsine-halide Complexes

$Rh(NO)(AsPh_3)_2Cl$. This complex is formed by reaction of $Rh(AsPh_3)_3Cl$ or $Rh(CO)(AsPh_3)_2Cl$ with NO in chloroform [1]. It is a dark brown solid; see table below for physical properties.

$Rh(NO)(AsPh_3)_2X_2$. The **dichloro** complex is obtained by reaction of $[Rh(NO)_2Cl]_n$ or $[Rh(NO)Cl_2]_n$ with $AsPh_3$ in tetrahydrofuran or chloroform, or by reaction of $RhCl_3 \cdot nH_2O$ in ethanol with $AsPh_3$ and $p\text{-}MeC_6H_4SO_2N(NO)Me$ in refluxing methoxyethanol [2, 3, 4], or NOCl in ethanol [7]. The corresponding **dibromo** complex is obtained from $RhBr_3$, and the **di-iodide** from a mixture of $RhCl_3$ and an excess of LiI or KI [4].

$Rh(NO)(AsPh_3)_2ClBr$ is obtained by reaction of $RhCl_3 \cdot 3H_2O$, $AsPh_3$ and NOCl in ethanol [7].

$Rh(NO)(AsPh_3)_3ClBr$ is obtained from the reaction between $RhCl_3 \cdot nH_2O$, $AsPh_3$ and $NOBr_3$ in ethanol [5]. It is isolated as a dark brown solid and, in refluxing ethanol, is converted into $[Rh(AsPh_3)_2Br]_n$ [5].

$Rh(NO)(AsPh_3)_2Cl_3$ is prepared by treatment of $Rh(NO)(AsPh_3)_2Cl$ in dimethylformamide (DMF) with Cl_2 in chloroform. It is a pale brown solid which reverts to the monochloride in boiling DMF [1].

$Rh(NO)(AsPh_3)_2(NO_3)_2$. This complex is obtained by reaction of $Rh(NO)(CO)(AsPh_3)_2XCl$ (X = Cl, Br or I) with $AgNO_3$ in chloroform, and is isolated as green crystals [6].

Rhodium nitrosyl-arsine complexes:

Complex	Colour	Melting Point in °C	ν_{NO} in cm^{-1}	Ref.
$Rh(NO)(AsPh_3)_2Cl$	dark brown		1628	[1]
$Rh(NO)(AsPh_3)_2Cl_2$	orange-brown	249 to 250	1630	[3]
		245 to 250	1628	[4]
$Rh(NO)(AsPh_3)_2Br_2$	red-brown	259 to 260	1629	[3]
		260	1630	[4]
$Rh(NO)(AsPh_3)_2I_2$	violet-brown	249 to 250	1626	[3]
		240 to 245	1621	[4]
$Rh(NO)(AsPh_3)_3ClBr$	dark brown	185 to 187	1600	[5]
$Rh(NO)(AsPh_3)_2Cl_3$	pale brown		1629	[1]
$Rh(NO)(AsPh_3)_2(NO_3)_2$	green		1655	[6]
$Rh(NO)(AsPh_3)_2ClBr$	brown	285 to 290		[7]

$Rh(NO)_2(AsPh_3)_2(NO_2)\cdot 1.5CH_2Cl_2$ is prepared by reaction of $Rh(CO)(AsPh_3)_2Cl$ with N_2O_3 in dichloromethane. It was isolated as a red-yellow solid, melting point 165 to 167°C, ν_{NO} = 1560 and 1530 cm^{-1} [8].

$Rh_2(NO)_3(AsPh_3)_4Cl_5$ is said to be formed when $RhCl_3\cdot 3H_2O$, $AsPh_3$ and NOCl react in ethanol. It is dark brown with a broad NO absorption in the IR at 1600 to 1640 cm^{-1} [7].

References:

[1] Yu. N. Kukushkin, L. E. Danilina, M. M. Singh (Zh. Neorgan. Khim. **16** [1971] 2718/21; Russ. J. Inorg. Chem. **16** [1971] 1449/50). – [2] W. Hieber, K. Heinicke (Z. Anorg. Allgem. Chem. **316** [1962] 321/6). – [3] G. R. Crooks, B. F. G. Johnson (J. Chem. Soc. Dalton Trans. **1970** 1662/5). – [4] S. D. Robinson, M. F. Uttley (J. Chem. Soc. A **1971** 1254/7). – [5] K. K. Pandey, S. Datta, U. C. Agarwala (Transition Metal Chem. [Weinheim] **4** [1979] 337/9).

[6] A. Dowerah, M. M. Singh (Transition Metal Chem. [Weinheim] **2** [1976] 74/5). – [7] K. K. Pandey, U. C. Agarwala (J. Inorg. Nucl. Chem. **42** [1980] 293/4). – [8] K. K. Pandey, U. C. Agarwala (Z. Anorg. Allgem. Chem. **457** [1979] 235/7).

2.1.1.9 Nitrosyl-tertiary-stibine Complex, $Rh(NO)(SbPh_3)_3$

This complex is prepared at room temperature by reaction of $[Rh(NO)_2Cl]_n$ with $SbPh_3$ in tetrahydrofuran. The compound forms dark red crystals, melting point ~110°C with decomposition. $Rh(NO)(SbPh_3)_2Cl_2$ is also produced in this reaction, but no details are available.

W. Hieber, K. Heinicke (Z. Anorg. Allgem. Chem. **316** [1962] 321/6).

2.1.2 Thionitrosyl Complexes

$[Rh(NS)(PPh_3)Cl_2]_2$. This compound is prepared by reaction of $Rh(NO)(PPh_3)_3$ with $(NSCl)_3$ in dichloromethane. It is formed as brown crystals, melting point 220°C. It was dimeric in solution, exhibiting ν(NS) at 840 cm^{-1}. It is believed to dimerise via NS bridges [1].

$Rh(NS)(PPh_3)_2Cl_2$ is prepared by addition of PPh_3 to $[Rh(NS)(PPh_3)Cl_2]_2$ in dichloromethane [1], reaction of $RhCl_3\cdot 3H_2O$ with PPh_3 and $(NSCl)_3$ in ethanol or treatment of $Rh(PPh_3)_3Cl$ with $(NSCl)_3$ in tetrahydrofuran [2]. It was obtained as red-brown crystals, melting point 160 or 165°C, and exhibited ν(NS) at 1120 cm^{-1} [1, 2].

$Rh(NS)(CO)(PPh_3)_2Cl_2$ is formed by reaction of $Rh(CO)(PPh_3)_2Cl$ with $(NSCl)_3$ in dichloromethane. It is a red-brown solid, melting point 128 to 130°C. It exhibited ν(NS) and ν(CO) at 1118 and 2090 cm^{-1}, respectively, and is believed to have an all-*trans* structure in which the Rh-N-S bond is bent [2].

$Rh(NS)(PPh_3)_2ClBr$. This complex is formed by reaction of $Rh(PPh_3)_3Br$ with $(NSCl)_3$ in tetrahydrofuran. It was isolated as a brown solid, melting point 160 to 162°C, with ν(NS) at 1120 cm^{-1} [2].

$[Rh(NS)(AsPh_3)Cl_2]_2$ is obtained by reaction of $Rh(NO)(AsPh_3)_2Cl_2$ with $(NSCl)_3$ in carbon tetrachloride. It is a red-brown solid, melting point 220°C, exhibited ν(NS) at 1120 cm^{-1}, and is believed to dimerise via NS bridges [1].

$Rh(NS)(AsPh_3)_2Cl_2$. This complex is prepared by treatment of $[Rh(NS)(AsPh_3)_3Cl_2]_2$ with $AsPh_3$ in dichloromethane. It is formed as a red-brown solid, melting point 169°C; ν(NS) occurred at 1116 cm^{-1} [1].

$Rh(NS)(CO)(AsPh_3)_2Cl_2$ is obtained by treatment of $Rh(CO)(AsPh_3)_2Cl$ with $(NSCl)_3$ in dichloromethane. It occurs as red-brown crystals, melting point 159°C, with ν(NS) and ν(CO) at 1115 and 2070 cm^{-1}, respectively. Its structure is believed to be similar to $Rh(NS)(CO)(PPh_3)_2Cl_2$ [2].

$Rh(NS)(PPh_3)(AsPh_3)Cl_2$ is prepared by treatment of $[Rh(NS)(AsPh_3)Cl_2]_2$ with PPh_3 in dichloromethane. It was isolated as a red-brown solid, melting point 165°C, and exhibited ν(NS) at 1116 cm^{-1} [1].

References:

[1] K. K. Pandey, U. C. Agarwala (Inorg. Chem. **20** [1981] 1308/10). – [2] K. K. Pandey, S. Datta, U. C. Agarwala (Z. Anorg. Allgem. Chem. **468** [1980] 228/30).

2.1.3 Nitro Complexes

In this section we shall consider both nitro and nitrito complexes (i.e., those in which the ligand may be bound via nitrogen or oxygen, respectively); in most cases, if not all, the complexes are likely to be of the nitro type.

2.1.3.1 Unsubstituted Complexes

See "Rhodium" 1938, pp. 51, 73, 78, 87, 94/5, 97, 99.

The $[Rh(NO_2)_6]^{3-}$ Ion

Salts of this are usually made from $[RhCl_6]^{3-}$ and NO_2^-; the potassium and ammonium salts are only very sparingly soluble in water. A study has been carried out on the formation of $[Rh(NO_2)_6]^{3-}$ from $[RhCl_6]^{3-}$ and $[RhCl_n(H_2O)_{6-n}]^{(3-n)}$ at different pH. The stability constant of $[Rh(NO_2)_6]^{3-}$, determined potentiometrically, is $(3.2 \pm 0.4) \times 10^{-25}$ [1].

Vibrational Spectra. See under individual salts for solid state spectra. The Raman spectrum of $[Rh(NO_2)_6]^{3-}$ has been studied in aqueous solution from 100 to 1300 cm^{-1} [4], from 800 to 1500 cm^{-1} in the infrared and Raman [2] and from 100 to 1500 cm^{-1} in the infrared and Raman [3].

Electronic Spectra. For solid state spectra see under individual salts. The electronic absorption spectrum of $[Rh(NO_2)_6]^{3-}$ in aqueous solution has been recorded (200 to 420 nm, reproduced in paper) [5], and the emission spectrum of solid $K_3[Rh(NO_2)_6]$ from 200 to 1050 nm at 77 K, was measured (reproduced in paper) [6]. The spectrum of $[Rh(NO_2)_6]^{3-}$ in water (200 to 400 nm, reproduced in paper) was measured as a function of time, and also measured in a $NaNO_3$-KNO_3 melt (300 to 600 nm, reproduced in paper) [7]. Assignments were proposed [5, 6, 7].

Polarography. No polarographic reduction could be observed for $Na_3[Rh(NO_2)_6]$ [8].

Chemical Reactions. With dimethylglyoxime (DMGH), $Na_3[Rh(NO_2)_6]$ in boiling aqueous solution with NH_4Cl gives $(NH_4)[Rh(DMG)_2(NO_2)_2] \cdot 2H_2O$ [9]. The kinetics of reaction of $[Rh(NO_2)_6]^{3-}$ with amidosulphuric acid have been studied [10]. Studies on the oxidation of $[Rh(NO_2)_6]^{3-}$, as the sodium salt, by cerium(IV) and MnO_4^- show that three of the NO_2^- groups only are oxidised (to NO_3^-), and this was ascribed to the *trans* influence of NO_2^- [11]. With NH_2SO_3H, $[Rh(NO_2)_4(H_2O)_2]^-$, $[Rh(NO_2)_5(H_2O)_5]^{2-}$ and $(Rh(NO_2)_3(SO_4)(H_2O)_2]^{2-}$ are formed [12].

References:

[1] T. E. Fomina, V. S. Chekushin (Izv. Vysshikh Uchebn. Zavedenii Khim. Khim. Tekhnol. **19** [1976] 190/3 from C.A. **85** [1976] No. 25986). – [2] M. Le Postollec, J.-P. Mathieu (Compt. Rend. **254** [1962] 1800/2). – [3] M. Le Postollec, J.-P. Mathieu, H. Poulet (J. Chim. Phys. **60** [1963] 1319/33, 1321). – [4] J.-P. Mathieu, S. Cornevin (J. Chim. Phys. **36** [1939] 271/9, 278). – [5] H.-H. Schmidtke (Z. Physik. Chem. [Frankfurt] **40** [1963] 96/108, 100).

[6] T. R. Thomas, G. A. Crosby (J. Mol. Spectrosc. **38** [1971] 118/29, 122). – [7] F. B. Ogilvie, O. G. Holmes (Can. J. Chem. **44** [1966] 447/50). – [8] S. A. Repin (J. Appl. Chem. [USSR] **20** [1947] 46/54 from C.A. **1947** 5814). – [9] V. V. Lebedinskii, I. A. Fedorov (Izv. Sekt. Platiny No. 15 [1938] 19/25 from C.A. **1939** 2060). – [10] Y. N. Kukushkin, O. V. Stefanova (Zh. Neorgan. Khim. **22** [1977] 3375/7; Russ. J. Inorg. Chem. **22** [1977] 1844/5).

[11] A. V. Belyaev (Zh. Neorgan. Khim. **12** [1967] 1097/9; Russ. J. Inorg. Chem. **12** [1967] 577/9). – [12] Yu. N. Kukushkin, O. V. Stefanova (Koord. Khim. **5** [1979] 1379/82; Soviet J. Coord. Chem. **5** [1979] 1075/7).

Salts of Alkali Metals and Ammonium

$Na_3[Rh(NO_2)_6]$ is made from $Na_3[RhCl_6]$ and $NaNO_2$ [1, 2].

The infrared spectrum has been measured (300 to 1500 cm^{-1}, reproduced in paper) and some assignments proposed [3]; data from 800 to 1420 cm^{-1} were also given [4]. The infrared and Raman spectra of the solid were measured and assigned (100 to 1500 cm^{-1}) [5]. The electronic spectrum was run in a KCl pellet (180 to 350 nm) and assignments proposed [6].

$Na_3[Rh(NO_2)_6]$, dissolved in H_2SO_4, has been used for rhodium plating [7].

References:

[1] J.-P. Mathieu, S. Cornevin (J. Chim. Phys. **36** [1939] 271/9, 278). – [2] V. V. Lebedinskii, E. V. Shenderetskaya, A. G. Maiorova (Zh. Prikl. Khim. **32** [1959] 928/9; J. Appl. Chem. [USSR] **32** [1959] 944/6). – [3] M. Le Postollec, J.-P. Mathieu, H. Poulet (J. Chim. Phys. **60** [1963] 1319/33, 1321). – [4] M. Le Postollec, J.-P. Mathieu (Compt. Rend. **254** [1962] 1800/2). – [5] M. Le Postollec, J.-P. Mathieu (Compt. Rend. B **265** [1967] 138/40).

[6] K. G. Caulton, R. F. Fenske (Inorg. Chem. **6** [1967] 562/8). – [7] S. A. Simonova, E. I. Maslov, G. N. Molodkina, O. T. Lokshtanova, Yu. N. Kukushkin (Zh. Prikl. Khim. **49** [1976] 2524/6; J. Appl. Chem. [USSR] **49** [1976] 2521/3).

$K_3[Rh(NO_2)_6]$ is made from the sodium salt with KCl [1].

The infrared spectrum has been measured (300 to 1500 cm^{-1}, reproduced in paper) and some assignments made [2]; also from 800 to 1500 cm^{-1} [3]; infrared and Raman spectra of the solid were measured and assigned (80 to 1500 cm^{-1}) [4].

The electronic absorption spectrum of the aqueous solution was measured (200 to 450 nm, reproduced in paper) and assigned [1]. The luminescence spectrum of the solid was measured at 85 K and the absorption spectrum at 295 K, and bands assigned [5]; the emission spectrum of the solid at 77 K was also recorded (300 to 800 nm, reproduced in paper) [7].

The salt is colourless and isotropic, with a refractive index of 1.686 [6]. The X-ray photoelectronic spectrum of $K_3[Rh(NO_2)_6]$ gave the following binding energies: Rh $3d_{5/2}$ = 310.7, N 1s = 404.3, K $2p_{3/2}$ = 293.3 eV [8].

Fusion of $K_3[Rh(NO_2)_6]$ with KHF_2 at 300°C gives $K_3[RhF_6]\cdot KHF_2$, and at 500°C $K_3[RhF_6]$ is formed [9].

References:

[1] H.-H. Schmidtke (Z. Physik. Chem. [Frankfurt] **40** [1964] 96/108, 100, 107). – [2] M. Le Postollec, J.-P. Mathieu, H. Poulet (J. Chim. Phys. **60** [1963] 1319/33, 1321). – [3] M. Le Postollec, J.-P. Mathieu (Compt. Rend. **254** [1962] 1800/2). – [4] M. Le Postollec, J.-P. Mathieu (Compt. Rend. B **265** [1967] 138/40). – [5] P. E. Hoggard, H.-H. Schmidtke (Ber. Bunsenges. Physik. Chem. **77** [1973] 1052/8).

[6] M. N. Lyashenko (Tr. Inst. Krist. Akad. Nauk SSSR No. 7 [1952] 67/72 from C.A. **1956** 2349). – [7] T. R. Thomas, G. A. Crosby (J. Mol. Spectrosc. **38** [1971] 118/29, 122). – [8] V. I. Nefedov, M. A. Porai-Koshits (Mater. Res. Bull. **7** [1972] 1543/52). – [9] R. D. Peacock (J. Chem. Soc. **1955** 3291/2).

$(NH_4)_3[Rh(NO_2)_6]$. The infrared spectrum (300 to 1400 cm^{-1}) has been measured [6].

$Rb_3[Rh(NO_2)_6]$. $Cs_3[Rh(NO_2)_6]$. The salts are cubic, with a = 10.83 and 11.30 Å, respectively [7].

$Na(NH_4)_2[Rh(NO_2)_6]$ is made by reaction of $Na_3[Rh(NO_2)_6]$ with NH_4Cl [1, 2].

The X-ray crystal structure shows it to be face-centred cubic, space group Fm3-T_h^3 with a = 10.52 Å [3].

The solubility in water (g per 100 g H_2O) at 10, 20 and 25°C is 0.095, 0.146 and 0.209 [4]. It is used, by virtue of its relative insolubility, for separation of rhodium from other platinum group metals [2]. When boiled with aqueous NH_4OH, $Rh(NO_2)_3(NH_3)_3$ and $K[Rh(NO_2)_4(NH_3)_2]$ are formed [5].

References:

[1] V. V. Lebedinskii, E. V. Shenderetskaya (Izv. Sekt. Platiny No. 29 [1955] 61/5 from C.A. **1956** 6243). – [2] V. V. Lebedinskii, E. V. Shenderetskaya, A. G. Maiorova (Zh. Prikl. Khim. **32** [1959] 928/9; J. Appl. Chem. [USSR] **32** [1959] 944/6). – [3] G. B. Bokii, L. A. Popova (Bull. Acad. Sci. USSR Div. Chem. Sci. **1945** 89/93 from C.A. **1945** 5140). – [4] I. A. Peskov (Zh. Prikl. Khim. **22** [1949] 290/3 from C.A. **1949** 6047). – [5] V. V. Lebedinskii, E. V. Shenderetskaya (Ann. Sect. Platine Inst. Chim. Chem. Gen. USSR No. 18 [1945] 19/22 from C.A. **1947** 5044).

[6] Y. Puget, C. Duval (Compt. Rend. **250** [1960] 4141/2). – [7] L. K. Frevel (Ind. Eng. Chem. Anal. Ed. **14** [1942] 687/93, 691).

Salts of Heavy Metals

$Ba_3[Rh(NO_2)_6]$ is cubic, with a = 10.70 Å [1].

$Tl_3[Rh(NO_2)_6]$ is made from $TlNO_3$ and rhodium residues in 16 M HNO_3, followed by treatment with SO_2 and $NaNO_2$ [2]. The salt has a cubic structure with a = 10.91 Å [1].

$[Co(NH_3)_6][Rh(NO_2)_6]$ is made from $Tl_3[Rh(NO_2)_6]$, HCl and HNO_3, $NaNO_2$ and $[Co(NH_3)_6]Cl_3$ [2].

References:

[1] L. K. Frevel (Ind. Eng. Chem. Anal. Ed. **14** [1942] 687/93, 691). – [2] S. J. Lyle, R. Wellum (Radiochim. Acta **8** [1967] 119/20).

2.1.3.2 Substituted Complexes

$[Rh(NO_2)_5(H_2O)]^{2-}$. This is made from $[Rh(NO_2)_6]^{3-}$ and NH_2SO_3H [1]. From pH titrations the pK of this species at 25°C is 8.86 [2]; $K_1 = (0.7 \pm 0.3) \times 10^{-7}$ [1].

$[Rh(NO_2)_4(H_2O)_2]^-$. This is made from $[Rh(NO_2)_6]^{3-}$ and NH_2SO_3H. The acid dissociation constants are $(1.0 \pm 0.1) \times 10^{-6}$ and $(3.1 \pm 0.2) \times 10^{-10}$ [1].

$[Rh(NO_2)_3(H_2O)_2(SO_4)]^{2-}$ is made from $[Rh(NO_2)_6]^{3-}$ and NH_2SO_3H. The sulphate ligand is believed to be bonded as a monodentate ligand. The acid dissociation constants are $(1.1 \pm 0.5) \times 10^{-7}$ and $(1.5 \pm 0.6) \times 10^{-11}$ [1].

$K_3[Rh(NO_2)_3Cl_3]$. The X-ray photoelectronic spectrum shows the following binding energies: Rh $3d_{5/2} = 310.6$, N 1s = 404.3, Cl $2p_{3/2} = 198.8$, K $2p_{3/2} = 293.2$ eV [3].

References:

[1] Yu. N. Kukushkin, O. V. Stefanova (Koord. Khim. **5** [1979] 1379/80; Soviet J. Coord. Chem. **5** [1979] 1075/7). – [2] A. V. Belyaev (Izv. Sibirsk. Otd. Akad. Nauk SSSR Ser. Khim. Nauk **1970** No. 4, pp. 11/8; Sib. Chem. J. **1970** 459/65). – [3] V. I. Nefedov, M. A. Porai-Koshits (Mater. Res. Bull. **7** [1972] 1543/52, 1545).

2.1.4 Organonitrile Complexes

General Literature:

R. A. Walton, The Reactions of Metal Halides with Alkyl Cyanides, Quart. Rev. [London] **19** [1965] 126/43.

2.1.4.1 Complexes with Acetonitrile

***trans*-$[RhCl_2(CH_3CN)_4]Y$** ($Y = ClO_4$, NO_3). These salts are prepared by heating a mixture of *mer*-$RhCl_3(CH_3CN)_3$ and the appropriate silver salt (AgY) under reflux in acetonitrile for 30 min, and are isolated as pale yellow crystals by concentration of the filtered reaction solution. The Raman active vibration $\nu(RhCl_2)$ occurs at 304 cm^{-1} [1]; a printing error renders other spectroscopic data unintelligible.

***trans*-$[RhBr_2(CH_3CN)_4]ClO_4$** is similarly prepared from *mer*-$RhBr_3(CH_3CN)_3$ and $AgClO_4$. The Raman active vibration $\nu(RhBr_2)$ occurs at 194 cm^{-1} [1].

***fac*-$RhCl_3(CH_3CN)_3$** is prepared by bringing a solution of $RhCl_3 \cdot 3H_2O$ in acetonitrile briefly to the boil, and is precipitated as an orange-brown solid by addition of diethyl ether [2, 3]. The electronic spectrum has been reported: $\lambda_{max} = 408$ nm (reflectance) [2]; $\lambda_{max} = 410$ nm (reflectance), $\lambda_{max} = 408$ nm (aqueous solution), and assigned ($^1A_{1g} \rightarrow {}^1T_{1g}$) [3]. Infrared data have been reported and assignments given (frequencies in cm^{-1}; s = strong, m = medium, w = weak): 2941(s), C-H stretch; 2874(m), C-H stretch; 2304(m), C≡N stretch; 1414(s), CH_3 deformation; 1366(w), CH_3 deformation; 1042(s), CH_3 rock; 954(w), C-C stretch; 456(w), 432(m), 415(w), no assignment; 356(s), Rh-Cl stretch and 346(s), Rh-Cl stretch [2], 456(w), 432(m), 415(w), 356(s) and 346(s) [3]. Proton NMR data have been recorded: $\delta(CH_3CN) = 2.60$(s) ppm in CD_3CN [3], $\delta(CH_3CN) = 2.58$ ppm in D_2O [2].

The heat of decomposition $RhCl_3(CH_3CN)_3(cryst) \rightarrow RhCl_3(cryst) + 3CH_3CN(gas)$ has been given as 81 ± 2 kJ/mol [4]. The complex isomerises to the more stable *mer* isomer on heating in acetonitrile [2].

***mer*-$RhCl_3(CH_3CN)_3$** is conveniently prepared by heating a solution of $RhCl_3 \cdot 3H_2O$ in 1/1 ethanol/acetonitrile under reflux for 4 h [3]. It can also be obtained by heating the less stable *fac* isomer in acetonitrile [2], and is described as a yellow precipitate [3]. The deuteriated complex *mer*-$RhCl_3(CD_3CN)_3$ is obtained by CH_3CN/CD_3CN exchange in boiling CD_3CN over a period of 1 h [1]. The product obtained by warming $RhCl_3$ with excess CH_3CN and originally formulated as *mer*-$RhCl_3(CH_3CN)_3$ [5] has been reformulated as the *fac* isomer [6]. Spectroscopic data reported in [5] therefore are presumably those of the *fac* isomer.

The electronic spectrum of the *mer* isomer has been reported, $\lambda_{max} = 433$ nm, CH_3NO_2 solution, and assigned ($^1A_{1g} \rightarrow {}^1T_{1g}$). The far infrared spectrum has been recorded and assignments given (frequencies in cm^{-1}; s = strong, m = medium, (v)w = (very) weak, sh = shoulder): ν(Rh-Cl) = 356(s), 350(s), 300(sh); other bands 456(m), 430(vw), 254(m), 201(w), 159(w). Proton NMR data have been reported: $\delta(CH_3) = 2.64$(s), 2.59(s) ppm [3]. The heat of decomposition *mer*-$RhCl_3(CH_3CN)_3$(cryst) → $RhCl_3$(cryst) + $3CH_3CN$(gas) has been given as 97.5 ± 0.2 kJ/mol [4].

***mer*-$RhBr_3(CH_3CN)_3H_2O$** is prepared by stirring a solution of $RhBr_3 \cdot nH_2O$ in 10:1 $CH_3CN/CH(OC_2H_5)_3$ overnight, and is precipitated as orange-red crystals by addition of dry diethyl ether. The electronic spectrum ($\lambda_{max} = 480$ nm) is consistent with a *mer* configuration but the proton NMR spectrum ($\delta(CH_3) = 2.84$ ppm) supports a *fac* arrangement of the CH_3CN ligands [1].

$[RhCl_3(CH_3CN)_2]_n$ is prepared by heating an acetone solution of *mer*-$RhCl_3(CH_3CN)_3$ under reflux for 1 h and deposits as a pink precipitate. Treatment with CH_3CN under reflux for 1 h regenerates the parent complex [1].

$RhBr_3(CH_3CN)_2(H_2O)$ is prepared by heating a solution of $RhBr_3 \cdot 2H_2O$ in CH_3CN under reflux for 7 h and is isolated as dark brown crystals by evaporation of the wine-red reaction solution. The electronic spectrum with $\lambda_{max} = 465$(sh) nm ($^1A_{1g} \rightarrow {}^1T_{1g}$ or $^1T_{2g}$) 334 and 266 nm (charge transfer), infrared spectrum with ν(CN) = 2315, ν(RhBr) = 321(s), 304(s), 296(s) cm^{-1} and proton NMR spectrum with $\delta(CH_3) = 2.58$(s), 2.55(s) ppm have all been reported [7].

***trans*-$[N(C_2H_5)_4][RhCl_4(CH_3CN)_2]$** is prepared by addition of equimolar amounts of $[N(C_2H_5)_4]Cl$ and $RhCl_3 \cdot 3H_2O$ to a 1:1 $CH_3CN:H_2O$ solvent mixture and precipitates out as orange plates. The electronic spectrum exhibits $\lambda_{max} = 490$ ($^1A_{1g} \rightarrow {}^1T_{1g}$) and 423 nm ($^1A_{1g} \rightarrow {}^1T_{2g}$) for CH_3NO_2 solution; 486 ($^1A_{1g} \rightarrow {}^1T_{1g}$) and 418 nm ($^1A_{1g} \rightarrow {}^1T_{2g}$) for aqueous solution, the infrared spectrum ν(CN) = 2320, ν(RhCl) = 338, 300(sh) cm^{-1} and the proton NMR spectrum $\delta(CH_3CN) = 2.59$(s) ppm. The far infrared spectrum (500 to 150 cm^{-1}) is reproduced [3].

***trans*-$Cs[RhCl_4(CH_3CN)_2]$** is similarly prepared using CsCl [3].

***trans*-$[N(C_2H_5)_4][RhBr_4(CH_3CN)_2]$** is prepared by addition of $[N(C_2H_5)_4]Br$ to a boiling solution of $RhBr_3 \cdot 2H_2O$ in acetonitrile and is precipitated as a brown solid by addition of diethyl ether to the cooled reaction solution. The electronic spectrum shows $\lambda_{max} = 490$(sh) ($^1A_{1g} \rightarrow {}^1T_{1g}$), 450(sh) ($^1A_{1g} \rightarrow {}^1T_{2g}$), 344, 304(sh) and 260 nm (all charge transfer), the infrared spectrum ν(CN) = 2315, ν(RhBr) = 307(s), 286(s) cm^{-1} and the proton NMR spectrum $\delta(CH_3CN) = 2.61$(s) ppm [7].

For **$[Rh(C_{28}H_{20}N_4)(CH_3CN)_2](ClO_4)_3 \cdot H_2O$** see p. 209.

$M_2[RhCl_5(CH_3CN)]$ (M = NH_4, $N(CH_3)_4$, Cs; see "Rhodium" 1938, pp. 152/3). These salts are prepared by adding the appropriate chloride (MCl) to a solution of $Na_3[RhCl_6]$ in a 1:1 mixture of $CH_3CN:H_2O$. They range in colour from brown to orange-brown depending upon the cation [3].

For **$K_3[Rh(CN)_5(CH_3CN)]$** see "Rhodium" Suppl. Vol. B 1, 1982, p. 197; for **$[Rh(CH_3CN)(NH_3)_5](ClO_4)_3$** and CD_3CN analogue see p. 138.

The following spectroscopic data have been recorded [3]:

M	λ_{max} in nm	Assignment	ν(CN) in cm^{-1}	ν(RhCl) in cm^{-1}
NH_4	465[a)]	$^1A_{1g} \rightarrow {}^1T_{1g}$	2304	
	382	$^1A_{1g} \rightarrow {}^1T_{2g}$		
Cs	462(464)[b)]	$^1A_{1g} \rightarrow {}^1T_{1g}$	2304	343(s)
	382(380)	$^1A_{1g} \rightarrow {}^1T_{2g}$		328(s)
				295(sh)

a) Spectra taken in 1:1 $CH_3CN:H_2O$ solution. – b) Spectra taken by reflectance (or in aqueous solution).

$RhCl(CH_3CN)(PPh_3)_2$ is prepared by dissolving $RhCl(C_2H_4)(PPh_3)_2$ in benzene containing the appropriate amount of CH_3CN. The kinetics of replacement of CH_3CN by triphenylphosphine have been investigated [8, 9].

For **nitrosyl acetonitrile** complexes see pp. 99, 107/8; for **$[Rh(PCy_3)_2(CH_3CN)_2H_2](PF_6)$** see "Rhodium" Suppl. Vol. B 3.

References:

[1] R. D. Gillard, B. T. Heaton, H. Shaw (Inorg. Chim. Acta **7** [1973] 102/4). – [2] B. D. Catsikis, M. L. Good (Inorg. Nucl. Chem. Letters **4** [1968] 529/31). – [3] B. D. Catsikis, M. L. Good (Inorg. Chem. **8** [1969] 1095/9). – [4] G. Beech, G. Marr, S. J. Ashcroft (J. Chem. Soc. A **1970** 2903/6). – [5] B. F. G. Johnson, R. A. Walton (J. Inorg. Nucl. Chem. **28** [1966] 1901/5).

[6] R. A. Walton (Can. J. Chem. **46** [1968] 2347/52, 2351). – [7] B. D. Catsikis, M. L. Good (Inorg. Chem. **10** [1971] 1522/4). – [8] Y. Ohtani, A. Yamagishi, M. Fujimoto (Bull. Chem. Soc. Japan **52** [1979] 1537/8). – [9] Y. Ohtani, A. Yamagishi, M. Fujimoto (Bull. Chem. Soc. Japan **52** [1979] 2149/50).

2.1.4.2 Complexes with Propionitrile

***trans*-$[RhCl_2(C_2H_5CN)_4][ClO_4]$** is prepared by heating a slurry of *trans*-$[RhCl_2(CH_3CN)_4]$-$[ClO_4]$ in C_2H_5CN at 100°C for 30 min and is precipitated by addition of diethyl ether. The electronic spectrum (λ_{max} = 410 nm) and selected infrared data (ν(CN) = 2320 cm^{-1}) have been recorded. Analytical data indicate possible hydration of the complex ($4H_2O$) [1].

***mer*-$RhCl_3(C_2H_5CN)_3$** is prepared by warming $RhCl_3$ with excess C_2H_5CN until a clear orange-yellow solution is obtained, and deposits from solution on cooling as yellow crystals. The electronic spectrum has been recorded: λ_{max} = 431 nm (reflectance); 429, 370 nm (CH_3NO_2 solution); it is reproduced from 600 to 350 nm in CH_3NO_2 solution. Infrared frequencies in cm^{-1}: 2310(s), 2305(sh), 1415(m), 1310(m), 1260(w), 1160(w, br), 1090(sh), 1076(s), 1005(w), 790(s) cm^{-1}; conductivity 4.2 $cm^2 \cdot \Omega^{-1} \cdot mol^{-1}$ for a 3.33×10^{-3} M solution in CH_3NO_2 [2]. This complex has also been formulated as a monohydrate, *mer*-$RhCl_3(CH_3CH_2CN)_3 \cdot H_2O$; ν(OH) = 3500 cm^{-1}, λ_{max} = 422 nm, δ(α-CH_2) = 3.00, 2.96 ppm [1].

For **$[Rh(C_2H_5CN)(NH_3)_5](ClO_4)_3$** see p. 138.

References:

[1] R. D. Gillard, B. T. Heaton, H. Shaw (Inorg. Chim. Acta **7** [1973] 102/4). – [2] B. F. G. Johnson, R. A. Walton (J. Inorg. Nucl. Chem. **28** [1966] 1901/5).

2.1.4.3 Complex with 2-Methoxypropionitrile, *mer*-$RhCl_3(CH_3OCH_2CH_2CN)_3$

The compound is prepared by heating $RhCl_3$ with $CH_3OCH_2CH_2CN$ in boiling chloroform and separates as orange crystals from the solution on evaporation [1]. The electronic spectrum has been reported: λ_{max} = 427 nm (reflectance); 425, 376 nm (CH_3NO_2 solution) [1]. Infrared data (frequencies in cm^{-1}): 2285(s), 1315(m), 1250(mw), 1218(ms), 1182(ms), 1150(w), 1105(vs), 1050(ms), 1010(mw), 990(ms), 942(m), 885(vvw), 845(w, br), 805(m) [1], 448(ms), 385(w), ~360(sh), 347(s), ~335(sh), 275(m, br) [2]; conductivity 4.3 $cm^2 \cdot \Omega^{-1} \cdot mol^{-1}$ for an 8.34×10^{-3} M solution in CH_3NO_2 [1].

References:

[1] B. F. G. Johnson, R. A. Walton (J. Inorg. Nucl. Chem. **28** [1966] 1901/5). – [2] R. A. Walton (Can. J. Chem. **44** [1966] 1480/2).

2.1.4.4 Complexes with n-Butyronitrile

***trans*-$[RhCl_2(CH_3CH_2CH_2CN)_4][ClO_4]$** is prepared by heating a slurry of *trans*-$[RhCl_2(CH_3CN)_4][ClO_4]$ in neat butyronitrile at 100°C for 30 min, and is precipitated from solution by addition of diethyl ether. The electronic spectrum (λ_{max} = 415 nm) and selected infrared bands (ν(CN) = 2325 cm^{-1}) have been reported.

***mer*-$RhCl_3(CH_3CH_2CH_2CN)_3$** is prepared by warming $RhCl_3 \cdot nH_2O$ with excess CH_3CH_2-CH_2CN in the presence of $CH(OC_2H_5)_3$ as a dehydrating agent. The electronic spectrum (λ_{max} = 429 nm) and proton NMR spectrum ($\delta(\alpha\text{-}CH_2)$ = 2.92, 2.88 ppm) have been reported.

R. D. Gillard, B. T. Heaton, H. Shaw (Inorg. Chim. Acta **7** [1973] 102/4).

2.1.4.5 Complexes with iso-Butyronitrile

***mer*-$RhCl_3\{(CH_3)_2CHCN\}_3$** is prepared by warming $RhCl_3 \cdot nH_2O$ with isobutyronitrile in the presence of triethylorthoformate as a dehydrating agent. The electronic spectrum (λ_{max} = 425 nm) and proton NMR spectrum ($\delta(CH_3)$ = 3.82 ppm) have been reported.

$[RhCl_3\{(CH_3)_2CHCN\}_2]_n$ is obtained as a pink precipitate by heating a solution of $RhCl_3\{(CH_3)_2CHCN\}_3$ in acetone under reflux for 1 h. This reaction can be reversed by boiling the product with excess $(CH_3)_2CHCN$ for 1 h.

R. D. Gillard, B. T. Heaton, H. Shaw (Inorg. Chim. Acta **7** [1973] 102/4).

For **nitrosyl** complexes containing Bu^tCN see p. 99.

2.1.4.6 Complex with Glutaronitrile, $[RhCl_3(NCCH_2CH_2CH_2CN)_{1.5}]_n$

The compound is obtained as an insoluble orange powder by warming a mixture of $RhCl_3$ with excess glutaronitrile. The electronic spectrum (diffuse reflectance) has been reported (λ_{max} = 425(sh) nm) and is reproduced (500 to 350 nm). The infrared spectrum (2400 to 700 cm^{-1}) has been recorded, bands at 2302 and 2240 cm^{-1} are attributed to bound and free cyanide groups, respectively.

B. F. G. Johnson, R. A. Walton (J. Inorg. Nucl. Chem. **28** [1966] 1901/5).

2.1.4.7 Complexes with Phenylacetonitrile

***trans*-[RhCl$_2$(C$_6$H$_5$CH$_2$CN)$_4$][ClO$_4$]** is prepared by heating a slurry of *trans*-[RhCl$_2$(CH$_3$CN)$_4$]-[ClO$_4$] and $C_6H_5CH_2CN$ at 100°C for 30 min and is precipitated by addition of diethyl ether. The electronic spectrum (λ_{max} = 415 nm) and selected infrared bands (ν(CN) = 2320 cm^{-1}) have been reported.

***mer*-RhCl$_3$(C$_6$H$_5$CH$_2$CN)$_3$** is prepared by warming $RhCl_3 \cdot nH_2O$ with phenylacetonitrile in the presence of CH(OC$_2$H$_5$)$_3$ as a dehydrating agent. The electronic spectrum (λ_{max} = 425 nm) and selected proton NMR data (δ(α-CH$_2$) = 4.75, 4.19 ppm) have been recorded.

[RhCl$_3$(C$_6$H$_5$CH$_2$CN)$_2$]$_n$ is prepared by heating a toluene solution of *mer*-RhCl$_3$(C$_6$H$_5$CH$_2$CN)$_3$ under reflux for 1 h.

R. D. Gillard, B. T. Heaton, H. Shaw (Inorg. Chim. Acta **7** [1973] 102/4).

2.1.4.8 Complexes with Benzonitrile

***trans*-[RhCl$_2$(C$_6$H$_5$CN)$_4$][ClO$_4$]** is prepared by heating a slurry of *trans*-[RhCl$_2$(CH$_3$CN)$_4$]-[ClO$_4$] and neat benzonitrile at 100°C for 30 min, and is precipitated by cautious addition of diethyl ether [1]. A printer's error has rendered the spectroscopic data unintelligible.

***mer*-RhCl$_3$(C$_6$H$_5$CN)$_3$** is prepared by warming a mixture of $RhCl_3$ and C_6H_5CN until a clear solution is obtained, and is isolated as a yellow solid from the cooled, concentrated solution [2]. The electronic (reflectance) spectrum (λ_{max} = 420 nm) [2] and the infrared spectrum (2300 to 650 cm^{-1}) with ν(CN) = 2280 cm^{-1} [2], (421 to 285 cm^{-1}) [3] have been reported. The heat of decomposition $RhCl_3(C_6H_5CN)_3(cryst) \rightarrow RhCl_3(cryst) + 3C_6H_5CN(gas)$ has been recorded as 128 ± 3 kJ/mol [4].

[N(C$_2$H$_5$)$_4$][RhCl$_4$(C$_6$H$_5$CN)$_2$]·H$_2$O is prepared by warming a suspension of [N(C$_2$H$_5$)$_4$]$_3$-[Rh$_2$Cl$_9$] in benzonitrile until a reddish orange solution forms, and is isolated as a pale orange (apricot) solid by slow addition of anhydrous diethyl ether. The electronic spectrum with λ_{max} = 490 nm ($^1A_{1g} \rightarrow {}^1T_{1g}$), 418 nm ($^1A_{1g} \rightarrow {}^1T_{2g}$) in CH_3NO_2 solution and selected infrared bands (ν(CN) = 2283 cm^{-1}; ν(RhCl) = 346, 312 cm^{-1}) have been reported [5].

References:

[1] R. D. Gillard, B. T. Heaton, H. Shaw (Inorg. Chim. Acta **7** [1973] 102/4). – [2] B. F. G. Johnson, R. A. Walton (J. Inorg. Nucl. Chem. **28** [1966] 1901/5). – [3] R. A. Walton (Can. J. Chem. **44** [1966] 1480/2). – [4] G. Beech, G. Marr, S. J. Ashcroft (J. Chem. Soc. A **1970** 2903/6). – [5] B. D. Catsikis, M. L. Good (Inorg. Chem. **10** [1971] 1522/4).

2.1.4.9 Complex with 3,4,4-Trimethyl-2-cyanobut-2-en-1,4-olide

3,4,4-Trimethyl-2-cyanobut-2-en-1,4-olide = 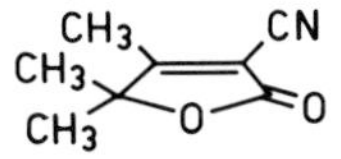 = RCN

[Rh(RCN)$_6$]Cl$_3$·H$_2$O is prepared by treatment of $RhCl_3 \cdot 3H_2O$ with aqueous RCN in the presence of KCl. The complex dehydrates at 160°C and decomposes above 250°C. Coordination through the cyanide group is proposed. No other details are given in the abstract.

S. N. Sarkisyan, A. A. Avetisyan, L. I. Sagradyan (Arm. Khim. Zh. **31** [1978] 539/41 from C.A. **89** [1978] No. 172841).

For complexes **[Rh(RCN)(NH$_3$)$_5$](ClO$_4$)$_3$** with R = CH-CH$_2$, CH=CH-CH$_3$, Ph, o-, m- or p-C$_6$H$_4$F, see p. 138.

2.1.5 Amide Complex, $Rh(NH_2)_3$

The existence of this compound has been tentatively demonstrated by potentiometric titration of $[Rh(NH_3)_5Br]Br_2$ with KNH_2 in liquid ammonia [1]. The pale yellow colour of the ammine decreases and is progressively replaced by green and yellow-green colours, the stoichiometry of the reaction being consistent with production of the tri-amide. There was apparently no further reaction with KNH_2 beyond this point. The compound is reported to have been isolated from the reaction $[Rh(NH_3)_5Br]Br_2$ with KNH_2 in liquid ammonia, but the preparation is not readily reproducible. Under slightly different conditions a dark brown solid is formed which, upon removal of NH_3 under vacuum, becomes black and pyrophoric. Similar treatment of $[Rh(NH_3)_5Cl]Cl_2$ afforded only Rh metal. The tri-amide is yellow but unstable and acts as a hydrogenation catalyst for olefins and aromatic nitro compounds [2].

References:

[1] G. W. Watt, G. R. Choppin, J. L. Hall (J. Electrochem. Soc. **101** [1954] 235/8). – [2] G. W. Watt, A. Broodo, W. A. Jenkins, S. G. Parker (J. Am. Chem. Soc. **76** [1954] 5989/93).

2.1.6 Hydrazine Complexes

A large number of very ill-defined hydrazine complexes of rhodium has been reported.

Chloro-Hydrazine Complexes. A number of these has been reported, made by reaction of $K_3[RhCl_6]$ with hydrazine in various molar proportions at temperatures from 15 [1, 3] to 60°C [2]. Proposed formulae and gram susceptibilities are listed in the following table.

Complex (proposed formula)	Colour	$\chi_{mol} \times 10^6$	μ_{eff} in B.M.	Ref.
$RhCl \cdot 2.5N_2H_4$	yellow	−80		[1]
$RhCl \cdot 2.75N_2H_4$		−14		[1]
$RhCl \cdot 3.25N_2H_4$	yellow	0		[1]
$RhCl \cdot 3.5N_2H_4$	yellow	−10		[1, 2]
$RhCl_2 \cdot 1.75N_2H_4$		1180	1.73	[3]
$RhCl_2 \cdot 2.25N_2H_4$		−16		[2]
$Rh_2Cl_3 \cdot 2N_2H_4$	ochre	4700		[2, 3]
		4890	3.38	[2, 3]
$Rh_2Cl_3 \cdot 2.25N_2H_4$	yellow	1715	2.15	[3]
$Rh_2Cl_4 \cdot 3.5N_2H_4$	ochre	1190	1.73	[3]
$Rh_2Cl_4 \cdot 4N_2H_4$		1187	1.73	[3]
$Rh_2Cl_5 \cdot 3N_2H_4$	ochre	195	0.96	[3]
$Rh_2Cl_5 \cdot 4N_2H_4$	ochre	−112		[2]
$Rh_3Cl \cdot 5.5N_2H_4$	yellow	25		[1]
$Rh_3Cl_2 \cdot 8.5N_2H_4$	orange-yellow	127		[1]
$Rh_3Cl_5 \cdot 4N_2H_4$	ochre	1045	1.76	[3]
		−20		[2]
$Rh_4Cl_3 \cdot 14.5N_2H_4$		−18		[1]
$Rh_4Cl_{7.5} \cdot 5.5N_2H_4$		−160		[2]

(Continued)

Complex (proposed formula)	Colour	$\chi_{mol} \times 10^6$	μ_{eff} in B.M.	Ref.
$Rh_4Cl_{7.5} \cdot 10.5\,N_2H_4$		−246		[2]
$Rh_4Cl_9 \cdot 5\,N_2H_4$		−225		[2]
$Rh_4Cl_9 \cdot 7.5\,N_2H_4$		−318		[2]
$Rh_6Cl_5 \cdot 23\,N_2H_4$		0		[1]

Iodo Hydrazine Complexes. A number of ill-defined complexes between RhI_3 and N_2H_4 has been claimed, made from RhI_3 and N_2H_4 in varying proportions [4].

Complex	Colour	$\chi_{mol} \times 10^6$
$RhI_2 \cdot 3.5\,N_2H_4$	orange-yellow	−66
$Rh_2I_3 \cdot 5\,N_2H_4$	orange-yellow	−89
$Rh_3I_5 \cdot 3.5\,N_2H_4$	yellow	−75
$RhI \cdot 3.5\,N_2H_4$	orange	19
$RhI \cdot 2.75\,N_2H_4$	orange	−6

References:

[1] L. Cambi, E. D. Paglia (Atti Accad. Nazl. Lincei Classe Sci. Fis. Mat. Nat. Rend. [8] **29** [1960] 488/90). – [2] L. Cambi, E. D. Paglia (Atti Accad. Nazl. Lincei Classe Sci. Fis. Mat. Nat. Rend. [8] **28** [1960] 770/6). – [3] L. Cambi, E. D. Paglia, C. Giuffrida (Atti Accad. Nazl. Lincei Classe Sci. Fis. Mat. Nat. Rend. [8] **29** [1960] 8/14). – [4] L. Cambi, E. D. Paglia (Atti Accad. Nazl. Lincei Classe Sci. Fis. Mat. Nat. Rend. [8] **29** [1960] 253/6).

For **cyano hydrazine** complexes, see "Rhodium" Suppl. Vol. B 1, 1982, pp. 197/8; for **acetato hydrazine** complexes, see p. 20, and for **miscellaneous carboxylate hydrazine** complexes, see p. 62.

2.1.7 Ammine Complexes

See also "Rhodium" 1938, pp. 102/53.

An extensive literature exists for this class of compound. The complexes are formed as hexammines, $[Rh(NH_3)_6]^{3+}$, as pentammines, $[Rh(NH_3)_5X]^{2+}$-, a particularly large class of compounds within the group, and as tetrammines, $[Rh(NH_3)_4XY]$ which exist as *cis* and *trans* isomers. The triammines, $Rh(NH_3)_3X_3$, which also form *mer* and *fac* isomers, diammines and monoammines are less numerous. There are several photochemical studies of the penta- and tetrammines which reveal photosolvation and, in the case of $[Rh(NH_3)_4XY]^+$, photoisomerisation processes. There are very few examples of Rh(II) ammine complexes.

2.1.7.1 Unsubstituted Ammines

$[Rh(NH_3)_6]Cl$. Very brief reference has been made to this compound, and no preparative details have been given. The Raman spectrum has been reported.

H. Poulet, J. P. Mathieu (Compt. Rend. **255** [1962] 1514/5).

Hexammine Complexes

$[Rh(NH_3)_6]Y_3$. The **chloride** is prepared by reaction of $RhCl_3 \cdot nH_2O$ with aqueous concentrated ammonia either at atmospheric pressure [1] or in an autoclave [2]. It is formed as a yellow powder or crystals and is soluble in water. The **nitrate** has been described in [3].

For structural and spectroscopic details, see below.

$[Rh(NH_3)_6]Y$ (Y = MnF_6, GaF_6, ScF_6, $CrCl_6$, $PbCl_6$, $SbCl_6$, $CuCl_5$, $CuBr_5$). The **hexafluoromanganate(III)** is prepared by treatment of $[Rh(NH_3)_6][NO_3]_3$ with aqueous HF containing $Mn(NO_3)_2$, followed by oxidation with $KMnO_4$ [4]; it is a violet solid. The related GaF_6^{3-} and ScF_6^{3-} have been mentioned only briefly [5]. The **hexachlorochromate(III)** is made by reaction of $[Rh(NH_3)_6]Cl_3$ with $CrCl_3$ in water, and is obtained as a rose-violet solid [6]. The **hexachloroplumbate(II/IV)** and **hexachloro-antimonate(III)** have been described only briefly, and no preparative details were reported [7, 8, 9]. The **pentachlorocuprate(II)** is prepared from $[Rh(NH_3)_6]Cl_3$ and $CuCl_2 \cdot 2H_2O$ in water containing HCl [10]; the related $[Rh(NH_3)_6][CuBr_5]$ has been referred to only briefly [11].

Structural Data on $[Rh(NH_3)_6]^{3+}$ Salts. $[Rh(NH_3)_6]Cl_3$ forms crystals of low birefringence, $n \approx 1.679$ [12], and its structure has been briefly described [13]. The cation is presumably octahedral. An X-ray structural study of $[Rh(NH_3)_6][MnF_6]$ indicated that the compound was cubic, space group $Pa3-T_h^6$, Z = 4, with a = 10.05 Å [4]. The species $[Rh(NH_3)_6][MF_6]$ (M = Ga or Sc) also crystallise in a cubic lattice and are isomorphous [5]. $[Rh(NH_3)_6][CrCl_6]$ is similarly cubic, space group $Pa3-T_h^6$, Z = 4, with a = 11.24 Å [6].

Semi-empirical Wolfsburg-Helmholtz molecular orbital calculations have been made using $[Rh(NH_3)_6]^{3+}$ and attempts were made using the results to interpret the electronic spectrum of this complex [14]. Molecular orbital calculations using the linear combination of atomic orbitals approach were used to demonstrate that, in $[Rh(NH_3)_6]^{3+}$, the metal forms strong σ-covalent bonds with the NH_3 ligands. The diamagnetic susceptibility term for $[Rh(NH_3)_6]^{3+}$ has been calculated [16].

References:

[1] R. Bramley, B. N. Figgis (J. Chem. Soc. A **1967** 861/3). – [2] H. H. Schmidtke (Z. Physik. Chem. [Frankfurt] **40** [1964] 96/108, 101, 107). – [3] Gmelin Handbuch „Rhodium" 1938, pp. 106/7. – [4] K. Wieghardt, H. Siebert (Z. Anorg. Allgem. Chem. **381** [1971) 12/20). – [5] K. Wieghardt, H. Siebert (J. Mol. Struct. **7** [1971] 305/13).

[6] H. H. Eysel (Z. Anorg. Allgem. Chem. **390** [1972] 210/6). – [7] P. Day, I. D. Hall (J. Chem. Soc. A **1970** 2679/82). – [8] N. S. Hush (Progr. Inorg. Chem. **8** [1967] 391). – [9] P. Burroughs, A. Hamnett, A. F. Orchard (J. Chem. Soc. Dalton Trans. **1974** 565/7). – [10] G. C. Allen, N. S. Hush (Inorg. Chem. **6** [1967] 4/8).

[11] G. C. Allen, G. A. M. El-Sharkawy (Inorg. Nucl. Chem. Letters **6** [1970] 281/4). – [12] M. N. Lyashenko (Tr. Inst. Krist. No. 7 [1952] 67/72 from C.A. **1956** 2439). – [13] N. Belov (Ann. Sect. Platine Inst. Chim. Gen [USSR] No. 18 [1945] 112/92 from C.A. **1947** 41). – [14] P. K. Mehrotra, P. T. Manoharan (Chem. Phys. Letters **39** [1976] 194/8). – [15] I. B. Bersuker, S. S. Budnikov, A. S. Kimoglo (Teor. Eksperim. Khim. **12** [1976] 621/32 from C.A. **86** [1976] No. 62903).

[16] C. J. Ballhausen, R. W. Asmussen (Acta Chem. Scand. **11** [1957] 479/83).

IR, UV and Electronic Spectral Data from $[Rh(NH_3)_6]^{3+}$ Salts

The infrared and Raman spectra of $[Rh(NH_3)_6]Cl_3$ and its deuteriate have been measured (4000 to 200 cm^{-1}): ν_s(Rh-N) is at 514 cm^{-1} (483 cm^{-1} in the deuteriate), ν_{as}(Rh-N) at 472 cm^{-1}; δ_{as}(NRhN) at 294.5 cm^{-1} (240 cm^{-1} in the deuteriate) [1, 2]. The stretching force constant for the

Rh-N bond was calculated to be 2.10 mdyn/Å, commensurate with a Rh-N bond order of 0.8 [1]. However, a Urey-Bradley Force Field calculation has given a force constant of 2.24 mdyn/Å [3]. Other measurements of the infrared and Raman spectra of $[Rh(NH_3)_6]Cl_3$ have been made over the ranges 3200 to 825 cm^{-1} [4], 3000 to 300 cm^{-1} [5 to 9] and 200 to 100 cm^{-1} [5]. The vibrational spectra of $[Rh(NH_3)_6][MnF_6]$ have been measured between 3320 and 285 cm^{-1} [10], and the resonance Raman spectrum of $[Rh(NH_3)_6][PbCl_6]$ between 850 and 200 cm^{-1} [11].

The electronic absorption spectrum of $[Rh(NH_3)_6]Cl_3$ has been measured in solution: two absorptions occurred at 32700 and 39100 cm^{-1} [12], at 32800 and 39200 cm^{-1} [13], or at 33300 and 40000 cm^{-1} [14]; assignment was made of these bands to the $^1A_{1g} \rightarrow {}^1T_{1g}$ and $^1A_{1g} \rightarrow {}^1T_{2g}$ transitions in octahedral symmetry [14, 15]. The solid state reflectance spectrum of $[Rh(NH_3)_6]Cl_3$ has also been recorded [16, 17] (reproduced in paper, 200 to 450 nm; for solid and aqueous solution [17]). From the electronic spectra, the crystal field ($\Delta = 10$ Dq, and electronic repulsion (B) parameters and nephelauxetic ratios (β) were calculated to be 34000, 430, and 0.60 cm^{-1}, respectively, and comparison of these was made with $[Co(NH_3)_6]^{3+}$ and $[Ir(NH_3)_6]^{3+}$ [15]. The electronic spectra of $[Rh(NH_3)_6]Y$ ($Y = MnF_6$, $CrCl_6$, $PbCl_6$, $CuCl_5$, $CuBr_5$) have also been reported [10, 18 to 21].

References:

[1] K. H. Schmidt, A. Mueller (Inorg. Chem. **14** [1975] 2183/7). – [2] W. P. Griffith (J. Chem. Soc. A **1966** 899/901). – [3] R. D. Hancock, A. Evers (J. Inorg. Nucl. Chem. **35** [1973] 2558/61). – [4] J. W. Palmer, F. Basolo (J. Inorg. Nucl. Chem. **15** [1960] 279/86). – [5] J. M. Terrasse, H. Poulet, J. P. Mathieu (Spectrochim. Acta **20** [1964] 305/15).

[6] M. Freymann (Ann. Chim. [Paris] [11] **11** [1939] 11/72, 47). – [7] J. P. Mathieu (J. Chim. Phys. **36** [1939] 308/25, 309, 322). – [8] H. Poulet, J. P. Mathieu (Compt. Rend. **255** [1962] 1514/5). – [9] I. M. Cheremisina (Zh. Strukt. Khim. **19** [1978] 336/51; J. Struct. Chem. USSR **19** [1978] 286/300, 297). – [10] K. Wieghardt, H. Siebert (Z. Anorg. Allgem. Chem. **381** [1971] 12/20).

[11] R. J. H. Clark, N. R. Trumble (J. Chem. Soc. Dalton Trans. **1976** 1145/9). – [12] C. K. Jørgensen (Acta Chem. Scand. **19** [1956] 500/17). – [13] H. H. Schmidtke (Z. Physik. Chem. [Frankfurt] **40** [1964] 96/108). – [14] T. R. Thomas, G. A. Crosby (J. Mol. Spectrosc. **38** [1971] 118/29). – [15] H. H. Schmidtke (J. Mol. Spectrosc. **11** [1963] 483/5).

[16] A. V. Babaeva, R. I. Rudyi (Izv. Akad. Nauk SSSR Ser. Fiz. **18** [1954] 729/30; Bull. Acad. Sci. USSR Phys. Ser. **18** [1954] 410/1). – [17] A. V. Babaeva, R. I. Rudyi (Zh. Neorgan. Khim. **1** [1956] 921/9; Russ. J. Inorg. Chem. **1** No. 5 [1956] 42/51, 45). – [18] H. H. Eysel (Z. Anorg. Allgem. Chem. **390** [1972] 210/6). – [19] P. Day, I. D. Hall (J. Chem. Soc. A **1970** 2679/82). – [20] G. C. Allen, N. S. Hush (Inorg. Chem. **6** [1967] 4/8).

[21] G. C. Allen, G. A. M. El-Sharkawy (Inorg. Nucl. Chem. Letters **6** [1970] 281/4).

Other Spectroscopic Studies of $[Rh(NH_3)_6]^{3+}$ Salts

The emission and excitation spectra of $[Rh(NH_3)_6]Cl_3$ have been recorded in solution at room temperature, in rigid glasses between 102 and 77 K, and as a solid. The ammine displayed a luminescence which could be assigned to a spin-forbidden d-d phosphorescence [1].

X-ray emission and absorption spectral studies have been made of $[Rh(NH_3)_6]Cl_3$ [2, 3]. The electronic structure of the central metal atom was investigated using the X-ray fluorescence emission $L\beta_2$ and edge-absorption L_{III} spectra of the Rh atom [2].

The X-ray photoelectron spectrum of $[Rh(NH_3)_6]Cl_3$ [4] and $[Rh(NH_3)_6][SbCl_6]$ [5] have been obtained, and the nitrogen 1s electron binding energies in $[Rh(NH_3)_6][NO_3]_3$ found to be 400.3 (NH_3) and 407.3 eV (NO_3), respectively [6].

The ^{14}N nuclear magnetic resonance spectrum of $[Rh(NH_3)_6]Cl_3$ exhibited a single resonance at $\delta = +295$ ppm (vs. CH_3CN) [7].

Ligand field photoexcitation of $[Rh(NH_3)_6]^{3+}$ in aqueous solution resulted in photo-dissociation and -aquation of one NH_3 group, giving $[Rh(NH_3)_5(H_2O)]^{3+}$. Similar studies were made of $[Rh(ND_3)_6]^{3+}$, and there was discussion of the mechanisms of non-radiative deactivation of the reactive excited state of $[Rh(NH_3)_6]^{3+}$ [8].

References:

[1] T. R. Thomas, G. A. Crosby (J. Mol. Spectrosc. **38** [1971] 118/29). – [2] L. Mazalov, A. P. Sadovskii, A. V. Belyaev, A. V. Chernyavskii, E. S. Gluskin, L. F. Berkhoer (Izv. Sibirsk. Otd. Akad. Nauk SSSR Ser. Khim. Nauk **1971** No. 2, pp. 51/9 from C.A. **76** [1972] No. 39710). – [3] A. P. Sadovskii, A. V. Belyaev (Izv. Sibirsk. Otd. Akad. Nauk SSSR Ser. Khim. Nauk **1967** No. 6, pp. 57/61; Sib. Chem. J. **1967** 703/6). – [4] N. I. Nefedov, M. A. Porai-Koshits (Mater. Res. Bull. **7** [1972] 1543/52). – [5] P. Burroughs, A. Hamnett, A. F. Orchard (J. Chem. Soc. Dalton Trans. **1974** 565/7).

[6] D. N. Hendrickson, J. M. Hollander, W. L. Jolly (Inorg. Chem. **8** [1969] 2642/7). – [7] R. Bramley, B. N. Figgis, R. S. Nyholm (J. Chem. Soc. A **1967** 861/3). – [8] J. P. Petersen, P. C. Ford (J. Phys. Chem. **78** [1974] 1144/9).

Chemical Reactions and Applications of $[Rh(NH_3)_6]^{3+}$ Salts

The rate of hydrogen exchange with $[Rh(NH_3)_6]^{3+}$ in 0.1M deuterioacetic acid-acetate buffer has been measured and the value compared with other hexammines [1]. Thermal decomposition of $[Rh(NH_3)_6]Cl_3$ afforded rhodium metal, NH_4Cl and N_2 [2]. The zeolite NaY has been exchanged with $[Rh(NH_3)_6]Cl_3$ to obtain a zeolite in which the Rh ammine was in the interior of the material. Treatment of this product with CO and hydrogen afforded a catalyst used in the hydroformylation of olefins [3]. The thermal decomposition of $[Rh(NH_3)_6](NO_2)_3$ has also been studied [4].

The effect of $[Rh(NH_3)_6]Cl_3$ on the sensitivity and gradation of photographic emulsions has been investigated [5].

References:

[1] T. W. Palmer, F. Basolo (J. Inorg. Nucl. Chem. **15** [1960] 279/86). – [2] W. W. Wendlandt, L. A. Funes (J. Inorg. Nucl. Chem. **26** [1964] 1879/84). – [3] E. Mantovani, N. Palladino, A. Zanobio (Ger. 2804307 [1978] from C.A. **89** [1978] No. 136452). – [4] A. V. Ablov, T. A. Mal'kova (Zh. Neorgan. Khim. **15** [1970] 3081/3; Russ. J. Inorg. Chem. **15** [1970] 1604/6). – [5] M. T. Beck, P. Kiss, T. Szalay, E. C. Porzsolt, G. Bazsa (J. Signalaufzeichnungsmaterialien **4** No. 2 [1976] 131/7 from C.A. **85** [1976] No. 70648).

2.1.7.2 Substituted Ammines

2.1.7.2.1 Pentammines (see "Rhodium" 1938, pp. 109/12, 129/41)

Pentammine Hydrides

$[Rh(NH_3)_5H]^{2+}$. The **sulphate** is prepared by treatment of $[Rh(NH_3)_5Cl]Cl_2$ with Zn dust at 60°C in aqueous $(NH_4)_2SO_4$ solution containing NH_4OH (10 to 20%) [1 to 5]. The **dibromide, di-iodide, tetrabromomercurate** and **perchlorate** are obtained by addition of KBr, KI, $K_2[HgBr_4]$ or $KClO_4$ to the sulphate in aqueous solution [3]. The deuteride, $[Rh(ND_3)_5D]^{2+}$, can be made by exchange with D_2O [1, 3]. The hydride can also be prepared by electrochemical reduction, at −1.19 V vs. standard calomel electrode, using a mercury cathode, of $[Rh(NH_3)_5(OH)]^{2+}$ in aqueous ammonia containing NH_4ClO_4 [6]. $[Rh(NH_3)_5H]^{2+}$ can also be formed when $[Rh(NH_3)_5Cl]^{2+}$ is treated with $NaBH_4$ in cold aqueous solution [10]. The ion $[Rh(NH_3)_5H]^{2+}$ is colourless in solution, the salts being pale yellow or white as solids, and are air-stable in the solid state. The salts are generally water-soluble although unstable towards aquation [1 to 5].

Structure. $[Rh(NH_3)_5H](ClO_4)_2$ is orthorhombic, space group $P2_12_12_1$-D_2^4 (Z=4) with a=10.2866(2), b=8.0689(10), c=15.0146(21) Å. The compound contained octahedral cations which were stacked in zig-zag chains of essentially linear Rh-N···N-Rh elements, whose Rh atoms were 6.013(1) Å apart. The ammine group which was directed towards the next cation in the chain was the closest to the Rh atom, with Rh-N=2.048(11) Å. The other three NH_3 groups *cis* to the hydride ligand had an average Rh-N distance of 2.079(7) Å. The NH_3 group which was *trans* to the hydride ligand had Rh-N=2.244(13) Å, indicating the strong *trans*-directing influence excerted by the H-ligand; the Rh-N bond length was 1.82(17) Å [6].

Spectra. The IR spectra of $[Rh(NH_3)_5H]^{2+}$ [1, 2, 3, 5] and $[Rh(NH_3)_5D]^{2+}$ have been recorded, and ν(N-H/D), ν(Rh-H), ν(Rh-N), $\delta(NH_3)N$ (asym. and sym.) and $\varrho(NH_3ND_3)$ reported [3]. The Rh-H stretching frequency (in liquid paraffin) was influenced by the nature of the counter-anion [3] (see table below). The electronic absorption spectra of $[Rh(NH_3)_5H]^{2+}$ have been measured between 36000 and 20000 cm^{-1} [1, 3, 7], and the position of the hydride ligand in the spectrochemical series established [7]. The ^1H NMR spectrum of $[Rh(NH_3)_5H]^{2+}$ in aqueous solution has been measured; δ(RhH)=−17.1 ppm, J(RhH)=14.5 Hz [3, 5].

Rh-H/D and Rh-N stretching frequencies in $[Rh(NH_3)_5H]^{2+}$ and $[Rh(ND_3)_5D]^{2+}$:

Compound	νRh-H/D (Nujol) in cm^{-1}	ν(Rh-N)	Ref.
$[Rh(NH_3)_5H](SO_4)$	2073		[1]
	2079	465, 494	[3, 5]
	2080		[2, 10]
$[Rh(ND_3)_5D](SO_4)$	1486	440, 472	[3]
$[Rh(NH_3)_5H](ClO_4)_2$	2126	499, 479	[3]
$[Rh(ND_3)_5D](ClO_4)$	1530	432, 400	[3]
$[Rh(NH_3)_5H]Br_2$	2015	483, 452	[3]
$[Rh(ND_3)_5D]Br_2$	1452[a)]	448, 413	[3]
$[Rh(NH_3)_5H]I_2$	2045	481, 452	[3]
$[Rh(NH_3)_5H][HgBr_4]$	2016	460, 457	[3]

[a)] Hexachlorobutadiene mull

Reactions. The pentammine hydride was unstable in aqueous solution, being aquated to give *trans*-$[Rh(NH_3)_4(H_2O)H]^{2+}$, and the equilibrium $[Rh(NH_3)_5H]^{2+} + H_2O \rightleftharpoons [Rh(NH_3)_4(H_2O)H]^{2+} + NH_3$ has been established [1, 5] (see tetrammine hydrides, p. 147). Ammonia could also be displaced by treatment of $[Rh(NH_3)_5H]^{2+}$ with aliphatic polyammines giving, for example, $[Rh(en)_2(NH_3)H]^{2+}$ and $[Rh(pn)_2(NH_3)H]^{2+}$ [8]; en=1,2-diaminoethane, pn=1,2-diaminopropane; these compounds are discussed in the section dealing with aliphatic diamino complexes, pp. 167, 187.

The hydride cation reacted with oxygen giving a blue or purple species which may be $[Rh(NH_3)_4(H_2O)OORh(NH_3)_5(H_2O)]^{5+}$ or $[Rh(NH_3)_4(OH)(H_2O)]^{+}$ [6], but which has not been fully characterised. It also reacted with alkenes and fluorinated alkynes affording species containing

Rh-C bonds, of the type $[Rh(NH_3)_5R]^{2+}$ where R = ethyl, n-propyl, n-butyl, C_4F_6H, C_2F_4H, $(CF_3)C_3(CF_3)H$ [3]. With CN^-, the hydride gave either $[Rh(CN)_5H]^{3-}$ or $[Rh(CN)_4(OH_2)H]^{2-}$ [8], whereas HCl caused precipitation of Rh metal, and 50% acetic acid afforded $[Rh(OCOCH_3)_2(H_2O)]_2$ [3].

Water-soluble unsaturated carbocyclic acids could be hydrogenated using hydrogen in the presence of $[Rh(NH_3)_5H]^{2+}$ in aqueous solution [9].

References:

[1] J. A. Osborn, A. R. Powell, G. Wilkinson (J. Chem. Soc. Chem. Commun. **1966** 461/2). – [2] A. P. Powell (Platinum Metals Rev. **11** [1967] 58/9). – [3] K. Thomas, J. A. Osborn, A. R. Powell, G. Wilkinson (J. Chem. Soc. A **1968** 1801/6). – [4] A. R. Powell (Brit. 1196583 [1970] from C.A. **73** [1970] No. 59620). – [5] J. A. Osborn, K. Thomas, G. Wilkinson (Inorg. Syn. **13** [1971] 213/5).

[6] B. A. Coyle, J. A. Ibers (Inorg. Chem. **11** [1972] 1105/9). – [7] J. A. Osborn, R. D. Gillard, G. Wilkinson (J. Chem. Soc. **1964** 3168/73). – [8] K. Thomas, G. Wilkinson (J. Chem. Soc. A **1970** 356/60). – [9] G. C. Bond, Johnson, Matthey & Co. Ltd. (Brit. 1197723 [1970] from C.A. **73** [1970] No. 9840). – [10] R. D. Gillard, G. Wilkinson (J. Chem. Soc. **1963** 3594/9).

Pentammine Cyanides

$[Rh(NH_3)_5CN]X_2$. Reaction of *trans*-$Na[Rh(NH_3)_4(SO_3)_2]$ with aqueous KCN affords $[Rh(NH_3)_4(SO_3)(CN)]\cdot 2H_2O$ which, on treatment with aqueous HCl, affords $[Rh(NH_3)_4(OH)_2CN]Cl_2$. In the presence of ammonia and ammonium chloride this last compound gives the **dichloride** $[Rh(NH_3)_5CN]Cl_2$. When this is dissolved in hot water and treated with conc. HNO_3, the **dinitrate,** $[Rh(NH_3)_5CN](NO_3)_2$ is formed [1].

Both the dichloride and dinitrate formed as white solids. The IR spectrum of $[Rh(NH_3)_5CN]Cl_2$ exhibited ν(CN) at 2121 cm^{-1}, and the compound was reduced polarographically at $E_{1/2} = -1.36$ V vs. standard calomel electrode [1]. The Rh-N stretching frequencies of the dichloride and dinitrate species have been reported [2].

Reaction with $AgNO_3$ afforded $[Rh(NH_3)_5CN]_2\cdot Ag(NO_3)_5\cdot 1.5H_2O$ [1].

References:

[1] I. B. Baranovskii, A. V. Babaeva (Zh. Neorgan. Khim. **13** [1968] 3148/50; Russ. J. Inorg. Chem. **13** [1968] 1624). – [2] I. B. Baranovskii, Yu. Ya. Kharitonov, G. Ya. Mazo (Zh. Neorgan. Khim. **15** [1970] 1715/7; Russ. J. Inorg. Chem. **15** [1970] 881/3).

Pentammines with Oxygen Donors

Aqua and Hydroxo Species

$[Ru(NH_3)_5(H_2O)]^{3+}$. The **perchlorate** $[Rh(NH_3)_5(H_2O)](ClO_4)_3$ is obtained by reaction of a hot aqueous solution of $[Rh(NH_3)_5Cl]Cl_2$ with NaOH, followed by recrystallisation from dilute $HClO_4$ [1, 2, 3], or with $AgClO_4$ in water [4]. The ^{18}O-labelled species $[Rh(NH_3)_5(H_2{}^{18}O)](ClO_4)_3$ is prepared by dissolving the unlabelled aqua cation in a 10^{-3}M $HClO_4/H_2{}^{18}O$ solution and refluxing the mixture [7]. The **bromide** $[Rh(NH_3)_5(H_2O)]Br_3$ is obtained by reaction of $[Rh(NH_3)_5Cl]Cl_2$ with NaOH, followed by careful treatment of the $[Rh(NH_3)_5OH]^{2+}$ with HBr [1].

The **trifluorosulphonate** $[Rh(NH_3)_5(H_2O)](OSO_2CF_3)_3$ is prepared by treatment of $[Rh(NH_3)_5Cl]Cl_2$ with $AgOSO_2CF_3$ in $(CH_3)_2NHCO$ containing water [5].

The **nitric acid adduct** $[Rh(NH_3)_5(H_2O)](NO_3)_3 \cdot HNO_3$ is obtained by recrystallisation of $[Rh(NH_3)_5(H_2O)](NO_3)_3$ from dilute nitric acid [6].

References:

[1] G. W. Bushnell, G. C. Lalor, E. A. Moelwyn-Hughes (J. Chem. Soc. A **1966** 712/23). – [2] G. C. Lalor, G. W. Bushnell (J. Chem. Soc. A **1968** 2520/2). – [3] S. C. Chan (Australian J. Chem. **20** [1967] 61/8). – [4] T. W. Swaddle, D. R. Stranks (J. Am. Chem. Soc. **94** [1972] 8357/60). – [5] L. M. Jackman, R. M. Scott, R. H. Portman, J. F. Dornish (Inorg. Chem. **18** [1979] 1497/502).

[6] R. Ugo, R. D. Gillard (Inorg. Chim. Acta **1** [1967] 311/4). – [7] F. Monacelli, E. Viel (Inorg. Chim. Acta **1** [1967] 467/70).

Spectra. The Rh-N stretching frequencies of $[Rh(NH_3)_5(H_2O)]Br_3$ have been determined (480 and 453 cm^{-1}) [2]. The electronic spectra of $[Rh(NH_3)_5(H_2O)]Cl_3$ have been obtained in water/methanol/ethyleneglycol glasses at 85 K [3], and in aqueous solution between 20000 and 45000 cm^{-1} [4, 5, 6]. The luminescence spectrum of $[Rh(NH_3)_5(H_2O)]Cl_3$ was measured in water/methanol/ethyleneglycol glasses at 85 K [3].

Electrochemistry. Polarographic studies have shown that $[Rh(NH_3)_5(H_2O)]^{3+}$ undergoes a two-electron irreversible reduction at −1.07 V vs. standard calomel electrode (0.1 M $NaClO_4$ solution) [7, 8].

References:

[1] A. D. Cunningham, D. A. House, H. J. K. Powell (Australian J. Chem. **23** [1970] 2375/8). – [2] I. B. Baranovskii, Yu. Ya. Kharitonov, G. Ya. Mazo (Zh. Neorgan. Khim. **15** [1970] 1715/7; Russ. J. Inorg. Chem. **15** [1970] 881/3). – [3] P. E. Hoggard, H.-H. Schmidtke (Ber. Bunsenges. Physik. Chem. **77** [1973] 1052/8). – [4] C. J. Jørgensen (Acta Chem. Scand. **10** [1956] 500/17). – [5] D. A. Palmer, G. M. Harris (Inorg. Chem. **13** [1974] 965/9).

[6] K. Nag, P. Banerjee (Indian J. Chem. A **14** [1976] 418/20). – [7] A. A. Vlcek (Discussion Faraday Soc. No. 26 [1958] 164/71). – [8] H.-H. Schmidtke (Z. Physik. Chem. [Frankfurt] **45** [1965] 306/16).

Chemical Reactions. The kinetics of displacement of water by halide ion in $[Rh(NH_3)_5(H_2O)]^{3+}$ have been studied extensively [1], as a function of ionic strength [2, 3] and nucleophile concentration [3, 4, 5], pressure [5], and the possibility of ion-pairing [3]. The activation parameters for the anation reactions of $[Rh(NH_3)_5(H_2O)]^{3+}$ by Cl^-, Br^-, I^-, NCS^-, N_3^-, $CH_3CO_2^-$, oxalate and carbonate have been reported [6].

The rate of exchange between bound water in $[Rh(NH_3)_5(H_2O)]^{3+}$ and solvent water have been determined using ^{18}O tracer techniques, and the mechanism is apparently associative [7, 8]. Photochemical excitation of $[Rh(NH_3)_5(H_2O)]^{3+}$ led to exchange of solvent and coordinated water, and if Cl^- or Br^- was present, $[Rh(NH_3)_5X]^{2+}$ (X = Cl, Br) was formed [9, 10].

The pulse radiolytic one-electron reduction of $[Rh(NH_3)_5(H_2O)]^{3+}$ afforded solvated $[Rh(NH_3)_4]^{2+}$ which reacted with O_2 apparently giving the relatively stable $[Rh(NH_3)_4(O_2)(H_2O)]^{2+}$, which was not isolated [11]. The rate constants for reaction of carbonate radical with $[Rh(NH_3)_5(H_2O)]^{3+}$ has been measured in aqueous solution [12].

The thermal deaquation of $[Rh(NH_3)_5(H_2O)]X_3$ (X = Br, I or ClO_4) has been studied by thermogravimetric, differential thermal, thermomagnetic and gas evolution analytical techniques [13, 14].

References:

[1] A. B. Lamb (J. Am. Chem. Soc. **61** [1939] 699/708). – [2] M. J. Pavelich, S. M. Maxey, C. Pfaff (Inorg. Chem. **17** [1978] 564/70). – [3] H. L. Bott, A. J. Poe, K. Shaw (J. Chem. Soc. A **1970** 1745/50). – [4] F. Monacelli (Inorg. Chim. Acta **2** [1968] 263/8). – [5] R. van Eldik, D. A. Palmer, H. Kelm (Inorg. Chem. **18** [1979] 1520/7).

[6] R. van Eldik (Z. Anorg. Allgem. Chem. **416** [1975] 88/96). – [7] T. W. Swaddle, D. R. Stranks (J. Am. Chem. Soc. **94** [1972] 8357/60). – [8] F. Monacelli, E. Viel (Inorg. Chim. Acta **1** [1967] 467/70). – [9] P. C. Ford, J. D. Petersen (Inorg. Chem. **14** [1975] 1404/8). – [10] G. A. Shagisultanova, V. V. Yasinetskii (Zh. Fiz. Khim. **50** [1976] 91/5; Russ. J. Phys. Chem. **50** [1976] 51/3).

[11] J. Lilie, M. G. Simic, J. F. Endicott (Inorg. Chem. **14** [1975] 2129/33). – [12] V. W. Cope, M. Z. Hoffman, N.-S. Chen (J. Phys. Chem. **82** [1978] 2665/9). – [13] W. W. Wendlandt, P. F. Franke (J. Inorg. Nucl. Chem. **26** [1964] 1885/93). – [14] J. P. Smith, W. W. Wendlandt (Nature **201** [1964] 291/2).

$[Rh(NH_3)_5(OH)]^{2+}$. The hydroxo species is obtained by heating $[Rh(NH_3)_5Cl]Cl_2$ with an excess of Ag_2O in water [1]. It is formed as a bright yellow solid which is soluble in water.

The heats of neutralization of $[Rh(NH_3)_5(H_2O)](ClO_4)_3$ by NaOH (which gave $[Rh(NH_3)_5OH]^{2+}$) in aqueous acidic solutions containing $HClO_4$ have been determined, and the data compared with that from similar studies of $[M(NH_3)_5(H_2O)]^{3+}$ (M = Cr, CO) [2].

The electronic spectra of $[Rh(NH_3)_5(OH)](ClO_4)_2$ have been obtained in water/methanol/ethyleneglycol glasses at 85 K [3], and in aqueous solution between 20000 and 45000 cm^{-1} [4, 5, 6]. The luminescence spectrum of $[Rh(NH_3)_5(OH)](ClO_4)_2$ was also measured [3].

From studies using ^{18}O tracer techniques, it was clear that no appreciable exchange occurred between $[Rh(NH_3)_5(OH)]^{2+}$ and 0.1M NaOH [7]. $[Rh(NH_3)_5(OH)]^{2+}$ did not undergo photo-exchange or photo-anation when undergoing ligand field excitation [8].

References:

[1] F. Monacelli, F. Basolo, R. G. Pearson (J. Inorg. Nucl. Chem. **24** [1962] 1241/50). – [2] A. J. Cunningham, D. A. House, H. K. J. Powell (Australian J. Chem. **23** [1970] 2375/8). – [3] P. E. Hoggard, H. H. Schmidtke (Ber. Bunsenges. Physik. Chem. **77** [1973] 1052/8). – [4] C. J. Jørgensen (Acta Chem. Scand. **10** [1956] 500/17). – [5] D. A. Palmer, G. M. Harris (Inorg. Chem. **13** [1974] 965/9).

[6] K. Nag, P. Banerjee (Indian J. Chem. A **14** [1976] 418/20). – [7] F. Monacelli, E. Viel (Inorg. Chim. Acta **1** [1967] 467/70). – [8] P. C. Ford, J. D. Petersen (Inorg. Chem. **14** [1975] 1404/8).

Alkoxy and Alcoholato Complexes

$[Rh(NH_3)_5(OCH_3)](ClO_4)_2$. This compound, isolated with ½ CH_3OH of crystallisation, is prepared by heating of $[Rh(NH_3)_5(OCOCCl_3)](ClO_4)_2$ with 1M $NaOCH_3$ in methanol [1]. The pure perchlorate can be obtained if the product is recrystallised in the presence of $LiClO_4$ [2].

$[Rh(NH_3)_5(OC_2H_5)](ClO_4)_2$ is obtained similarly, using $NaOC_2H_5$ and dry ethanol [2]. The compound forms yellow crystals and is soluble in water.

The electronic spectrum of the methoxide has been measured in the range 240 to 340 nm (shown in paper) [1]. The ^1H NMR spectra of the complexes were recorded in D_2O (see table below). The pK_a values of $[Rh(NH_3)_5(OCH_3)]^{2+}$ and $[Rh(NH_3)_5(OC_2H_5)]^{2+}$ were determined to be 5.8 and 6.0, respectively [2].

[Rh(NH$_3$)$_5$(ROH)](ClO$_4$)$_3$ (R = CH$_3$ or C$_2$H$_5$) are prepared by reaction of [Rh(NH$_3$)$_5$(OR)](ClO$_4$)$_2$ with the stoichiometric amount of HClO$_4$, but the compounds have apparently not been isolated. Their electronic spectra have been recorded in the range 240 to 340 nm (shown in paper), and details of their ^{1}H NMR spectra given in the table. The kinetics of aquation of [Rh(NH$_3$)$_5$(ROH)]$^{3+}$ have been determined [2].

^{1}H NMR spectrum obtained from [Rh(NH$_3$)$_5$(OR)]$^{2+}$ and [Rh(NH$_3$)$_5$(ROH)]$^{3+}$:

Complex	ppm vs. TMS	τ(Rh-H) in Hz
[Rh(NH$_3$)$_5$(OCH$_3$)]$^{2+}$	3.06	0.6 ± 0.1
[Rh(NH$_3$)$_5$(CH$_3$OD)]$^{3+}$	3.39	0.6 ± 0.1
[Rh(NH$_3$)$_5$(OC$_2$H$_5$)]$^{2+}$	3.28q(CH$_2$)$^{a)}$	6.8 ± 0.2
[Rh(NH$_3$)$_5$(C$_2$H$_5$OD)]$^{3+}$	3.66q(CH$_2$)$^{a)}$	7.2 ± 0.2

$^{a)}$ CH$_3$ signal obscured by t-butyl alcohol internal reference in D$_2$O.

[Rh(NH$_3$)$_5$(OCHNH$_2$)](ClO$_4$)$_3$ is prepared by photolysis of [Rh(NH$_3$)$_5$(OH$_2$)](ClO$_4$)$_3$ in formamide. It forms yellow crystals and the IR spectrum (Nujol) exhibited bands at 1715 and 1390 cm^{-1} characteristic of O-coordinated formamide. The electronic spectrum of the complex has also been measured [3]. The kinetics and mechanism of formamide between solvent and bound ligand in [Rh(NH$_3$)$_5$(OCHNH$_2$)](ClO$_4$)$_3$ have been determined [4].

[Rh(NH$_3$)$_5$(OSMe$_2$)](ClO$_4$)$_3$ is obtained by heating [Rh(NH$_3$)$_5$(OH$_2$)](ClO$_4$)$_3$ in dimethylsulphoxide, followed by addition of aqueous HClO$_4$. The complex is a yellow solid which is soluble in water and (CH$_3$)$_2$SO. The IR spectrum (Nujol) exhibited bands at 982 and 943 cm^{-1} characteristic of O-bound dimethylsulphoxide, and the electronic absorption spectrum of the complex was measured [3].

References:

[1] F. Monacelli (Inorg. Chim. Acta **1** [1967] 271/4, 271). – [2] E. Borghi, F. Monacelli (Inorg. Chim. Acta **23** [1977] 53/7, 53/4). – [3] M. A. Bergkamp, R. J. Watts, P. C. Ford (J. Am. Chem. Soc. **102** [1980] 2627/31, 2628). – [4] S. T. D. Lo, M. J. Sisley, T. W. Swaddle (Can. J. Chem. **56** [1978] 2609/15).

Carboxylato-pentammine Complexes

[Rh(NH$_3$)$_5$(OCOR)](ClO$_4$)$_2$ (R = H, CH$_3$, CH$_2$F, CHF$_2$, CF$_3$, CCl$_3$). The formate, acetate, pivaloate (R = C(CH$_3$)$_3$) and trifluoroacetate are prepared by reaction of [Rh(NH$_3$)$_5$(OH)](OH)$_2$ with an excess of the appropriate carboxylic acid [1, 2, 8]. The acetate may also be made, as the trifluoromethanesulphonate salt [Rh(NH$_3$)$_5$(OCOCH$_3$)](OSO$_2$CF$_3$)$_2$, by treatment of [Rh(NH$_3$)$_5$-(OH$_2$)](OSO$_2$CF$_3$)$_3$ with acetic anhydride and PhCH$_2$N(CH$_3$)$_2$ [3]. The monofluoro-, difluoro- and trichloro-acetato species are prepared by treatment of [Rh(NH$_3$)$_5$(OH)](ClO$_4$)$_3$ with hot 50% aqueous carboxylic acid [4].

The perchlorate salts where R = H, CH$_3$, C(CH$_3$)$_3$ and CF$_3$ were obtained as white crystals [1, 2], while the remaining compounds formed yellow crystals [3, 4].

[Rh(NH$_3$)$_5$(OCOC$_2$H$_5$)]$^{2+}$ is obtained by reaction of [Rh(NH$_3$)$_5$(OH)]$^{2+}$ with propionic anhydride [5], but no preparative details were given.

From the IR spectra of $[Rh(NX_3)_5(OCOR)]Br_2$, (X = H or D; R = H, D, CH_3, CD_3, the ν(RhN) has been obtained [6, 8]. The electronic spectra of $[Rh(NH_3)_5(OCOR)]^{2+}$ (R = CH_3, $C(CH_3)_3$, CF_3) have been reported [1, 2] (spectra shown between 20000 and 50000 cm^{-1} for R = H, CH_3 [2]), and the luminescence spectra of (R = H or CH_3) measured [7].

$[Rh(NH_3)_5(OCOCH_2NH_3)](ClO_4)_3$ is formed when $[Rh(NH_3)_5(H_2O)](ClO_4)_3$ is refluxed in an aqueous solution containing an excess of glycine. The electronic spectrum has been measured (region 250 to 350 nm shown in paper), and it was clear that the glycine is monodentate and bound to the Rh via an O atom. The kinetics and mechanism of formation of the complex were determined [9].

$[Rh(NH_3)_5(C_2O_4H)](ClO_4)_2$ is obtained by reaction of $[Rh(NH_3)_5(H_2O)](ClO_4)_3$ with oxalic acid [10, 11], or from $[Rh(NH_3)_5OH](OH)_2$ and H_2O_4 and $HClO_4$. The electronic absorption spectrum was reported [10, 11], also the infrared spectrum [12], and the rate constants for the anation of $[Rh(NH_3)_5(H_2O)]^{3+}$ by $H_2C_2O_4$, $HC_2O_4^-$ and $C_2O_4^{2-}$ were determined [10].

$[Rh(NH_3)_5(C_2O_4)]\cdot 2NaClO_4\cdot 3H_2O$ is made from $[Rh(NH_3)_5(OH)](OH)_2$, $H_2C_2O_4$, $NaClO_4$ and $HClO_4$ [12].

For **$[(NH_3)_3Co(\mu\text{-}OH)_2(\mu^3\text{-}C_2O_4)Rh(NH_3)_5][Co(NH_3)_3](ClO_4)_5\cdot 2H_2O$** see p. 8.

References:

[1] F. Monacelli, F. Basolo, R. G. Pearson (J. Inorg. Nucl. Chem. **24** [1962] 1241/7). – [2] H.-H. Schmidtke (Z. Physik. Chem. [Frankfurt] **45** [1965] 305/16, 313). – [3] L. M. Jackman, R. M. Scott, R. H. Portman, J. F. Dornish (Inorg. Chem. **18** [1979] 1497/502). – [4] F. Monacelli (J. Inorg. Nucl. Chem. **29** [1967] 1079/87). – [5] D. A. Buckingham, L. M. Engelhardt (J. Am. Chem. Soc. **97** [1975] 5915/7).

[6] I. B. Baranovskii, Yu. Ya. Kharitonov, G. Ya. Mazo (Zh. Neorgan. Khim. **15** [1970] 1715/7; Russ. J. Inorg. Chem. **15** [1970] 881/3). – [7] P. E. Hoggard, H.-H. Schmidtke (Ber. Bunsenges. Physik. Chem. **77** [1973] 1052/8). – [8] Yu. Ya. Kharitonov, G. Ya. Mazo, I. B. Baranovskii (Zh. Neorgan. Khim. **15** [1970] 2305/8; Russ. J. Inorg. Chem. **15** [1970] 1193/4). – [9] C. Chatterjee, A. K. Basak (Bull. Chem. Soc. Japan **52** [1979] 2710/2). – [10] R. van Eldik (Z. Anorg. Allgem. Chem. **416** [1975] 88/96).

[11] R. van Eldik (React. Kinet. Catal. Letters **2** [1975] 251/6 from C.A. **83** [1975] No. 103950). – [12] K. Wieghardt (Z. Naturforsch. **29b** [1974] 809/10).

Nitrito- and Nitro-pentammine Complexes

$[Rh(NH_3)_5(ONO)]^{2+}$. The **dichloride** is obtained by reaction of $[Rh(NH_3)_5(OH)]Cl_2$ with sodium nitrite at ice-bath temperatures with a small amount of HCl [1]. If $HClO_4$ (90%) is used instead of HCl, the corresponding **perchlorate** is formed [6]. Both the chloride and perchlorate were isolated as white crystals, which were water-soluble, and readily isomerised to the nitro isomer, $[Rh(NH_3)_5(NO_2)]^{2+}$ in the solid state and in solution [1, 2, 3].

The IR spectrum of $[Rh(NH_3)_5(ONO)]Cl_2$ between 2000 and 800 cm^{-1} has been measured, ν(ONO) ≈ 1460 and 1065 cm^{-1} [2, 4]. An assignment of the fundamental modes associated with the ONO group in $[Rh(NR_3)_5(ONO)]Cl_2$ (R = H or D) and $[Rh(NH_3)_5(O^{15}NO)]Cl_2$ has been made [5]. The electronic spectrum of the nitrito complex has been measured in aqueous solution (spectrum shown in paper between 20000 and 50000 cm^{-1}) [8].

The kinetics of base-catalysed isomerisation of the nitrito- to the nitro-form have been measured: $k_s = 9.6(\pm 0.2) \times 10^{-4}\ s^{-1}$ [3]; rate of isomerisation (in $10^{-4}\ s^{-1}$) in the solid state: k(KBr disc) = 0.75 (25°C), 1.7(35°C), 5.5(51°C); k(Nujol) = $1.6 \times 10^{-4}\ s^{-1}$ [2]. Activation volumes for the linkage isomerisation of $[Rh(NH_3)_5(ONO)](NO_3)_2$ in aqueous solution were also measured [6]. The kinetics of acid-catalysed hydrolysis of $[Rh(NH_3)_5(ONO)]^{2+}$ have been determined [7].

$[Rh(NH_3)_5(NO_2)]^{2+}$. The dichloride is obtained by reaction of $[Rh(NH_3)_5(H_2O)]^{3+}$ with HNO_2 in the presence of HCl at room temperature [2], or by base-catalysed isomerisation of the corresponding nitrito isomer [3].

The stability constants of $[Rh(NH_3)_5(NO_2)]X$ ($X = SO_4^{2-}$, SeO_3^{2-}, $S_2O_3^{2-}$, TeO_3^{2-}, CO_3^{2-}) were determined spectrophotometrically and by electrical conductivity measurements [14, 18]. The diamagnetic susceptibility of $[Rh(NH_3)_5(NO_2)]^{2+}$ has been calculated [19].

The IR spectrum of $[Rh(NH_3)_5(NO_2)]Cl_2$ between 2000 and 800 cm^{-1} has been measured, and bands at 1420 and 830 cm^{-1} associated with $\nu(RhNO_2)$ [2, 4]. An assignment of the fundamental modes of the NO_2 group in $[Rh(NR_3)_5(NO_2)]Cl_2$ (R = H or D) and $[Rh(NH_3)_5(^{15}NO_2)]Cl_2$ has been made [5]. Assignments of the Rh-N vibrations in $[Rh(NR_3)_5(NO_2)]Cl_2$ (R = H or D) have also been made [20]. The electronic spectrum of $[Rh(NH_3)_5(NO_2)]^{2+}$ in aqueous solution has been measured (spectrum shown between 20000 and 45000 cm^{-1}) [9], and the luminescence spectrum determined at 85 K [10]. The N(1s) binding energies for $[Rh(NH_3)_5(NO_2)]Br_2$ have been obtained from the photoelectron spectrum of the compound: $-NO_2 = 404.4$ eV, $NH_3 = 400.3$ eV [11].

$[Rh(NH_3)_5(NO_2)]^{2+}$ is reduced polarographically in an irreversible two-electron step at −1.10 [12] or −1.26 V [13] (0.1M $NaClO_4$ solution, standard calomel electrode).

The kinetics of acid- and base-catalysed hydrolysis of $[Rh(NH_3)_5(NO_2)]^{2+}$ have been determined [7, 15] and the volumes of activation of the base catalysed process measured [16]. Thermogravimetric curves have been obtained from $[Rh(NH_3)_5(NO_2)](NO_3)_2$ [12], and on heating $[Rh(NH_3)_5(NO_2)](NO_2)_2$ at 190 to 200°C, $Rh(NH_3)_3(NO_2)_3$ was formed [17].

References:

[1] F. Basolo, G. S. Hammaker (J. Am. Chem. Soc. **82** [1966] 1001/2). – [2] F. Basolo, G. S. Hammaker (Inorg. Chem. **1** [1962] 1/5). – [3] W. G. Jackson, G. A. Lawrence, P. A. Lay, A. M. Sargeson (Inorg. Chem. **19** [1980] 904/10). – [4] J. L. Burmeister (Coord. Chem. Rev. **3** [1968] 225/45, 226/7). – [5] M. J. Cleare, W. P. Griffith (J. Chem. Soc. A **1967** 1144/7).

[6] M. Mares, D. A. Palmer, H. Kelm (Inorg. Chim. Acta **27** [1978] 153/6). – [7] B. E. Crossland, P. J. Staples (J. Chem. Soc. A **1971** 2853/6). – [8] H.-H. Schmidtke (Z. Physik. Chem. [Frankfurt] **45** [1965] 305/16). – [9] C. K. Jørgensen (Acta Chem. Scand. **10** [1956] 500/17). – [10] P. E. Hoggard, H.-H. Schmidtke (Ber. Bunsenges. Physik. Chem. **77** [1973] 1052/8).

[11] D. N. Hendrickson, J. M. Hollander, P. L. Jolly (Inorg. Chem. **8** [1969] 2642/7). – [12] I. B. Baranovskii (Zh. Neorgan. Khim. **12** [1967] 2860/3; Russ. J. Inorg. Chem. **12** [1967] 1512/3). – [13] K. Nag, P. Banerjee (Indian J. Chem. A **14** [1976] 418/20). – [14] A. K. Pyartman, V. E. Mironov, M. A. Chugunnikova (Zh. Neorgan. Khim. **23** [1978] 260/2; Russ. J. Inorg. Chem. **23** [1978] 147/8). – [15] S. Balt, A. Jelsma (J. Inorg. Nucl. Chem. **43** [1981] 1287/91).

[16] D. A. Palmer (Australian J. Chem. **32** [1979] 2589/95). – [17] A. V. Ablov, T. A. Mel'kova (Zh. Neorgan. Khim. **15** [1970] 3081/3; Russ. J. Inorg. Chem. **15** [1970] 1604/6). – [18] M. V. Sof'in, A. K. Pyartman, M. A. Chugunnikova, V. E. Mironov (Zh. Fiz. Khim. **51** [1977] 1281; Russ. J. Phys. Chem. **51** [1977] 760). – [19] C. J. Ballhausen, R. W. Asmussen (Acta Chem. Scand. **11** [1959] 479/83). – [20] I. B. Baranovskii, Yu. Ya. Kharitonov, G. Ya. Mazo (Zh. Neorgan. Khim. **15** [1970] 1715/7; Russ. J. Inorg. Chem. **15** [1970] 881/3).

Nitrato-pentammine Complexes

$[Rh(NH_3)_5(NO_3)]X_2$. The **nitrate** ($X = NO_3$) is prepared by reaction of $[Rh(NH_3)_5(H_2O)](ClO_4)_3$, dissolved in the minimum of water, with conc. HNO_3 [1]. It can also be obtained by heating $[Rh(NH_3)_5(H_2O)](NO_3)_3$ [2].

The **perchlorate** is formed by treatment of a saturated aqueous solution of $[Rh(NH_3)_5(NO_3)](NO_3)_2$ with conc. $HClO_4$ [2]. Both salts were yellow and were soluble in water, although the dinitrate was less so after drying at 120°C for several hours [1]; see also "Rhodium" Suppl. Vol. B 1, 1982, p. 66.

The simultaneous determination of thermally stimulated light emission and differential thermal analysis curves of $[Rh(NH_3)_5(NO_3)](NO_3)_2$ has been made [3]. The acid and base hydrolysis of $[Rh(NH_3)_5(NO_3)]^{2+}$ and the ammoniation of $[Rh(NH_3)_5(NO_3)]^{2+}$ have been studied kinetically [1].

References:

[1] F. Monacelli, S. Viticoli (Inorg. Chim. Acta **7** [1973] 231/4, 231). – [2] W. P. Griffith (The Chemistry of Rarer Platinum Metals, Interscience, New York 1967, p. 340). – [3] W. W. Wendlandt (Thermochim. Acta **34** [1980] 313/20). – [4] S. Balt, A. Jelsma (Inorg. Chem. **20** [1981] 733/7).

Sulphato-pentammine Complex $[Rh(NH_3)_5(SO_4)](ClO_4)$

This complex is prepared by treatment of $[Rh(NH_3)_5(H_2O)](ClO_4)_3$ with a strong anion exchanger in the OH^- form, the resulting solution being neutralised by H_2SO_4. After evaporation the yellow solid, which may have been $[Rh(NH_3)_5(H_2O)](SO_4)_{1.5}$, was dehydrated in vacuo, dissolved in water and treated with $Ba(ClO_4)_2$, when the perchlorate precipitated. The compound was obtained as a white solid which was soluble in water. The electronic spectrum has been measured in aqueous solution (spectrum from 240 to 360 mμ shown in paper) [1]. The ^1H NMR spectrum of $[Rh(NH_3)_5(HSO_4)]^{2+}$ was determined in H_2SO_4 solution [2]. The acid and base hydrolysis of $[Rh(NH_3)_5(SO_4)]^+$ has been measured [3].

References:

[1] F. Monacelli (Inorg. Chim. Acta **2** [1968] 263/8, 264). – [2] D. N. Henderson, W. L. Jolly (Inorg. Chem. **9** [1970] 1197/201). – [3] F. Monacelli (Inorg. Chim. Acta **7** [1973] 65/73).

For **$[Rh(NH_3)_5SO_3]^+$**, see "Rhodium" Suppl. Vol. B 1, 1982, p. 151; for **$[Rh(NH_3)_5(NCS)]^{2+}$** and **$[Rh(NH_3)_5(SCN)]^{2+}$**, see p. 137.

Carbonato- and Phosphato-pentammine Complexes

$[Rh(NH_3)_5(CO_3)](ClO_4) \cdot H_2O$ is prepared by reaction of $[Rh(NH_3)_5(H_2O)]^{3+}$ in aqueous solution at pH = 8 to 8.5 either with Li_2CO_3 or with LiOH followed by passage of CO_2 [1, 2]. The carbonate is also formed in the reaction of $[Rh(NH_3)_5(OH)]^{2+}$ with CO_2 [3], and is isolated as a pale yellow solid [1]. The kinetics of the acid-catalysed decarboxylation of $[Rh(NH_3)_5(CO_3)]^{2+}$ [2] and of the uptake of CO_2 by $[Rh(NH_3)_5(OH)]^{2+}$ [2, 3] have been investigated. Mechanistic information has been obtained on the effect of pressure on the formation and acid-catalysed aquation of $[Rh(NH_3)_5(CO_3)]^+$ [5].

$Rh(NH_3)_5PO_4$. This complex is obtained by reaction of $[Rh(NH_3)_5Cl]Cl_2$ with aqueous $AgNO_3$ followed by addition of $NaH_2PO_4 \cdot H_2O$. The complex was purified by ion exchange and was isolated as a yellow solid. The ^{31}P NMR spectrum of this compound ($\delta(P) = -13.8$ ppm vs. external H_3PO_4 at pH = 12) was dependent on pH, and there was evidence for the existence of $[Rh(NH_3)_5(H_2PO_4)]^{2+}$ ($\delta(P) = -7.9$ ppm at pH = 1.8) and $[Rh(NH_3)_5(HPO_4)]^+$ ($\delta(P) = -11.9$ ppm at pH = 6.2). The ^{31}P NMR spectra of $[Rh(NH_3)_5(PO_3F)]^{2-}$ ($\delta(P) = -7.8$ ppm at pH = 7.8) and $[Rh(NH_3)_5(HPO_3)]^{2-}$ ($\delta(P) = -13.9$ ppm at pH = 7.0) were also recorded [4].

References:

[1] S. Ficner, D. A. Palmer, T. P. Dasgupta, G. M. Harris (Inorg. Syn. **17** [1977] 152/5). – [2] D. A. Palmer, G. M. Harris (Inorg. Chem. **13** [1974] 965/9). – [3] G. M. Harris, T. P. Dasgupta (J. Indian Chem. Soc. **54** [1977] 62/7). – [4] P. Seel, G. Bohnstedt (Z. Anorg. Allgem. Chem. **435** [1977] 257/67). – [5] U. Spitzer, R. van Eldik, H. Kelm (Inorg. Chem. **21** [1982] 2821/3).

Pentammine Complexes with N- or S-Donors

For **$[Rh(NH_3)_5NO_2]^{2+}$**, see p. 134.

Thiocyanato and Isothiocyanato Complexes

$[Rh(NH_3)_5(NCS)]X_2$. The **perchlorate** ($X = ClO_4$) is obtained by heating $[Rh(NH_3)_5(H_2O)](ClO_4)_3$ with NaCNS in aqueous solution [1, 2]. The **bromide** (X = Br) is prepared by reaction of $[Rh(NH_3)_5(H_2O)]Br_3$ with an excess of KCNS in water, and the deuteride, $[Rh(ND_3)_5(NCS)]Br_2$ is prepared similarly from $[Rh(ND_3)_5(H_2O)]Br_3$ [3].

$[Rh(NH_3)_5(SCN)]X_2$. The **perchlorate** is isolated when $[Rh(NH_3)_5(H_2O)](ClO_4)_3$ is allowed to stand in aqueous solution containing NaSCN [2].

The **isothiocyanate** (-NCS) perchlorate was light yellow whereas the **thiocyanate** (-SCN) was deep yellow; both are water soluble [2]. There was negligible interconversion of the two isomers at 40°C over 48 h, but at 78°C, the S-bonded form converted to the N-bonded isomer over 3 h [1].

From IR spectral studies of $[Rh(NH_3)_5(NCS)]^{2+}$, ν(CN) has been reported at 2145 [1, 4] or 2148 cm^{-1} [3], while ν(CS) occurred at 815 [1, 4] or 843 cm^{-1} [3]. The Rh-N stretching frequencies for $[Rh(NH_3)_5(NCS)]Br_2$ were reported to occur at ~495 and ~468 cm^{-1}, and at ~451 and 468 cm^{-1} in the corresponding deuteride [4]. The IR spectrum of $[Rh(NH_3)_5(SCN)](ClO_4)_2$ showed ν(CN) at 2115 and ν(CS) at 730 cm^{-1} [1]. The electronic spectra of $[Rh(NH_3)_5(NCS)]^{2+}$ in the range 200 to 400 nm in solution [1, 2] and in D_2O/methanol/ethyleneglycol glasses at 77 K [5] (spectra shown in the paper) have been reported, and some assignments made [5, 6]. Luminescence spectra have been reported and interpreted for $[Rh(NH_3)_5(NCS)]^{2+}$ [5, 6]. The electronic spectrum of $[Rh(NH_3)_5(SCN)]^{2+}$ has also been described in aqueous solution in the range 200 to 400 nm [1].

$[Rh(NH_3)_5(NCS)]Br_2$ underwent an irreversible two-electron reduction in aqueous solution at −1.04 [3] or −1.14 V [7] vs. standard calomel electrode ($NaClO_4$ as base electrolyte). The thermogravimetric curves for the isothiocyanate have been determined [3].

The kinetics and mechanism of base hydrolysis of $[Rh(NH_3)_5(NCS)]^{2+}$ have been investigated [8] and the equilibrium constant for the reaction of Ag^+ with the isothiocyanate complex (log K = 3.38) was determined [9]. The photochemical aquation of $[Rh(NH_3)_5(NCS)]^{2+}$ and $[Rh(NH_3)_5(SCN)]^{2+}$ has been studied [10].

References:

[1] H.-H. Schmidtke (J. Am. Chem. Soc. **87** [1965] 2522/3). – [2] H.-H. Schmidtke (Z. Physik. Chem. [Frankfurt] **45** [1965] 305/16). – [3] I. B. Baranovskii (Zh. Neorgan. Khim. **12** [1967] 2860/3; Russ. J. Inorg. Chem. **12** [1967] 1512/3). – [4] I. B. Baranovskii, Yu. Ya. Kharitonov, G. Ya. Mazo (Zh. Neorgan. Khim. **15** [1970] 1715/7; Russ. J. Inorg. Chem. **15** [1970] 881/3). – [5] T. R. Thomas, G. A. Crosby (J. Mol. Spectrosc. **38** [1971] 118/29).

[6] P. E. Hoggard, H.-H. Schmidtke (Ber. Bunsenges. Physik. Chem. **77** [1973] 1052/8). – [7] K. Nag, P. Banerjee (Indian J. Chem. A **14** [1976] 418/29). – [8] C. Chatterjee, P. Chaudhuri, D. Banerjee (Indian J. Chem. **8** [1970] 1123/5). – [9] G. C. Lalor, H. Miller (J. Inorg. Nucl. Chem. **37** [1975] 1832/3). – [10] J. F. Endicott, G. J. Guillermo (J. Phys. Chem. **80** [1976] 949/53).

Cyanato and Alkylcyanato Complexes

$[Rh(NH_3)_5(NCO)](ClO_4)_2$ is obtained by heating $[Rh(NH_3)_5(H_2O)](ClO_4)_3$ with urea in dimethylacetamide. An intermediate in this reaction appears to be $[Rh(NH_3)_5(NH_2CO_2H)]^{3+}$. The N-bonded cyanate forms a pale yellow solid which is water soluble [1].

The IR spectrum of the compound exhibited (CN), (CO) and the NCO bonding mode at 2264 (very strong), 1355 (medium, shoulder) and 585 (weak) cm^{-1}, respectively. The electronic spectrum of the compound has been reported in the range 200 to 400 nm [1].

The $[Rh(NH_3)_5(NCO)](ClO_4)_2$ salt underwent acid-catalysed hydrolysis to give $[Rh(NH_3)_6]^{3+}$ and CO_2 [1], and the photochemical behaviour of the compound has been investigated [2].

$[Rh(NH_3)_5(NH_2CO_2H)]^{3+}$ is formed as an intermediate in the reaction between $[Rh(NH_3)_5(H_2O)]^{3+}$ and urea in dimethylacetamide, and during the acid-catalysed hydrolysis of $[Rh(NH_3)_5(NCO)]^{2+}$. It has been isolated and characterised by its electronic spectrum [1].

References:

[1] P. C. Ford (Inorg. Chem. **10** [1971] 2153/8). – [2] A. S. El-Awady, F. Basolo (Inorg. Chim. Acta **31** [1978] L363/L366).

Nitrile Complexes

$[Rh(NH_3)_5(NCR)](ClO_4)_3$ (R = CH_3, C_2H_5, CD_3, CH=CH_2; CH_2=CCH_3, C_6H_5, o-, m- and p-FC_6H_4). These complexes are prepared by heating $[Rh(NH_3)_5(H_2O](ClO_4)_3$ with the appropriate nitrile and molecular sieves in dimethylacetamide in a sealed tube at 110°C for 5 h. They were obtained as white crystals [1].

The IR spectra of the compounds revealed a band in the region 2280 to 2330 cm^{-1} due to ν(CN) which varied with the nature of R and which was 50 to 70 cm^{-1} higher than the corresponding frequency in the free nitrile [1]. The electronic spectra of the complexes were recorded in the range 200 to 400 nm [1, 6] (spectra shown in [6]).

The ^{1}H NMR spectra of $[Rh(NH_3)_5(NCCH_3)]^{3+}$ exhibited $\delta(CH_3)$ = 2.546 ppm, and of $[Rh(NH_3)_5(NCC_2H_5)]^{3+}$, $\delta(CH_2)$ = 2.981 (J(RhH) = 0.4 Hz) and $\delta(CH_3)$ = 1.264 ppm. The ^{1}H NMR spectra of $[Rh(NH_3)_5(NCR)]^{3+}$ where R = CCH_3=CH_2 were shown in the paper in the range 180 to 220 and 570 to 650 Hz, and R = 3,5-$D_2C_6H_3CN$ and 4-DC_6H_4CN, in the range 740 to 820 Hz. The ^{19}F NMR spectra of these complexes with R = o-, m- and p-FC_6H_4CN were also reported [2].

The kinetics of base hydrolysis of $[Rh(NH_3)_5(NCR)]^{3+}$ (R = CH_3 and C_6H_5), which gave $[Rh(NH_3)_5(NHCOR)]^{2+}$, have been determined [3, 4, 5].

References:

[1] R. D. Foust, P. C. Ford (Inorg. Chem. **11** [1972] 899/901). – [2] R. D. Foust, P. C. Ford (J. Am. Chem. Soc. **94** [1972] 5686/96). – [3] A. W. Zanella, P. C. Ford (J. Chem. Soc. Chem. Commun. **1974** 795/6). – [4] A. W. Zanella, P. C. Ford (Inorg. Chem. **14** [1975] 42/7). – [5] A. W. Zanella, P. C. Ford (Inorg. Chem. **14** [1975] 700/1).

[6] J. D. Petersen, R. J. Watts, P. C. Ford (J. Am. Chem. Soc. **98** [1976] 3188/94).

Ketamide Complexes

$[Rh(NH_3)_5(NHCOR)]^{2+}$ (R = CH_3 or C_6H_5). Reaction of $[Rh(NH_3)_5(H_2O)](ClO_4)_3$ with acetamide and molecular species, or of $[Rh(NH_3)_5(NCCH_3)](ClO_4)_3$ with aqueous NaOH and $NaClO_4$ gives $[Rh(NH_3)_5(NHCOCH_3)](ClO_4)_2$ [1, 2]. The corresponding $[Rh(NH_3)_5(NHCOC_6H_5)]^{2+}$ is produced by the base hydrolysis of $[Rh(NH_3)_5(NCC_6H_5)]^{2+}$ [3].

These complexes are obtained as white solids, and can be protonated to give $[Rh(NH_3)_5(NH_2COR)]^{3+}$ (R = CH_3, C_6H_5). The electronic spectra of the $[Rh(NH_3)_5(NHCOR)]^{2+}$ and their protonated derivatives have been recorded [1, 3].

References:

[1] A. W. Zanella, P. C. Ford (Inorg. Chem. **14** [1975] 42/7). – [2] A. W. Zanella, P. C. Ford (J. Chem. Soc. Chem. Commun. **1974** 795/6). – [3] A. W. Zanella, P. C. Ford (Inorg. Chem. **14** [1975] 700/1).

Azide Complexes

$[Rh(NH_3)_5(N_3)]X_2$. The **perchlorate** is prepared by reaction of $[Rh(NH_3)_5(H_2O)](ClO_4)_3$ with an excess of NaN_3 in aqueous solution [1, 2, 3]. The **chloride** may be obtained from the perchlorate by ion-exchange (in the Cl^- form), and the **tetraphenylborate** by addition of $NaBPh_4$ to the chloride in aqueous solution [4]. The **azide** was isolated as a bright yellow solid [3, 4].

Molecular orbital calculations were made for $[Rh(NH_3)_5(N_3)]^{2+}$ and interpreted in terms of the photochemical behaviour of the coordinated azide [5].

The electronic spectrum of $[Rh(NH_3)_5(N_3)]^{2+}$ has been measured in aqueous solution in the range 20000 to 50000 cm^{-1} (spectrum shown in paper) [3] and as a glass in D_2O/methanol/ethyleneglycol at 79°C (spectrum shown from 8000 to 44000 cm^{-1}) [6].

The photochemical aquation of $[Rh(NH_3)_5(N_3)]^{2+}$ in acidic media gave $[Rh(NH_3)_5(H_2O)]^{3+}$ [7], but irradiation in the range 350 to 400 nm in neutral solution did not cause aquation, either of N_3^- or NH_3 [8]. Photolysis of the azide in dilute HCl solution gave $[Rh(NH_3)_5(NH_2Cl)]^{3+}$ and $[Rh(NH_3)_5(H_2O)]^{3+}$ [3, 9, 10], and it was proposed that reaction intermediates might include $[Rh(NH_3)_5(N)]^{2+}$ and $[Rh(NH_3)_5(NH)]^{3+}$ [9, 11]. The acid- and base-catalysed hydrolysis reactions of $[Rh(NH_3)_5(N_3)]^{2+}$ have been investigated kinetically [12, 13].

References:

[1] C. S. Davis, G. C. Lalor (J. Chem. Soc. A **1968** 1095/7). – [2] C. S. Davis, G. C. Lalor (J. Chem. Soc. A **1970** 445/9). – [3] H.-H. Schmidtke (Z. Physik. Chem. [Frankfurt] **45** [1965] 305/16). – [4] J. L. Reed, H. D. Gafney, F. Basolo (J. Am. Chem. Soc. **96** [1974] 1363/9). – [5] J. I. Zink (Inorg. Chem. **14** [1975] 446/8).

[6] T. R. Thomas, G. A. Crosby (J. Mol. Spectrosc. **38** [1971] 118/29). – [7] J. F. Endicott, G. J. Ferrandi (J. Phys. Chem. **80** [1976] 949/53). – [8] J. F. Endicott, G. J. Ferrandi (J. Chem. Soc.

Chem. Commun. **1973** 674/5). – [9] J. L. Reed, F. Wang, F. Basolo (J. Am. Chem. Soc. **94** [1972] 7173/4). – [10] T. Inoue, J. F. Endicott, G. J. Ferrandi (Inorg. Chem. **15** [1976] 3098/104).

[11] G. J. Ferrandi, J. F. Endicott (J. Am. Chem. Soc. **95** [1973] 2371/2). – [12] P. J. Staples (J. Chem. Soc. A **1968** 2731/3). – [13] C. S. Davis, G. C. Lalor (J. Chem. Soc. A **1968** 2328/31).

Amido Complexes

$[Rh(NH_3)_5(NH_2OH)]Cl_2$. This compound is obtained by photolysis of $[Rh(NH_3)_5(N_3)](ClO_4)_2$ in $HClO_4$ followed by ion exchange and treatment with HCl. It was isolated as an off-white solid. The IR spectrum revealed bands at 2850 and 960 cm^{-1} typical of NH_2OH, and the electronic spectrum of the compound was measured [1].

$[Rh(NH_3)_5(NH_2Cl)]Cl_2$ is prepared by photolysis of $[Rh(NH_3)_5(N_3)](ClO_4)_2$ in dilute aqueous HCl containing NaCl, followed by ion exchange (acid form), addition of further HCl and finally $HClO_4$. The compound was obtained as a white solid, and its electronic spectrum was measured between 20000 and 50000 cm^{-1} (shown in paper) [2].

References:

[1] J. L. Reed, H. D. Gafney, F. Basolo (J. Am. Chem. Soc. **96** [1974] 1363/9). – [2] T. Inoue, J. F. Endicott, G. J. Ferrandi (Inorg. Chem. **15** [1976] 3098/104, 3098).

Heterocycle N-Donor Complexes

$[Rh(NH_3)_5(NC_5H_4R)](ClO_4)_3$ (NC_5H_4R = pyridines; R = H, 3-Cl, 4-CH_3 or 4-CN). These complexes are obtained by heating $[Rh(NH_3)_5(H_2O)](ClO_4)_3$ with the appropriate pyridine in $(CH_3)_2NCHO$ [1, 2].

$[Rh(ND_3)_5(NC_5D_5)](ClO_4)_3$ is prepared similarly from $[Rh(ND_3)_5(D_2O)](ClO_4)_3$ and deuteriopyridine [1].

The electronic spectra of these complexes were measured in the range 200 to 350 nm [1 to 3], and the luminescence spectra and lifetimes at 77 K and photoaquation quantum yields measured [1, 3]. The ^1H and ^{13}C NMR spectra have also been reported [6, 7].

$[Rh(NH_3)_5(pz)]X_3$ (pz=pyrazine). The **perchlorate** is obtained by heating $[Rh(NH_3)_5(H_2O)]$-$(ClO_4)_3$ with pyrazine in $(CH_3)_2NCHO$, and the **chloride** is formed from it by passing the salt through an ion-exchange column in the Cl^- form. The complexes were isolated as white solids, and the electronic spectrum of $[Rh(NH_3)_5(pz)]^{3+}$ has been measured in aqueous solution [4].

$Rh(NH_3)_5(pz)Fe(CN)_5$ has been briefly described and is formed by reaction of $[Rh(NH_3)_5(pz)]^{3+}$ with $Na_3[Fe(CN)_5(NH_3)]\cdot 3H_2O$. The electronic spectrum revealed a metal-ligand charge transfer transition at 570 nm (in KBr disc) [5].

$[Rh(NH_3)_5(pz)Ru(NH_3)_5](ClO_4)_5$ is prepared by reaction of $[Rh(NH_3)_5(pz)]Cl_3$ with a zinc-reduced aqueous solution derived from $[Ru(NH_3)_5Cl]Cl_2$. Aerial oxidation of this pink solution followed by addition of $NaClO_4$ gives the purple solid $[Rh(NH_3)_5(pz)Ru(NH_3)_5](ClO_4)_5$. This compound, which is believed to contain Rh^{III} and Ru^{II}, can be oxidised using Ce^{4+} or polarographically ($E_{1/2} = 0.71 \pm 0.02$ V vs. standard hydrogen electrode in 0.1M Cl^- solution) to give $[Rh(NH_3)_5(pz)Ru(NH_3)_5]^{6+}$, but this has not been isolated [4]. The electronic spectra of the 5+ and 6+ cations were measured in aqueous solution, between 250 and 600 nm, and the 5+ species exhibited a metal-ligand charge transfer band at 528 nm; extinction coefficient $\varepsilon = 1.8\times10^4$ $mol^{-1}\cdot cm^{-1}$ [2, 4].

$[Rh(NH_3)_5(NC_5H_4CN)Ru(NH_3)_5]^{5+}$ has been mentioned only briefly, and exhibited a metal-ligand charge transfer transition at 458 nm. Photolysis of this complex in aqueous solution afforded $[Rh(NH_3)_5(H_2O)]^{3+}$ and $[Ru(NH_3)_5(NCC_5H_4N)]^{2+}$ [2].

$[Rh(NH_3)_5(bpH)](ClO_4)_4$ (bp = 4,4′-bipyridyl) is obtained by zinc-reduction of $[Rh(NH_3)_5Cl]Cl_2$ in the presence of 4,4′-bipyridyl, followed by addition of HBr and $HClO_4$ in the absence of air. The electronic spectrum of the complex was measured in aqueous solution [5].

$[Rh(NH_3)_5(bp)Fe(CN)_5]\cdot xH_2O$. This compound is formed by treatment of equimolar amounts of $[Rh(NH_3)_5(bp)]^{3+}$ with $[Fe(CH)_5(H_2O)]^{3-}$. It exhibited a metal-ligand charge transfer band (in KBr discs) at 515 nm [5].

References:

[1] J. D. Petersen, R. J. Watts, P. C. Ford (J. Am. Chem. Soc. **98** [1976] 3188/94). – [2] J. A. Gelroth, J. E. Figard, J. D. Petersen (J. Am. Chem. Soc. **101** [1979] 3649/51). – [3] P. C. Ford, G. Malouf, J. D. Petersen (Advan. Chem. Ser. No. 150 [1976] 187/200 from C.A. **84** [1976] No. 102256]. – [4] C. Creutz, H. Taube (J. Am. Chem. Soc. **95** [1973] 1086/94, 1088, 1090). – [5] A. Yeh, A. Haim, M. Tanner, A. Ludi (Inorg. Chim. Acta **33** [1979] 51/6).

[6] J. D. Doi (Inorg. Chem. **20** [1981] 3345/9). – [7] D. K. Lavalee, M. D. Baughman, M. P. Phillips (J. Am. Chem. Soc. **99** [1977] 718/24).

Pentammine Halides, $[Rh(NH_3)_5X]^{2+}$ (X = Cl, Br, I)

The **chloro** complex $[Rh(NH_3)_5Cl]Cl_2$ is prepared by reaction of $RhCl_3\cdot 3H_2O$ either with aqueous ammonium carbonate followed by extraction with HCl [1] or with aqueous ammonia in ethanol [2]. It may also be obtained by reaction of $Na_3[RhCl_6]$ with aqueous ammonia in methanol [3] and by treatment of $[Rh(NH_3)_5(H_2O)](ClO_4)_3$ with chloride [4, 5]. The **bromo** complex $[Rh(NH_3)_5Br]Br_2$ is obtained by reaction of $[Rh(NH_3)_5(OH)]^{2+}$ with HBr [6] and both it and the **iodide** species $[Rh(NH_3)_5I]I_2$ can be prepared by treatment of $[Rh(NH_3)_5(H_2O)](ClO_4)_3$ with halide ion [5, 6, 7]. The corresponding **perchlorates** (X = Cl, Br or I, Y = ClO_4) can be obtained by recrystallisation of $[Rh(NH_3)_5X]X_2$ from dilute $HClO_4$ solutions [5, 6, 7]. The complexes form yellow crystals which are soluble in water.

References:

[1] S. N. Anderson, F. Basolo (Inorg. Syn. **7** [1963] 214/20). – [2] J. A. Osborn, K. Thomas, G. Wilkinson (Inorg. Syn. **13** [1971] 213/5). – [3] C. Ouannes (Compt. Rend. **247** [1958] 1202/4). – [4] G. C. Lalor, G. W. Bushnell (J. Chem. Soc. A **1968** 2820/2). – [5] S. C. Chan (Australian J. Chem. **20** [1967] 61/8).

[6] G. W. Bushnell, G. C. Lalor, E. A. Moelwyn-Hughes (J. Chem. Soc. A **1966** 719/23). – [7] C. S. E. Boyce, G. C. Lalor, H. Miller (J. Inorg. Nucl. Chem. **41** [1979] 857/61).

Properties of $[Rh(NH_3)_5X]^{2+}$ Salts

Stability Constants. The stability constants for the formation of the outer-sphere complexes $[Rh(NH_3)_5X]Y_2$ (X = Cl or Br, Y = SO_4^{2-}, $S_2O_3^{2-}$, SeO_3^{2-}, TeO_3^{2-} and CO_3^{2-}) have been determined using calorimetric, spectrophotometric and electrical conductivity measurements [1]. At zero ionic strength, the order of complex stability was $SO_4^{2-} > CO_3^{2-} > TeO_3^{2-} = SeO_3^{2-} > S_2O_3^{2-}$ [1, 2]. The values of log β_I for $[Rh(NH_3)_5Cl](SO_4)$ and $[Rh(NH_3)_5Cl](S_2O_3)$ at 25°C and zero ionic strength were 2.90±0.05 and 2.50±0.05, respectively [3].

References:

[1] A. K. Pyartman, V. E. Mironov, M. A. Chugunnikova (Koord. Khim. **6** [1980] 114/6). – [2] M. V. Sof'in, A. K. Pyartman, M. A. Chugunnikova, V. E. Mironov (Zh. Fiz. Khim. **51** [1977] 1281; Russ. J. Phys. Chem. **51** [1977] 760). – [3] A. K. Pyartman, V. E. Mironov, M. A. Chugunnikova (Zh. Neorgan. Khim. **23** [1978] 260/2; Russ. J. Inorg. Chem. **23** [1978] 147/8).

Crystallographic Properties. The X-ray crystal structure of $[Rh(NH_3)_5Cl]Cl_2$ is orthorhombic, space group Pnma-D_{2h}^{16}, Z=4, with a=13.40(1), b=10.50(1), c=6.751(7) Å [1]; a=13.36(1), b=10.46(1), c=5.741(8) Å [2]. $[Rh(NH_3)_5Br]Br_2$ is also orthorhombic, space group Pnma-D_{2h}^{16}, Z=4, with a=13.80(1), b=10.83(1), c=6.974(4) Å [1]. Both compounds contain an octahedral cation, with Rh-N distances in the range 2.071(5) to 2.085(8) Å [1] or 2.051(4) to 2.061(4) Å [2] and an Rh-Cl bond length of 2.355(2) [1] or 2.356(1) Å [2]. The bromo complex has Rh-N distances in the range 2.052(9) to 2.060(10) Å and an Rh-Br bond length of 2.491(2) Å. There is no significant difference between the Rh-N distances *trans* and *cis* to the halide atom [2]. The chloro complex is isomorphous with $[Co(NH_3)_5Cl]Cl_2$ [3].

References:

[1] R. S. Evans, E. A. Hopcus, J. Bordner, A. F. Schreiner (J. Cryst. Mol. Struct. **3** [1973] 235/45). – [2] M. Weishaupt, H. Bezter, J. Straehle (Z. Anorg. Allgem. Chem. **440** [1978] 52/64). – [3] M. N. Lyashenko (Tr. Inst. Krist. Akad. Nauk SSSR **7** [1952] 67/72 from C.A. **1956** 2349).

Vibrational Spectra. An analysis of the normal vibrational modes of $[Rh(NH_3)_5X]X_2$ and $[Rh(ND_3)_5X]X_2$ (X=Cl, Br) has been performed, and the IR frequencies assigned. The spectra, in liquid paraffin, were shown between 4000 and 400 cm^{-1} [1]. Assignment has been made of ν(RhX) [2, 3, 4] and ν(RhN) [3, 4] in these complexes, and other frequencies, including δ(NRhN), have also been recorded. Raman and far IR spectral data have also been obtained from these halo complexes, and the spectrum for $[Rh(NH_3)_5Cl]Cl_2$ and $[Rh(ND_3)_5Cl]Cl_2$ shown between 550 and 50 cm^{-1} [5]. The values for $\nu(Rh^{35}Cl)$ and $\nu(Rh^{37}Cl)$ were quoted as 309 and 303 cm^{-1}, with a corresponding stretching force constant of 1.47 md/Å. The stretching force constant for the Rh-N bond was 2.34 md/Å [3]. However, the Rh-X stretching frequencies in $[Rh(NH_3)_5X]^{2+}$ (X=Cl or Br) were given as 318 (Raman, X=Cl) and 309, 303 (IR, X=Cl) [5] or 274 and 282 (strong and broad, X=Cl) and 199 (Raman) and 206 cm^{-1} (IR, X=Br) or 207 cm^{-1} (X=Br) [4].

References:

[1] Yu. Ya. Kharitonov, N. A. Knyazeva, G. Ya. Mazo, I. B. Baranovskii, N. B. Generalova (Zh. Neorgan. Khim. **16** [1971] 1974/80; Russ. J. Inorg. Chem. **16** [1971] 1050/4). – [2] I. B. Baranovskii, G. Ya. Mazo (Zh. Neorgan. Khim. **17** [1972] 1682/5; Russ. J. Inorg. Chem. **17** [1972] 870/2). – [3] I. B. Baranovskii, Yu. Ya. Kharitonov, G. Ya. Mazo (Zh. Neorgan. Khim. **15** [1970] 1715/7; Russ. J. Inorg. Chem. **15** [1970] 881/3). – [4] K. W. Bowker, E. R. Gardner, J. Burgess (Inorg. Chim. Acta **4** [1970] 626/8). – [5] M. W. Bee, S. F. A. Kettle, D. B. Powell (Spectrochim. Acta A **30** [1974] 139/50).

Electronic Spectra. Electronic spectral measurements have been made of $[Rh(NH_3)_5Cl]^{2+}$ [1], in particular in the range 250 to 350 mμ in aqueous acidic solution, and in aqueous solution and in the solid state [2, 3]. The electronic spectra of $[Rh(NH_3)_5X]^{2+}$ (X=Cl, Br, I) have been recorded in 0.01 M $HClO_4$ solution, and the data interpreted in terms of ligand field excited states, and ligand to metal charge transfer excited states [4]. The spectra have also been recorded in the range 20000 to 45000 cm^{-1} in water [5] and 8000 to 44000 cm^{-1} in water/methanol/ethyleneglycol glasses [6], assignments being made in both cases. The electronic spectrum of $[Rh(NH_3)_5I]^{2+}$ in aqueous solution has been reported [7] and shown from 200 to 500 nm [8]. A study has been made of the solvent shifts in the UV absorption spectra of solutions of $[Rh(NH_3)_5X]^{2+}$ (X = Br, I) [9].

Luminescence spectra of $[Rh(NH_3)_5X]^{2+}$ (X = Cl, Br, I) have been measured as glasses (water/methanol/ethyleneglycol) over the range 8000 to 44000 cm^{-1} at −77°C [6] and at 85 and 295 K [10]. The luminescence lifetimes of $[Rh(NH_3)_5Cl]^{2+}$, $[Rh(ND_3)_5Cl]^{2+}$ and the corresponding bromo species have been measured [11 to 14]. The quenching of emission from photo-excited $[Rh(NH_3)_5Cl]Cl_2$ has been observed [15].

References:

[1] L. Moggi (Gazz. Chim. Ital. **97** [1967] 1089/96). – [2] A. V. Babaeva, R. I. Rudyi (Izv. Akad. Nauk SSSR Ser. Fiz. **18** [1954] 729/30; C.A. **1956** 106/7). – [3] A. V. Babaeva, R. I. Rudyi (Zh. Neorgan. Khim. **1** [1956] 921/9; Russ. J. Inorg. Chem. **1** No. 5 [1956] 42/51, 46). – [4] W. D. Blanchard, W. R. Mason (Inorg. Chim. Acta **78** [1978] 159/68, 163). – [5] C. K. Jørgensen (Acta Chem. Scand. **10** [1956] 500/17).

[6] T. R. Thomas, G. A. Crosby (J. Mol. Spectrosc. **38** [1971] 118/29). – [7] T. L. Kelly, J. F. Endicott (J. Chem. Soc. Chem. Commun. **1971** 1061/2). – [8] T. L. Kelly, J. F. Endicott (J. Am. Chem. Soc. **94** [1972] 1797/804). – [9] C. St. E. Boyce, G. C. Lalor (Rev. Latinoamer. Quim. **5** [1974] 88/93 from C.A. **81** [1974] No. 43581). – [10] P. E. Hoggard, H.-H. Schmidtke (Ber. Bunsenges. Physik. Chem. **77** [1973] 1052/8).

[11] M. A. Bergkamp, R. J. Watts, P. C. Ford, J. Brannon, D. Magde (Chem. Phys. Letters **58** [1978] 125/8). – [12] M. A. Bergkamp, J. Brannon, D. Magde, R. J. Watts, P. C. Ford (J. Am. Chem. Soc. **101** [1979] 4549/54). – [13] M. A. Bergkamp, R. J. Watts, P. C. Ford (J. Am. Chem. Soc. **102** [1980] 2627/31). – [14] M. A. Bergkamp, R. J. Watts, P. C. Ford (J. Phys. Chem. **85** [1981] 684/6). – [15] A. W. Adamson, R. C. Fakuda, M. Larson, H. Macke, J. P. Puaux (Inorg. Chim. Acta **44** [1980] L13/L15).

Other Physical Properties. The X-ray photoelectron spectra of $[Rh(NH_3)_5Cl]Cl_2$ provided the following binding energies: N1s = 390.8 (1.6) eV, C1 $2p_{3/2}$ = 199.2 and 197.8 (1.4) eV [1]. Similar studies of $[Rh(NH_3)_5Cl](NO_3)_2$ gave the N1s binding energies as 400.4 (NH_3) and 407.2 eV (NO_3), while the Cl $2p_{3/2}$ binding energy was 198.9 eV [2].

The X-ray K spectrum of $[Rh(NH_3)_5Cl](NO_3)_2$ showed a narrow absorption maximum at ~10 eV which is due to transfer of a 1s electron due to Cl to a free molecular orbital [3]. This absorption band has been used for analysing the electronic structure of $[Rh(NH_3)_5Cl]Cl_2$ [4].

The ^1H NMR spectra of $[Rh(NH_3)_5X]^{2+}$ have been measured in conc. H_2SO_4. Resonances due to the NH_3 protons occurred at $\delta = -3.78$ (Cl) and −3.80 (X = Br) ppm, occurring as singlets because of rapid exchange with the solvent [5].

References:

[1] J. R. Ebner, D. L. McFadden, D. R. Tyler, R. A. Walton (Inorg. Chem. **15** [1976] 3014/8). – [2] I. B. Baranovskii, G. Ya. Mazo (Zh. Neorgan. Khim. **17** [1972] 1682/5; Russ. J. Inorg. Chem. **17** [1972] 870/2). – [3] A. P. Sadovskii, A. V. Belyaev (Izv. Sibirsk. Otd. Akad. Nauk SSSR Ser. Khim. Nauk **1967** No. 6, pp. 57/61; Sib. Chem. J. **1967** 703/6). – [4] K. I. Narbutt, V. I. Nefedov, M. A. Porai-Koshits, A. P. Kochetkova (Zh. Strukt. Khim. **13** [1972] 451/7; J. Struct. Chem. [USSR] **13** [1972] 422/7). – [5] D. N. Hendrickson, W. L. Jolly (Inorg. Chem. **9** [1970] 1197/201).

Electrochemical Behaviour

$[Rh(NH_3)_5Cl]^{2+}$ is reduced polarographically in a two-electron step which has been described as reversible, with $E_{1/2} = -0.93$ V vs. standard calomel electrode [1], but more extensive studies have suggested it is irreversible [2 to 5]. In aqueous solution (0.001 M) with $NaClO_4$ as base-electrolyte (0.1M) $E_{1/2}$ was reported as −0.87 [3] or −0.92 V [4] vs. standard calomel electrode. However, $E_{1/2}$ varied with the base electrolyte: −0.93 (1 M NH_4Cl), −0.93 (1M NH_4Cl/1M NH_4OH), −0.80 (0.02 M NH_4Cl), −0.96 (1M KNO_3 with gelatin) and −0.96 V (1M K_2SO_4) [5]. Polarographic data have been compared with those obtained from other related species $[Rh(NH_3)_5X]^{n+}$ (n = 2, X = NO_2, NCS; n = 3, X = H_2O) and a correlation made between $E_{1/2}$ and the position of the first d-d transition in the electronic spectra of the complexes [3]. The mechanism of electrodeposition of Rh from electrochemically reduced $[Rh(NH_3)_5Cl](NO_3)_2$ in liquid NH_3 has been investigated [6, 7].

References:

[1] D. Cozzi, F. Pantani (J. Inorg. Nucl. Chem. **8** [1958] 385/98). – [2] A. A. Vlcek (Discussions Faraday Soc. No. 26 [1958] 164/71). – [3] K. Nag, P. Banerjee (Indian J. Chem. A **14** [1976] 418/20). – [4] I. B. Baranovskii (Zh. Neorgan. Khim. **12** [1967] 2860/3; Russ. J. Inorg. Chem. **12** [1967] 1512/3). – [5] J. B. Willis (J. Am. Chem. Soc. **66** [1944] 1067/9).

[6] G. W. Watt, J. W. Vaughan (J. Electrochem. Soc. **110** [1963] 723/6). – [7] G. W. Watt, A. Broodo, W. A. Jenkins, S. G. Parker (J. Am. Chem. Soc. **76** [1954] 5989/93).

Chemical Reactions

Thermal Decomposition. The stoichiometry of the thermal dissociation reactions of $[Rh(NH_3)_5X]X_2$ (X = Cl, Br, I) have been studied by thermogravimetry, differential thermal analysis, gas evolution analysis, and chemical analysis of the products [1, 2]. Decomposition afforded the metal, NH_4X, halogen (X = Br or I), N_2 and ammonia. Under hydrogen at 200 to 340°C, $[Rh(NH_3)_5Cl]Cl_2$ decomposed to Rh, NH_3 and NH_4Cl [2]. The simultaneous determination of thermally stimulated light emission and differential thermal analysis of $[Rh(NH_3)_5X]^{2+}$ (X = Cl, Br) has been made [3]. The kinetics of thermal deammoniation of $[Rh(NH_3)_5X]X_2$ (X = Cl, Br, I) indicated that in the solid state $[Rh(NH_3)_4X_2]X$ was formed [4].

References:

[1] W. W. Wendlandt, P. H. Franke (J. Inorg. Nucl. Chem. **26** [1964] 1885/93). – [2] P. D. Perkins, G. T. Kerr (Inorg. Chem. **70** [1981] 3581/2). – [3] W. W. Wendlandt (Thermochim. Acta **34** [1980] 313/20). – [4] S. Kohata, H. Kawaguchi, N. Itoh, A. Ohyoshi (Bull. Chem. Soc. Japan **53** [1980] 807/8).

Photolysis. The photolabilisation of $[Rh(NH_3)_5X]^{2+}$ (X = Cl, Br or I) may give either $[Rh(NH_3)_5(H_2O)]^{3+}$ or *trans*-$[Rh(NH_3)_4(H_2O)X]^{2+}$ [1 to 6]. The mechanism of this reaction when X = Cl has been investigated [6] and appears to be a function of solvent [7, 8], Cl substitution predominating in water and formamide, while NH_3 substitution occurs in $(CH_3)_2NCHO$, methanol and $(CH_3)_2SO$ [8]. The product of substitution is also determined by the irradiation frequency [9], and a ligand-field-based interpretation of this photosolvation process has been given [10]. If photolysis of $[Rh(NH_3)_5I]^{2+}$ occurred at 254 nm in the presence of small amounts of I^-, then ammonia and equal amounts of *trans*-$[Rh(NH_3)_4(H_2O)I]^+$ and *trans*-$[Rh(NH_3)_4I_2]^+$ were formed, but in the absence of I^-, only small amounts of the *trans* di-iodide were detected [3].

References:

[1] L. Moggi (Gazz. Chim. Ital. **7** [1967] 1089/96). – [2] T. L. Kelly, J. F. Endicott (J. Am. Chem. Soc. **94** [1972] 278/9). – [3] T. L. Kelly, J. F. Endicott (J. Am. Chem. Soc. **92** [1970] 5733/4). – [4] T. L. Kelly, J. F. Endicott (J. Am. Chem. Soc. **94** [1972] 1797/804). – [5] G. A. Shagisultanova, V. V. Yasinetskii (Zh. Fiz. Khim. **50** [1976] 91/5; Russ. J. Phys. Chem. **50** [1976] 51/3).

[6] C. Kutal, A. W. Adamson (Inorg. Chem. **12** [1973] 1454/6). – [7] M. A. Bergkamp, R. J. Watts, P. C. Ford (J. Am. Chem. Soc. **102** [1980] 2627/31). – [8] M. A. Bergkamp, R. J. Watts, P. C. Ford (J. Chem. Soc. Chem. Commun. **1979** 623/4). – [9] J. I. Zink (Inorg. Chem. **12** [1973] 1018/24). – [10] M. J. Incorvia, J. I. Zink (Inorg. Chem. **13** [1974] 2489/94).

Hydrolysis. Kinetic studies showed that reaction of $[Rh(NH_3)_5X]^{2+}$ (X = Cl, Br, I) with OH^-, which gave $[Rh(NH_3)_5(OH)]^{2+}$, followed 2nd order rate laws [1], the rate increasing in the order Cl < Br < I [2]. The kinetic data were interpreted in terms of the electronegativity of X [3] and solvation effects were important in the control of reactivity [3]. The deuterium isotope effect on the base hydrolysis of $[Rh(NR_3)_5X]^{2+}$ (R = H or D; X = Cl, Br, I) has also been investigated [4], and the pressure dependencies of the rates of base hydrolysis of $[Rh(NH_3)_5X]^{2+}$, leading to calculations of volumes of activation, investigated [5]. From mechanistic studies of the base hydrolysis reaction, it was proposed that $[Rh(NH_3)_5(NH_2)Cl]^+$ was formed as an intermediate [6]. It appeared that other nucleophiles, e.g. NO_2^-, I^-, NH_3 or thiourea, had little or no effect on the rate of Cl^- release from $[Rh(NH_3)_5Cl]^{2+}$ during hydrolysis [7, 10]. The kinetics of the forward and reverse reactions in the system $[Rh(NH_3)_5Cl]^{2+} + H_2O \rightleftharpoons [Rh(NH_3)_5(H_2O)]^{3+} + Cl^-$ in aqueous solution have been determined [8]. An unusual kinetic and thermodynamic *trans* effect due to NH_3 was observed in the hydrolysis of $[Rh(NH_3)_5X]^{2+}$ (X = Cl, Br or I) [9, 10]. A study has been made of the relation of the heat of activation to the entropy of activation for the hydrolysis of $[Rh(NH_3)_5X]^{2+}$ [11].

Mercury(II) catalysed hydrolyses of $[Rh(NH_3)_5Cl]^{2+}$ in water [12, 13] and other solvents [14] have been described, and there is evidence for the intermediacy of species of the type $[Rh(NH_3)_5XHg]^{4+}$ (X = Cl, Br and I) [15 to 18]. The pressure dependence on rates of Hg^{2+}-induced aquation of $[Rh(NH_3)_5Cl]^{2+}$ has been measured [19] and the volume of activation of the mercury(II)-catalysed aquation of $[Rh(NH_3)_5I]^{2+}$ determined [20]. Aquation of the iodo-pentammine rhodium(III) species by Ag^+ has also been observed, and there is evidence for the intermediacy of $[\{Rh(NH_3)_5I\}_2Ag]^{5+}$ and $[Rh(NH_3)_5IAg_n]^{(2+n)+}$ (n = 1, 2, 3) [21, 22]. The theory of kinetics of ligand substitution in $[Rh(NH_3)_5X]^{2+}$ has been analysed by quantum mechanical methods [23].

References:

[1] A. B. Lamb (J. Am. Chem. Soc. **61** [1939] 699/708). – [2] G. W. Bushnell, G. C. Lalor, E. A. Moelwyn-Hughes (J. Chem. Soc. A **1966** 719/23). – [3] S. C. Chan (Australian J. Chem. **20** [1967] 61/8). – [4] S. C. Chan (J. Inorg. Nucl. Chem. **34** [1972] 793/6). – [5] D. A. Palmer (Australian J. Chem. **32** [1979] 2589/95).

[6] J. A. Broomhead, F. Basolo, R. G. Pearson (Inorg. Chem. **3** [1964] 826/32). – [7] S. A. Johnson, F. Basolo, R. G. Pearson (J. Am. Chem. Soc. **85** [1963] 1741/7). – [8] G. C. Lalor, G. W. Bushnell (J. Chem. Soc. A **1968** 2520/2). – [9] A. J. Pöe, K. Shaw (J. Chem. Soc. Chem. Commun. **1967** 52/4). – [10] A. J. Pöe, K. Shaw, M. J. Wendt (Inorg. Chim. Acta **1** [1967] 371/7).

[11] A. V. Belyaev, V. P. Kazakov, B. V. Ptitsyn (Dokl. Akad. Nauk SSSR **160** [1965] 149/50; Dokl. Phys. Chem. Proc. Acad. Sci. USSR **160/165** [1965] 5/7). – [12] A. B. Venediktov, A. V. Belyaev (Zh. Neorgan. Khim. **17** [1972] 2222/6; Russ. J. Inorg. Chem. **17** [1972] 1158/60). – [13] I. V. Kozhevnikov, E. S. Rudakov (Inorg. Nucl. Chem. Letters **8** [1972] 571/6). – [14] J. Burgess,

M. G. Price (J. Chem. Soc. A **1971** 3108/12). – [15] S. C. Chan, S. F. Chan (J. Inorg. Nucl. Chem. **34** [1972] 2311/4).

[16] A. B. Venediktov, A. V. Belyaev (Izv. Sibirsk. Otd. Akad. Nauk SSSR Ser. Khim. Nauk **1973** 49/52 from C.A. **79** [1973] No. 118790). – [17] A. B. Venediktov, A. V. Belyaev (Izv. Sibirsk. Otd. Akad. Nauk SSSR Ser. Khim. Nauk **1972** 148/53 from C.A. **78** [1973] No. 62786). – [18] A. B. Venediktov, A. V. Belyaev (Izv. Sibirsk. Otd. Akad. Nauk SSSR Ser. Khim. Nauk **1973** 42/8 from C.A. **79** [1973] No. 46192). – [19] D. A. Palmer, R. van Eldik, T. P. Dasgupta, H. Kelm (Inorg. Chim. Acta **34** [1979] 91/5). – [20] W. Weber, D. A. Palmer, H. Kelm (Inorg. Chim. Acta Letters **54** [1981] L177/L180).

[21] C. St. E. Boyce, G. C. Lalor, H. Miller (J. Inorg. Nucl. Chem. **41** [1979] 857/61). – [22] G. C. Lalor, H. Miller (Rev. Latinoamer. Quim. **7** [1976] 102/6 from C.A. **86** [1977] No. 9243). – [23] E. D. German, R. R. Dogonadze (J. Res. Inst. Catal. Hokkaido Univ. **20** [1972] 34/49 from C.A. **78** [1973] No. 48469).

Miscellaneous Reactions

The chemical effects of (n,γ) reactions on $[Rh(NH_3)_5Cl]Cl_2$ [1, 2] and $[Rh(NH_3)_5Br](NO_3)_2$ [3] have been investigated. The chemical effects of ^{80m}Br isomeric transitions in crystalline $[Rh(NH_3)_5Br](NO_3)_2$ have been described [4, 5, 6].

Radiolysis of aqueous solutions of $[Rh(NH_3)_5Cl]Cl_2$ saturated with air caused formation of metallic rhodium [7]. However, pulse radiolytic one-electron reduction of $[Rh(NH_3)_5Cl]^{2+}$ indicated that $[Rh(NH_3)_5Cl]^+$ and $[Rh(NH_3)_4]^{2+}$ (possibly solvated) were formed [8]. The reactions of $[Rh(NH_3)_5Cl]^{2+}$ with carbonate radicals have been studied [9].

Reduction of $[Rh(NH_3)_5X]^{2+}$ (X = Cl, Br, I) with Cr^{II} gave, eventually, Rh metal, but Rh^{II} intermediates may have been formed [10, 11]. It has been reported that in this reaction $[Rh_2(H_2O)_{10}]^{4+}$ is formed [11]. Reduction of $[Rh(NH_3)_5Br]^{2+}$ by KNH_2 in liquid ammonia, giving $Rh(NH_2)_3$, has been described in Section 2.1.5, p. 124.

Rh^{III}-exchanged zeolites have been prepared by reaction of NaY zeolite with aqueous $[Rh(NH_3)_5Cl]^{2+}$. At 600 to 770 K, lattice-bound Rh^{II} was formed and was investigated by ESR spectroscopy [12]. Formation of rhodium carbonyl species from $[Rh(NH_3)_5Cl]^{2+}$ deposited in zeolite centres has also been observed [13]. The chemical conditions for the photographic activity of $[Rh(NH_3)_5Cl]Cl_2$ have been studied [14].

The isotope exchange reactions of X-labelled $[Rh(NH_3)_5X]^{2+}$ (X = ^{36}Cl, ^{82}Br, ^{131}I) and its iridium analogues have been examined in aqueous acidic solutions [15, 16]. The Rh complex exchanged via a first-order reaction and nearly 60 times faster than its Ir analogue [15].

It has been proposed, from kinetic studies, that the ammoniation of $[Rh(NH_3)_5Br](ClO_4)_2$, giving $[Rh(NH_3)_6](ClO_4)_3$, proceeds via a conjugate-base mechanism involving $[Rh(NH_3)_4(NH_2)Br]^+$ and $[Rh(NH_3)_4(NH_2)]^{2+}$ as intermediates [17].

References:

[1] E. R. Gardner, M. E. Gravenor, R. D. Harding, J. B. Raynor, R. Autchakit (Radiochim. Acta **13** [1970] 100/4). – [2] E. R. Gardner, M. E. Wilson, R. D. Harding, J. B. Raynor (Radiochim. Acta **17** [1972] 41/5). – [3] G. B. Schmidt, K. Heine, W. Herr (Radiat. Damage Solids Part. Proc. Symp., Venice 1962 [1962/63], Vol. 3, pp. 93/102 from C.A. **61** [1964] 11402). – [4] W. Herr, G. B. Schmidt (Z. Naturforsch. **17a** [1962] 309/14). – [5] G. B. Schmidt, W. Herr (Z. Naturforsch. **18a** [1963] 505/9).

[6] G. B. Schmidt (Radiochim. Acta **5** [1966] 178/80). – [7] L. I. Barsova, A. K. Pikaev, V. I. Spitsyn, A. A. Balandin (Dokl. Akad. Nauk SSSR **144** [1962] 344/6; Proc. Acad. Sci. USSR Chem. Sect. **142/147** [1962] 417/9). – [8] J. Lilie, M. G. Simic, J. F. Endicott (Inorg. Chem. **14** [1975] 2129/33). – [9] V. W. Cope, M. Z. Hoffman, N.-S. Chen (J. Phys. Chem. **82** [1978] 2615/9). – [10] G. T. Takali, R. T. M. Fraser (Proc. Chem. Soc. **1964** 116).

[11] J. J. Ziolkowski, H. Taube (Bull. Acad. Polon. Sci. Ser. Sci. Chim. **21** [1973] 113/7). – [12] C. Naccache, Y. Ben Taarit, M. Boudart (ACS Symp. Ser. 40 **1977** 156/65 from C.A. **86** [1977] No. 128079). – [13] M. Primet, J. C. Vedrine, C. Naccache (J. Mol. Catal. **4** [1978] 411/2). – [14] M. T. Beck, P. Kiss, T. Szalay, E. C. Porzsolt, G. Bazsa (J. Signalaufzeichnungsmaterialien **4** [1976] 131/7 from C.A. **85** [1976] No. 70648). – [15] G. B. Schmidt (Z. Physik. Chem. [Frankfurt] **41** [1964] 26/32).

[16] G. B. Schmidt (Z. Physik. Chem. [Frankfurt] **50** [1966] 222/4). – [17] S. Balt, A. Jelsma (Inorg. Chem. **20** [1981] 733/7).

2.1.7.2.2 Tetrammine Complexes

See "Rhodium" 1938, pp. 141/2.

Rhodium(II) Tetrammine Complex, $[Rh(NH_3)_4]^{2+}$

There is a brief report that this species may be formed in the pulse radiolytic one-electron reduction of $[Rh(NH_3)_5X]^{n+}$ (n = 2, X = Cl; n = 3, X = H_2O) and $[Rh(NH_3)_4Br_2]^+$ in aqueous solution. It was probably solvated and underwent exchange of NH_3 by H_2O. The species also apparently disproportionated, and it reacted with O_2 giving $[Rh(NH_3)_4(OH_2)(O_2)]^{2+}$. The reduction of $[Rh(NH_3)_4Br_2]^+$ by $[Rh(NH_3)_4]^{2+}$ resulted in a chain reaction which catalysed the exchange of Br^- for H_2O in the Rh^{III} species.

J. Lilie, M. G. Simic, J. F. Endicott (Inorg. Chem. **14** [1975] 2129/33).

Hydrido Rhodium(III) Tetrammine Complexes

***trans*-$[Rh(NH_3)_4HCl]^+$.** This compound is apparently formed by reaction of sodium borohydride with *trans*-$[Rh(NH_3)_4Cl_2]^+$ in ice-cold water, but no preparative details have been given [1, 2]. The electronic spectrum has been reported in aqueous solution in the range 250 to 450 nm [2].

***trans*-$[Rh(NH_3)_4H(H_2O)]^{2+}$** exists in aqueous solution in equilibrium with $[Rh(NH_3)_5H]^{2+}$ when the latter is dissolved in water [3 to 6]. The **sulphate** is prepared by dissolving $[Rh(NH_3)_5H](SO_4)$ in dilute aqueous solution and precipitation with acetone [3, 4, 5], and is isolated as an air-stable white solid.

The IR spectrum of *trans*-$[Rh(NH_3)_4H(H_2O)]^{2+}$ exhibited ν(RhH) at 2145 [3] or 2146 cm^{-1} [5]; ν(RhN) = 479 (weak), 463 (medium), 445 (weak) cm^{-1}; ν(NH), δ(NH_3, asymmetric and symmetric), ϱ(NH_3) have also been reported. The ^1H NMR spectrum revealed τ(RhH) = 32.0, J(RhH) = 25 Hz [5].

The compound reacted with ethylene, tetrafluoroethylene and acetylene forming Rh-C bonded species of the type $[Rh(NH_3)_4R(H_2O)]^{2+}$ (R = C_2H_5, C_2F_4H and CH=CH_2; isolated as SO_4^{2-} salts when R = C_2H_5 and C_2F_4H [5, 7]. It also reacted photolytically with O_2, in acidic aqueous solutions giving **$[Rh(NH_3)_4(O_2H)(H_2O)]^{2+}$** [8] which was also apparently formed by reaction of electrolytically reduced $[Rh(NH_3)_5H]^{2+}$ with O_2 [9]. Water-soluble unsaturated carboxylic acids could be hydrogenated with hydrogen in the presence of $[Rh(NH_3)_4H(H_2O)]^{2+}$ [10].

References:

[1] R. D. Gillard, G. Wilkinson (J. Chem. Soc. **1963** 3594/9). – [2] J. A. Osborn, R. D. Gillard, G. Wilkinson (J. Chem. Soc. **1964** 3168/73). – [3] J. A. Osborn, A. R. Powell, G. Wilkinson (J. Chem. Soc. Chem. Commun. **1966** 461/2). – [4] A. R. Powell (Platinum Metals Rev. **11** [1967] 58/9). – [5] K. Thomas, J. A. Osborn, A. R. Powell, G. Wilkinson (J. Chem. Soc. A **1968** 1801/6).

[6] J. A. Osborn, K. Thomas, G. Wilkinson (Inorg. Syn. **13** [1971] 213/5). – [7] A. R. Powell (Brit. 1196583 from C.A. **73** [1970] No. 59620). – [8] J. F. Endicott, C.-L. Wong, T. Inoue, P. Natarajan (Inorg. Chem. **18** [1979] 450/4). – [9] L. E. Johnston, J. A. Page (Can. J. Chem. **47** [1969] 4241/6). – [10] G. C. Bond (Brit. 11197723 from C.A. **73** [1970] No. 98401).

Cyano-tetrammine Complexes

$[Rh(NH_3)_4(CN)(H_2O)]Cl_2$ is obtained by reaction of $Rh(NH_3)_4(CN)(SO_3)$ with hot conc. HCl. On cooling, the compound was formed as a white powder. The IR spectrum exhibited ν(CN) at 2189 and 2130 cm^{-1}, consistent with terminal and bridging CN groups.

I. B. Baranovskii, A. V. Babaeva (Zh. Neorgan. Khim. **13** [1968] 3148/50; Russ. J. Inorg. Chem. **13** [1968] 1624).

For **$[Rh(NH_3)_4(CN)(SO_3)]\cdot 2H_2O$** see "Rhodium" Suppl. Vol. B 1, 1982, p. 151.

Hydroxo- and Aquo-tetrammine Complexes

$[Rh(NH_3)_4(OH)_2]^+$. The *cis* isomer is obtained by photolysis of *trans*-$[Rh(NH_3)_4(OH)X]^+$ (X = Cl, Br or I) or *trans*-$[Rh(NH_3)_4(OH)(H_2O)]^{2+}$ in base [1, 2] or by deprotonation of basic *cis*-$[Rh(NH_3)_4(OH)(H_2O)]^{2+}$ or *cis*-$[Rh(NH_3)_4(H_2O)_2]^{3+}$ in aqueous solution [2, 3, 4]. The *trans* isomer is prepared in solution by deprotonation of *trans*-$[Rh(NH_3)_4(OH)(H_2O)]^{2+}$ in base [1, 2, 4]. However, the salts have not apparently been isolated, although their electronic spectra have been measured in the range 200 to 500 nm in aqueous basic solution (0.04 to 0.1M NaOH) [2, 3, 4].

$[Rh(NH_3)_4(OH)(H_2O)]^{2+}$. The *cis* isomer may be prepared by photolysis of *trans*-$[Rh(NH_3)_4(OH)X]^{2+}$ (X = Cl or Br) in neutral aqueous solution [2].

***cis*-$[Rh(NH_3)_4(OH)(H_2O)](S_2O_6)\cdot 0.2H_2O$** is formed by refluxing *cis*-$[Rh(NH_3)_4Cl_2]Cl\cdot 0.5H_2O$ with $AgNO_3$ in water in the absence of light, then adding $Na_2S_2O_6\cdot 2H_2O$ in the presence of some pyridine [3]. *Trans*-$[Rh(NH_3)_4(OH)(H_2O)]S_2O_6$ is prepared by reaction of $[Rh(NH_3)_5H](SO_4)$ with hot aqueous $HClO_4$, followed by addition of H_2O_2 and then $Na_2S_2O_6\cdot 2H_2O$ in the presence of some NaOH but in the dark [4]. It is also formed by dissolving *trans*-$[Rh(NH_3)_4(H_2O)_2](ClO_4)_3$ in aqueous solution at pH = 7 in the presence of $Na_2S_2O_6\cdot 2H_2O$ [4].

The *cis* dithionate was isolated as yellow crystals whereas the *trans* form was obtained as cream-coloured crystals. The electronic spectra of the two isomers have been measured in the range 200 to 500 nm in aqueous solution [2, 3, 4].

$[Rh(NH_3)_4(H_2O)_2]^{3+}$. The *cis* isomer may be obtained by protonation of *cis*-$[Rh(NH_3)_4(OH)_2]^+$ or *cis*-$[Rh(NH_3)_4(OH)(H_2O)]^{2+}$ in aqueous solution [1 to 4]. The ***cis*** and ***trans*** **perchlorates** are isolated when *cis*- and *trans*-$[Rh(NH_3)_4(OH)(H_2O)](S_2O_6)$ are treated with $HClO_4$. Both species were isolated as white solids [4].

The pK_a values of the *cis* and *trans* isomers were determined as follows: *cis* form $pK_a^1 = 6.40$, $pK_a^2 = 8.32$; *trans* form $pK_a^1 = 4.92$, $pK_a^2 = 8.26$ [2, 4]. The electronic spectra of the complexes were measured in the range 200 to 500 nm in aqueous $HClO_4$ solutions [2, 3, 4]. Isomerisation of both forms occurred on photolysis in aqueous solution [2].

$[Rh(NH_3)_4(\mu\text{-}OH)_2Rh(NH_3)_4]Br_4 \cdot 4H_2O$ is formed when *cis*-$[Rh(NH_3)_4(OH)(H_2O)](S_2O_6)$ is heated at 120°C for 15 h. The dithionate so formed was not isolated but can be converted to the bromide by addition of NH_4Br in aqueous solution. The complex was obtained as yellow crystals, and its electronic spectrum has been reported in neutral and acid (0.012 M $HClO_4$) aqueous solution [3].

***trans*-$[Rh(NH_3)_4(OH)(O_2H)]^+$** is apparently formed when $[Rh(NH_3)_5H]^{2+}$ reacts with O_2 in ammoniacal NH_4ClO_4 solutions. The formulation of the species was based on electrochemical studies, and it was reported that electrochemical reduction afforded *trans*-$[Rh(NH_3)_4(OH)_2]^+$. The electronic spectrum of the compound was briefly reported, and addition of $HClO_4$ afforded coloured paramagnetic but uncharacterised species [5].

$[Rh(NH_3)_4(H_2O)(O_2H)]^{2+}$ has been briefly described as the product of the photo-induced oxygenation of *trans*-$[Rh(NH_3)_4H(OH_2)]^{2+}$. It decomposed slowly in aqueous solution forming paramagnetic coloured species which were not characterised [6].

$[Rh(NH_3)_4(C_2O_4)](ClO_4) \cdot H_2O$ is obtained by heating $[Rh(NH_3)_5Cl]^{2+}$ with a mixture of sodium oxalate and oxalic acid in water in an autoclave at 120°C for 14 h. On addition of $HClO_4$, the complex was obtained as pale-yellow crystals. It is a useful intermediate for the preparation of *cis*-$[Rh(NH_3)_4X_2]^+$ (X = Cl or Br) by addition of HX [7].

***trans*-$[Rh(NH_3)_4(H_2O)(Me_2NCHO)](ClO_4)_2$, *trans*-$[Rh(NH_3)_4(H_2O)(Me_2SO)](ClO_4)_2$.** These compounds are prepared by photolysis of *trans*-$[Rh(NH_3)_4(H_2O)Cl](ClO_4)_2$ in either dimethylformamide or dimethylsulphoxide. The IR spectra of the $(CH_3)_2NCHO$-containing complexes exhibited ν(CO) at 1650 and 1370 cm^{-1}, and the $(CH_3)_2SO$-containing complex showed ν(SO) at 982 and 943 cm^{-1}; both spectra indicating O-bonded ligands. The electronic spectra of the complexes in $(CH_3)_2NCHO$ and $(CH_3)_2SO$ were also reported [8].

$[Rh(NH_3)_4(OH)X]^+$ (X = Cl, Br). The *cis* and *trans* isomers of these species are made by treatment of *cis*- and *trans*-$[Rh(NH_3)_4(H_2O)X]^{2+}$ with OH^-. The electronic spectra of both isomers have been reported in basic aqueous solution [1 to 4].

$[Rh(NH_3)_4(H_2O)X](S_2O_6)$. The *cis* chloride is prepared by treatment of *cis*-$[Rh(NH_3)_4Cl_2]Cl \cdot 0.5H_2O$ with Ag_2O in aqueous solution containing a small amount of $HClO_4$, followed by addition of $Na_2S_2O_6 \cdot 2H_2O$. The *cis* bromide, *trans* chloride and *trans* bromide are similarly prepared from cis-$[Rh(NH_3)_4Br_2]Br$ and *trans*-$[Rh(NH_3)_4X_2]X$ (X = Cl or Br), respectively. *Trans*-$[Rh(NH_3)_4(H_2O)X]^{2+}$ can also be obtained by photolysis of *cis*- or *trans*-$[Rh(NH_3)_4X_2]^+$ (X = Cl, Br or I) in aqueous solution. The dithionate salts of the *cis* chloride, *cis* bromide, *trans* chloride and *trans* bromide were isolated as pale-yellow, light-orange, orange-brown and red-brown crystals, respectively [9]. The electronic spectra of these complexes have been recorded in the range 200 to 500 nm in neutral and acidic aqueous solutions [2, 4, 9, 10]. The pK_a values of $[Rh(NH_3)_4(H_2O)X]^{2+}$ were determined as follows: X = Cl, *cis* = 7.84, *trans* = 6.75; X = Br, *cis* = 7.89, *trans* = 6.87 [2, 4].

References:

[1] L. H. Skibsted, P. C. Ford (J. Chem. Soc. Chem. Commun. **1979** 853/4). – [2] L. H. Skibsted, P. C. Ford (Inorg. Chem. **19** [1980] 1828/34). – [3] M. P. Hancock (Acta Chem. Scand. A **33** [1979] 499/502). – [4] L. H. Skibsted, P. C. Ford (Acta Chem. Scand. A **34** [1980] 109/13). – [5] L. E. Johnston, J. A. Page (Can. J. Chem. **47** [1969] 4241/6).

[6] J. F. Endicott, C.-L. Wong, T. Inoue, P. Natarajan (Inorg. Chem. **18** [1979] 450/4). – [7] M. P. Hancock (Acta Chem. Scand. A **29** [1975] 468/70). – [8] M. A. Bergkamp, R. J. Watts, P. C. Ford (J. Am. Chem. Soc. **102** [1980] 2627/31). – [9] L. H. Skibsted, D. Strauss, P. C. Ford (Inorg. Chem. **18** [1979] 3171/7, 3171/2). – [10] T. L. Kelly, J. F. Endicott (J. Phys. Chem. **76** [1972] 1937/46).

Tetrammine-azido Complex, *cis*-$[Rh(NH_3)_4(N_3)Cl]_2(S_2O_6)$

This compound is prepared by reaction of *cis*-$[Rh(NH_3)_4(H_2O)Cl]S_2O_6$ with NaN_3 in hot water. It was isolated as bright yellow crystals, but decomposition with HNO_2 revealed that the compound contained ~30% of *cis*-$[Rh(NH_3)_4Cl_2](S_2O_6)$. The electronic spectrum of the compound has been reported.

L. H. Skibsted, D. Strauss, P. C. Ford (Inorg. Chem. **18** [1979] 3171/7, 3171/2).

Tetrammine-dihalide Complexes, $[Rh(NH_3)_4X_2]Y$

X = Cl, Br, I. For convenience the chloro, bromo and iodo complexes are combined together.

***trans*-$[Rh(NH_3)_4Cl_2]Cl$** is obtained by reaction of $RhCl_3 \cdot 3H_2O$ with NH_4Cl and $(NH_4)_2CO_3$ by heating in aqueous solution for 3 h followed by dissolution of the precipitate hot HCl and cooling the solution [1, 2]. It may also be prepared by reaction of $[Rh(NH_3)_5H]Cl_2$ at pH = 6 to 7 with KCl, followed by acidification to pH < 1 and addition of H_2O_2 and HCl [3]. The complex was isolated as yellow crystals [2, 3]. The **perchlorate** is prepared by recrystallisation of *trans*-$[Rh(NH_3)_4Cl_2]Cl$ from aqueous $NaClO_4$ [3].

***trans*-$[Rh(NH_3)_4Cl_2](NO_3) \cdot H_2O$.** This compound is obtained either by recrystallisation of *trans*-$[Rh(NH_3)_4Cl_2]Cl$ from cold HNO_3 [1, 2], or by reaction of $[NH_4]_3[RhCl_6] \cdot H_2O$ with NH_4Cl, $(N_2H_6)Cl_2$, CH_3COONH_4 and HNO_3 [4]. It was isolated as golden-yellow crystals [1, 2].

***cis*-$[Rh(NH_3)_4Cl_2]Y$.** The **chloride,** as a hemihydrate, is prepared by treatment of $[Rh(NH_3)_4(C_2O_4)](ClO)_4 \cdot H_2O$ with HCl, and was isolated as bright yellow crystals [5]. The **dithionate** (Y = ½ S_2O_6) is formed from the chloride by metathesis with $Na_2S_2O_6 \cdot 2H_2O$ in water, and was obtained as pale yellow crystals [5].

***trans*-$[Rh(NH_3)_4Br_2]Y$.** The **bromide** is prepared by treating *trans*-$[Rh(NH_3)_4Cl_2]Cl$ with NaBr in boiling water, and was obtained as orange crystals [3]. The **perchlorate** is formed by photolysis of $[Rh(NH_3)_5Br](ClO_4)_2$ in aqueous $HClO_4$, followed by addition of NaBr and recrystallisation of the product from aqueous solution containing $NaClO_4$ [6].

***cis*-$[Rh(NH_3)_4Br_2]Br$** is obtained by reaction of $[Rh(NH_3)_4(C_2O_4)](ClO_4) \cdot H_2O$ with HBr, and the **dithionate,** *cis*-$[Rh(NH_3)_4Br_2](S_2O_6)$ is prepared by metathesis of the bromide with $Na_2S_2O_6 \cdot 2H_2O$. Both compounds were isolated as orange crystals [5].

***trans*-$[Rh(NH_3)_4I_2]I$** is prepared either by heating *trans*-$[Rh(NH_3)_4Cl_2]Cl$ with aqueous NaI at 90°C in the absence of light, or by reaction of the *cis*-dichloride with NaI in warm water, followed by removal of $Rh(NH_3)_3I_3$ by filtration and concentration of the filtrate. The compound was isolated as red-brown crystals [3].

Properties of $[Rh(NH_3)_4X_2]^+$ Salts. The structure of *trans*-$[Rh(NH_3)_4Cl_2]Cl$ has been determined by X-ray crystallography: triclinic, space group P1-C_i^1, Z = 2; a = 6.35(2), b = 5.77(2), c = 13.81(5) Å, α = 115.3(2)°, β = 99.6(2)°, γ = 77.5(2)°. The cation has an octahedral geometry with an average Rh-N distance of 2.07 Å, and an average Rh-Cl distance of 2.31 Å. The structure of $[Rh(NH_3)_4Cl_2]^+$ and Cl^- ions is linked by N-H···Cl hydrogen bonds [7]. X-ray powder diffraction data have been obtained from trans-$[Rh(NH_3)_4Cl_2](NO_3)\cdot H_2O$ and d spacings recorded [8].

A molecular orbital treatment has been made of the electronic structure of *trans*-$[Rh(NH_3)_4Cl_2]^+$ [9]. The electronic spectra of *trans*-$[Rh(NH_3)_4Cl_2]^+$ [1, 3, 8, 10,11] (spectrum shown in paper from 340 to 500 nm [1]), *cis*-$[Rh(NH_3)_4Cl_2]^+$ [4, 10], *trans*-$[Rh(NH_3)_4Br_2]^+$ [3, 5, 6, 10, 11], *cis*-$[Rh(NH_3)_4Br_2]^+$ [5, 10] and *trans*-$[Rh(NH_3)_4I_2]^+$ [3, 11] have been reported in the range 200 to 500 nm, and in aqueous solution. The luminescence spectrum of *trans*-$[Rh(NH_3)_4Cl_2]^+$ has been measured as a glass (in D_2O/methanol/ethyleneglycol) at 77 K, and the spectrum shown from 8000 to 44 000 cm^{-1} [12].

The X-ray photoelectron spectrum of *trans*-$[Rh(NH_3)_4Cl_2]Cl$ has been reported. The binding energies: $Cl\,2p_{3/2}$ = 199.2 and 197.8 (1.4) eV; N1s = 399.6 (1.4) eV [13]. The X-ray K spectra of *trans*-$[Rh(NH_3)_4Cl_2]Y$ (Y = Cl or NO_3) have been measured and have been used to demonstrate the covalent character of the Rh-Cl bond [14]. Use has been made of X-ray emission and absorption spectroscopy to probe the electronic structure of the metal in *trans*-$[Rh(NH_3)_4Cl_2](NO_3)$ [5]. The ^1H NMR spectrum of *trans*-$[Rh(NH_3)_4Cl_2]^+$ in conc. H_2SO_4 has been measured ($\tau = -3.85$) [16].

References:

[1] S. A. Johnson, F. Basolo (Inorg. Chem. **1** [1962] 925/32). – [2] S. N. Anderson, F. Basolo (Inorg. Syn. **7** [1963] 214/20, 216, 217). – [3] A. J. Pöe, M. V. Twigg (Can. J. Chem. **50** [1972] 1089/92). – [4] B. I. Peshchevitskii, A. V. Belyaev, S. Ya. Dvurechenskaya (Izv. Sibirsk. Otd. Akad. Nauk SSSR Ser. Khim. Nauk **1972** 138/40 from C.A. **77** [1973] No. 159615). – [5] M. P. Hancock (Acta Chem. Scand. A **29** [1975] 468/70).

[6] T. L. Kelly, J. F. Endicott (J. Phys. Chem. **76** [1972] 1937/46). – [7] I. A. Baidina, N. V. Podberezskaya, L. P. Solov'eva (Zh. Strukt. Khim. **15** [1974] 940/1; J. Struct. Chem. [USSR] **15** [1974] 834/5). – [8] M. M. Muir, W.-L. Huang (Inorg. Chem. **12** [1973] 1831/5). – [9] V. I. Baranovskii, Yu. N. Kukushkin, N. S. Panina, K. A. Khokhryakov (Koord. Khim. **7** [1981] 415/20; Soviet J. Coord. Chem. **7** [1981] 195/200). – [10] L. H. Skibsted, D. Strauss, P. C. Ford (Inorg. Chem. **18** [1974] 3171/7).

[11] W. D. Blanchard, W. R. Mason (Inorg. Chim. Acta **78** [1978] 159/68). – [12] T. R. Thomas, G. A. Crosby (J. Mol. Spectrosc. **38** [1971] 118/29). – [13] J. R. Ebner, D. M. McFadden, D. R. Tyler, R. A. Walton (Inorg. Chem. **15** [1976] 3014/8). – [14] A. P. Sadovskii, A. V. Belyaev (Izv. Sibirsk. Otd. Akad. Nauk SSSR Ser. Khim. Nauk **1967** No. 6, pp. 57/61; Sib. Chem. J. **1967** 703/6). – [15] L. N. Mazalov, A. P. Sadovskii, A. V. Belyaev, L. I. Chernayavskii, E. S. Gluskin, L. F. Berkoer (Izv. Sibirsk. Otd. Akad. Nauk SSSR Ser. Khim. Nauk **1971** No. 2, pp. 51/9 from C.A. **76** [1972] No. 39710).

[16] D. N. Hendrickson, W. L. Jolly (Inorg. Chem. **9** [1970] 1197/201).

Studies have been made of the photoaquation of *cis*- and *trans*-$[Rh(NH_3)_4X_2]^+$ (X = Cl, Br). This process led to the formation of *trans*-$[Rh(NH_3)_4(OH_2)X]^{2+}$ (see section on hydroxy and aquo tetrammine complexes, p. 148) [1 to 8].

The kinetics and mechanism of hydrolysis of *trans*-$[Rh(NH_3)_4Cl_2]^+$ in aqueous solution have been studied [9], and the rate of exchange of bound Cl with Cl^- in solution investigated [10]. The kinetics of aquation of *trans*-$[Rh(NH_3)_4Cl_2]^+$ in the presence of Hg^{2+} have been reported, and $[Rh(NH_3)_4Cl_2Hg]^{3+}$ detected as an intermediate [11]. A study of the chemical effects of (n,γ) reactions in *trans*-$[Rh(NH_3)_4Cl_2]^+$, using ^{36}Cl-labelled compounds, showed that at least one Rh-Cl bond was broken [12].

References:

[1] D. Strauss, P. C. Ford (J. Chem. Soc. Chem. Commun. **1977** 194/5). – [2] L. H. Skibsted, D. Strauss, P. C. Ford (Inorg. Chem. **18** [1979] 3171/7). – [3] T. L. Kelly, J. F. Endicott (J. Phys. Chem. **76** [1972] 1937/46). – [4] M. M. Muir, W.-L. Huang (Inorg. Chem. **12** [1973] 1831/5). – [5] C. Kutal, A. W. Adamson (Inorg. Chem. **12** [1973] 1454/6).

[6] J. Sellan, R. Rumfeldt (Can. J. Chem. **54** [1976] 519/25). – [7] J. I. Zink (Inorg. Chem. **12** [1973] 1018/24). – [8] M. J. Incorvia, J. I. Zink (Inorg. Chem. **13** [1974] 2489/94). – [9] S. A. Johnson, F. Basolo, R. G. Pearson (J. Am. Chem. Soc. **85** [1963] 1741/7). – [10] J. Burgess, K. W. Bowker, E. R. Gardner, F. M. Mekhail (J. Inorg. Nucl. Chem. **14** [1979] 1215/7).

[11] A. B. Venediktov, A. V. Belyaev (Koord. Khim. **2** [1976] 1414/21; Soviet J. Coord. Chem. **2** [1976] 1085/91). – [12] E. R. Gardner, M. E. Wilson, R. D. Harding, J. B. Raynor (Radiochim. Acta **17** [1972] 41/5).

2.1.7.2.3 Triammine Complexes

See „Rhodium“ 1938, p. 146.

$Rh(NH_3)_3(NO_2)_3$. This is prepared either by heating $Na[NH_4]_2[Rh(NO_2)_6]$ with aqueous ammonia [1], or by reacting $Na_3Rh(NO_2)_6$ in a stepwise fashion with NH_4Cl, HCl and K_2SO_4, and ammonia [2]. It was said that the compound was isolated as a yellow solid [1], but one report describes it as colourless crystals [3]. The optical properties of crystals of the compound have been measured [3], and X-ray photoelectron spectral measurements made [4].

$Rh(NH_3)_3(OH)_3$ is said to be formed by the action of ammonia on a mixture of boiling $H_3[RhCl_6]$ and H_2O_2 [5]. However, the compound has been reformulated as $Rh(H_2O)_2(NH_3)(OH)_3$ [6] (see section on mono-ammine complexes, p. 154). The compound was said to decompose thermally under nitrogen giving Rh and Rh_2O_3, under hydrogen giving Rh metal, and under oxygen giving pure Rh_2O_3 [5]. It was also said to react readily with HCl, H_2SO_4, HNO_3, $HClO_4$ and CH_3COOH giving compounds of the type $Rh(NH_3)_3X_3$ [5], but some of these have been reformulated (see section on mono-ammine complexes, p. 154).

$[Rh(NH_3)_3(H_2O)_3]Cl_3$ has been briefly described but few preparative details given. The acid dissociation constants were determined as $pK_1 = 5.54$, $pK_2 = 7.55$ and $pK_3 = 9.53$ [7].

$Rh(NH_3)_3Cl_3$. This is said to be formed by the reaction of $Rh(NH_3)_3(OH)_3$ with conc. HCl. It was isolated as an orange-yellow compound which behaved as a non-electrolyte in aqueous solution. The IR spectrum indicated that the complex was a mixture of two isomers, probably containing *mer* and *fac* Cl_3 arrangements [8].

The optical properties of crystals of $Rh(NH_3)_3Cl_3$ have been determined [3], and experimental data on the refractive indices and crystal dimensions used to calculate the Langevin diamagnetic susceptibility of the compound [9]. The electronic spectrum has been measured (shown in paper from 20000 to 50000 cm^{-1} in aqueous solution) [10]. X-ray photoelectron spectra of $Rh(NH_3)_3Cl_3$ have been obtained, and compared with data from related complexes [4]. The X-ray K spectrum of $Rh(NH_3)_3Cl_3$ has been obtained and revealed a narrow maximum at ~10 eV [11]. $Rh(NH_3)_3Cl_3$ reacted with boiling saturated $(NH_4)_2SO_3$ solution giving $[NH_4]_3$-$[Rh(SO_3)_3]$ [12].

$Rh(NH_3)_3I_3$. The luminescence spectrum of this compound was reported in glasses (D_2O/methanol/ethyleneglycol) at 85 K [13], and the photoelectron spectrum has been reported [4].

References:

[1] V. V. Lebedinskii, E. V. Shenderetskaya (Izv. Sekt. Platiny **18** [1945] 19/22 from C.A. **1947** 5044). – [2] L. M. Gerasimovich, S. I. Khotyanovich (Issled. Obl. Elektroosazhdeniya Metal. Mater. 15th Resp. Konf. Elektrokhim. Lit.SSR, Vilnius 1977, pp. 76/9 from C.A. **89** [1973] No. 13913). – [3] M. N. Lyashenko (Tr. Inst. Krist. Akad. Nauk SSSR **7** [1952] 67/72 from C.A. **1956** 2439). – [4] V. I. Nefedov, M. A. Porai-Koshits (Mater. Res. Bull. **7** [1972] 1543/52). – [5] G. Pannetier, R. Bonnaire, P. Alepee, P. Davous, V. Huynh (J. Less-Common Metals **18** [1969] 275/84).

[6] G. Pannetier, R. Bonnaire, P. Alepee, P. Davous, V. Huynh (J. Less-Common Metals **21** [1970] 81/3). – [7] S. Ya. Dvurechenskaya, A. V. Belyaev, B. I. Peshchevitskii (Izv. Sibirsk. Otd. Akad. Nauk SSSR Ser. Khim. Nauk **1973** 49/53 from C.A. **79** [1973] No. 35692). – [8] G. Pannetier, R. Bonnaire, P. Alepee, P. Davous, V. Huynh (J. Less-Common Metals **18** [1969] 285/94). – [9] G. B. Avetikyan, Yu. N. Kukushkin, K. A. Khokhryakov, L. I. Danilina, E. F. Strizhev (Izv. Vysshikh Uchebn. Zavedenii Khim. Khim. Tekhnol **21** [1978] 7/10 from C.A. **88** [1978] No. 181380). – [10] H. H. Schmidtke (Z. Anorg. Allgem. Chem. **339** [1965] 103/12).

[11] A. P. Sadovskii, A. V. Belyaev (Izv. Sibirsk. Otd. Akad. Nauk SSSR. Ser. Khim. Nauk **1967** No. 6, pp. 57/61; Sib. J. Chem. **1967** 57/62). – [12] V. V. Lebedinskii, E. V. Shenderetskaya, A. G. Maiorova (Zh. Prikl. Khim. **32** [1959] 928/9; J. Appl. Chem. [USSR] **32** [1959] 944/6). – [13] P. E. Hoggard, H. H. Schmidtke (Ber. Bunsenges. Physik. Chem. **77** [1973] 1052/8).

For **$[Rh(NH_3)_3(SO_3)_3]^{3-}$**, see "Rhodium" Suppl. Vol. B 1, 1982, pp. 149/50.

2.1.7.2.4 Diammine Complexes

$[Rh(NH_3)_2(NO_2)_4]^-$. The **potassium** salt is prepared by boiling $Na(NH_4)_2[Rh(NO_2)_6]$ in aqueous ammonia, followed by removal of $Rh(NH_3)_3(NO_2)_3$ by filtration, and addition of KCl. The **caesium** salt is obtained similarly using CsCl. Both salts were isolated as white crystals, the K^+ species being isolated as monohydrate [1].

$[Rh(NH_3)_2(H_2O)_4]Cl_3$ has been mentioned briefly but no preparative details were given. The acid dissociation constants were determined: $pK_1 \approx 4.0$, $pK_2 \approx 6.7$. At pH > 7, the compound was apparently polymeric [2].

$Na[Rh(NH_3)_2(S_2O_3)_2]$ is obtained by decomposition of $Na_{12}(NH_4)_2[Rh_4 \cdot 13S_2O_3 \cdot 10NH_3] \cdot 2.5H_2O$ or $Na_8[Rh_2 \cdot 7S_2O_3 \cdot 6NH_3]$, said to be formed by reaction of $(NH_4)_3RhCl_3$, and $Rh(NH_3)_3Cl_3$ with $Na_2S_2O_3 \cdot 5H_2O$ in aqueous ammonia. It reacted with aqueous ammonia giving $Na_3[Rh(NH_3)_3(S_2O_3)_3]$. The nature of this species is uncertain [3].

References:

[1] V. V. Lebedinskii, E. V. Shenderetskaya (Izv. Sekt. Platiny **31** [1955] 53/5 from C.A. **1956** 16525). – [2] S. Ya. Dvurechenskaya, A. V. Belyaev, B. I. Peshchevitskii (Izv. Sibirsk. Otd. Akad. Nauk SSSR Ser. Khim. Nauk **1973** 49/53 from C.A. **79** [1973] No. 35692). – [3] I. I. Chernyaev, A. G. Maiorova (Zh. Neorgan. Khim.**5** [1960] 1208/20; Russ. J. Inorg. Chem. **5** [1960] 583/8).

For **$[Rh(NH_3)_2(SO_3)_4]^{5+}$**, see "Rhodium" Suppl. Vol. B 1, 1982, p. 149.

2.1.7.2.5 Mono-ammino Complexes

See "Rhodium" 1938, pp. 151/2.

$Rh(H_2O)_2(NH_3)(OH)_3$. This compound, originally formulated as $Rh(NH_3)_3(OH)_3$ [1], is prepared by reaction of ammonia with a boiling aqueous solution of $H_3[RhCl_6]$ and H_2O_2. It is isolated as a yellow solid [2].

$[Rh(NH_3)(H_2O)_5]Cl_3$ has been briefly reported, but no preparative details were given. The acid dissociation constants were measured, $pK_1 \approx 3.8$, $pK_2 \approx 6.3$. The salt is apparently polymeric when pH>7 in aqueous solution [3].

$K[Rh(NH_3)(NO_2)_3(OH)]$ is obtained by reaction of $K[Rh(NH_3)_2(NO_2)_4] \cdot 0.5H_2O$ with $(NH_4)_2SO_4$ at pH = 1 to 2, and heating $K_2[Rh_2(NH_3)_2(N_2)(NO_2)_6(OH)_2]$ so formed in a CO_2 flow [4].

References:

[1] G. Pannetier, R. Bonnaire, P. Alepee, P. Davous, V. Huynh (J. Less-Common Metals **18** [1969] 275/84). – [2] G. Pannetier, R. Bonnaire, P. Alepee, P. Davous, V. Huynh (J. Less-Common Metals **21** [1970] 81/3). – [3] S. Ya. Dvurechenskaya, A. V. Belyaev, B. I. Peshchevitskii (Izv. Sibirsk. Otd. Akad. Nauk SSSR Ser. Khim. Nauk **1973** 49/53 from C.A. **79** [1973] No. 35692). – [4] L. S. Volkova, V. M. Volkov, S. S. Chernikov (Zh. Neorgan. Khim. **16** [1971] 2594/5; Russ. J. Inorg. Chem. **16** [1971] 1383.

2.1.7.2.6 Binuclear Complexes

$[Rh(NH_3)_4(H_2O)(CH_3CO_2)_2(OH)]_2$. This binuclear Rh^{III} compound is obtained by reaction of $Rh(H_2O)_2(NH_3)(OH)_3$ with acetic acid [1, 2]. It is reduced by ethanol in acetic acid to give a binuclear Rh^{II} complex $[Rh(NH_3)(H_2O)(CH_3CO_2)_2]_2$. The compound is formulated as $[Rh(CH_3CO_2)(HOH \cdots CH_3CO_2)(OH)(NH_3)]_2$ [2].

$[Rh(NH_3)(H_2O)(CH_3CO_2)_2]_2$ was shown to be diamagnetic, and its IR (measured from 4000 to 800 cm^{-1} in KBr) and electronic spectrum (150 to 600 nm) measured. It was formulated as $[Rh(NH_3)(CH_3CO_2)(HOH \cdots CH_3CO_2)]_2$, containing μ-CH_3CO_2 bridges and a Rh-Rh bond. A crystallographic examination established that the compound was monoclinic, space group either Cc-C_s^4 or C2/c-C_{2h}^6, Z = 4, with a = 13.281, b = 8.623, c = 14.218 Å and β = 118.71° [3]. The compound underwent hydrolysis to give a species formulated as $(NH_3)(H_2O)(OH)Rh(CH_3CO_2)_2$-$Rh(CH_3CO_2)(H_2O)(NH_3)$, and its thermogravimetric properties were examined [2].

$[Rh(NH_3)(H_2O)_3(C_2O_4)]C_2O_4$ is prepared by reaction of an aqueous suspension of $Rh(NH_3)(H_2O)_2(OH)_3$ with oxalic acid. The compound was isolated as an orange-yellow solid, and was soluble in ethanol, acetone and benzene. The IR spectrum was measured in KBr discs (shown in paper from 4000 to 500 cm^{-1}) and ^1H NMR spectral studies confirmed the presence of NH_3. It was suggested that the compound contains two oxalate bridging groups. The conductivity of the saturated aqueous solution (195 g/l at 21°C) is 9×10^{-2} $\Omega^{-1} \cdot cm^{-1}$. Thermal studies show decomposition at 70 to 80°C and formation of metallic rhodium below 225°C [4].

$[Rh(NH_3)(H_2O)_3(O_2CCH_2CO_2)](O_2CCH_2CO_2)$ and **$[Rh(NH_3)(H_2O)_3(O_2CCH_2CH_2CO_2)]$-$(O_2CCH_2CH_2CO_2)$** are prepared in the same way as the oxalate above, using malonic and succinic acids. The IR spectra (4000 to 500 cm^{-1} shown in paper) and ^1H NMR spectra were said to be consistent with their formulation. The compounds decomposed on standing, and on heating under hydrogen in nitrogen afforded Rh metal or Rh_2O_3 [5].

$[Rh(NH_3)(OH)_2X]_2$ ($X=NO_3$ or ClO_4). These compounds are prepared by treatment of $Rh(NH_3)(H_2O)_2(OH)_3$ with cold HNO_3 or hot $HClO_4$ [6]. They were isolated as orange-yellow solids, and their IR spectra were measured (4000 to 500 cm^{-1}, shown in paper). They were thought to contain two bridging nitrate or perchlorate groups [6].

$K_2[Rh_2(NH_3)_2(N_2)(NO_2)_6(OH)_2]$. This complex is said to be formed by reaction of $K[Rh(NH_3)_2(NO_2)_4]\cdot 0.5H_2O$ with $(NH_4)_2SO_4$ at pH=1 to 2 in aqueous solution. The Raman spectrum revealed a band at 2070 cm^{-1} which shifted to 2045 cm^{-1} when ^{15}N-labelled $(NH_4)_2SO_4$ was used in the synthesis. It was suggested that the complex contained a linear RhN-NRh group [3].

References:

[1] G. Pannetier, R. Bonnaire, P. Alepee, P. Davous, V. Huynh (J. Less-Common Metals **18** [1969] 275/84). – [2] G. Pannetier, J. Segall (J. Less-Common Metals **22** [1970] 305/16). – [3] L. S. Volkova, V. M. Volkov, S. S. Chernikov (Zh. Neorgan. Khim. **16** [1971] 2594/5; Russ. J. Inorg. Chem. **16** [1971] 1383). – [4] G. Pannetier, A. Mbomé, N. Platzer (J. Less-Common Metals **22** [1970] 61/70, 62). – [5] G. Pannetier, A. Mbomé, N. Platzer (J. Less-Common Metals **23** [1971] 403/14).

[6] G. Pannetier, A. Mbomé, C. Laurent (J. Less-Common Metals **25** [1971] 27/38).

2.1.7.2.7 Other Polynuclear Complexes

$Na_{14}(NH_4)_2[Rh_4\cdot 13S_2O_3\cdot 2NaS_2O_3\cdot 12NH_3]$. This species is apparently formed by reaction of $(NH_4)_2RhCl_6\cdot H_2O$ with $Na_2S_2O_3\cdot 5H_2O$ in aqueous ammonia. It can be converted into $Na_6(NH_4)[Rh_2\cdot 6S_2O_3\cdot NaS_2O_3\cdot 6NH_3]$ and decomposed in the presence of ammonia giving $Na_4[Rh_2\cdot 5S_2O_3\cdot 6NH_3]\cdot 3H_2O$. The nature of these species is uncertain.

$Na_{12}(NH_4)_2[Rh_{14}\cdot 13S_2O_3]\cdot 2.5H_2O$ is also obtained by reaction of $(NH_4)_3RhCl_6$ with sodium thiosulphate in aqueous ammonia, and decomposed into $Na_3[Rh(NH_3)_2(S_2O_3)_3]$.

$Na_8[Rh_2\cdot 7S_2O_3\cdot 6NH_3]$ is prepared by reaction of $Rh(NH_3)_3Cl_3$ with $Na_2S_2O_3\cdot 5H_2O$ in aqueous ammonia. It reacted with ammonia, or decomposed in boiling water, to give $Na_3[Rh(NH_3)_2(S_2O_3)_3]$.

The nature of the above species is uncertain.

I. I. Chernyaev, A. G. Maiorova (Zh. Neorgan. Khim. **5** [1960] 1208/20; Russ. J. Inorg. Chem. **5** [1960] 583/8).

2.1.8 Alkyl-, Allyl- and Aryl-amine Complexes

2.1.8.1 Rhodium(I) Complex, $Rh(\overline{NHCH_2CH_2})_3I$ ($\overline{NHCH_2CH_2}$ = ethyleneimine, aziridine)

It is claimed that this complex is prepared either by reaction of $Rh(\overline{NHCH_2CH_2})_3(CO)Cl$ with an excess of HI in methanol [1], or by treatment of $Rh(\overline{NHCH_2CH_2})_3Cl_3$ with Ag_2O in water followed by addition of KI [2]. It is described as a red-brown [1] or dark brown [2] solid. However, an X-ray crystallographic examination of a sample of this material revealed it to be *trans*-$[Rh(\overline{NHCH_2CH_2})_4I_2]I$ [3].

References:

[1] M. R. Hoffmann, J. O. Edwards (Inorg. Nucl. Chem. Letters **10** [1974] 837/43). – [2] J. Scherzer, P. K. Phillips, L. B. Clapp, J. O. Edwards (Inorg. Chem. **5** [1966] 847/51). – [3] K. Lussier, J. O. Edwards, R. Eisenberg (Inorg. Chim. Acta **3** [1969] 468/70).

2.1.8.2 Rhodium(III) Alkyl- and Allyl-amine Complexes

2.1.8.2.1 Unsubstituted Amine, $[Rh(NHCH_2CH_2)_6]Cl_3$ ($NHCH_2CH_2$ = ethyleneimine, aziridine)

This compound is prepared by reaction of $Rh(NHCH_2CH_2)_3Cl_3$ in an excess of ethyleneimine. It is obtained as white crystals, and its formation is difficult to reproduce.

J. Scherzer, P. K. Phillips, L. B. Clapp, J. O. Edwards (Inorg. Chem. **5** [1966] 847/51).

2.1.8.2.2 Substituted Amines

Pentamine Complexes

$[Rh(NH_2CH_3)_5(OH_2)](ClO_4)_3$ is obtained by heating $[Rh(NH_2CH_3)_5Cl](ClO_4)_2$ with $AgClO_4$ in perchloric acid solution for 24 h. It is isolated as pale yellow crystals, and its UV spectrum has been recorded in dilute $HClO_4$ in the range 190 to 530 nm. Acid dissociation constants have been reported for $[Rh(NH_2CH_3)_5(OH_2)]^{3+}$, and it has been shown that NH_2CH_3 is a poorer electron donor than NH_3 in this type of complex [1].

$[Rh(NH_2R)_5Cl]Cl_2$ (R = CH_3, C_2H_5, n-C_3H_7, n- and iso-C_4H_9). These complexes are formed when $RhCl_3 \cdot 3H_2O$ is treated with $(NH_3R)Cl$ in aqueous ethanol to which NH_2R is then added; R = CH_3 [1]; R = C_2H_5, n-C_3H_7, n- and iso-C_4H_9 [2]. The **allylamine** complex (R = $NH_2C_3H_5$) is prepared by treatment of $RhCl_3 \cdot 3H_2O$ in ethanol with an excess of allylamine [3], or by reaction of dry $RhCl_3$ with allylamine [4].

$[Rh(NH_2CH_3)_5Cl](ClO_4)_2$ is obtained by dissolving $[Rh(NH_2CH_3)_5Cl]Cl_2$ in aqueous $HClO_4$ and adding perchlorate ion [1].

$[Rh(NH_2R)_5Br]Br_2$ (R = CH_3, C_2H_5, n-C_3H_7, n- and iso-C_4H_9). These complexes are prepared either by heating a mixture of finely ground $RhBr_3 \cdot 3H_2O$, NaBr and $(NH_3R)Cl$ in aqueous ethanol to which NH_2R had been added, or by reacting $RhCl_3 \cdot 3H_2O$ with NaBr and $(NH_3R)Br$ under similar conditions [2].

The **chloro** complexes were obtained as yellow crystals and the melting point of the allylamine complex was 170°C [1, 2, 4].

The conductivities of these complexes have been measured in aqueous solutions, and are consistent with 1:2 electrolytes [2].

The IR spectra of $[Rh(NH_2R)_5Cl]^{2+}$ and $[Rh(NH_2R)_5Br]^{2+}$ have been obtained in Nujol mulls or KBr discs [2, 3, 4]. The values of ν(NH) and δ(as-NH_2) were reported for the chloro complexes where R = CH_3, C_2H_5, n-C_3H_7 or C_3H_5 [4], and it was established that in $[Rh(NH_2C_3H_5)_5Cl]Cl_2$, the amine was coordinated via the N atom and the C=C bond did not bind to the metal [3]. From Raman and IR spectral studies the values of ν(Rh-X; X = halide) were found to be in the range 290 to 343 (Raman) and 313 to 343 cm^{-1} (IR) for X = Cl, and 171 to 214 (Raman) and 170 to 217 cm^{-1} (IR) for X = Br. The electronic spectra of $[Rh(NH_2R)_5X]^{2+}$ (X = Cl or Br) have been measured in the range 200 to 500 nm in aqueous solution, ethanol, methanol or aqueous ethanol (1:4) mixtures [2].

Polarographic studies of $[Rh(NH_2R)_5X]^{2+}$ in aqueous solution revealed that the complexes underwent a reduction in the range −0.69 to −0.70 V for X = Cl (vs. standard calomel electrode with KCl supplementary electrode) and −0.58 to −0.61 V for X = Br (vs. standard calomel electrode with KBr supplementary electrode). A correlation was made between the reduction potential and ν(max) in the electronic spectra of the complexes [2].

The kinetics of base hydrolysis of $[Rh(NH_2CH_3)_5Cl]^{2+}$ have been measured [1] and the rate constants and activation parameters for the substitution of Cl by Br in $[Rh(NH_2R)_5Cl]^{2+}$ have been determined. For R = CH_3, C_2H_5 and n-C_3H_7 at 353 K, 10^5 k = 1.64, 4.50 and 5.40 s^{-1}; ΔH = 32.1 ± 4.6, 24.1 ± 0.7 and 23.7 ± 1.0 kcal/mol; ΔS = 10.2 ± 12.9, −11 ± 2 and −11 ± 3 cal $\cdot mol^{-1} \cdot K^{-1}$ [2].

$[Rh(NH_2C_3H_5)_5Cl]Cl_2$ was stable towards thermal decomposition in ethanol [3].

$[Rh(NHCH_2CH_2)_5Cl]X_2$. The **chloride** (X = Cl) is obtained in the reaction between $RhCl_3 \cdot 3H_2O$ with an excess of ethyleneimine after removal of $Rh(NHCH_2CH_2)_3Cl$ by filtration. The **tetraiodomercurate** (X_2 = HgI_4) is formed after treatment of the dichloride with K_2HgI_4. Both compounds were isolated as yellow salts [5].

References:

[1] T. W. Swaddle (Can. J. Chem. **55** [1977] 3166/71). – [2] S. P. Dagnall, M. P. Hancock, B. T. Heaton, D. H. Vaughan (J. Chem. Soc. Dalton Trans. **1977** 1111/6). – [3] H. Sawai, H. Hirai (Inorg. Chem. **10** [1971] 2068/70). – [4] E. F. Shubochkina, M. A. Seifer, O. E. Zvyagintsev (Zh. Neorgan. Khim. **13** [1968] 1382/5; Russ. J. Inorg. Chem. **13** [1968] 724/6). – [5] J. Scherzer, P. K. Phillips, '.. B. Clapp, J. O. Edwards (Inorg. Chem. **5** [1966] 847/51).

Tetra-amine Dihalide Complexes

***cis*-$[Rh(NH_2R)_4Cl_2]X$** (R = CH_3 or C_2H_5). The chloride (X = Cl) is obtained by filtration of the solution after reaction of $RhCl_3 \cdot 3H_2O$ with NH_2R in aqueous alkaline media containing traces of methanol or ethanol [1]. The corresponding **nitrates** are prepared by addition of conc. HNO_3 to the reaction solutions, and are collected by filtration [1].

***cis*-$[Rh(NH_2C_3H_5)_4Cl_2]X$.** The **chloride** (X = Cl) is produced by treatment of $RhCl_3 \cdot 3H_2O$ with allylamine in aqueous ethanol. The corresponding **nitrate** is formed when a small amount of HNO_3 is present in the reaction mixtures. The complexes are probably *cis* but this has not been confirmed [1].

All of the above complexes were isolated as yellow crystals [1].

***trans*-$[Rh(NH_2R)_4Cl_2]X \cdot nH_2O$** (R = CH_3, C_2H_5, n-C_3H_7, n-C_4H_9, i-C_4H_9). The **chlorides** (X = Cl) are prepared by reaction of $[Rh(NH_2R)_5Cl]Cl_2$ with zinc dust in the presence of NH_2R in warm water (50°C). Addition of KCl and H_2O_2 afforded the desired complexes. The corresponding **nitrates** (X = NO_3) are formed when the chlorides are dissolved in and recrystallised from dilute HNO_3 [2, 4].

***trans*-$[Rh(NH_2C_2H_5)_4Cl_2]Cl$** may also be prepared by reaction of $RhCl_3 \cdot 3H_2O$ with $NH_2C_2H_5$ under basic conditions and evaporation of the filtrate of the reaction mixture. The **nitrate** can be produced also by recrystallisation from dilute HNO_3 [3, 4].

***trans*-$[Rh(NH_2R)_4Cl_2]X$** (R = i-C_3H_7 or C_6H_{11}). The **chlorides** (X = Cl) are prepared by reaction of $RhCl_3 \cdot 3H_2O$ with the appropriate amine in ethanol. The corresponding **perchlorates** are obtained by treatment of the chloride with $LiClO_4$ or $NaClO_4$ [2].

***trans*-[Rh(NH$_2$R)$_4$Br$_2$]X.** ***trans*-[Rh(NH$_2$CH$_3$)$_4$Br$_2$]Br** is obtained by heating *trans*-[Rh(NH$_2$CH$_3$)$_4$Cl$_2$]Cl with HBr and recrystallising the product from aqueous HBr [2]. ***trans*-[Rh(NH$_2$C$_2$H$_5$)$_4$Br$_2$]Br** is formed by treatment of *trans*-[Rh(NH$_2$C$_2$H$_5$)$_4$Cl$_2$](NO$_3$) with a large excess of NaBr in water. The corresponding **perchlorate** (X = ClO$_4$) is produced by treating the filtrate after removal of [Rh(NH$_2$C$_2$H$_5$)$_4$Br$_2$]Br with HClO$_4$. ***trans*-[Rh(NH$_2$C$_2$H$_5$)$_4$I$_2$]I** is similarly prepared by reaction of the nitrate with an excess of NaI in water containing $[IrCl_6]^{2-}$. Reaction of *trans*-[Rh(NH$_2$C$_2$H$_5$)$_4$I$_2$]I with aqueous NaCl, followed by addition of HClO$_4$, gives ***trans*-[Rh(NH$_2$C$_2$H$_5$)$_4$ClI](ClO$_4$)**, and with NaBr and HClO$_4$, ***trans*-[Rh(NH$_2$C$_2$H$_5$)$_4$BrI](ClO$_4$)** is formed [3].

The dichloro complexes are yellow or orange-yellow, the dibromo complexes orange, the diiodo species brown, the chloro-iodo and bromo-iodo cations orange and orange-red, respectively [2, 3].

The IR spectra of *cis*- and *trans*-[Rh(NH$_2$R)$_4$X$_2$]Y have been reported. Both ν(RhN) and ν(NH$_2$), as well as δ(cis-NH$_2$), have been assigned [1, 3, 4], and the spectra shown in the ranges 3800 to 2200, 2000 to 800 and 700 to 400 cm^{-1} [3]. The Rh-X stretching frequency has been observed in the Raman spectra of *trans*-$[Rh(NH_2R)_4Cl_2]^+$ in the range 289 to 307 cm^{-1}. The electronic spectra of *cis*- and *trans*-$[Rh(NH_2R)_4X_2]^+$ have been reported in the range 350 to 280 nm [1] and 200 to 500 nm [2]. The spectra of $[Rh(NH_2C_2H_5)_4XY]^+$ (X = Y = Cl, Br or I; X = Br, Y = I) have been measured and the spectra in the range 20000 to 50000 cm^{-1} shown in the paper [3].

Polarographic studies of *trans*-$[Rh(NH_2R)_4X_2]^+$ (X = Cl or Br; R = CH$_3$, C$_2$H$_5$, n-C$_3$H$_7$, i-C$_3$H$_7$, n-C$_4$H$_9$, i-C$_4$H$_9$, C$_6$H$_{11}$) have established that the complexes were reduced in the potential range −0.49 to −0.56 V (vs. standard calomel electrode with KCl or KBr supplementary electrodes) [2].

The thermal stability of *trans*-[Rh(NH$_2$R)$_4$X$_2$]X (R = CH$_3$, X = Cl; R = C$_2$H$_5$, X = Cl, Br or I) has been investigated by thermal gravimetry. The thermal stability of the ethylamine complexes decreased in the order Cl > Br > I. The thermogravimetric curve of [Rh(NH$_2$C$_2$H$_5$)$_4$Cl$_2$]Cl was shown in the paper [4].

The rate of exchange of amino protons in *trans*-$[Rh(NH_2CH_3)_4Cl_2]^+$ and the rate constants and activation parameters for substitution of Cl by Br in *trans*-$[Rh(NH_2R)_4Cl_2]^+$ (R = CH$_3$, C$_2$H$_5$ and n-C$_3$H$_7$) have been determined and compared with other Rh^{III}, Co^{III} and Cr^{III} complexes. In this reaction, with R = CH$_3$, C$_2$H$_5$ and n-C$_3$H$_7$ at 353 K, 10^5 k = 2.0, 4.59 and 5.75 s^{-1}, ΔH = 29.2 ± 1.0, 24.5 ± 0.7 and 24.9 ± 3.3 kcal/mol, ΔS = 2.3 ± 2.3, −9.4 ± 2.1 and −7.8 ± 9.3 cal · $mol^{-1} \cdot K^{-1}$ [2].

[Rh(NHCH$_2$CH$_2$)$_4$X$_2$]X (NHCH$_2$CH$_2$ = ethyleneimine, aziridine). The **chloro** complex (X = Cl) is obtained by fractional crystallisation of the evaporated filtrate obtained after removal of Rh(NHCH$_2$CH$_2$)$_3$Cl$_3$ from the reaction between RhCl$_3$·3H$_2$O and an excess of ethyleneimine. The **bromo** complex (X = Br) is prepared from an aqueous solution of RhBr$_3$ and ethyleneimine, and the **iodo** species (X = I) from RhI$_3$ and ethyleneimine in ethanol [5].

[Rh(NHCH$_2$CH$_2$)$_4$Cl$_2$]Br and **[Rh(NHCH$_2$CH$_2$)$_4$Cl$_2$]I** are obtained by treatment of [Rh(NHCH$_2$CH$_2$)$_4$Cl$_2$]Cl with KBr [5] and KI [5, 6], respectively.

The colours of [Rh(NHCH$_2$CH$_2$)$_4$Cl$_2$]Cl and the corresponding bromo and iodo species were bright yellow, brown and yellow-brown, respectively, while [Rh(NHCH$_2$CH$_2$)$_4$Cl$_2$]Br was yellow [5].

The crystal and molecular structure of *trans*-Rh(NHCH$_2$CH$_2$)$_4$I$_2$]I has been determined by X-ray methods. The complex was tetragonal, space group $P\bar{4}n2$-D^8_{2d} (Z = 2) with a = 7.81(1), c = 14.26(2) Å. The $[Rh(NHCH_2CH_2)_4I_2]^+$ group is crystallographically required to have D$_2$ symmetry, the coordination geometry about the metal being a tetragonally distorted octahedron. The Rh-N and Rh-I distances were 1.99(2) and 2.681(5) Å, respectively, and the ethyleneimine N atoms alternated slightly above and below the equatorial plane of the distorted octahedron; $D_{exp} = 2.48 \pm 0.04$ g/cm³ [6].

The electronic spectra of *trans*-$[Rh(\overline{NHCH_2CH_2})_4X_2]^+$ have been measured in water in the range 20000 to 46000 cm^{-1}, and spectral assignments tentatively made. The spectrum of *trans*-$[Rh(\overline{NHCH_2CH_2})_4Cl_2]Cl$ was shown in the paper [5].

***trans*-$[Rh(NHC_4H_8)_4Cl_2]X$** (NHC_4H_8 = pyrrolidine). The **chloride** (X = Cl) is prepared by reaction of $RhCl_3 \cdot 3H_2O$ with pyrrolidine in ethanol at 40°C, and the **perchlorate** (X = ClO_4) is obtained from it by reaction with $NaClO_4$. Both salts formed yellow crystals and developed the odour of pyrrolidine on storage. The conductivity of the complex in aqueous solution was consistent with its formulation as a 1:1 electrolyte, and the Raman (289 to 307 cm^{-1}) and electronic spectra (200 to 500 nm in water, methanol and ethanol) have been measured [2].

References:

[1] E. F. Shubochkina, M. A. Seifer, O. E. Zvyagintsev (Zh. Neorgan. Khim. **13** [1968] 1382/5; Russ. J. Inorg. Chem. **13** [1968] 724/6). – [2] S. P. Dagnall, M. P. Hancock, B. T. Heaton, D. H. Vaughan (J. Chem. Soc. Dalton Trans. **1977** 1111/6). – [3] E. F. Shubochkina, M. A. Golubnichaya (Zh. Neorgan. Khim. **15** [1970] 2752/6; Russ. J. Inorg. Chem. **15** [1970] 1430/3). – [4] L. K. Shubochkin, E. F. Shubochkina, M. A. Golubnichaya (Zh. Neorgan. Khim. **19** [1974] 3300/3; Russ. J. Inorg. Chem. **19** [1974] 1807/9). – [5] J. Scherzer, P. K. Phillips, L. B. Clapp, J. O. Edwards (Inorg. Chem. **5** [1966] 847/51).

[6] R. Lussier, J. O. Edwards, R. Eisenberg (Inorg. Chim. Acta **3** [1969] 468/70).

Tris-amine Complexes

$Rh(NH_2\text{-n-}C_3H_7)_3Cl_3$ is prepared by reaction of $RhCl_3 \cdot 3H_2O$ with four molar equivalents of n-propylamine. The allylamine complex **$Rh(NH_2C_3H_5)_3Cl_3$** is made similarly. These complexes are yellow (melting point 200°C) and orange-yellow (melting point 180°C), respectively [1].

$Rh(\overline{NHCH_2CH_2})_3Cl_3$. This compound is obtained by reaction of $RhCl_3 \cdot 3H_2O$ with ethyleneimine in ice-cold methanol solution. The corresponding bromide, **$Rh(\overline{NHCH_2CH_2})_3Br_3$**, is prepared from the trichloride on treatment with aqueous HBr. The chloride and bromide were bright yellow and red-orange, respectively [2].

The IR spectra (Nujol mulls) of the allylamine complexes indicated that the C=C bond was not coordinated to the metal [1]. The electronic spectra of $Rh(\overline{NHCH_2CH_2})_3X_3$ have been measured in water in the range 20000 to 46000 cm^{-1}, and tentative spectral assignments made; the reflectance spectrum of $Rh(\overline{NHCH_2CH_2})_3Cl_3$, and the solution spectrum in water, dimethylformamide and methanol were shown in the paper [2].

$Rh(NH_2C_3H_5)_3Cl_3$ was thermally stable in hot ethanol, but reacted with HCl at 50°C to give unidentified (organic) carbonyl compounds [1].

$[Rh(\overline{NHCH_2CH_2})_3(H_2O)_2(OH)](HgI_4)$. Treatment of $Rh(\overline{NHCH_2CH_2})_3Cl_3$ with Ag_2O and water affords a green-yellow species which is precipitated by addition of $K_2[HgI_4]$ [2].

$Rh(NHC_5H_{10})_3Cl_3$ (NHC_5H_{10} = piperidine). This compound is prepared by refluxing $[NH_2C_5H_{10}]_3RhCl_6 \cdot 3H_2O$ with a large excess of piperidine in benzene. It was isolated as orange-yellow crystals. X-ray powder diffraction data were collected and the ten most intense lines reported. From these, and from IR spectral measurements (360 to 150 cm^{-1} in Nujol mulls), it was established that a mixture of 1,2,6-(mer) and 1,2,3-(fac) isomers was present, the former being predominant [3].

$Rh(C_6H_{11}NO)_3Cl_3$ ($C_6H_{11}NO$ = N-formylpiperidine). On heating $RhCl_3 \cdot 3H_2O$ with an excess of piperidine in ethanol, this compound is formed, together with Rh metal [4]. It decomposed on treatment with $P(C_6H_5)_3$ or $NaBH_4$, releasing N-formylpiperidine. In dimethylformamide it is a useful catalyst for the hydrogenation of oct-1-ene, cyclohexane and other olefins, and the reduction of nitrobenzene to aniline [4].

References:

[1] H. Sawai, H. Hirai (Inorg. Chem. **10** [1971] 2068/70). – [2] J. Scherzer, P. K. Phillips, L. B. Clapp, J. O. Edwards (Inorg. Chem. **5** [1966] 847/51). – [3] E. R. Birnbaum (J. Inorg. Nucl. Chem. **35** [1973] 3145/54). – [4] I. Jardine, F. J. McQuillin (Tetrahedron Letters **1972** 173/4).

Other Complexes

$Rh(NH_2R)_2Cl_3$. The **n-propylamine** (R = n-C_3H_7) complex is made by treating $RhCl_3 \cdot 3H_2O$ in ethanol with two molar equivalents of n-propylamine. The related **allylamine** (R = C_3H_5) complex is obtained similarly. These compounds are light brown in colour (melting points 180 and 175°C, respectively). The allylamine complex readily decomposed in ethanol at 50°C giving various organic products [1].

$[NH_2(CH_3)_2][Rh(NH_2CH_3)_2Cl_4]$. This complex is formed by heating in air a mixture of $RhCl_3$ in dimethylformamide which had been carbonylated with CO. It was a brown-red complex. Conductivity measurements established that the salt was a 1:1 electrolyte in water, and its IR spectrum was recorded between 1640 and 820 cm^{-1} [2].

$Rh(NH_2R)(H_2O)Cl_3$ (R = n-C_3H_7, C_3H_5). The n-propyl- and allyl-amine complexes are prepared by reacting $RhCl_3 \cdot 3H_2O$ with one mole equivalent of the appropriate amine. They were red-brown solids, melting points 155 and 120°C, respectively. They decomposed readily in ethanol at 50°C [1].

$[Rh(NH_3)_5(\overline{NHCH_2CH_2})]Cl_3$ is prepared from $[Rh(NH_3)_5Cl]Cl_2$ and ethyleneimine. It was a white solid whose electronic spectrum was measured in the range 20000 to 46000 cm^{-1} [3].

References:

[1] H. Sawai, H. Hirai (Inorg. Chem. **10** [1971] 2068/70). – [2] I. B. Bondarenko, N. A. Buzina, Yu. S. Varshavskii, M. I. Gel'fman, V. V. Razumovskii, T. G. Cherkasova (Zh. Neorgan. Khim. **16** [1971] 3071/8; Russ. J. Inorg. Chem. **16** [1971] 1615/7). – [3] J. Scherzer, P. K. Phillips, L. B. Clapp, J. O. Edwards (Inorg. Chem. **5** [1966] 847/51).

2.1.8.3 Rhodium(III) Arylamine Complexes

***trans*-$[Rh(NH_2Ph)_4Cl_2]Cl$.** This complex is prepared from $RhCl_3 \cdot 3H_2O$ and aniline in ethanol at 40°C. It is isolated as yellow-brown needles. The Raman spectrum of the compound has been recorded in the range 289 to 307 cm^{-1}, and the electronic spectrum from 200 to 560 nm in alcohol/water solutions [1].

Rh(3- or 4-$NH_2C_6H_4NO_2)_3Cl_3$. Treatment of $Na_3[RhCl_6]$ with three molar equivalents 3- or 4-nitroaniline afforded these compounds. The 4-nitro species can also be prepared by heating $[NH_3C_6H_4NO_2]_3[RhCl_6]$ at 130°C. The IR spectrum of the compounds indicated that the 3-nitroaniline product had *fac* geometry while the 4-nitro product had *mer* geometry [2].

$[Rh(NH_2C_6H_5)_2Cl_3]_n$ is obtained from the reaction between azobenzene and $RhCl_3 \cdot 3H_2O$ in refluxing ethanol. It was a buff solid, melting point 275°C (with decomposition) [3].

Rh(3- or 4-$NH_2C_6H_4NO_2)_2(H_2O)Cl_3$. These compounds are obtained by reaction of $Na_3[RhCl_6]$ with two molar equivalents of 3- or 4-nitroaniline [2].

$Rh(NH_2C_6H_4C_6H_4NH_2)_2Cl_3$ ($NH_2C_6H_4C_6H_4NH_2$ = benzidine = 4,4'-diamino biphenyl). The complex is prepared by reaction of $RhCl_3 \cdot 3H_2O$ with a slight excess over two molar equivalents of benzidine in ethanol. The compound is described as dirty brown. It was apparently a 1:2 electrolyte in dimethylsulphoxide but thought to be non-ionic with a hexacoordinate metal in the solid state. The complex was diamagnetic, and IR spectral bands due to the benzidine ligand were detected [4].

References:

[1] S. P. Dagnall, M. P. Hancock, B. T. Heaton, D. H. Vaughan (J. Chem. Soc. Dalton Trans. **1977** 1111/6). – [2] Yu. N. Kukushkin, G. N. Portnov, T. O. Blyumental', S. A. Simanova (Zh. Obshch. Khim. **50** [1980] 1108/10; J. Gen. Chem. [USSR] **50** [1980] 895/7). – [3] M. I. Bruce, M. Z. Iqbal, F. G. A. Stone (J. Organometal. Chem. **40** [1972] 393/401). – [4] S. M. F. Rahman, N. Ahmad, V. Kumar (Indian J. Chem. **13** [1975] 86/8).

2.2 Complexes with Saturated Bidentate Nitrogen Donors

2.2.1 Ethylenediamine Complexes

$NH_2CH_2CH_2NH_2$ = en = ethylenediamine, 1,2-diaminoethane

This is a substantial aspect of the chemistry of Rhodium. The complexes $[Rh(en)_3]^{3+}$, *cis*-$[Rh(en)_2X_2]^+$, and their C-alkylated analogues exist as optical isomers, and there has been much study of the optical properties of the compounds as related to the Cotton effect. Photochemical studies of *cis*- and *trans*-$[Rh(en)_2X_2]^+$ have also been performed. The mechanism of substitution and interconversion of these species has been much investigated.

General Literature:

R. D. Gillard, Some Aspects of Catalytic Syntheses in Rhodium Chemistry, Coord. Chem. Rev. **8** [1972] 149/57.

K. Garbett, R. D. Gillard, Optical Configuration of Werner Complexes of Diamines, Coord. Chem. Rev. **1** [1966] 179/86.

R. D. Gillard, The Cotton Effect in Coordination Compounds, Progr. Inorg. Chem. **7** [1966] 215/76.

2.2.1.1 Unsubstituted Tris-ethylenediamine Complexes

2.2.1.1.1 Rhodium(I) Complexes

$[Rh(en)_2]^+$. This species has been implicated in the hydrazine-catalysed substitution of *trans*-$[Rh(en)_2Cl_2]^+$ by Br^- or I^- [1]. In this reaction the *trans* dichloride may be slowly reduced to $[Rh(en)_2]^+$ which then forms an intermediate $[Rh(en)_2\text{-}ClRh(en)_2Cl]^{2+}$ with *trans*-$[Rh(en)_2Cl_2]^+$. The existence of $[Rh(en)_2]^+$ is also implicated in the electrochemical reduction of *trans*-$[Rh(en)_2Cl_2]^+$ at a Hg electrode [2, 3]. The complex *trans*-$[Rh(en)_2H(OH_2)]^{2+}$ may engage in a pH-dependent equilibrium of the type $[Rh(en)_2H(OH_2)]^{2+} + OH^- \rightleftharpoons [Rh(en)_2]^+ + 2H_2O$ [3, 4]. The species $[Rh(en)_2]^+$ may react with $HgCl_2$ giving $[Rh(en)_2HgCl]^{2+}$ and possibly $[\{Rh(en)_2\}_2Hg]^{4+}$ and $[\{Rh(en)_2\}_2HgCl]^{3+}$; some electronic spectral data were presented to support these suggestions [2].

References:

[1] D. J. Baker, R. D. Gillard (Chem. Commun. **1967** 520/1). – [2] J. Gulens, F. C. Anson (Inorg. Chem. **12** [1973] 2568/74). – [3] J. Gulens, D. Konrad, F. C. Anson (J. Electrochem. Soc. **121** [1974] 1421/9). – [4] R. D. Gillard, B. T. Heaton, D. H. Vaughan (J. Chem. Soc. A **1970** 3126/30).

2.2.1.1.2 Rhodium(III) Complexes

Tris-ethylenediamine Complexes

$[Rh(en)_3]Y_3$. The **chloride**, as a trihydrate, is prepared by boiling an aqueous solution of $RhCl_3 \cdot 3H_2O$ with ethylenediamine [1]. The mono-hydrated **perchlorate** is obtained by treating *trans*-$[Rh(en)_2Cl_2]NO_3$ with ethylenediamine in aqueous ethanol containing a trace of $NaBH_4$ [2]. The **iodide** may be prepared by reaction of $RhCl_3$ with 95% ethylenediamine in isopropanol [3]. This affords a solid which is dissolved in water and treated with KI. The monohydrated iodide is also isolated from the reaction between RhI_3 and ethylenediamine in aqueous ethanol [4].

The chloride was obtained as white needles [1], the perchlorate and iodides as white solids [2, 3, 4].

The $[Rh(en)_3]^{3+}$ ion has been optically resolved using lithium tartrate in aqueous solution [1]; by using electrophoretic methods in an electrolyte of Na^+-d-tartrate and $AlCl_3$ [5]; and by chromatographic methods using ion-exchange resins [6]. Optically pure forms of $[Rh(en)_3]^{3+}$ are not affected by boiling them in solution with activated charcoal, Pt black or silica gel [7].

$[Rh(en)_3][M(C_2O_4)_3] \cdot xH_2O$ (C_2O_4 = oxalate). Addition of Λ- or Δ-$[Rh(en)_3]^{3+}$ to solutions of $[M(C_2O_4)_3]^{3-}$ (M = V, Cr, Mn, Fe, Co) affords these complexes where M = V, x = 0.5 (0 after drying in vacuo); M = Cr, x = 0.5 (0); M = Mn, Fe or Co, x = 1 (0.5) [8].

References:

[1] F. Galsbøl (Inorg. Syn. **12** [1970] 269/80). – [2] A. W. Addison, R. D. Gillard, P. S. Sheridan, L. R. H. Tipping (J. Chem. Soc. Dalton Trans. **1974** 709/19). – [3] G. W. Watt, J. K. Crum (J. Am. Chem. Soc. **87** [1965] 5366/70). – [4] H.-H. Schmidtke (Z. Physik. Chem. [Frankfurt] **38** [1963] 170/83). – [5] Y. Yoneda, T. Miura (Bull. Chem. Soc. Japan **45** [1972] 2126/9).

[6] R. D. Gillard, P. C. H. Mitchell (Transition Metal Chem. [Weinheim] **1** [1976] 223/5). – [7] D. Sen, W. C. Fernelius (J. Inorg. Nucl. Chem. **10** [1959] 269/74). – [8] R. D. Gillard, D. J. Shepherd, D. A. Tarr (J. Chem. Soc. Dalton Trans. **1976** 594/9).

Structures

The structure of (±)-$[Rh(en)_3]Cl_3 \cdot 3H_2O$ has been determined by X-ray crystallography [1] (the crystal characteristics are given in the following table). The $[Rh(en)_3]^{3+}$ cation has trigonal symmetry, with Rh-N = 2.056(7) and 2.067(6) Å, the N-Rh-N bond angles being 83.6(2)°. Unit cell dimensions have been obtained from X-ray diffraction studies of (±)-$[Rh(en)_3]Cl_3 \cdot 3H_2O$ [1 to 4], (−)-$[Rh(en)_3]Cl_3 \cdot 2H_2O$ [2], (−)-$[Rh(en)_3]Br_3 \cdot H_2O$ [4], (±)- and (−)-$[Rh(en)_3](SCN)_3$ [5], the mixed active racemate [(+)-{$Rh(en)_3$}-(+)-{$Cr(en)_3$}]$Cl_6 \cdot 6H_2O$ [2, 7], and [(+)-{$Rh(en)_3$}-(+)-{$Co(en)_3$}]$Cl_6 \cdot$aq [4]. The structure of this last complex was determined by X-ray methods, the geometry of the Rh-containing cation being the same as that described above, Rh-N = 2.054(8) and 2.072(7) Å, N-Rh-N = 83.4(3)° [4]. It was established that (+)-$[Rh(en)_3]^{3+}$ and (−)-$[Cr(en)_3]^{3+}$ had the same absolute configuration [7] and that (−)-$[Rh(en)_3]^{3+}$ and (+)-$[M(en)_3]^{3+}$ (M = Cr, Co) had the same absolute configuration [4]. The salt (−)-$[Rh(en)_3]Br_3 \cdot H_2O$ was isomorphous with (+)-$[Co(en)_3]X_3 \cdot H_2O$ (X = Cl, Br) [5].

X-ray data obtained from $[Rh(en)_3]X_3$:

Complex	Crystal Form	Space Group	Dimensions in Å	Ref.
(±)-$[Rh(en)_3]Cl_3 \cdot 3H_2O$	trigonal	$P\bar{3}c1$-D^4_{3d}	a = 11.614(2) c = 15.492(4) γ = 120° Z = 4	[1, 2]
$[Rh(en)_3]Cl_3 \cdot$aq	hexagonal		a = 11.60 c = 15.48	[4]
(−)-$[Rh(en)_3]Cl_3 \cdot 2H_2O$	cubic	$F4_132$-O^4	a = 21.675(1)	[2]
(−)-$[Rh(en)_3]Br_3 \cdot H_2O$	tetragonal	$P4_32_12$-D^8_4	a = 10.070(2) c = 16.688(5) Z = 4	[5]
(±)-$[Rh(en)_3](SCN)_3$	ortho-rhombic	Pbca-D^{15}_{2h}	a = 14.652(5) b = 14.268(6) c = 17.480(6)	[6]
(−)-$[Rh(en)_3](SCN)_3$	ortho-rhombic	$P2_12_12_1$-D^4_2	a = 14.711(4) b = 13.481(3) c = 9.166(2)	[6]
[(+)-{$Rh(en)_3$}-(+){$Cr(en)_3$}]-$(SCN)_6 \cdot 1.5H_2O$	monoclinic	$P2_1$-C^2_2	a = 15.707(5) b = 17.739(4) c = 9.426(3) γ = 131.04(2)°	[6]
[(+)-{$Rh(en)_3$}-(+)-{$Cr(en)_3$}]-$Cl_6 \cdot 6H_2O$	trigonal	P321-D^2_3	a = 11.587(3) c = 15.522(6) γ = 120°	[2, 7]
[(+)-{$Rh(en)_3$}-(+)-{$Co(en)_3$}]-$Cl_6 \cdot$aq	hexagonal		a = 11.54 c = 15.50	[4]

Calculations have been performed of the electronic structure of $[Rh(en)_3]^{3+}$ and it was shown that Rh forms stronger covalent bonds with ethylenediamine than does Co. The experimental data on the MO levels were used in the interpretation of the electronic spectrum of the cation [8].

References:

[1] A. Whuler, C. Brouty, P. Spinat, P. Herpin (Acta Cryst. B **32** [1976] 2238/9). – [2] A. Whuler, C. Brouty, P. Spinat, P. Herpin (Compt. Rend. C **284** [1977] 117/9). – [3] J. ter Berg (Rec. Trav. Chim. **58** [1939] 93/8). – [4] P. Andersen, F. Galsbøl, S. E. Harnung (Acta Chem. Scand. **23** [1969] 3027/37). – [5] P. Spinat, A. Whuler, C. Brouty (J. Appl. Cryst. **13** [1980] 616/7).

[6] C. Brouty, A. Whuler, P. Spinat (J. Appl. Cryst. **13** [1980] 452/3). – [7] A. Whuler, C. Brouty, P. Spinat, P. Herpin (Acta Cryst. B **32** [1976] 2542/4). – [8] I. B. Bersuker, S. S. Budnikov, A. S. Kimoglo (Teor. Eksperim. Khim. **12** [1976] 621/32; Theor. Exptl. Chem. [USSR] **12** [1976] 481/91).

Thermodynamic Functions

The thermodynamics of formation of outer-sphere complexes of $[Rh(en)_3]^{3+}$ with SO_3^{2-}, SeO_3^{2-}, TeO_3^{2-}, $S_2O_3^{2-}$, CO_3^{2-} and SO_4^{2-} [2 to 4], with F^-, Cl^- and Br^- [5], and with ClO_4^- [1], have been measured spectrophotometrically and by electrical conductivity methods.

References:

[1] A. K. Pyartman, N. P. Kolobov, L. E. Merkul'eva, V. E. Mironov (Zh. Neorgan. Khim. **19** [1974] 1691/2; Russ. J. Inorg. Chem. **19** [1974] 920/1). – [2] V. E. Mironov, A. K. Pyartman, Yu. B. Solov'ev, V. P. Ivankova, M. V. Sof'in (Koord. Khim. **6** [1980] 1233/6; Soviet J. Coord. Chem. **6** [1980] 621/4). – [3] V. E. Mironov, G. K. Ragulin, I. E. Umova, Yu. B. Solov'ev, V. P. Mikhailova, Phan Tham Dong (Zh. Fiz. Khim. **46** [1972] 257/8; Russ. J. Phys. Chem. **46** [1972] 155/6). – [4] V. E. Mironov, V. P. Mikhailova, Yu. B. Solov'ev (Zh. Fiz. Khim. **49** [1975] 1298/300; Russ. J. Phys. Chem. **44** [1975] 763/4). – [5] A. K. Pyartman, M. V. Sof'in, V. E. Mironov (Koord. Khim. **7** [1981] 1877/9 from C.A. **96** [1982] No. 75350).

Spectra

The IR spectrum of $[Rh(en)_3]Cl_3 \cdot H_2O$ in the range 1750 to 450 cm^{-1} has been interpreted in terms of coordination of the ethylenediamine in the gauche form [1, 2]. A correlation was also made between ν(Rh-N) and the stability of the Rh-N bond in comparison with other related ethylenediamine complexes [2]. Extensive IR and Raman spectral studies have been made of (±)-$[Rh(en)_3]Cl_3 \cdot xH_2O$ in the ranges 4000 to 40 cm^{-1} [3, 4, 8], 3800 to 200 cm^{-1} [5] and 3000 to 200 cm^{-1} [6], of (+)-$[Rh(en)_3]Cl_3 \cdot 3H_2O$ (3800 to 600 cm^{-1}) [5], of C- and N-deuterated $[Rh(en)_3]Cl_3 \cdot xH_2O$ [4, 7, 8], and of ^{15}N-substituted analogues [6, 7, 8] (ν(Rh-N) in the following table). A normal coordinate analysis of $[Rh(en)_3]^{3+}$ has been made [7, 8], and it was suggested that the (±)-isomer could be differentiated from the (+)- and (−)-isomers by examination of the vibrations in the region 200 to 40 cm^{-1} [5]. The IR and Raman spectra of $[Rh(en)_3]I_3$ [8] and of (+)- and (±)-$[Rh(en)_3]I_3 \cdot H_2O$ were measured in the range 3800 to 400 cm^{-1} [5].

Rh-N stretching frequencies in $[Rh(en)_3]^{3+}$:

Complex	ν(RhN)	Remarks	Ref.
$[Rh(en)_3]_2[PtCl_4]_3$	580	IR (solid)	[2]
$[Rh(en)_3]Cl_3 \cdot H_2O$	550 535 sh	IR (solution)	[4]
	550 (N/D) 537 sh (N/D)	IR (solution)	[4]

Complex	ν(RhN)	Remarks	Ref.
$[Rh(en)_3]Cl_3 \cdot H_2O$	555 sh 546 508	Raman (solid)	[4]
	544 vs 507 m	Raman (solution)	[4]
$[Rh(en)_3]Cl_3$	445	IR (solution)	[5]
	447	Raman (solid)	
$[Rh(en)_3]I_3$	570 w	IR (solid)	[8]
	558 w	IR (solid)	[8]

References:

[1] D. B. Powell, N. Sheppard (J. Chem. Soc. **1959** 791/5). – [2] D. P. Powell, N. Sheppard (J. Chem. Soc. **1961** 1112/4). – [3] J. Gouteron-Vaissermann (Compt. Rend. B **275** [1972] 149/52). – [4] D. W. James, M. J. Nolan (Inorg. Nucl. Chem. Letters **9** [1973] 319/29). – [5] J. Gouteron (J. Inorg. Nucl. Chem. **38** [1976] 63/71).

[6] G. Borch, P. Klaeboe, P. H. Nielsen (Spectrochim. Acta A **34** [1978] 87/91). – [7] G. Borch, P. H. Nielsen, P. Klaeboe (Acta Chem. Scand. A **31** [1977] 109/19). – [8] G. Borch, J. Gastavsen, P. Klaeboe, P. H. Nielsen (Spectrochim. Acta A **34** [1978] 93/9).

The electronic spectrum of $[Rh(en)_3]^{3+}$ has been measured in aqueous solution in the range 20000 to 45000 cm^{-1} [1, 2], and that of $[Rh(en)_3]I_3 \cdot H_2O$ in the range 25000 to 50000 cm^{-1} (in water, spectrum shown in the paper) [3, 4].

The Cotton effect as applied to $[Rh(en)_3]^+$, and the electronic spectrum of D-(−)-$[Rh(en)_3]^{3+}$, have been discussed [5], and the influence of the ligands on the residual rotational strength of the first absorption band in the electronic spectrum of $[Rh(en)_3]^{3+}$ described [6]. The circular dichroism spectra of (+)- and (−)-$[Rh(en)_3]X_3 \cdot H_2O$ where X = Cl [7 to 10], X = ClO_4 [11, 12], SeO_3 [12], TeO_3 [12], and $X_3 = [M(C_2O_4)_3]^{3-}$ (M = V, Mn, Rh) [9, 13], have been reported. In all papers the spectra were measured in aqueous solution, and specific and/or molar rotations were given.

The luminescence spectra of $[Rh(en)_3]Cl_3$ and $[Rh(en)_3]I_3$ at 85 K and at room temperature have been reported [14]. Lifetime and quantum yields for non-radiative processes in $[Rh(en)_3]^{3+}$ were measured [15].

References:

[1] C. K. Jørgensen (Acta Chem. Scand. **10** [1956] 500/17). – [2] C. K. Jørgensen (J. Chim. Phys. **56** [1959] 889/96). – [3] H.-H. Schmidtke (Z. Physik. Chem. [Frankfurt] **38** [1963] 170/83). – [4] M. Billardon (Compt. Rend. **251** [1966] 2320/2). – [5] R. D. Gillard (Progr. Inorg. Chem. **7** [1966] 215/76, 260/3).

[6] F. Woldbye (Proc. Roy. Soc. [London] A **297** [1967] 79/87). – [7] F. Galsbøl (Inorg. Syn. **12** [1970] 269/80). – [8] A. J. McCafferty, S. F. Mason, R. E. Ballard (J. Chem. Soc. **1965** 2883/92). – [9] R. D. Gillard, D. J. Shepherd, D. A. Tarr (J. Chem. Soc. Dalton Trans. **1976** 594/9). – [10] S. K. Hall, B. E. Douglas (Inorg. Chem. **7** [1968] 533/6).

[11] S. F. Mason, B. J. Norman (Chem. Commun. **1965** 73/5). – [12] S. F. Mason, B. J. Norman (J. Chem. Soc. A **1966** 307/12). – [13] R. D. Gillard (J. Chem. Soc. **1963** 2092/5). – [14] P. E.

Hoggard, H.-H. Schmidtke (Ber. Bunsenges. Physik. Chem. **77** [1973] 1052/8). – [15] J. E. Hillis, M. K. de Armond (J. Lumin. **4** [1971] 273/90 from C.A. **76** [1972] No. 147219).

The ^{1}H NMR spectra of $[Rh(en)_3]Cl_3$ in aqueous solution [1], in the presence of NaOD [2], in D_2O containing PO_4^{3-} [3, 4], and in the presence of SeO_4^{2-} or SeO_3^{2-} ions [4] were measured. The ^{1}H NMR spectrum of $[Rh(en)_3](ClO_4)_3$ in D_2O in the presence of D_2SO_4 has also been reported [4]. The ^{15}N (natural abundance) NMR spectrum of $[Rh(en)_3]Cl_3$ has been measured and J(Rh-N) assigned [5].

References:

[1] D. B. Powell, N. Sheppard (J. Chem. Soc. **1959** 791/5). – [2] T. G. Appleton, J. R. Hall, C. J. Hawkins (Inorg. Chem. **9** [1970] 1299/300). – [3] J. L. Sudmeier, G. L. Blackmer (Inorg. Chem. **10** [1971] 2010/8). – [4] L. R. Froebe, B. E. Douglas (Inorg. Chem. **9** [1970] 1513/6). – [5] K. S. Bose, E. H. Abbott (Inorg. Chem. **16** [1977] 3190/3).

Miscellaneous Properties

The acidity of $[Rh(en)_3]I_3$ is weak; upper limit for H^+ dissociation constant 10^{-12} [1]. The rate of exchange of N-H protons in $[Rh(en)_3]^{3+}$ in deuterioacetic acid/acetate buffer has been measured [2], and the rate of exchange of ^{14}C-labelled ethylenediamine with $[Rh(en)_3]^{3+}$ estimated [3].

γ-Radiation of $[Rh(en)_3]Cl_3 \cdot 2.5\,H_2O$ has been investigated by electron spin resonance (ESR), and a species containing O_2^- may have been formed [4].

The utility of $[Rh(en)_3]X_3$ as a material for second harmonic generation in crystals, and for laser action [5], and as a measure of concentration and separation of radioactive mixtures containing rhodium [6], have been investigated.

References:

[1] A. A. Grinberg, L. V. Vrublevskaya, Kh. I. Gil'dengerskel, A. I. Stetsenko (Zh. Neorgan. Khim. **4** [1959] 1018/27; Russ. J. Inorg. Chem. **4** [1959] 462/6). – [2] J. W. Palmer, F. Basolo (J. Inorg. Nucl. Chem. **15** [1960] 279/86). – [3] R. G. Wilkins, D. S. Popplewell (Rec. Trav. Chim. **75** [1956] 815/8). – [4] Yu. V. Glazhov, N. I. Zotov, L. A. Il'yukevich, L. N. Neokladnova (Khim. Vysokikh Energ. **9** [1975] 88/90 from C.A. **82** [1975] No. 148395). – [5] C. B. van den Berg (Phys. Status Solidi **11** [1965] 617/28).

[6] J. Steigmann (Phys. Rev. [2] **59** [1941] 498/501).

2.2.1.2 Substituted Ethylenediamine Complexes

2.2.1.2.1 Bis-ethylenediamine Complexes

2.2.1.2.1.1 Hydrido Complexes

***cis*-$[Rh(en)_2H_2]BPh_4$.** This complex is prepared by reaction of *cis*- or *trans*-$[Rh(en)_2Cl_2]Cl$ in ice-cold water with $NaBH_4$, followed by addition of $NaBPh_4$ [1]. It is a white solid which is unstable in solution, decomposing to Rh metal, and which is reconverted to *cis*-$[Rh(en)_2Cl_2]^+$ by oxidation in aqueous solution in the presence of Cl^- [1]. The IR spectrum (Nujol mull) was said to exhibit ν(RhH) at 2100 cm^{-1} [1], but this was reassigned to 1969 cm^{-1} [2]. The electronic spectrum has also been recorded in aqueous solution, and the position of the H^- ion in the spectrochemical series assigned. The ^{1}H NMR spectrum of the complex in aqueous, slightly alkaline solution was measured: τ(RhH) = 31, J(RhH) = 31 Hz [1] or 31.6 or 32.0, J(RhH) = 27 ± 0.1 Hz [2].

***trans*-[Rh(en)$_2$HX]$^+$** (X = Cl, Br or I). These species are obtained as intermediates in the reaction of *trans*-[Rh(en)$_2$X$_2$]$^+$ (X = Cl, Br or I) with $NaBH_4$ in ice-cold aqueous solution. Attempts to isolate them as BPh_4^- salts were only partially successful, as the products after addition of $NaBPh_4$ were mixtures of mono- and di-hydrido compounds, and were also explosive. The IR and NMR spectra of these compounds afforded ν(RhH) and τ(RhH) (see table) [1, 2], and their electronic spectra were recorded [2].

[Rh(en)$_2$H(OH$_2$)](ClO$_4$)$_2$. The *cis* isomer is obtained by addition of dilute $HClO_4$ to *trans*-[Rh(en)$_2$(NH$_3$)H](ClO$_4$)$_2$ with gentle heating [3]. It is isolated as fine colourless needles which are apparently explosive. The IR spectrum exhibited ν(RhH) at 2135 cm^{-1}, and ν(OH) at 3565 and 3470 cm^{-1}. The ^{1}H NMR spectrum has been measured: τ(RhH) = 32.6, J(RhH) = 30 Hz [3]. Mixtures of *cis* and *trans* isomers of [Rh(en)$_2$H(OH$_2$)]$^{2+}$ are apparently formed when a fresh ice-cold solution of *trans*-[Rh(en)$_2$H(OH$_2$)](ClO$_4$)$_2$ is treated with conc. $HClO_4$, or when a stream of N_2 is passed through the solution of the perchlorate in water [3]. The ^{1}H NMR spectrum of the *trans* isomer which exists in equilibrium with the *cis* form exhibited τ(RhH) = 32.2, J(RhH) = 30 Hz [3].

[Rh(en)$_2$H(NH$_3$)](ClO$_4$)$_2$. The *trans* isomer is obtained when [Rh(NH$_3$)$_5$H]SO$_4$ is treated with ethylenediamine in gently warmed water [3]. Addition of $NaClO_4$ affords the complex as white needles which are stable in air, and can be recrystallised from boiling water without decomposition. The IR spectrum was measured in the range 4000 to 600 cm^{-1}; ν(RhH) = 2056 cm^{-1}. The ^{1}H NMR spectrum gave τ(RhH) = 27.8, J(RhH) = 22 Hz [3]. The *cis* isomer, preparative details for which are not available, is formed together with the *trans* isomer, but is lost in the purification of the latter. The IR spectrum of the *cis* isomer showed ν(RhH) at 2025 cm^{-1}, and the ^{1}H NMR spectrum was measured: τ(RhH) = 26.3, J(RhH) = 19 Hz [3]. In aqueous solution, the *cis* and *trans* isomers existed in an equilibrium which was established by NMR spectroscopy [3].

***trans*-[Rh(en)$_2$H(OH)]BPh$_4$** is obtained by electrolytic reduction of *trans*-[Rh(en)$_2$Cl$_2$]$^+$ in aqueous solution under Ar over a mercury pool [4]. It is a white solid whose IR spectrum exhibited ν(RhH) = 2058 cm^{-1}. The ^{1}H NMR spectrum in aqueous solution showed τ(RhH) = 30.6 ± 0.2, J(RhH) = 30 Hz. Addition of HCl to the compound appears to give a mixture of [Rh(en)$_2$H(OH$_2$)]Cl$_2$ and [Rh(en)$_2$H(OH$_2$)]Cl$_3$·H$_5$O$_2$. The hydrido hydroxide apparently reacted with O_2 giving *trans*-[Rh(en)$_2$(OH)(OOH)]$^+$ and *trans*-[Rh(en)$_2$(OH)$_2$]$^+$ [4].

IR and NMR spectral data from hydrido complexes:

Complex	ν(RhH) in cm^{-1}	τ(RhH)	J(RhH) in Hz	Ref.
cis-[Rh(en)$_2$H$_2$]$^+$	2100 (as BPh_4^-)	31	31	[1]
	1969	31.6		[2]
		32.0	27 ± 0.1	
	2093	31	31	[1]
trans-[Rh(en)$_2$HCl]$^+$	2100	31	31	[2]
trans-[Rh(en)$_2$HBr]$^+$	2120			[2]
trans-[Rh(en)$_2$HI]$^+$	2140	30.2		[2]
cis-[Rh(en)$_2$HCl]$^+$	2105			[2]
trans-[Rh(en)$_2$H(OH$_2$)]$^{2+}$		32.2	30	[3]

Complex	ν(RhH) in cm^{-1}	τ(RhH)	J(RhH) in Hz	Ref.
cis-$[Rh(en)_2H(OH_2)]^{2+}$	2135 (as ClO_4)	32.6	30	[3]
trans-$[Rh(en)_2H(NH_3)]^{2+}$	2056 (as ClO_4)	27.8	22	[3]
cis-$[Rh(en)_2H(NH_3)]^{2+}$	2025 (as ClO_4)	26.3	19	[3]
trans-$[Rh(en)_2H(OH)]^{+}$	2058 (as BPh_4^-)	30.6 ± 0.2	30	[4]

References:

[1] R. D. Gillard, G. Wilkinson (J. Chem. Soc. **1963** 3594/9). – [2] J. A. Osborn, R. D. Gillard, G. Wilkinson (J. Chem. Soc. **1964** 3168/73). – [3] K. Thomas, G. Wilkinson (J. Chem. Soc. A **1970** 356/60). – [4] R. D. Gillard, B. T. Heaton, D. H. Vaughan (J. Chem. Soc. A **1970** 3126/30).

2.2.1.2.1.2 Complexes with Oxygen Donors

Rhodium(II) Complexes

$[\{Rh(en)_2(OH_2)\}_2]^{4+}$. This species is thought to be formed on controlled-potential electrolytic reduction of *trans*-$[Rh(en)_2Cl_2]^+$ in aqueous solution at −1.0 V vs. standard calomel electrode. The compound was formulated as containing a Rh-Rh bond. It was diamagnetic and its electronic spectrum in solution was reported. The species reacted with X^- (X=Cl, Br or I) affording $[\{Rh(en)_2X\}_2]^{2+}$, whose electronic spectrum was recorded [1]. However, it has been suggested that the aquo species may in fact be a Hg^{II} bridged complex of Rh^I, $[\{Rh(en)_2\}_2Hg]^{4+}$ [2].

References:

[1] R. D. Gillard, B. T. Heaton, D. H. Vaughan (J. Chem. Soc. A **1971** 734/5). – [2] J. Gulens, F. C. Anson (Inorg. Chem. **12** [1973] 2568/74).

Rhodium(III) Complexes

***trans*-$[Rh(en)_2(OH)(OH_2)](ClO_4)_2$** is prepared by reaction of *trans*-$[Rh(en)_2Cl_2]Cl$ with NaOH in the pH range 5 to 8 [1], with LiOH in the pH range 4 to 6 [2], or of *trans*-$[Rh(en)_2Cl_2](ClO_4)$ with NaOH and $AgClO_4$ at pH=7 [5].

The compound is isolated as yellow needles [1 to 3] whose electronic spectrum was reported in the ranges 250 to 350 nm [1, 3] and 240 to 430 nm [2]. The pK_a value for loss of a proton from the water molecules was 7.81 ± 0.02 [3]. The kinetics of Cl^- anation of the complex have been determined [2, 3]. The compound reacted in acid solution with CO_2 giving *trans*-$[Rh(en)_2(OCO_2)(OH_2)]ClO_4$ [5].

***cis*-$[Rh(en)_2(OH)(OH_2)]^{2+}$.** The **perchlorate** salt is prepared by treatment of *cis*-$[Rh(en)_2Cl_2]ClO_4$ with aqueous LiOH and $AgClO_4$ [4]. The **dithionate** is obtained by refluxing *cis*-$[Rh(en)_2Cl_2]Cl \cdot 1.5\,H_2O$ in aqueous solution followed by addition of $Na_2S_2O_6 \cdot 2H_2O$ and a small amount of pyridine [6]. Both the perchlorate and dithionate were obtained as yellow solids. The electronic spectrum of *cis*-$[Rh(en)_2(OH)(OH_2)]^{2+}$ has been measured [4] (spectrum shown in paper) [6]. The pK_a value for the dissociation of H^+ from the water molecules was 7.64 [7]. The compound reacted with CO_2 in acid solution forming *cis*-$[Rh(en)_2(OCO_2)(OH_2)]^+$ [6].

$[Rh(en)_2(OH)_2]^+$. The *trans* isomer is obtained by deprotonation of *trans*-$[Rh(en)_2(OH)(OH_2)]^{2+}$ by base in aqueous solution [1], or by dissolving *trans*-$[Rh(en)_2(OH_2)_2](ClO_4)_3$ in aqueous NaOH [8]. The *cis* isomer is obtained by treatment of *cis*-$[Rh(en)_2(OH)(OH_2)]^{2+}$ with aqueous base [4]. There are few preparative details, but the electronic spectra of the species have been recorded: *trans* in the range 250 to 350 nm [1, 3, 8] and *cis* tabulated and shown in the paper [4].

$[Rh(en)_2(OH_2)_2]^{3+}$. The *trans* **perchlorate** is obtained by reaction of *trans*-$[Rh(en)_2I_2]I$ with aqueous $AgClO_4$ and $HClO_4$. It is formed as pale yellow crystals [8]. The *trans* isomer may also be obtained by acidification of *trans*-$[Rh(en)_2(OH)(OH_2)]^{2+}$ in aqueous solution using $HClO_4$ [1, 2]. The *cis* isomer is obtained similarly from *cis*-$[Rh(en)_2(OH)(OH_2)]^{2+}$ [4].

The electronic spectra of the *trans* isomer has been measured in aqueous solution in the range 250 to 350 nm [1] or 240 to 430 nm [2, 3, 8], whereas that of the *cis* isomer has been mentioned [4]. The $pK_a(1)$ and $pK_a(2)$ values of the *cis* isomer were 6.72 and 7.64, respectively [9]. A correlation has been made between the polarographic half-wave reduction potential of *cis*- and *trans*-$[Rh(en)_2(OH_2)_2]^{3+}$ and the first d-d transition in their electronic spectra [10]. The kinetics and mechanism of Cl^- [2] and Br^- [11, 12] anation of *trans*-$[Rh(en)_2(OH_2)_2]^{3+}$ have been determined.

$[\{Rh(en)_2\}_2(\mu\text{-}OH)_2]^{4+}$. The **bis-dithionate** salt of the *meso* isomer is obtained by heating *cis*-$[Rh(en)_2(OH)(OH_2)]S_2O_6$ at 120°C for 10 h [6]. It is converted to the pale yellow **bromide,** *meso*-$[\{Rh(en)_2\}_2(\mu\text{-}OH)_2]Br_4 \cdot 2H_2O$ by treatment of the dithionate salt in water with saturated aqueous NH_4Br [6, 13]. The **perchlorate,** Δ,Λ-$[\{Rh(en)_2\}_2(\mu\text{-}OH)_2](ClO_4)_4$ is prepared by treatment of the tetrabromide with aqueous $NaClO_4$ [13].

The electronic spectra of these three salts in aqueous solution have been tabulated [6, 13] and an equilibrium between four species, $[\{Rh(en)_2\}_2(\mu\text{-}OH)_2]^{4+}$, $[\{Rh(en)_2(OH_2)\}_2(\mu\text{-}OH)]^{5+}$, $[Rh(en)_2(OH_2)(\mu\text{-}OH)Rh(en)_2(OH)]^{4+}$ and $[\{Rh(en)_2\}_2(\mu\text{-}OH)(\mu\text{-}OH_2)]^{5+}$ has been established [13].

References:

[1] R. van Eldik, D. A. Palmer, G. M. Harris (Inorg. Chem. **19** [1980] 3673/9). – [2] M. J. Pavelich (Inorg. Chem. **14** [1975] 982/8). – [3] A. J. Pöe, K. Shaw (J. Chem. Soc. A **1970** 393/6). – [4] D. A. Palmer, R. van Eldik, H. Kelm, G. M. Harris (Inorg. Chem. **19** [1980] 1009/13). – [5] N. S. Rowan, R. M. Milburn, T. P. Dasgupta (Inorg. Chem. **15** [1976] 1477/84).

[6] M. P. Hancock (Acta Chem. Scand. A **33** [1979] 499/502). – [7] S. Ya. Dvurechenskaya, A. V. Belyaev, B. I. Peshchevitskii (Zh. Neorgan. Khim. **19** [1974] 788/93; Russ. J. Inorg. Chem. **19** [1974] 427/30). – [8] C. Burgess, F. R. Hartley, D. E. Rogers (Inorg. Chim. Acta **13** [1975] 35/42). – [9] S. Ya. Dvurechenskaya, A. V. Belyaev, B. I. Peshchevitskii (Izv. Sibirsk. Otd. Akad. Nauk SSR Ser. Khim. Nauk **1972** 154/61 from C.A. **78** [1973] No. 63027). – [10] H.-H. Wei, T.-T. Li (J. Chinese Chem. Soc. [Taipei] [2] **24** [1977] 147/56 from C.A. **88** [1977] No. 143430).

[11] H. L. Bott, A. J. Pöe, K. Shaw (Chem. Commun. **1968** 793/4). – [12] H. L. Bott, A. J. Pöe, K. Shaw (J. Chem. Soc. A **1970** 1745/50). – [13] M. P. Hancock, B. Nielsen, J. Springborg (Acta Chem. Scand. A **36** [1982] 313/22).

Carbonato Complexes

***trans*-$[Rh(en)_2(OCO_2)(OH_2)]ClO_4$.** This complex is obtained by dissolving *trans*-$[Rh(en)_2(OH)(OH_2)](ClO_4)_2$ in dilute $HClO_4$, adding $NaHCO_3$ and adjusting the pH to the range 4 to 5.8. The electronic spectrum in aqueous solution has been measured (250 to 350 nm). The kinetics and mechanism of formation of the complex have been determined [1].

***cis*-[Rh(en)$_2$(OCO$_2$)(OH$_2$)]$^+$** is formed by treatment of *cis*-[Rh(en)$_2$(OH)(OH$_2$)](ClO$_4$)$_2$ in dilute HClO$_4$ solution with solid Li$_2$CO$_3$ until the pH reaches 7. It has not been isolated [2].

Its electronic spectrum, and that of *cis*-Rh(en)$_2$(OCO$_2$)OH which is apparently obtained from the aquo species by treatment with aqueous base, have been measured. The rates of formation of these carbonato species have been measured [2].

***cis*-[Rh(en)$_2$(O$_2$CO)]ClO$_4$** is prepared by refluxing an aqueous solution containing *cis*-[Rh(en)$_2$(OCO$_2$)(OH$_2$)]$^+$ at pH = 7, followed by addition of HClO$_4$. It is formed as a cream-coloured precipitate whose electronic spectrum has been briefly reported (aqueous solution, shown in paper). The kinetics and mechanism of formation of the complex were determined [2].

Na[*trans*-Rh(en)$_2$(OCO$_2$)$_2$]. This compound is prepared from *trans*-[Rh(en)$_2$(OH)(OH$_2$)]-(ClO$_4$)$_2$ by treatment with dilute aqueous NaOH (pH 9) and an excess of Na$_2$CO$_3$ and NaHCO$_3$. It is isolated as an impure solid contaminated with NaHCO$_3$ and Na$_2$CO$_3$. Its electronic spectrum (aqueous solution) has been measured in the range 250 to 350 nm [1]. It was decarbonylated at pH ≦ 6 [2].

References:

[1] R. van Eldik, D. A. Palmer, G. M. Harris (Inorg. Chem. **19** [1980] 3673/9). – [2] D. A. Palmer, R. van Eldik, H. Kelm, G. M. Harris (Inorg. Chem. **19** [1980] 1009/13).

Oxalato Complexes

[Rh(en)$_2$(C$_2$O$_4$)]X. The **nitrate** (X = NO$_3$) is prepared by refluxing *cis*- or *trans*-[Rh(en)$_2$Cl$_2$]NO$_3$ with Na$_2$C$_2$O$_4$ in aqueous solution [1]. The **perchlorate** (X = ClO$_4$) is obtained by treating *trans*-[Rh(en)$_2$Cl$_2$]NO$_3$·H$_2$O [2] or *cis*- or *trans*-[Rh(en)$_2$Cl$_2$]ClO$_4$ [3] with an aqueous solution of Na$_2$C$_2$O$_4$ and NaOH containing a trace of NaBH$_4$, followed by addition of LiClO$_4$ [2] or NaClO$_4$ [3]. The **bromide** (X = Br) is obtained by reaction of RhCl$_3$·3H$_2$O with en·2HCl, adjustment of the pH to 6.5 to 7 (using NaOH) and adding a trace of NaBH$_4$ and Na$_2$C$_2$O$_4$ followed by NaBr [3]. It may also be obtained directly from *cis*- or *trans*-[Rh(en)$_2$Cl$_2$]ClO$_4$ with Na$_2$C$_2$O$_4$, a trace of NaBH$_4$ and NaBr [3]. The nitrate and bromide are isolated as pale yellow crystals [1, 3], whereas the perchlorate is formed as a white solid [2].

Racemic (±)-[Rh(en)$_2$(C$_2$O$_4$)]$^+$ has been resolved using (+)-[Co(EDTA)]$^-$ (EDTA = ethylenediaminetetraacetate) [4]. Optically impure [Rh(en)$_2$(C$_2$O$_4$)]Cl underwent spontaneous resolution [4].

X-ray powder diffraction data (d-spacings) have been obtained for (+)-[Rh(en)$_2$(C$_2$O$_4$)]Cl, (±)-[Rh(en)$_2$(C$_2$O$_4$)]ClO$_4$·0.5H$_2$O and [(+)-{Rh(en)$_2$(C$_2$O$_4$)}][(+)-{Co(EDTA)}]·H$_2$O. It was established that [(+)-{Rh(en)$_2$(C$_2$O$_4$)}][(+)-{Co(EDTA)}] and [(−)-{Co(en)$_2$(C$_2$O$_4$)}][(+)-{Co(EDTA)}] were isomorphous [4].

The IR (1850 to 750 cm^{-1}, KBr disc) [1] and electronic spectra [1, 3] of [Rh(en)$_2$(C$_2$O$_4$)]NO$_3$ have been briefly reported, and it was established that the oxalate ligand was bidentate [1]. The electronic and CD spectra of (+)-[Rh(en)$_2$(C$_2$O$_4$)]$^+$ have been measured [4].

The reactivity of [Rh(en)$_2$(C$_2$O$_4$)]$^+$ towards O-exchange with solvent, and aquation to *cis*-[Rh(en)$_2$(OH$_2$)$_2$]$^{3+}$, has been examined [5], and it was shown that the oxalate readily reacted with HX (X = Cl or Br) forming *cis*-[Rh(en)$_2$X$_2$]$^+$ [2].

References:

[1] R. M. Milburn, T. P. Dasgupta, L. Damrauer (Inorg. Chem. **9** [1970] 2789/91). – [2] A. W. Addison, R. D. Gillard, P. S. Sheridan, L. R. H. Tipping (J. Chem. Soc. Dalton Trans. **1974**

709/19). – [3] M. P. Hancock (Acta. Chem. Scand. A **33** [1979] 15/8). – [4] R. D. Gillard, L. R. H. Tipping (J. Chem. Soc. Dalton Trans. **1977** 1241/7). – [5] N. S. Rowan, R. M. Milburn, T. P. Dasgupta (Inorg. Chem. **15** [1976] 1477/84).

Acetato and Malonato Complexes

***trans*-[Rh(en)$_2$(OCOCH$_3$)$_2$]ClO$_4$.** The chromatographic behaviour of this compound in acetone/water/acetic acid and acetone/water/HCl mixtures has been investigated [1].

[Rh(en)$_2$(malonate)]NO$_3$. Reaction of *trans*-[Rh(en)$_2$Cl$_2$]NO$_3$·H$_2$O with sodium malonate in aqueous solution in the presence of a trace of $NaBH_4$, followed by addition of HNO_3, affords this complex [2].

References:

[1] L. Ossicini (Ric. Sci. Rend. A [2] **3** [1963] 913/8 from C.A. **60** [1964] 7436). – [2] A. W. Addison, R. D. Gillard, P. S. Sheridan, L. R. H. Tipping (J. Chem. Soc. Dalton Trans. **1974** 709/19).

Amino-acid Complexes

[Rh(en)$_2$(glycinato-O]I$_2$. This complex is prepared by reaction of *cis-* or *trans-*[Rh(en)$_2$Cl$_2$]NO$_3$ with an aqueous ethanol solution of glycine and NaOH, followed by addition of NaI [1, 2].

The corresponding **(S)-alaninate** ($C_3H_6NO_2$) [1, 2], **(S)-valinate** ($C_5H_{10}NO_2$) [1, 2], **(S)-leucinate** ($C_6H_{12}NO_2$) [1, 2], **(S)-serinate** ($C_3H_6NO_3$) [1], **(2S,3R)-threoninate** ($C_4H_8NO_3$) [1] and **(S)-methioninate** ($C_5H_{10}NO_2S$) [1, 2], as iodides, are prepared similarly. The **L-phenylalaninate** and **L-tyrosinate,** as iodides are obtained only from *trans*-[Rh(en)$_2$Cl$_2$]NO$_3$ [2].

The complexes except the phenylalaninate and tyrosinate have been resolved into two diastereoisomers using silver antimonyltartrate in the absence of light, the pure optical forms being obtained by treatment of the antimonyltartrate with $AgNO_3$, followed by addition of NaI [1].

[Rh(en)$_2$(glycinato-O]S$_2$O$_6$, [Rh(en)$_2$(alaninato-O)]S$_2$O$_6$. These complexes are obtained by reacting an aqueous solution of *trans*-[Rh(en)$_2$Cl$_2$]NO$_3$·H$_2$O with sodium glycinate or L-alaninate and a trace of $NaBH_4$, followed by addition of $Na_2S_2O_6 \cdot 2H_2O$. They are isolated as white solids [3].

Spectroscopic Properties. The IR spectra of the glycinate, alaninate, phenylalaninate, valinate, leucinate, methioninate and tyrosinate were briefly reported, and their electronic spectra recorded [2]. The electronic and CD spectra of these complexes in the region 250 to 400 nm (aqueous solution) have been reported and all the (−)-D isomers assigned the Λ absolute configuration relative to the absolute configuration of (+)-D-[Co(en)$_3$]$^{3+}$ [1].

References:

[1] S. K. Hall, B. E. Douglas (Inorg. Chem. **7** [1968] 530/2). – [2] J. F. Waller, J. Hu, B. E. Bryant (J. Inorg. Nucl. Chem. **27** [1965] 2371/4). – [3] A. W. Addison, R. D. Gillard, P. S. Sheridan, L. R. H. Tipping (J. Chem. Soc. Dalton Trans. **1974** 709/19).

Peroxo Complexes

***trans*-[Rh(en)$_2$(OH)(OOH)]$^+$** is thought to be formed when *trans*-[Rh(en)$_2$H(OH)]$^+$ reacts with oxygen. The presence of the hydroperoxy group was identified by an iodide titration [1].

$[Rh(en)_2(NO_2)(O_2)]^+$. The complex (X = NO_2) is formed when colourless *cis*-$[Rh(en)_2(NO_2)_2]^+$ becomes red in daylight in the presence of O_2. The complex is not optically active and has been isolated as a nitrate and a chloride (preparative details are not given). The complex was paramagnetic, exhibited ν(OO) at 1052 cm^{-1} in its Raman spectrum, and its electronic spectrum was briefly reported. It reacted with aquo-Rh^{III} complexes giving $[\{Rh(en)_2L\}_2(O_2)]^{n+}$ (L = Cl, NO_2 or H_2O) [2].

$[Rh(en)_2(H_2O)(O_2)]^{2+}$ and **$[Rh(en)_2Cl(NO_2)]^+$** have been only very briefly mentioned [2] and the ESR spectra of *cis* and *trans* isomers of the chloro complex analysed in terms of an MO scheme [3].

$[(H_2O)(en)_2RhO_2Rh(en)_2(OH_2)]^{4+}$ is believed to be formed when a mixture of *trans*-$[Rh(en)_2(OH_4)_2]^+$ and *trans*-$[Rh(en)_2(OH)(OOH)]^+$ is concentrated in aqueous solution at pH = 9. This gives an intermediate which, on addition of conc. $HClO_4$, affords the binuclear tetra-cation as a blue species whose electronic spectrum was recorded. If the intermediate is allowed to stand, a red species is produced which may be oxidised by Ce^{IV} to give blue $[Cl(en)_2Rh(O_2)$-$Rh(en)_2Cl]^{3+}$ [1]. The electronic spectrum of this binuclear dichloride has been reported [1], and its ESR spectrum analysed in terms of a simple MO scheme which places the unpaired electron largely on the O_2 ligand [3]. Treatment of the red species mentioned above with conc. $HClO_4$ afforded *trans*-$[Rh(en)_2Cl(OH_2)]^{2+}$, and with Cl_2, *trans*-$[Rh(en)_2Cl_2]^+$ was formed. The complex equilibria involving these red and blue species have been discussed [1].

References:

[1] R. D. Gillard, B. T. Heaton, D. H. Vaughan (J. Chem. Soc. A **1970** 3126/30). – [2] R. D. Gillard, J. D. Pedrosa de Jesus, L. R. H. Tipping (J. Chem. Soc. Chem. Commun. **1977** 58). – [3] J. B. Rayner, R. D. Gillard, J. D. Pedrosa de Jesus (J. Chem. Soc. Dalton Trans. **1982** 1165/6).

2.2.1.2.1.3 Complexes Containing Oxygen Donor Atoms and Halides

***trans*-$[Rh(en)_2(OH_2)F](ClO_4)_2$** is made by reaction of *trans*-$[Rh(en)_2Cl_2]ClO_4$ with AgF in refluxing aqueous solution. It is obtained as very pale yellow crystals whose electronic spectrum in aqueous solution has been reported briefly [1].

***trans*-$[Rh(en)_2(OH_2)Cl](ClO_4)_2$** is prepared by reaction of *trans*-$[Rh(en)_2Cl_2]ClO_4$ with AgCl in water [1] or in dilute $HClO_4$ [2]. The electronic spectrum of the complex has been reported [1], and reproduced in the range 240 to 430 nm [2]. The pK_a value of the complex was 7.53 [3] and its reaction with CO_2, which afforded a carbonato complex, has been studied kinetically [2]. The anation of the complex with Cl^-, Br^- and I^- has also been investigated kinetically [4].

***cis*-$[Rh(en)_2(OH_2)Cl]^{2+}$** is obtained by aquation of *cis*-$[Rh(en)_2Cl_2]^+$ in aqueous solutions of KCl. The pK_a value of the complex was 7.53 [5]. The ^{13}C NMR spectrum of *cis*- and *trans*-$[Rh(en)_2(OH_2)Cl]^{2+}$ allowed distinction between the two isomers [6].

***trans*-$[Rh(en)_2(OH_2)Br](ClO_4)_2$** is formed by reaction of *trans*-$[Rh(en)_2Br_2]ClO_4$ with $AgClO_4$ in water [1, 6], in dilute $HClO_4$[2], or by refluxing *trans*-$[Rh(en)_2Cl_2]ClO_4$ in aqueous solution [4].

The electronic spectrum of the complex has been reported [1], in the ranges 300 to 500 nm (aqueous solution, shown in paper) [2] and 200 to 500 nm (aqueous solution) [7, 8]. The pK_a value of the complex was 6.30 [3]. The anation of the complex with Cl^- and Br^- has been studied [9, 10], and the photolabilisation and photoaquation of the compound discussed [8, 11]. The kinetics of uptake of CO_2 by the compound have been measured [2].

***cis*-[Rh(en)$_2$(OH$_2$)Br](ClO$_4$)$_2$.** This is prepared by reaction of *cis*-[Rh(en)$_2$Br$_2$]ClO$_4$ with AgClO$_4$ in water [7], or aquation of *cis*-[Rh(en)$_2$Br$_2$]$^+$ with aqueous KBr [5].

The electronic spectrum has been measured in the range 200 to 500 nm (aqueous solution) [7], and the pK$_a$ value of the complex was 7.44 [5].

***trans*-[Rh(en)$_2$(OH$_2$)I]X$_2$.** The **perchlorate** (X = ClO$_4$) is formed by reaction of *trans*-[Rh(en)$_2$I$_2$]ClO$_4$ with one mole equivalent of AgClO$_4$ in water [7, 12], or with two mole equivalents of AgClO$_4$ in dilute HClO$_4$ [2]. The **tetraphenylborate** (X = BPh$_4$) is obtained by treating the perchlorate with NaBPh$_4$ [1]. The electronic spectrum of the complex has been reported [1], and shown in the range 300 to 500 nm [2, 8, 12] (aqueous solution). The pK$_a$ value of the compound was >6.30 [5]. The photolysis of the compound in aqueous solution has been investigated [12], and the kinetics of uptake of CO$_2$ by the complex have been determined [2].

***cis*-[Rh(en)$_2$(OH$_2$)I]$^{2+}$** is formed by heating as [Rh(en)$_2$I$_2$]$^+$ in aqueous KI. The pK$_a$ value of the species was 7.54 [5].

***trans*-[Rh(en)$_2$(OH)X]$^+$.** The complexes where X = F [1], Cl, Br or I [1, 2, 4] are obtained by reacting *trans*-[Rh(en)$_2$(OH$_2$)X]$^{2+}$ with OH$^-$ in aqueous solution. The species have not apparently been isolated, but their electronic spectra have been obtained in aqueous solution [1, 4] in the ranges 250 to 430 (X = Cl) or 300 to 500 nm [2]. The kinetics of base hydrolyses of *trans*-[Rh(en)$_2$(OH)X]$^+$ where X = Cl or Br [13] and I [14] have been determined.

***trans*-Rh(en)$_2$(O$_2$CO)X.** The **chloride** (X = Cl) is obtained by reaction of *trans*-[Rh(en)$_2$-(OH$_2$)Cl]$^{2+}$ with NaHCO$_3$ in water followed by addition of ethanol. The compound was isolated but was contaminated with [Rh(en)$_2$(OH)Cl]$^+$ or [Rh(en)$_2$(OH$_2$)Cl]$^{2+}$. The related **bromide** (X = Br) and **iodide** (X = I) were made similarly in solution, but were not isolated. The electronic spectra of these complexes were measured (250 to 430 nm for X = Cl, 300 to 500 nm for X = Br or I in aqueous solution). The formation and decarboxylation of these species was studied kinetically [2].

References:

[1] C. Burgess, F. R. Hartley, D. E. Rogers (Inorg. Chim. Acta **13** [1975] 35/42). – [2] R. van Eldik, D. A. Palmer, H. Kelm, G. M. Harris (Inorg. Chem. **19** [1980] 3679/83). – [3] S. Ya. Dvurechenskaya, A. V. Belyaev, B. I. Peshchevitskii (Izv. Sibirsk. Otd. Akad. Nauk SSSR Ser. Khim. Nauk **1972** 63/9 from C. A. **77** [1972] No. 157134). – [4] H. L. Bott, A. J. Pöe (J. Chem. Soc. A **1967** 205/12). – [5] S. Ya. Dvurechenskaya, A. V. Belyaev, B. I. Peshchevitskii (Zh. Neorgan. Khim. **19** [1974] 788/93; Russ. J. Inorg. Chem. **19** [1974] 427/30).

[6] F. P. Jakse, J. P. Paukstelis, J. D. Petersen (Inorg. Chim. Acta **27** [1978] 225/31). – [7] S. F. Clark, J. D. Petersen (Inorg. Chem. **18** [1979] 3394/9). – [8] T. L. Kelly, J. F. Endicott (J. Phys. Chem. **76** [1972] 1937/46). – [9] H. L. Bott, A. J. Pöe (Chem. Commun. **1968** 793/4). – [10] H. L. Bott, A. J. Pöe, K. Shaw (J. Chem. Soc. A **1970** 1745/50).

[11] L. G. Vanquickenborne, A. Ceulemans (Inorg. Chem. **20** [1981] 110/3). – [12] S. F. Clark, J. D. Petersen (Inorg. Chem. **19** [1980] 2917/21). – [13] A. J. Pöe, C. P. J. Vuik (Can. J. Chem. **53** [1975] 1842/8). – [14] A. J. Pöe, C. P. J. Vuik (J. Chem. Soc. Dalton Trans. **1972** 2250/3).

2.2.1.2.1.4 Complexes Containing N-Donor Atoms

***trans*-[Rh(en)$_2$(NH$_3$)(OH$_2$)]X$_3$.** The **nitrate** (X = NO$_3$) is obtained by heating *trans*-[Rh(en)$_2$Cl$_2$]NO$_3$ with conc. aqueous ammonia followed by addition of HNO$_3$. The **perchlorate** (X = ClO$_4$) is prepared similarly, using HClO$_4$. The electronic spectrum of the complex was obtained at pH = 4 in aqueous solution, the species probably being *trans*-$[Rh(en)_2(NH_3)(OH)]^{2+}$ [1]. The kinetics of reaction of this species with Cl^-, which afforded *trans*-$[Rh(en)_2(NH_3)Cl]^{2+}$, have been investigated [1, 2].

References:

[1] A. J. Pöe, K. Shaw (J. Chem. Soc. A **1970** 393/6). – [2] A. J. Pöe, K. Shaw (Chem. Commun. **1967** 52/4).

$[Rh(en)_2(NH_3)_2](ClO_4)_3$. The *cis* isomer is prepared by reacting *cis*-$[Rh(en)_2X_2](ClO_4)$ (X = Cl or Br) with Na in liquid NH_3. The residue from evaporation of the ammonia was treated with $HClO_4$ and passed down an ion-exchange column (HCl as eluant). The colourless band was collected and treated with $HClO_4$, giving the complex as white crystals. The *trans* isomer is obtained similarly from *trans*-$[Rh(en)_2Cl_2]ClO_4$, and is also formed as white crystals. The electronic and CD spectra of the complexes were measured (2000 to 45000 cm^{-1}, aqueous solution), and those for *cis*-(+)-$[Rh(en)_2(NH_3)_2]^{3+}$ were shown in the range 15000 to 41000 cm^{-1}. The ^1H NMR spectra of the complexes were measured in CF_3CO_2H solution [1].

***trans*-$[Rh(en)_2(N_2H_4)_2]^{3+}$** is apparently formed when *trans*-$[Rh(en)_2Cl_2]^+$ or *trans*-$[Rh(en)_2(N_2H_4)Cl]^{2+}$ is treated with hydrazine. The electronic spectrum has been briefly reported, and the complex reacted with HX rapidly affording *trans*-$[Rh(en)_2X_2]^+$ (X = Cl, Br or I) [10].

***cis*-$[Rh(en)_2(NH_3)Cl]Cl$** is obtained by treatment of *trans*-$[Rh(en)_2Cl_2]Cl$ with conc. aqueous ammonia, followed by addition of conc. HCl. The complex has been resolved into the (+)- and (−)-isomers using silver (+)-bromocamphorsulphonate. The electronic and CD spectra of this complex have been measured (aqueous solution, 15000 to 45000 cm^{-1}, region 15000 to 41000 cm^{-1} shown in paper) [1]. The ^1H NMR spectrum of the compound has been obtained, and ^{13}C NMR spectra used to establish the difference between the compound and its *trans* isomer in solution [2]. Photolysis of *cis*-$[Rh(en)_2(NH_3)Cl]^{2+}$ in aqueous solution afforded *cis*-$[Rh(en)_2(NH_3)(OH_2)]^{3+}$ [3].

$[Rh(en)_2(NH_3)Cl](NO_3)_2$. The *trans* isomer is prepared by refluxing *trans*-$[Rh(en)_2Cl_2]NO_3$ in conc. aqueous ammonia until no further colour change takes place, and then adding HNO_3. The *cis* isomer is obtained similarly, and both are isolated as pale yellow crystals. The electronic spectra of the complexes have been reported briefly [4].

***trans*-$[Rh(en)_2(NH_3)Br](NO_3)_2$.** The compound is formed by reacting *trans*-$[Rh(en)_2Br_2]NO_3$ with conc. aqueous ammonia, followed by refluxing and then addition of HNO_3 [1, 5]. It is isolated as yellow crystals [1].

The electronic spectrum in aqueous solution has been reported in the range 20000 to 45000 cm^{-1} [1] and 200 to 500 nm [5]. The CD spectrum has also been obtained [1], and the ^{13}C NMR spectrum in water measured [5]. Ligand field photoexcitation of the complex in water gave exclusively *trans*-$[Rh(en)_2(OH_2)Br]^{2+}$ [5].

***cis*-$[Rh(en)_2(NH_3)Br](NO_3)_2$** is prepared by reacting *cis*-$[Rh(en)_2Br_2]ClO_4$ with conc. aqueous ammonia, followed by addition of HNO_3 [5].

The electronic spectrum has been measured (200 to 500 nm, aqueous solution) [5], and the ^{13}C NMR spectrum of the complex obtained in water [5]. Ligand field photoexcitation of the complex in water afforded a mixture of *cis*-$[Rh(en)_2(NH_3)(OH_2)]^{3+}$ and *trans*-$[Rh(en)_2(OH_2)Br]^{2+}$ [5, 6].

$[Rh(en)_2(NH_3)I]I_2$. The *trans* isomer is obtained by reaction of *trans*-$[Rh(en)_2I_2]I$ with conc. aqueous ammonia, followed by refluxing, then addition of HI [1, 6]. It formed as yellow-orange crystals. The *cis* isomer is similarly obtained from *cis*-$[Rh(en)_2I_2]I$ [6].

$[Rh(en)_2(NH_3)I]S_2O_6$. The *cis* and *trans* isomers are obtained from the corresponding iodide salts by treatment with $Na_2S_2O_6$ in water [6].

The electronic spectra have been measured [6] and reported in the range 20000 to 45000 cm^{-1} [1]; the CD spectra have also been described [1]. The ^{13}C NMR spectra of both isomers have been measured in water. Ligand field photoexcitation of the *trans* isomer in water afforded only *trans*-$[Rh(en)_2(OH_2)I]^{2+}$, whereas that of the *cis* isomer gave both as $[Rh(en)_2(NH_3)(OH_2)]^{3+}$ and *trans*-$[Rh(en)_2(OH_2)I]^{2+}$ [6].

***cis*-$[Rh(en)_2(NH_2CH_2CH_2NH_3)Cl]Cl_3 \cdot 2H_2O$.** This complex is formed, together with some starting material and *trans*-$[Rh(en)_2Cl_2]^+$, by photolysis of $[Rh(en)_3]Cl_3 \cdot 3H_2O$ in deoxygenated HCl solutions over 4d at 5°C [7]. The electronic spectrum has been briefly reported [7] and the *cis*-configuration by ^{13}C NMR spectral studies [2, 7].

***trans*-$[Rh(en)_2(N_2H_4)Cl]^{2+}$** is apparently formed when *trans*-$[Rh(en)_2Cl_2]^+$ is treated with hydrazine. The electronic spectrum has been briefly reported [8].

***trans*-$[Rh(en)_2(NH_2CH_3)X](ClO_4)_2$.** The complexes where X = Br, I, NO_2 or N_3 are obtained by treatment of *trans*-$[Rh(en)_2(NH_2CH_3)Cl](ClO_4)_2$ (preparative details of this species were not given) with aqueous NaOH, followed by anion-exchange (OH^- form). The eluate was treated with dilute $HClO_4$ and NaX, followed by further anion-exchange (ClO_4^- form). The complexes are obtained as yellow crystals. The electronic spectra of the chloride (X = Cl) and azide (X = N_3) were measured in aqueous media (spectra shown in the range 280 to 380 nm) [9].

***trans*-$[Rh(en)_2(NH_2CH_3)(NO_2)](NO_3)_2$** is prepared by reaction of *trans*-$[Rh(en)_2(NHCH_3)Cl](ClO_4)_2$ with $[N(CH_2Ph)(CH_3)_3]OH$ and $NaNO_2$ in $(CH_3)_2SO$. The precipitate is treated with $NaBPh_4$, affording the tetraphenylborate salt, which is then dissolved in methanol/acetone to which $NaNO_2$ is added [9].

References:

[1] H. Ogino, J. C. Bailar (Inorg. Chem. **17** [1978] 1118/24). – [2] F. P. Jakse, J. V. Paukstelis, J. D. Petersen (Inorg. Chim. Acta **27** [1978] 225/31). – [3] J. D. Petersen, F. P. Jakse (Inorg. Chem. **18** [1979] 1818/21). – [4] S. A. Johnson, F. Basolo (Inorg. Chem. **1** [1962] 925/32). – [5] S. F. Clark, J. D. Petersen (Inorg. Chem. **18** [1979] 3394/9).

[6] S. F. Clark, J. D. Petersen (Inorg. Chem. **19** [1980] 2917/21). – [7] J. D. Petersen, F. P. Jakse (Inorg. Chem. **16** [1977] 2845/8). – [8] D. J. Baker, R. D. Gillard (Chem. Commun. **1967** 520/1). – [9] A. Panunzi, F. Basolo (Inorg. Chim. Acta **1** [1967] 223/7). – [10] D. J. Baker, R. D. Gillard (Chem. Commun. **1967** 520/1).

Nitro Complexes

$[Rh(en)_2(NO_2)_2]NO_3$. The *trans* isomer is prepared by refluxing *trans*-$[Rh(en)_2Cl_2]NO_3$ with an excess of $NaNO_2$ in water. On addition of HNO_3 the complex was obtained as white crystals. The *cis* isomer is prepared similarly [1].

The IR spectra of both isomers were measured: $\nu(NO_2)$ for the *trans* form was 825 cm^{-1}; for the *cis* isomer 830 and 840 cm^{-1}. The electronic spectra of both isomers in aqueous solution have been reported [1], and the acid-catalysed hydrolysis of the *trans* isomer has been investigated [2].

***trans*-[Rh(en)$_2$(ONO)Cl]ClO$_4$.** Reaction of *trans*-[Rh(en)$_2$Cl$_2$]ClO$_4$ with AgClO$_4$ followed by NaNO$_2$ affords this compound. This nitrito species rapidly rearranged to the **nitro** isomer at room temperature in aqueous solution ($t_{1/2} \approx 10$ min). The IR spectrum of the nitrito species was briefly reported [2].

***trans*-[Rh(en)$_2$(NO$_2$)Cl]NO$_3$.** Reaction of *trans*-[Rh(en)$_2$Cl$_2$]NO$_3$ with carefully controlled amounts of NaNO$_2$ in aqueous solution affords this complex as pale yellow crystals. The IR ($\nu(NO_2) = 830$ cm^{-1}) and electronic spectra of this complex have been reported [1].

***cis*-[Rh(en)$_2$(NO$_2$)Cl]ClO$_4$** is obtained from the *trans* isomer in aqueous solution, and is formed on heating *trans*-[Rh(en)$_2$Cl$_2$]$^+$ with NaNO$_2$. The IR spectrum was briefly described [2].

***cis*-[Rh(en)$_2$(NO$_2$)(N$_3$)]ClO$_4$.** Treatment of an aqueous solution of *cis*-[Rh(en)$_2$(NO$_2$)Cl]ClO$_4$ with NaOH, followed by anion-exchange (OH^- form) and addition of dilute HClO$_4$ and NaN$_3$, gives this complex as pale yellow crystals [3].

***cis*-[Rh(en)$_2$(NO$_2$)$_2$]NO$_3$.** Reaction of *cis*-[Rh(en)$_2$(NO$_2$)Cl]ClO$_4$ with [N(CH$_2$Ph)(CH$_3$)$_3$]OH and NaNO$_2$, followed by treatment of the precipitate with NaBPh$_4$, and addition to the resulting tetraphenylborate salt in acetone/methanol and NaNO$_3$, gives this compound as white crystals. The *cis* configuration was established by an IR spectrum (not reported) [3].

References:

[1] S. A. Johnson, F. Basolo (Inorg. Chem. **1** [1962] 925/32). – [2] P. J. Staples (J. Inorg. Nucl. Chem. **30** [1968] 3119/21). – [3] R. A. Bauer, F. Basolo (J. Am. Chem. Soc. **90** [1968] 2437/8). – [4] A. Panunzi, F. Basolo (Inorg. Chim. Acta **1** [1967] 233/7).

Azido Complexes

***trans*-[Rh(en)$_2$(N$_3$)$_2$]X.** The **azide** (X = N$_3$) is obtained by refluxing *trans*-[Rh(en)$_2$Cl$_2$]NO$_3$ with a large excess of NaN$_3$ in water. It forms deep golden light-sensitive crystals. The **chloride** (X = Cl) is obtained by anion-exchange (Cl^- form) of the azide, and is slightly light-sensitive. The IR spectrum of the chloride exhibited $\nu(N_3)$ at 2060 cm^{-1}, and the electronic spectrum of *trans*-[Rh(en)$_2$(N$_3$)$_2$]$^+$ in water was recorded [1].

***trans*-[Rh(en)$_2$(N$_3$)X]ClO$_4$.** The **chloride** (X = Cl) and **bromide** (X = Br) are prepared by treating *trans*-[Rh(en)$_2$(N$_3$)$_2$]ClO$_4$ with HClO$_4$ followed by NaX. The complexes are obtained as bright yellow crystals whose electronic spectra were measured in the range 280 to 390 nm (water) [2].

***trans*-[Rh(en)$_2$(N$_3$)(OH$_2$)]$^{2+}$** is formed on treatment of *trans*-[Rh(en)$_2$(N$_3$)$_2$]ClO$_4$ with aqueous HClO$_4$ (1 mole equivalent). It has not been isolated, but its electronic spectrum was recorded in the range 280 to 390 nm [2].

References:

[1] S. A. Johnson, F. Basolo (Inorg. Chem. **1** [1962] 925/32). – [2] E. J. Bounsall, A. J. Pöe (J. Chem. Soc. A **1966** 286/92).

2.2.1.2.1.5 Complexes Containing Halogeno Ligands

In this important section, we consider the preparations of the halo complexes *cis*- and *trans*-$[Rh(en)_2X_2]^+$ (X = Cl, Br, I) first, and then consider the properties of these as a group.

Preparation of Chloro Complexes

***trans*-$[Rh(en)_2Cl_2]X$.** The **nitrate** ($X = NO_3$) may be prepared by refluxing a mixture of $RhCl_3 \cdot 3H_2O$, en·2HCl and KOH in aqueous solution, followed by cooling and addition of conc. HNO_3 [1, 2]. The *trans* isomer is precipitated whereas the *cis* isomer remains in solution. The yield of the *trans* nitrate can be increased by adding $N_2H_4 \cdot 2HCl$ to the reaction mixture [3]. The complex is formed as golden-yellow crystals [1, 2]. The **chloride** (X = Cl) is obtained by passing a solution of the nitrate through an anion-exchange column (Cl^- form) [2].

***cis*-$[Rh(en)_2Cl_2]X$.** The **nitrate** ($X = NO_3$) may be obtained from the filtrate of the reaction between $RhCl_3 \cdot 3H_2O$, en·2HCl, KOH and HNO_3 once *trans*-$[Rh(en)_2Cl_2]NO_3$ has been removed [1, 2]. It may also be formed using $Na_3RuCl_6 \cdot 12H_2O$, en·2HCl and NaOH followed by addition of HNO_3 and prior removal of the *trans* nitrate salt [4], or from $Na_3Rh(NO_2)_6$ in aqueous solution containing ethylenediamine, HCl and HNO_3 [5]. It is isolated as bright yellow crystals [1, 2, 5].

The **chloride,** as a monohydrate, may be obtained from the *cis* nitrate by passage through an anion-exchange column (Cl^- form) [2], or by reaction of $[Rh(en)_2(C_2O_4)]ClO_4$ with conc. HCl [4]. It is isolated as a yellow solid [2, 4], and may be dehydrated by heating at 100°C for 30 min [6].

The *cis* dichloride may be resolved via d-α-bromocamphorsulphonate [1, 2], affording the l-*cis* isomer, and optically pure (+)-*cis*-$[Rh(en)_2Cl_2]^+$ can be formed by treating (+)-*cis*-$[Rh(en)_2(C_2O_4)]^+$ with HCl [7].

***trans*-$[Rh(en)_2Cl_2]Cl_2 \cdot H_5O_2$** is obtained by recrystallisation of *trans*-$[Rh(en)_2Cl_2]Cl$ from conc. HCl, and is obtained as shiny yellow plates. The IR spectrum of this complex has been measured and compared with that of *trans*-$[Co(en)_2Cl_2]Cl_2 \cdot H_5O_2$ and *trans*-$[Rh(bipy)_2Cl_2]Cl_2 \cdot H_5O_2$ [8].

***cis*-$[Rh(en)_2Cl_2]Cl \cdot HgCl_2$** is prepared by reacting *cis*-$[Rh(en)_2Cl_2]NO_3$ with $HgNO_3$ in dilute HNO_3, followed by addition of HCl [9].

References:

[1] S. A. Johnson, F. Basolo (Inorg. Chem. **1** [1962] 925/32). – [2] S. N. Anderson, F. Basolo (Inorg. Syn. **7** [1963] 214/20, 216, 217). – [3] B. I. Peshchevitskii, A. V. Belyaev, S. Ya. Dvurechenskaya (Izv. Sibirsk. Otd. Akad. Nauk SSSR Ser. Khim. Nauk **1972** 138/40 from C. A. **77** [1972] No. 159615). – [4] A. W. Addison, R. D. Gillard, P. S. Sheridan, L. R. H. Tipping (J. Chem. Soc. Dalton Trans. **1974** 709/19). – [5] S. Ya. Dvurechenskaya, A. V. Belyaev, B. I. Peshchevitskii (Zh. Neorgan. Khim. **19** [1974] 788/93; Russ. J. Inorg. Chem. **19** [1974] 427/30).

[6] H. Ogino, J. C. Bailar (Inorg. Chem. **17** [1978] 1118/24). – [7] R. D. Gillard, L. R. H. Tipping (J. Chem. Soc. Dalton Trans. **1977** 1241/7). – [8] R. D. Gillard, G. Wilkinson (J. Chem. Soc. **1964** 1640/6). – [9] N. V. Podberezskaya, A. V. Belyaev, V. V. Bakakin, I. A. Baidina (Zh. Strukt. Khim. **22** No. 6 [1981] 32/6; J. Struct. Chem. [USSR] **22** [1981] 827/31).

Preparation of Bromo Complexes

***trans*-$[Rh(en)_2Br_2]X$.** The **perchlorate** ($X = ClO_4$) is prepared by heating *trans*-$[Rh(en)_2Cl_2]NO_3$ with a large excess of NaBr in water containing a trace of HBr, followed by addition of HNO_3, and pouring the mixture into methanol containing $NaClO_4$ [1, 2]. The **nitrate**

($X = NO_3$) is formed by refluxing *trans*-[Rh(en)$_2$Cl$_2$]NO$_3$ with NaBr in water, followed by adding HNO$_3$. The nitrate was isolated as orange crystals [3].

***cis*-[Rh(en)$_2$Br$_2$]X.** The **perchlorate** ($X = ClO_4$) is formed by treating [Rh(en)$_2$(C$_2$O$_4$)]-ClO$_4$·0.5H$_2$O with dilute HBr and NaClO$_4$ [1], or by reacting Na$_3$[Rh(NO$_2$)$_6$] with ethylenediamine in dilute HClO$_4$ followed by treatment with HBr and recrystallisation from dilute HClO$_4$ [4]. The **bromide** (X = Br) is obtained by refluxing an aqueous solution of *cis*-[Rh(en)$_2$Cl$_2$]Cl with NaBr. The bromide is obtained as yellow crystals, and was resolved using silver (+)-bromocamphorsulphonate [5]. Optically pure (+)-*cis*-[Rh(en)$_2$Br$_2$]$^+$ could be obtained by treating (+)-[Rh(en)$_2$(C$_2$O$_4$)]$^+$ with HBr [6].

***trans*-[Rh(en)$_2$Br$_2$]Br$_2$·H$_5$O$_2$** is formed when *trans*-[Rh(en)$_2$Br$_2$]Br is recrystallised from conc. HBr solutions. The IR spectrum of the compound has been briefly mentioned [7].

References:

[1] S. F. Clark, J. D. Petersen (Inorg. Chem. **18** [1979] 3394/9). – [2] H. L. Bott, A. J. Pöe (J. Chem. Soc. **1965** 5931/4). – [3] S. A. Johnson, F. Basolo (Inorg. Chem. **1** [1962] 925/32). – [4] S. Ya. Dvurechenskaya, A. V. Belyaev, B. I. Peshchevitskii (Zh. Neorgan. Khim. **19** [1974] 788/93; Russ. J. Inorg. Chem. **19** [1974] 427/30). – [5] H. Ogino, J. C. Bailar (Inorg. Chem. **17** [1978] 1118/24).

[6] R. D. Gillard, L. R. H. Tipping (J. Chem. Soc. Dalton Trans. **1977** 1241/7). – [7] R. D. Gillard, G. Wilkinson (J. Chem. Soc. **1964** 1640/6).

Preparation of Iodo Complexes

***trans*-[Rh(en)$_2$I$_2$]X.** The **iodide** (X = I) can be made by refluxing *trans*-[Rh(en)$_2$Cl$_2$]NO$_3$ with NaI in aqueous solution. The product was obtained as rust-brown crystals after recrystallisation from aqueous NaI solution [1]. The **perchlorate** ($X = ClO_4$) is obtained by refluxing *trans*-[Rh(en)$_2$Cl$_2$]NO$_3$ with an excess of NaI in water. A precipitate of *trans*-[Rh(en)$_2$I$_2$]I which is formed is recrystallised from aqueous NaClO$_4$ solutions [2].

***cis*-[Rh(en)$_2$I$_2$]X.** The **iodide** is prepared by reacting [Rh(en)$_2$(C$_2$O$_4$)]ClO$_4$·0.5H$_2$O with HI in water [2] or by reacting Na$_3$[Rh(NO$_2$)$_6$] with aqueous ethylenediamine, dilute HClO$_4$ and NaI, the product being recrystallised from aqueous NaI solutions [3]. It could also be isolated from a reaction of *cis*-[Rh(en)$_2$Cl$_2$]NO$_3$ with NaI, followed by redissolving the product in warm water containing NaI [1]. The iodide is described as forming brick red [1] or orange [3] crystals. The **perchlorate** is obtained from the iodide by recrystallisation from aqueous NaClO$_4$ [2].

References:

[1] S. A. Johnson, F. Basolo (Inorg. Chem. **1** [1962] 925/32). – [2] S. F. Clark, J. D. Petersen (Inorg. Chem. **19** [1980] 2917/21). – [3] S. Ya. Dvurechenskaya, A. V. Belyaev, B. I. Peshchevitskii (Zh. Neorgan. Khim. **19** [1974] 788/93; Russ. J. Inorg. Chem. **19** [1974] 427/30.

Preparation of Mixed Dihalo Species

***trans*-[Rh(en)$_2$ClBr]X.** The **perchlorate** is formed either by irradiation of an aqueous solution of *trans*-[Rh(en)$_2$Cl$_2$]ClO$_4$, followed by addition of Br$^-$ [1], or by treatment of *trans*-[Rh(en)$_2$-Br$_2$]ClO$_4$ with AgNO$_3$, followed by addition of an excess of HCl and HClO$_4$ [2]. The **nitrate** is obtained by the latter route using HNO$_3$ in place of HClO$_4$. The relative stability of this complex was investigated kinetically [2].

***trans*-[Rh(en)$_2$ClI]ClO$_4$** is formed by treating *trans*-[Rh(en)$_2$I$_2$]I in water with a large excess of NaCl, followed by NaClO$_4$. It is isolated as a bright orange-red solid [3].

***trans*-[Rh(en)$_2$BrI]ClO$_4$** is prepared by reacting *trans*-[Rh(en)$_2$I$_2$]I in water with an excess of NaI, followed by NaClO$_4$ [3].

References:

[1] R. A. Bauer, F. Basolo (J. Am. Chem. Soc. **90** [1968] 2437/8). – [2] H. L. Bott, A. J. Pöe (J. Chem. Soc. **1965** 5931/4). – [3] E. J. Bounsall, A. J. Pöe (J. Chem. Soc. A **1966** 286/92).

Structures of *cis*- and *trans*-[Rh(en)$_2$X$_2$]$^+$ (X = Cl, Br, I)

According to investigations by X-rays, *cis*-[Rh(en)$_2$Cl$_2$]NO$_3$ forms triclinic crystals, space group $P\bar{1}$-C_i^1, a = 8.619(2), b = 6.528(2), c = 11.687(3) Å, α = 102.18(2)°, β = 90.98(2)°, γ = 112.46(2)°, Z = 2; D_{calc} = 2.00 g/cm^3. The *trans* isomer is monoclinic, space group $P2_1/n$, a = 6.441(2), b = 9.281(4), c = 10.010(4) Å, β = 102.42(3)°, Z = 2; D_{calc} = 2.02 g/cm^3. The complexes have the expected geometries, and bond lengths and angles were reported [1].

cis-[Rh(en)$_2$Cl$_2$]Cl·HgCl$_2$ is orthorhombic, space group Pcab-D_{2h}^{15}, a = 17.259(4), b = 11.953(2), c = 13.593 Å, Z = 8; D_{calc} = 2.85 g/cm^3. The complex has a layered structure with each Hg atom having an octahedral environment of Cl atoms, two each from two adjacent *cis*-[Rh(en)$_2$Cl$_2$]$^+$ groups [2].

X-ray powder diffraction data (d-spacings) have been reported for *cis*- and *trans*-[Rh(en)$_2$Cl$_2$]NO$_3$, and *trans*-[Rh(en)$_2$X$_2$]X (X = Br and I) [3].

References:

[1] I. A. Baidina, N. V. Podberezskaya, A. V. Belyaev, V. V. Bakakin (Zh. Strukt. Khim. **20** [1979] 1096/102; J. Struct. Chem. [USSR] **20** [1979] 934/8). – [2] N. V. Podberezskaya, A. V. Belyaev, V. V. Bakakin, I. A. Baidina (Zh. Strukt. Khim. **22** No. 6 [1981] 32/6; J. Struct. Chem. [USSR] **22** [1981] 827/31). – [3] M. M. Muir, W.-L. Huang (Inorg. Chem. **12** [1973] 1831/5).

Bonding and Thermodynamic Studies

The nature of the Rh-ligand bond has been probed in an MO LCAO study of [Rh(en)$_2$X$_2$]$^+$ and related complexes [1].

Thermodynamic parameters have been calculated from electronic, spectral and liquid chromatographic data for the displacement of Cl$^-$ from *cis*- and *trans*-[Rh(en)$_2$Cl$_2$]$^+$ [2]. The thermodynamics of formation of outer-sphere complexes of SO_4^{2-} and SeO_4^{2-} with *trans*-[Rh(en)$_2$Cl$_2$]$^+$ have been determined [3].

Equilibrium constants have been obtained spectrophotometrically for the stepwise replacement of one halide by another in the species *trans*-[Rh(en)$_2$X$_2$]$^+$. This replacement of Cl from *trans*-[Rh(en)$_2$Cl$_2$]$^+$ by I$^-$ gave K_1 = 7.0 and K_2 was estimated to be 2.2. For replacement of Br from *trans*-[Rh(en)$_2$Br$_2$]$^+$ by I$^-$, K_1 = 7.0 and K_2 was estimated to be 1.9. The enthalpies (in kcal/mol) for the reactions were ΔH_1° (Cl replaced by I) = −0.1 ± 0.9, ΔH_2° (Cl by I) = −6.2 ± 9.4, ΔH_1° (Br by I) = −2.6 ± 0.4 and ΔH_2° (Br by I) = 3.4 ± 0.3 [4].

References:

[1] I. B. Bersuker, S. S. Budnikov, A. S. Kimoglo (Teor. Eksperim. Khim. **12** [1976] 621/32; Theor. Exptl. Chem. [USSR] **12** [1976] 481/91). – [2] A. V. Belyaev, A. B. Venediktov (Koord.

Khim. **7** [1981] 1505/21; Soviet J. Coord. Chem. **7** [1981] 747/60). – [3] V. E. Mironov, L. A. Obozova, A. K. Pyartman, N. P. Kolobov (Koord. Khim. **7** [1981] 396/9 from C.A. **94** [1981] No. 215404). – [4] E. J. Bounsall, A. J. Pöe (J. Chem. Soc. A **1966** 286/92).

Spectral Studies

IR spectral studies have been used to distinguish between the *cis* and *trans* isomers of $[Rh(en)_2Cl_2]^+$. The frequencies of the wagging vibrations of the NH_2 groups in *trans*-$[Rh(en)_2Cl_2]_2(HCl_4)$ and *trans*-$[Rh(en)_2Cl_2]NO_3$ have been assigned, and the values compared with related Cu, Pd, Pt and Au compounds [2]. The far IR spectra of *trans*-$[Rh(en)_2X_2]X$ (X = Cl or Br) have been measured, and assignments made of the bands in the region 400 to 30 cm^{-1} (spectra shown in paper) [3]. For IR spectral studies of *trans*-$[Rh(en)_2XY]^+$ where X = Y = Cl, Br [4, 5], and X = Cl, Y = Br or I [5] have been made. Values of ν(Rh-X) (X = Cl or Br) are given in the following table; assignments of δ(NMN) were also reported [4].

IR spectral data from *trans*-$[Rh(en)_2XY]^+$:

Complex	ν(Rh-Cl) in cm^{-1}	ν(Rh-Br) in cm^{-1}	Ref.
$[Rh(en)_2Cl_2]^+$	350 s, sh		[4]
	343		[5]
$[Rh(en)_2Br_2]^+$		221 s, br	[4]
		222.7	[5]
$[Rh(en)_2ClBr]^+$	343 or 333.4	212.7	[5]
$[Rh(en)_2ClI]^+$	311.4		[5]

br = broad, s = strong, sh = shoulder

References:

[1] S. N. Anderson, F. Basolo (Inorg. Syn. **7** [1963] 214/20). – [2] A. A. Grinberg, Yu. S. Varshavskii (Dokl. Akad. Nauk SSSR **163** [1965] 646/9; Dokl. Chem. Proc. Acad. Sci. USSR **160/165** [1965] 692/4). – [3] I. B. Baranovskii, Ya. G. Mazo (Zh. Neorgan. Khim. **20** [1975] 444/51; Russ. J. Inorg. Chem. **20** [1975] 244). – [4] K. W. Bowker, E. R. Gardner, J. Burgess (Inorg. Chim. Acta **4** [1970] 626/8). – [5] E. J. Bounsall, A. J. Pöe (J. Chem. Soc. A **1966** 286/92).

The **electronic spectra** of *trans*-$[Rh(en)_2X_2]^+$ (X = Cl, Br and I) have been measured in aqueous solution in the ranges 200 to 500 nm [1], 240 to 450 nm (shown in paper) [2], and 20000 to 50000 cm^{-1} [3]. Assignments of the absorptions were made in terms of an MO treatment [3]. The electronic spectrum of *trans*-$[Rh(en)_2Cl_2]Cl$ was measured in the range 200 to 800 nm at 77 K in a glassy solution, and at room temperature [4]. The spectrum of *trans*-$[Rh(en)_2Br_2]^+$ was obtained in aqueous solution in the ranges 200 to 500 nm [5, 6], and of *trans*-$[Rh(en)_2I_2]^+$ similarly in the range 200 to 500 nm [5] and 280 to 390 nm (spectrum shown in paper). The electronic spectrum of *trans*-$[Rh(en)_2BrI]^+$ was also measured (shown in paper) [7].

The electronic spectra of *cis*-$[Rh(en)_2XY]^+$ (X = Cl, Br and I) have been obtained in aqueous solution in the ranges 200 to 500 nm [1] and 240 to 450 nm (spectra shown in paper) [2]. The absorption spectrum of *cis*-$[Rh(en)_2Cl_2]Cl$ in dilute HCl solution in the range 200 to 400 nm has been reported [8] and that of *cis*-$[Rh(en)_2Br_2]^+$ tabulated in the range 200 to 500 nm [6]. The electronic and CD spectra of (−)-*cis*-$[Rh(en)_2X_2]^+$ (X = Cl and Br) have been measured in the range 20000 to 45000 cm^{-1} (see paper) [9]; the electronic spectra of (+)-*cis*-$[Rh(en)_2X_2]^+$ (X = Cl and Br) have been briefly discussed [10].

References:

[1] S. A. Johnson, F. Basolo (Inorg. Chem. **1** [1962] 925/32). – [2] S. Ya. Dvurechenskaya, A. V. Belyaev, B. I. Peshchevitskii (Zh. Neorgan. Khim. **19** [1974] 788/93; Russ. J. Inorg. Chem. **19** [1974] 427/30). – [3] W. D. Blanchard, W. R. Mason (Inorg. Chim. Acta **28** [1978] 159/68). – [4] M. K. De Armond, J. E. Hillis (J. Chem. Phys. **54** [1971] 2247/53). – [5] T. L. Kelly, J. F. Endicott (J. Phys. Chem. **76** [1972] 1937/46).

[6] S. F. Clark, J. D. Petersen (Inorg. Chem. **18** [1979] 3344/9). – [7] E. J. Bounsall, A. J. Pöe (J. Chem. Soc. A **1966** 286/92). – [8] M. P. Hancock (Acta Chem. Scand. A **33** [1979] 15/8). – [9] H. Ogino, J. C. Bailar (Inorg. Chem. **17** [1978] 1118/24). – [10] R. D. Gillard, L. R. H. Tipping (J. Chem. Soc. Dalton Trans. **1977** 1241/7).

The **luminescence spectrum** of *trans*-$[Rh(en)_2Cl_2]Cl$ has been measured in glassy solution at 77 K and at room temperature [1]. Similar studies have been made of *trans*-$[Rh(en)_2Cl_2]NO_3$ [2]. Lifetime and quantum yields have been reported for non-radiative processes in *trans*-$[Rh(en)_2Cl_2]Cl$ [3].

References:

[1] M. K. De Armond, J. E. Hills (J. Chem. Phys. **54** [1971] 2247/53). – [2] P. E. Hoggard, H.-H. Schmidtke (Ber. Bunsenges. Physik. Chem. **77** [1973] 1052/8). – [37] J. E. Hillis, M. K. De Armond (J. Lumin. **4** [1971] 273/90 from C.A. **76** [1972] No. 147219).

X-ray fluorescence and edge-absorption spectroscopic studies have been made of $[Rh(en)_2Cl_2](NO_3)\cdot H_2O$ [1].

The ^{1}**H NMR spectra** of *trans*-$[Rh(en)_2Cl_2]^+$ in conc. H_2SO_4 [2] and of *cis*-$[Rh(en)_2X_2]^+$ (X = Cl, Br and I) in trifluoroacetic acid [3] have been reported. The ^{13}C NMR spectrum of *trans*-$[Rh(en)_2I_2]^+$ has been mentioned [4] and the use of ^{13}C NMR spectroscopy in identifying and differentiating between the *cis* and *trans* isomers of $[Rh(en)_2X_2]^+$ (X = Cl or Br) discussed [5, 6]. The natural abundance ^{15}N NMR spectrum of *trans*-$[Rh(en)_2Cl_2]Cl$ has been reported; J(Rh-N) = 17.1 Hz [7].

NQR spectroscopic measurements were made using *trans*-$[Rh(en)_2Cl_2]Cl_2\cdot H_5O_2$; ^{35}Cl NQR was observed at 17.089 and 13.460 MHz [8].

References:

[1] L. M. Mazalov, A. P. Sadovskii, A. V. Belyaev, L. I. Chernyavskii, E. S. Gluskin, L. F. Berkhoer (Izv. Sibirsk. Otd. Akad. Nauk SSSR Ser. Khim. Nauk **1971** 51/9 from C.A. **76** [1972] No. 39710). – [2] D. N. Hendrickson, W. L. Jolly (Inorg. Chem. **9** [1970] 1197/201). – [3] H. Ogino, J. C. Bailar (Inorg. Chem. **17** [1978] 1118/24). – [4] S. F. Clark, J. D. Petersen (Inorg. Chem. **19** [1980] 2917/21). – [5] F. P. Jakse, J. V. Paukstelis, J. D. Petersen (Inorg. Chim. Acta **27** [1978] 225/31).

[6] C. Burgess, F. R. Hartley (Inorg. Chim. Acta **14** [1975] L37/L38). – [7] K. S. Bose, E. H. Abbott (Inorg. Chem. **16** [1977] 3190/3). – [8] A. F. Schreiner, T. B. Brill (Inorg. Nucl. Chem. Letters **6** [1970] 355/8).

Electrochemical Properties

The molar conductances of *cis*- and *trans*-$[Rh(en)_2Cl_2]Cl \cdot H_2O$ have been measured in $(CH_3)_2SO$ solution: *cis* 37.55 to 33.73 and *trans* 32.98 to 30.51 $cm^2 \cdot \Omega^{-1} \cdot mol^{-1}$, depending on concentration. Limiting molar conductances were also determined, and the standard free energies of association at 25°C found to be −3.29 (cis) and 1.30 (trans) kcal/mol [1].

trans-$[Rh(en)_2XY]^+$ is reduced polarographically in aqueous solution in a two-electron step. The $E_{1/2}$ values (in V, vs. standard calomel electrode) were −0.79 (X = Y = Cl), −0.70 (X = Y = I) and −0.77 (X = Br, Y = Cl). The *cis* isomers were also examined polarographically but exhibited ill-defined electrode behaviour [2]. Correlations have been made between polarographic $E_{1/2}$ values and the d-d electronic transitions of *cis*- and *trans*-$[Rh(en)_2X_2]^+$, where X = Cl, Br and I [3] and where X = F, Cl, Br and I [4].

References:

[1] D. W. Watts, D. A. Palmer (Inorg. Chem. **10** [1971] 281/6). – [2] R. D. Gillard, J. A. Osborn, G. Wilkinson (J. Chem. Soc. **1965** 4107/10). – [3] D. R. Crow (Inorg. Nucl. Chem. Letters **5** [1969] 291/3). – [4] H.-H. Wei, T.-T. Li (J. Chinese Chem. Soc. [Taipei] [2] **24** [1977] 147/56 from C.A. **88** [1978] No. 143430).

Photochemical Properties

It has been established that ligand field photoexcitation in aqueous solution of *trans*-$[Rh(en)_2X_2]^+$, where X = Cl [1, 2], Br [3, 4] and I [5], led exclusively to aquation and formation of *trans*-$[Rh(en)_2(OH_2)X]^{2+}$. A model for predicting the photoreactions of and quantum yields from *trans*-$[Rh(en)_2X_2]^+$ (X = Cl, Br and I) has been developed [6], and ligand-field interpretations of these quantum yields have been offered [7].

The photolysis of *cis*-$[Rh(en)_2X_2]^+$, where X = Cl [2, 8, 9], and Br [3, 4, 9], led to the formation of *trans*-$[Rh(en)_2(OH_2)X]^{2+}$ although amine aquation was also observed and may be the principal reaction [8, 9].

References:

[1] C. Kutal, A. W. Adamson (Inorg. Chem. **12** [1973] 1454/6). – [2] J. D. Petersen, F. P. Jakse (Inorg. Chem. **18** [1979] 18/21). – [3] S. D. Clark, J. D. Petersen (Inorg. Chem. **18** [1979] 3394/9). – [4] T. L. Kelly, J. F. Endicott (J. Phys. Chem. **76** [1972] 1937/46). – [5] S. F. Clark, J. D. Petersen (Inorg. Chem. **19** [1980] 2917/21).

[6] J. I. Zink (Inorg. Chem. **12** [1973] 1018/24). – [7] M. J. Incorvia, J. I. Zink (Inorg. Chem. **13** [1974] 2489/94). – [8] J. Sellan, R. Rumfeldt (Can. J. Chem. **54** [1976] 519/25). – [9] J. Sellan, R. Rumfeldt (Can. J. Chem. **54** [1976] 1061/5).

Kinetic Studies

The kinetics of hydrolysis of *trans*-$[Rh(en)_2XY]^+$, where X = Y = Cl [1, 2], X = Cl, Y = Br, I, OH, N_3, NCS, NO_2 and CN [2], have been determined and discussed. The kinetics of base hydrolysis of *trans*-$[Rh(en)_2X_2]^+$ (X = Cl, Br and I) have been measured [3] and the acid hydrolysis and electron transfer reactions of *trans*-$[Rh(en)_2Cl_2]^+$ discussed in terms of the symmetry of this and related complexes [4]. The Hg^{2+} assisted aquations of *cis*- and *trans*-$[Rh(en)_2Cl_2]^+$ have been discussed in terms of Hg-containing intermediates [5 to 8].

The kinetics and mechanism of ammoniation of *cis*-$[Rh(en)_2Cl_2]ClO_4$ in liquid NH_3, giving *cis*-$[Rh(en)_2(NH_3)Cl]^{2+}$ and *cis*-$[Rh(en)_2(NH_3)_2]^{3+}$, have been determined [9].

The *trans* effect as applied to halide substitution in *trans*-$[Rh(en)_2X_2]^+$ has been investigated kinetically, [10, 11, 12], and the kinetics of the binuclear substitution of *trans*-$[Rh(en)_2I_2]^+$ by Cl^- and Br^-, and of *trans*-$[Rh(en)_2ClI]^+$ by OH^- and Br^-, studied [13].

References:

[1] S. A. Johnson, F. Basolo (J. Am. Chem. Soc. **85** [1963] 1741/7). – [2] M. L. Tobe (Inorg. Chem. **7** [1968] 1260/2). – [3] A. J. Pöe, C. P. J. Vuik (J. Chem. Soc. Dalton Trans. **1976** 661/6). – [4] K. B. Yatsimirskii (Teor. Eksperim. Khim. **8** [1972] 617/20; Theor. Exptl. Chem. [USSR] **8** [1972] 509/11). – [5] A. B. Venediktov, A. V. Belyaev (Koord. Khim. **2** [1976] 1414/21; Soviet J. Coord. Chem. **2** [1976] 1085/91).

[6] A. V. Belyaev, A. B. Venediktov, B. I. Peshchevitskii (Koord. Khim. **5** [1979] 1071/84; Soviet J. Coord. Chem. **5** [1976] 845/56). – [7] J. L. Armstrong, M. J. Blandamer, J. Burgess, A. Chen (J. Inorg. Nucl. Chem. **43** [1981] 173/4). – [8] A. O. Adeniran, G. J. Baker, G. J. Bennett, M. J. Blandamer, J. Burgess, N. K. Dhammi, P. P. Duce (Transition Metal Chem. [Weinheim] **7** [1982] 183/7). – [9] S. Salt, A. Jelsma (Transition Metal Chem. [Weinheim] **6** [1981] 119/22). – [10] F. Basolo, E. J. Bounsall, A. J. Pöe (Proc. Chem. Soc. **1963** 336).

[11] E. J. Bounsall, A. J. Pöe (Proc. 8th Intern. Conf. Coord. Chem., Vienna 1964, pp. 313/5 from C.A. **67** [1967] No. 26174). – [12] H. L. Bott, E. J. Bounsall, A. J. Pöe (J. Chem. Soc. A **1960** 1275/82). – [13] A. J. Pöe, C. P. J. Vuik (Inorg. Chem. **19** [1980] 1771/5).

Other Properties

The reaction enthalpy and activation energy for the dehydration/dehydrohalogenation of *trans*-$[Rh(en)_2Cl_2]Cl_2 \cdot H_5O_2$ were 3.7 ± 0.6 and 36 ± 3 kcal/mol, respectively [1].

The kinetics of chloride exchange in *cis*- and *trans*-$[Rh(en)_2Cl_2]^+$ have been determined [2]. The effect of Cl^- and H^+ ion concentration [3] and of temperature, X- and γ-rays [4], on the rate of ^{36}Cl exchange with *trans*-$[Rh(en)_2Cl_2]Cl$ has been investigated. The chemical effects of (n,γ) reactions in *trans*-$[Rh(en)_2Cl_2]Cl$ have been determined using ^{36}Cl-labelled complexes [5].

References:

[1] H. E. Le May, P. S. Nolan (Inorg. Nucl. Chem. Letters **7** [1971] 355/8). – [2] J. Burgess, K. W. Bowker, E. R. Gardner, F. M. McKhail (J. Inorg. Nucl. Chem. **41** [1979] 1215/7). – [3] K. W. Bowker, E. R. Gardner, J. Burgess (Trans. Faraday Soc. **67** [1971] 3076/80). – [4] G. B. Schmidt, K. Roessler (Radiochim. Acta **5** [1966] 123/6). – [5] E. R. Gardner, M. E. Wilson, R. D. Harding, J. B. Raynor (Radiochim. Acta **17** [1972] 41/5).

2.2.1.2.1.6 Complexes with Sulphur-Containing Ligands

Thiosulphato Complex, Na[*trans*-$Rh(en)_2(S_2O_3)_2$]

The compound is obtained by reaction of *trans*-$[Rh(en)_2Cl_2]NO_3$ with an excess of $Na_2S_2O_3 \cdot 5H_2O$ in aqueous solution. It is isolated as a pale green-yellow solid whose electronic spectrum was measured in the range 280 to 390 nm (aqueous solution).

E. J. Bounsall, A. J. Pöe (J. Chem. Soc. A **1966** 286/92).

Thiocyanato Complexes

***trans*-$[Rh(en)_2(SCN)_2]SCN$.** Reaction of $Na_3[Rh(SCN)_6]$ with ethylenediamine in refluxing acetonitrile, followed by chromatography on alumina, affords this complex as a yellow solid.

The IR spectrum in the range 4000 to 400 cm^{-1} was measured; ν(CN) = 2118, 2060 cm^{-1}; ν(CS) = 722, 708 cm^{-1}. The electronic spectrum in acetonitrile was reported in the range 190 to 300 nm. It was suggested that the thiocyanate was S-bonded to the metal [1].

***trans*-[Rh(en)$_2$(SCN)Cl]SCN** is prepared by refluxing *trans*-[Rh(en)$_2$Cl$_2$]NO$_3$ with an excess of NaSCN in water. It is isolated as a bright yellow solid whose IR spectrum exhibited bands due to thiocyanate at 2156, 2060, 835 and 735 cm^{-1}. The electronic spectrum of the complex in water was reported [2]. Correlations have been made between the polarographic reduction of this compound with its d-d-electronic transitions [3].

***trans*-[Rh(en)$_2$(OH$_2$)(NCS)]$^{2+}$, [Rh(en)$_2$(OH$_2$)(SCN)]$^{2+}$.** The acid dissociation constants of these species were determined potentiometrically to be >6.30 (SCN form) and <5.90 (NCS form) [4].

References:

[1] A. Bofar, E. Blasius, G. Klemm (Z. Anorg. Allgem. Chem. **449** [1979] 174/6). – [2] S. A. Johnson, F. Basolo (Inorg. Chem. **1** [1962] 925/32). – [3] D. R. Crow (Inorg. Nucl. Chem. Letters **5** [1969] 291/3). – [4] S. Ya. Dvurechenskaya, A. V. Belyaev, B. I. Peshchevitskii (Izv. Sibirsk. Otd. Akad. Nauk SSSR Ser. Khim. Nauk **1977** 63/9 from C.A. **77** [1972] No. 157134).

2.2.1.2.2 Mono-ethylenediamine and Related Complexes

M[Rh(en)Cl$_4$]. The **ethylenediamonium** salt (M = enH$_2$) is obtained when RhCl$_3$·3H$_2$O is refluxed with en·2HCl in aqueous solution in the absence of KOH [1]. The **potassium** salt (M = K), as a dihydrate, is formed by reacting an aqueous solution of RhCl$_3$·3H$_2$O and KCl with a 1:1 mixture of en·H$_2$O and acetic acid. It is isolated as brown prisms, and may be dehydrated at 115°C. K[Rh(NH$_2$CD$_2$CD$_2$NH$_2$)Cl$_4$] is similarly prepared using the C-deuteriated diamine. The **caesium** salt (M = Cs) is obtained by treating K[Rh(en)Cl$_4$]·2H$_2$O with CsCl in water, and is isolated as yellow crystals [2].

The far IR spectrum (400 to 30 cm^{-1}) of these compounds have been reported [2].

K[Rh(en)Br$_4$]·2H$_2$O is prepared by heating K[Rh(en)Cl$_4$]·2H$_2$O with KBr in water, and is isolated as red crystals. The far IR spectrum of this compound (400 to 30 cm^{-1}) has been recorded [2].

M[Rh(en)(SCN)$_4$]. The **sodium** salt (M = Na) is obtained by refluxing Na$_3$[Rh(SCN)$_6$] with ethylenediamine in acetonitrile, followed by chromatography on alumina. The orange-yellow hygroscopic species has been converted to the **tetra-n-butylammonium** salt (M = Bu$^n_4$N), which is not hygroscopic. The IR spectrum of the Na$^+$ salt was recorded: ν(CN) = 2115, 2062 cm^{-1}; ν(CS) = 744, 690, 672 cm^{-1}. The electronic spectrum of the compound was recorded in acetonitrile in the range 190 to 300 nm, and it was suggested that the thiocyanate was S-bonded to the metal [3].

Na$_2$(NH$_4$)$_3$[Rh(en)(SO$_4$)$_4$]·3H$_2$O is formed when Na$_{10}$(NH$_4$)$_6$[Rh$_4$(S$_2$O$_3$)$_{13}$(NaS$_2$O$_3$)$_2$(en)$_8$]·11.5H$_2$O is treated with H$_2$O$_2$. It is a yellow-white solid [4].

Rh(en)X$_3$. The chloride and bromide are prepared by boiling K[Rh(en)X$_4$]·2H$_2$O in water. The chloride is isolated as a yellow powder [2].

***trans*-[Rh(en)(py)$_2$Cl$_2$]Cl·3H$_2$O** is prepared by refluxing an ethanol solution of *trans*-[Rh(py)$_4$Cl$_2$]Cl·5H$_2$O with en·2HCl. The complex is isolated as yellow solids. The X-ray powder diffraction (d-spacings) data were reported together with IR and Raman spectral information [5].

***Δ-cis-α*-[Rh(en)(SS-edpp)]Cl** (edppH$_2$ = ethylenediamine-N,N′-di-S-α-propionic acid, HO$_2$CCHCH$_3$NHCH$_2$CH$_2$NHCHCH$_3$CO$_2$H). This complex is formed by refluxing a mixture of [Rh(pyr)$_4$Cl$_2$]Cl·5H$_2$O, (S,S)-edppH$_2$ and LiOH in water. The product is extracted with ether and, after filtration, the filtrate is treated with ethylenediamine in dimethylformamide. After refluxing, the complex is formed as a white solid [6].

***Δ-cis-β*-[Rh(en)(SS-edpp)]Cl·H$_2$O** is obtained by treating the residue after extraction of the Δ-*cis*-α isomer above, which is *Δ-cis-β*-H[Rh(SS-edpp)Cl$_2$]·3H$_2$O, with ethylenediamine in refluxing dimethylformamide, and is obtained as a creamy white solid [6].

The electronic and CD spectra of the α and β forms were measured (200 to 500 nm), and the ^{1}H NMR spectra reported [6].

References:

[1] S. A. Johnson, F. Basolo (Inorg. Chem. **1** [1962] 925/32). – [2] I. B. Baranovskii, R. E. Sevast'yanova, G. Ya. Mazo (Zh. Neorgan. Khim. **19** [1974] 3340/3; Russ. J. Inorg. Chem. **19** [1974] 1830/1). – [3] A. Bofar, E. Blasius, G. Klemm (Z. Anorg. Allgem. Chem. **449** [1979] 174/6). – [4] I. I. Chernyaev, A. G. Maiorova (Zh. Neorgan. Khim. **5** [1960] 1074/84; Russ. J. Inorg. Chem. **5** [1960] 517/21). – [5] A. W. Addison, R. D. Gillard, P. S. Sheridan, L. R. H. Tipping (J. Chem. Soc. Dalton Trans. **1974** 709/19).

[6] M. E. F. Sheridan, M.-J. Jun, C. F. Liu (Polyhedron **1** [1982] 659/63).

2.2.1.2.3 Miscellaneous Ethylenediamine Complexes

Na$_{10}$(NH$_4$)$_6$[Rh$_4$(S$_2$O$_3$)$_{13}$(NaS$_2$O$_3$)$_2$(en)$_8$]·11.5H$_2$O is apparently obtained by reaction of (NH$_4$)$_3$[RhCl$_6$]·H$_2$O in water with Na$_2$S$_2$O$_3$·5H$_2$O and ethylenediamine, followed by drying in vacuo over conc. H$_2$SO$_4$. It reacted with H$_2$O$_2$ giving Na$_2$(NH$_4$)$_3$[Rh(en)(SO$_4$)$_4$]·3H$_2$O, and was thought to contain Rh^{II} and Rh^{III} as well as S$_2$O$_3$ bridges.

Na$_4$(NH$_4$)$_3$[Rh(S$_2$O$_3$)$_6$(NaS$_2$O$_3$)(en)$_4$]·3H$_2$O and **Na$_2$(NH$_4$)$_3$[Rh(S$_2$O$_3$)$_3$(en)$_2$]·3.5H$_2$O** are claimed to be produced when the compound above is heated in water for 15 to 20 min and 60 min, respectively. The first species is also apparently formed by treating the compound above with NH$_3$, and was believed to contain S$_2$O$_3$ bridges.

Na$_8$[Rh$_2$(S$_2$O$_3$)$_7$(en)$_4$]. This compound is apparently prepared by heating Rh(NH$_3$)Cl$_3$ with Na$_2$S$_2$O$_3$·5H$_2$O and ethylenediamine. It was thought to contain S$_2$O$_3$ bridges.

Na$_3$[Rh(S$_2$O$_3$)$_3$(en)$_2$] is believed to be formed by heating aqueous Na$_8$[Rh$_2$(S$_2$O$_3$)$_7$(en)$_4$], or treating it with NH$_3$ at room temperature.

The above complexes were diamagnetic, and their formulations were established by cryoscopic molecular weight determinations, conductivity measurements and oxidation and reduction reactions.

I. I. Chernyaev, A. G. Maiorova (Zh. Neorgan. Khim. **5** [1960] 1074/84; Russ. J. Inorg. Chem. **5** [1960] 517/21).

2.2.1.3 Complexes with Deprotonated Ethylenediamine

Complexes containing $NHCH_2CH_2NH_2$ are designated as (en-H).

$Rh(en\text{-}H)_3$ is prepared by reaction of $[Rh(en)_3]I_3$ with three mole equivalents of K in liquid ammonia at −33.5°C [1, 2]. It may also be obtained by the freeze-drying of $[Rh(en)_3](OH)_3$, itself produced by anion exchange of Cl^- in $[Rh(en)_3]Cl_3$ by OH^- [3]. The compound forms a yellow hygroscopic, diamagnetic solid which is unstable in air, but stable in dry helium [1,2]. It dissolved in water forming a colourless basic solution and reacted with NH_4I in liquid NH_3 reforming $[Rh(en)_3]I_3$. X-ray powder diffraction data (d-spacings) were measured, and the IR spectrum in the range 4000 to 200 cm^{-1} (ν(Rh-N) = 585 cm^{-1}) reported [1].

$[Rh(en\text{-}H)_2(en)]I$ is obtained by reaction of $[Rh(en)_3]I_3$ with two mole equivalents of K in liquid NH_3 at −33.5°C [1, 2]. It is a pale yellow, hygroscopic, diamagnetic solid which dissolved in water forming clear light yellow basic solutions. X-ray powder diffraction data were given [1]. The IR spectrum in the range 4000 to 200 cm^{-1} (ν(Rh-N) = 582 cm^{-1}) [1] and the electronic spectrum of the compound [4] have been reported.

The complex reacted with NH_4I in liquid NH_3 regenerating $[Rh(en)_3]I_3$. It also reacted with gaseous BF_3 in chloroform giving, after five days, **$[Rh\{(en\text{-}H)BF_3\}_2(en)]I$** as an air-stable, slightly hygroscopic orange-yellow solid. The compound reacted with SO_2Cl_2 over one day at 25°C to give the air-stable, slightly hygroscopic bright yellow **$[Rh\{(en\text{-}H)_2SO_2\}(en)]Cl_2I$**, and with $SOCl_2$ giving the unstable yellow **$[Rh\{(en\text{-}H)_2SO\}(en)]Cl_2I$**. The X-ray powder diffraction data (d-spacings) and IR spectra (4000 to 200 cm^{-1}) of $[Rh\{(en\text{-}H)BF_3\}_2(en)]I$, $[Rh\{(en\text{-}H)_2SO_2\}$-$(en)]Cl_2I$ and $[Rh\{(en\text{-}H)_2SO\}(en)]Cl_2I$ were measured (IR spectra of the first two shown in paper) [2].

$[Rh(en\text{-}H)(en)_2]I_2$ is prepared by reaction of $[Rh(en)_3]I_3$ with one equivalent of K in liquid NH_3 at −76°C. It is a white diamagnetic solid whose X-ray powder diffraction data (d-spacings) and IR spectrum (4000 to 200 cm^{-1}) were obtained, ν(Rh-N) = 572 cm^{-1} [1].

$K[Rh(en\text{-}H)_2(en\text{-}2H)]$. Reaction of $[Rh(en)_3]I_3$ with five mole equivalents of K in liquid NH_3 for 5 d at 25°C afforded this compound. It formed dark green hygroscopic crystals which were stable in dry air and dry He. The compound was paramagnetic, $\mu_{eff} = 1.5 \pm 0.1$ B.M. X-ray powder diffraction data (d-spacings) and the IR spectrum (4000 to 200 cm^{-1}) have been reported. The complex reacted with aqueous HI giving $[Rh(en)_3]I_3$, and an oxidation state determination established the presence of Rh^{III} [5].

References:

[1] G. W. Watt, J. K. Crum (J. Am. Chem. Soc. **87** [1965] 5366/70). – [2] G. W. Watt, P. W. Alexander (Inorg. Chem. **7** [1968] 537/42). – [3] L. Labouvie, L. Heck (Ann. Univ. Saraviensis Math. Naturw. Fak. No. 15 [1980] 29/4 from C. A. **94** [1980] No. 9503). – [4] G. A. Moczygemba, J. J. Lagowski (J. Coord. Chem. **5** [1976] 71/6). – [5] G. W. Watt, J. K. Crum, J. T. Summers (J. Am. Chem. Soc. **87** [1965] 4641/2).

2.2.2 C-Alkylated Ethylenediamine and Related Diamine Complexes

This section contains complexes derived from 1,2-diamino-propane ($NH_2CH_2CHMeNH_2$ = pn), 2,3-diaminobutane ($NH_2CHMeCHMeNH_2$ = bn), 2,3-dimethyl-2,3-diaminobutane (NH_2C-$Me_2CMe_2NH_2$), *trans*-cyclopentane-1,2-diamine (cptn) and *trans*-cyclohexane-1,2-diamine (chxn).

2.2.2.1 Complexes of 1,2-Diaminopropane

2.2.2.1.1 Unsubstituted Complexes

$[Rh(pn)_3]Cl_3 \cdot 2H_2O$. Reaction of $RhCl_3 \cdot 3H_2O$ with (+)- or (−)-pn affords white crystals of $[Rh\{(+)\text{-pn}\}_3]Cl_3 \cdot nH_2O$ or $[Rh\{(-)\text{-pn}\}_3]Cl_3 \cdot nH_2O$, respectively [1,2]. The optical isomers obtained in this way from (R)-(−)-pn can be separated by fractional crystallisation via conversion to the tri-iodide or chloro-(+) tartrate salts giving Δ-$[Rh(R\text{-pn})_3]^{3+}$ and Λ-$[Rh(R\text{-pn})_3]^{3+}$, or via the (+)-tris[L-cysteine-sulphinato(2−)-SN] cobaltate(III) ion (this affords geometrical isomers in their optical forms) [8].

$[Rh(pn)_3]Br_3$. Treatment of $RhCl_3 \cdot 3H_2O$ with (−)-pn in aqueous solution followed by addition of HBr affords (+)- or (−)-$[Rh\{(-)\text{-pn}\}_3]Br_3$. The corresponding (+)- or (−)-$[Rh\{(+)\text{-pn}\}_3]Br_3$ are obtained similarly using (+)-pn [2].

$[Rh\{(-)\text{-pn}\}_3][M\{(+)\text{-pn}\}_3]Br_6 \cdot nH_2O$ (M = Rh or Co). These racemates are obtained by mixing almost saturated solutions of $[Rh\{(-)\text{-pn}\}_3]Br_3 \cdot nH_2O$ with $[M\{(+)\text{-pn}\}_3]Br_3 \cdot nH_2O$ [2].

$[Rh(pn)_3]I_3 \cdot H_2O$. This compound is prepared by reaction of RhI_3 with pn in aqueous ethanol [4]. (R)-$[Rh\{(-)\text{-pn}\}_3]I_3 \cdot H_2O$ is formed by reaction of $RhCl_3 \cdot 3H_2O$ with (R)-(−)-pn in aqueous solution followed by addition of NaI [5].

X-ray Diffraction and Spectra

X-ray diffraction data (d-line spacings) have been obtained from $[Rh\{(-)\text{-pn}\}_3]Br_3$ and $[Rh\{(-)\text{-pn}\}_3][M\{(+)\text{-pn}\}_3]Br_6$. It was established that (−)-$[Rh\{(+)\text{-pn}\}_3]^{3+}$ has the same absolute configuration as (+)-$[M\{(+)\text{-pn}\}_3]^{3+}$ (M = Cr or Co) [2].

The electronic spectrum of (+)-$[Rh\{(-)\text{-pn}\}_3]^{3+}$ (spectrum shown in paper) [6] and of $[Rh(pn)_3]I_3 \cdot H_2O$ in aqueous solution in the range 25000 to 50000 [4] or 28000 to 38000 cm^{-1} [5] have been measured. The circular dichroism spectra of (+)-$[Rh\{(-)\text{-pn}\}_3]^{3+}$ (shown in paper) [6], the isomers of $[Rh(R\text{-pn})_3]Cl_3$ (spectra shown in paper) [3], $[Rh(pn)_3]I_3 \cdot H_2O$ [5, 9] and (+)-$[Rh(pn)_3](ClO_4)_3$ [7, 8] have also been measured in aqueous solutions. The influence of SeO_3^{2-} and ClO_4^- on the circular dichroism spectrum of (+)-$[Rh(pn)_3]^{3+}$ has been investigated [7, 8].

References:

[1] F. Galsbøl (Inorg. Syn. **12** [1970] 269/80). – [2] P. Andersen, F. Galsbøl, S. E. Harnung (Acta Chem. Scand. **23** [1969] 3027/37). – [3] J. A. Hearson, S. F. Mason, J. W. Wood (Inorg. Chim. Acta **23** [1977] 95/6). – [4] H.-H. Schmidtke (Z. Physik. Chem. [Frankfurt] **38** [1963] 170/83). – [5] S. K. Hall, B. E. Douglas (Inorg. Chem. **7** [1968] 533/6).

[6] A. J. McCafferty, S. F. Mason, R. E. Ballard (J. Chem. Soc. **1965** 2883/92). – [7] S. F. Mason, B. J. Norman (Chem. Commun. **1965** 73/5). – [8] S. F. Mason, B. J. Norman (J. Chem. Soc. A **1966** 307/12). – [9] J. H. Dunlop, R. D. Gillard, G. Wilkinson (J. Chem. Soc. **1964** 3160/3).

2.2.2.1.2 Substituted Complexes

Hydrido Compounds

***trans*-$[Rh(pn)_2(NH_3)H](ClO_4)_2$.** The compound is formed when $[Rh(NH_3)_5H]SO_4$ in water is treated with pn and $NaClO_4$. It is isolated as colourless crystals whose IR spectrum was measured (4000 to 600 cm^{-1}, Nujol; ν(Rh-H) = 2060 cm^{-1}). The ^1H NMR spectrum in aqueous solution exhibited τ(RhH) = 27.4, J(RhH) = 24 Hz. In aqueous solution a small amount of the *cis* isomer was detected, τ(RhH) = 26.3, J(RhH) = 21 Hz [1].

$[Rh(pn)_2H(OH_2)]^{2+}$. The *cis* and *trans* isomers of this species are apparently formed when *trans*-$[Rh(pn)_2(NH_3)H]^{2+}$ is treated with $HClO_4$ in aqueous solution. The high-field ^1H NMR spectrum afforded two signals which were not assigned to particular isomers, at $\tau = 31.7$, J(RhH) = 32 Hz and $\tau = 32.3$, J(RhH) = 30 Hz [1].

Halo Complexes

***trans*-$[Rh\{(R)\text{-}pn\}_2Cl_2]X$.** The **nitrate** ($X = NO_3$) is prepared by refluxing $RhCl_3 \cdot 3H_2O$ with (R)-pn in water, then adding conc. HNO_3. The **chloride** (X = Cl) is obtained by passing the nitrate down an anion-exchange column (Cl^- form), and is isolated as a yellow solid [2]. The electronic and CD spectra of *trans*-$[Rh\{(R)\text{-}pn\}_2Cl_2]^+$ were measured in the ranges 280 to 480 nm (aqueous solution, spectrum shown in paper) [2] and 19000 to 45000 cm^{-1} (in methanol, spectrum shown in paper) [3].

***trans*-$[Rh\{(l)\text{-}pn\}_3Br_2]ClO_4$** is formed by refluxing $RhBr_3 \cdot 3H_2O$, l-pn·2HCl and KOH in aqueous solution. The mixture is passed down an ion-exchange column (H^+ form) and treated with $HClO_4$. The electronic and CD spectra were measured (19000 to 45000 cm^{-1} in methanol, shown in paper) [3].

$Na[Rh\{(R)\text{-}pn\}Cl_4] \cdot H_2O$ is prepared by reacting $RhCl_3 \cdot 3H_2O$ with (R)-pn in water. The yellow solution is passed down a cation-exchange column (Na^+ form), affording the complex as an orange-red precipitate. The electronic and CD spectra were measured in aqueous solution in the range 20000 to 33000 cm^{-1} (shown in paper) [3].

References:

[1] K. Thomas, G. Wilkinson (J. Chem. Soc. A **1970** 356/60). – [2] S. K. Hall, B. E. Douglas (Inorg. Chem. **7** [1968] 533/6). – [3] H. Ho, J. Fujita, K. Saito (Bull. Chem. Soc. Japan **42** [1969] 1286/91).

2.2.2.2 Complexes of 2,3-Butanediamine

Unsubstituted Complex, $[Rh(R,R\text{-}bn)_3]I_3 \cdot H_2O$

This is prepared by reaction of $RhCl_3 \cdot 3H_2O$ with (R, R)-bn in water [1]. The yellow complex is precipitated by addition of NaI. The absorption spectrum in aqueous solution in the range 350 to 400 nm, and the circular dichroism spectrum and optical rotation of the complex were reported. The effect of the chelating ligand on the rotational strength of the first d-d transition has been investigated [2].

References:

[1] S. K. Hall, B. E. Douglas (Inorg. Chem. **7** [1968] 533/6). – [2] F. Woldbye (Proc. Roy. Soc. [London] A **297** [1967] 79/87).

Substituted Complexes

***trans*-$[Rh(bn)_2Cl_2]X$.** The **nitrates** ($X = NO_3$) of the meso and dl-bn complexes are obtained by refluxing $RhCl_3 \cdot 3H_2O$ with meso and dl-bn·2HCl in aqueous KOH, followed by addition of HNO_3 [2]. They are obtained as golden-yellow crystals. The nitrate ($X = NO_3$) of (R, R)-bn is obtained similarly, and is converted to the **chloride** (X = Cl) by passage down an anion-exchange column (Cl^- form) [1]. The electronic spectra of the meso and dl-bn species have been reported [2], and that of the (R, R)-bn complex, together with its CD spectrum in aqueous solution, recorded in the range 280 to 480 nm [1].

$Na[Rh\{(R,R)\text{-bn}\}Cl_4]\cdot H_2O$ is obtained by heating $Na_3[RhCl_6]\cdot 12H_2O$ with (R, R)-bn in water, followed by addition of NaCl and HCl. The electronic and CD spectra were reported in the range 20 000 to 33 000 cm^{-1} [1].

References:

[1] S. K. Hall, B. E. Douglas (Inorg. Chem. **7** [1968] 533/6). – [2] S. A. Johnson, F. Basolo (Inorg. Chem. **1** [1962] 925/32).

2.2.2.3 Complexes of 2,3-Dimethyl-2,3-diaminobutane

***trans*-$[Rh(NH_2CMe_2CMe_2NH_2)_2Cl_2]Cl$** is prepared by refluxing a mixture of $RhCl_3\cdot 3H_2O$, $NH_2CMe_2CMe_2NH_2\cdot 2HCl$, and KOH in water. After removal of $[Rh(NH_2CMe_2CMe_2NH_2)Cl_3]_2$ by filtration, concentration of the solution afforded the complex; its electronic spectrum in aqueous solution has been reported.

***cis*-$[Rh(NH_2CMe_2CMe_2NH_2)_2Cl_2]Cl$** has been apparently formed on only one occasion after prolonged (10 h) refluxing of the reaction mixture which afforded the *trans* isomer. Its electronic spectrum was reported.

$Rh(NH_2CMe_2CMe_2NH_2)Cl_3$ is formed when $RhCl_3\cdot 3H_2O$, $NH_2CMe_2CMe_2NH_2$ and KOH are refluxed in aqueous solution. It is isolated as a bright yellow solid and was thought to be a dimer with Cl bridges.

S. A. Johnson, F. Basolo (Inorg. Chem. **1** [1962] 925/32).

2.2.2.4 Complexes Containing Ligands Related to C-Alkylated Ethylenediamine

Complexes of *trans*-Cyclopentane-1,2-diamine

$[Rh(cptn)_3]^{3+}$. The absolute configuration of $(+)\text{-}[Rh\{(+)\text{-cptn}\}_3]^{3+}$ has been established as D on the basis of stereospecific inducement of configuration at the metal atom by optically active ligands and a comparison of Cotton effect curves [1,2].

References:

[1] R. D. Gillard, G. Wilkinson (J. Chem. Soc. **1964** 1368/72). – [2] J. H. Dunlop, R. D. Gillard, G. Wilkinson (J. Chem. Soc. **1964** 3160/3).

Complexes of *trans*-Cyclohexane-1,2-diamine

$[Rh(chxn)_3]X_3$. The white **nitrates** $\Lambda\text{-}(+)\text{-}[Rh\{(-)\text{-chxn}\}_3](NO_3)_3\cdot 3H_2O$ and $\Delta\text{-}(+)\text{-}[Rh\{(-)\text{-chxn}\}_3](NO_3)_3$ are obtained by reaction of $RhCl_3\cdot 3H_2O$ with (−)-chxn in aqueous ethanol followed by addition of HNO_3. The isomers are separated by fractional crystallisation [1], the Λ-(+)-form being less soluble in water. The tri-nitrates may also be obtained from the tri-iodides (see below) by reaction with $AgNO_3$ in aqueous solution [2]. The **chloride**, $\Lambda\text{-}(+)\text{-}[Rh\{(+)\text{-chxn}\}_3]Cl_3$ is prepared from the nitrate by treatment with $FeCl_2\cdot 4H_2O$ in aqueous ammonia followed by addition of HCl. It is isolated as a pale yellow solid. The Δ-(+)-isomer is prepared from the corresponding nitrate by addition of HCl, and is isolated as a white solid [1]. The

iodides, described as L-$[Rh(d\text{-}chxn)_3]I_3$ and D-$[Rh(l\text{-}chxn)_3]I_3$, are obtained by reaction of $Na_3RhCl_6 \cdot 12H_2O$ with d- and l-forms of chxn in ethanol, followed by addition of NaI. They are converted to the corresponding **bromides**, **chlorates** and **perchlorates** by treatment with the appropriate Ag^+ salt in water [2].

Racemic $[Rh(chxn)_3]Cl_3$ is obtained from equimolar mixtures of Δ-$[Rh(chxn)_3]Cl_3$ and Λ-$[Rh(chxn)_3]Cl_3$ in water. The racemic $[\Delta\text{-}Rh(chxn)_3][\Lambda\text{-}Co(chxn)_3]Cl_6$ is similarly prepared from mixtures of Δ-$[Rh(chxn)_3]Cl_3 \cdot 4H_2O$ and Λ-$[Co(chxn)_3]Cl_3 \cdot 4H_2O$. The complexes are precipitated by addition of ethanol followed by dilute HCl [3].

(+)-$[Rh\{(-)\text{-}chxn\}_3](NO_3)_3 \cdot 3H_2O$ is hexagonal, space group $P6_3\text{-}C_6^6$, a = 13.339(2), c = 9.848(2)Å, Z = 2. The Rh atom is nearly octahedrally surrounded by the 6 N atoms, with average Rh-N = 2.082(3) Å. D_{calc} = 1.500, D_{exp} = 1.498 g/cm³. This is the "ob_3 isomer" whose absolute configuration is designated as Λ [4]. (+)-$[Rh(\textit{trans}\text{-}1(R)\text{-}2(R)\text{-}chxn)_3](NO_3)_3 \cdot 3H_2O$ is also hexagonal, space group $P6_3\text{-}C_6^6$, a = 13.008(2), c = 10.150(1) Å, Z = 2. The Rh atom is again octahedrally surrounded by N atoms, with average Rh-N = 1.076(4) and N-Rh-N bond angles of 83.6(8)°. This is the "lel isomer"; D_{calc} = 1.534, D_{exp} = 1.524 g/cm³ [5].

X-ray powder diffraction data obtained from (−)-$[Rh\{(+)\text{-}chxn\}_3]Cl_3$ have established that the species probably has the same absolute configuration as (−)-$[M\{(+)\text{-}chxn\}_3]Cl_3$ (M = Co, Cr and Ir), and that this is Λ [3].

The electronic and circular dichroism spectra, and optical rotations of Δ-(+) and Λ-(−)-$[Rh\{(-)\text{-}chxn\}_3]^{3+}$, and of Δ-(−) and Λ-(+)-$[Rh\{(+)\text{-}chxn\}_3]^{3+}$, as chlorides and nitrates, have been measured in the range 28000 to 45000 cm⁻¹ [1]. It has been suggested that (−)-$[Rh(+)\text{-}chxn_3]^{3+}$ has a D and (+)-$[Rh\{(-)\text{-}chxn\}_3]^{3+}$ an L absolute configuration on the basis of Cotton effects on the d-d electronic spectral transitions [6, 7].

References:

[1] F. Galsbøl, P. Steenbøl, B. S. Sørensen (Acta Chem. Scand. **26** [1972] 3605/11). – [2] F. M. Jaeger, L. Bijkerk (Z. Anorg. Allgem. Chem. **233** [1937] 97/139). – [3] P. Andersen, F. Galsbøl, S. E. Harning, T. Laier (Acta Chem. Scand. **27** [1973] 3973/8). – [4] R. Kuroda, Y. Sasaki, Y. Saito (Acta Cryst. B **30** [1974] 2053/5). – [5] H. Miyamae, S. Sato, Y. Saito (Acta Cryst. B **33** [1977] 3391/6).

[6] J. H. Dunlop, R. D. Gillard, G. Wilkinson (J. Chem. Soc. **1964** 3160/3). – [7] R. D. Gillard, G. Wilkinson (J. Chem. Soc. **1964** 1368/72).

2.2.3 N-Alkylated Ethylenediamine Complexes

A series of complexes have been made with $NH_2CH_2CH_2NHCH_3$ (men), $NH_2CH_2CH_2N(CH_3)_2$ (udmen), $CH_3NHCH_2CH_2NHCH_3$ (sdmen), $CH_3NHCH_2CH_2N(CH_3)_2$ (trimen) and $(CH_3)_2NHC_2CH_2N(CH_3)_2$ (tetmen).

Unsubstituted Complex, $[Rh(men)_3]I_3$

This complex is prepared either by reaction of $RhCl_3 \cdot 3H_2O$ with a large excess of men in aqueous ethanol followed by addition of KI [1], or by methylation of $Rh(en\text{-}H)_3$ with a large excess of methyl iodide in a sealed vessel at 25°C for one month [2]. It is described as a white solid [1] or a tan-yellow air-stable but slightly hygroscopic solid [2]. X-ray diffraction data (d-spacings) were reported [1, 2]. The IR spectrum was measured in the range 4000 to 200 cm⁻¹ [1,2]; ν(Rh-N) = 570, 558 cm⁻¹ [1]. The electronic spectrum was measured in water in the range 200 to 400 nm [1, 2].

References:

[1] G. W. Watt, P. W. Alexander (J. Am. Chem. Soc. **89** [1967] 1814/6). – [2] G. W. Watt, P. W. Alexander (Inorg. Chem. **7** [1978] 537/42).

Hydrido Complexes

***trans*-[Rh(tetmen)$_2$H(OH$_2$)](BPh$_4$)$_2$·3H$_2$O.** This complex is prepared by reaction of *trans*-[Rh(tetmen)$_2$Cl$_2$]ClO$_4$ with NaBH$_4$ in de-oxygenated water/diglyme mixtures at 0°C. The yellow mixture which forms is treated with HCl (to pH = 6) and NaBPh$_4$, affording the complex as a white precipitate. The electronic spectrum of the complex was measured in the range 280 to 460 nm. The ^{1}H NMR spectrum was obtained in D$_2$O solution; τ(RhH) = 32.2, J(RhH) = 30 Hz.

***trans*-[Rh(tetmen)$_2$HCl]PF$_6$.** Reaction of *trans*-[Rh(tetmen)$_2$Cl$_2$]Cl·2H$_2$O with NaBH$_4$ in a de-oxygenated water/diglyme mixture, followed by addition of ethanol saturated with HCl, and NaPF$_6$, affords this complex as pale yellow crystals. The electronic spectrum has been measured in the range 280 to 460 nm, and the ^{1}H NMR spectrum recorded in ethanol solution; τ(RhH) = 28.9, J(RhH) = 13 Hz. The IR spectrum exhibited ν(RhH) at 2158 cm^{-1}.

M. P. Hancock, B. T. Heaton, D. H. Vaughan (J. Chem. Soc. Dalton Trans. **1976** 931/6).

Complexes Containing Ethylenediamine

[Rh(men)$_2$(en)]I$_2$ and **[Rh(men)$_2$(en)]Br$_2$I** are prepared by reaction of [Rh(en-H)$_2$(en)]I with methyl iodide and methyl bromide at 25°C in a sealed vessel over three weeks and two months, respectively. They are formed as tan-yellow solids which were slightly hygroscopic. The IR spectra in the range 4000 to 200 cm^{-1} and the electronic spectra in water (200 to 400 nm) were reported [1, 2].

References:

[1] G. W. Watt, P. W. Alexander (J. Am. Chem. Soc. **89** [1967] 1814/8). – [2] G. W. Watt, P. W. Alexander (Inorg. Chem. **7** [1968] 537/42).

See also mono-ethylenediamine complexes, Section 2.2.1.2.2, pp. 184/5.

Halo Complexes

***trans*-[Rh(men)$_2$Cl$_2$]ClO$_4$** is formed when RhCl$_3$·3H$_2$O is treated with men·2HCl in aqueous solution containing NaOH. Adjustment of the pH to 7, followed by addition of NaClO$_4$, affords the complex [1].

***cis*-[Rh(udmen)$_2$Cl$_2$]X.** The **chloride** (X = Cl), described as the *trans* isomer, is obtained by warming RhCl$_3$·3H$_2$O with the ligand in water, the complex being precipitated by ethanol. The white **iodide** (X = I) is obtained by dissolving the chloride in water and adding KI [2]. The **perchlorate** (X = ClO$_4$) is obtained by the route for the men complex described above, using LiClO$_4$ instead of NaClO$_4$, and is isolated as feathery yellow crystals. The *cis* isomer has been resolved by cation-exchange chromatography of the perchlorate salt [1].

***trans*-[Rh(sdmen)$_2$Cl$_2$]X.** The yellow **chloride** (X = Cl), as a hemi-hydrate, is obtained by warming RhCl$_3$·3H$_2$O in water with the ligand, followed by addition of ethanol [1,2]. The **iodide** (X = I) is obtained by addition of KI to aqueous solutions of the chloride, and is recrystallised from aqueous ethanol, being formed as orange crystals [2]. The **perchlorate** (X = ClO$_4$) is formed from the chloride salt by treatment with LiClO$_4$, and was isolated as a yellow solid [1].

***trans*-[Rh(trimen)$_2$Cl$_2$]X.** The **chloride** (X = Cl) is obtained as yellow crystals by heating $RhCl_3 \cdot 3H_2O$ under N_2 with the ligand in ethanol containing only a small amount of water. The yellow-orange **iodide** (X = I) is obtained from it by treatment with KI and recrystallisation from ethanol/ether mixtures [2]. The **perchlorate** (X = ClO_4) is synthesised by refluxing $RhCl_3$ with trimen in ethanol (ether mixtures, addition of $LiClO_4$ and recrystallisation from ethanol). It was obtained as orange crystals [1].

***trans*-[Rh(tetmen)$_2$Cl$_2$]X.** The **chloride** (X = Cl) is prepared by heating $RhCl_3 \cdot 3H_2O$ with the ligand in ether/ethanol mixtures, and recrystallisation from the same solvent mixture. It forms yellow-orange crystals, and is hygroscopic [2]. The chloride, as a dihydrate, may also be obtained by passing the perchlorate in aqueous solution down an anion-exchange column (Cl^- form), and is isolated as orange flakes. The **perchlorate** (X = ClO_4) is formed by reaction of $RhCl_3 \cdot 3H_2O$ with the ligand in absolute ethanol, followed by addition of $LiClO_4$ [1].

Spectroscopic and electrochemical data from N-alkylated complexes *trans*-[Rh(diamine)$_2$Cl$_2$]X:

Diamine	X	ν(Rh-Cl) in cm^{-1}	ν(Rh-N) in cm^{-1}	Λ[a] in 10^{-3} mol/dm	$E_{1/2}$[b] in V	Ref.
men	ClO_4	303		85	−0.58	[1]
udmen	ClO_4	297		79	−0.47	[1]
	I		575, 531			[2]
sdmen	Cl	301		96	−0.54	[1]
			498, 474			[2]
	I		505, 475			[2]
trimen	ClO_4	290		80	−0.49	[1]
	Cl		518, 490			[2]
	I		515, 497			[2]
tetmen	ClO_4	283		82	−0.34	[1]
	Cl		537, 517			[2]

[a] In aqueous solution. – [b] In aqueous solution, using dropping Hg electrode and saturated calomel reference electrode.

Properties

X-ray diffraction data (d-spacings) have been obtained for the chloride and iodide salts [2].

The IR spectra of the complexes (ν(RhCl), ν(RhN)) have been reported in the range 400 to 250 cm^{-1} (see table above) [1, 2]. Electronic spectra of the complexes were measured in water and in ethanol in the range 280 to 460 nm [1, 2]. From these data the N-alkylated ligands could be placed in order in the spectrochemical series [2].

The conductances of the complexes were measured in water, and the polarographic reduction potentials obtained (see table above) [1]. The kinetics of base hydrolysis of the complexes have been determined [3], and it was established that *trans*-[Rh(sdmen)$_2$Cl$_2$]Cl is a useful reagent in the polarographic determination of serum proteins [4, 5].

References:

[1] M. P. Hancock, B. T. Heaton, D. H. Vaughan (J. Chem. Soc. Dalton Trans. **1976** 931/6). – [2] G. W. Watt, P. W. Alexander (J. Am. Chem. Soc. **89** [1967] 1814/8). – [3] M. P. Hancock, B. T.

Heaton, D. H. Vaughan (J. Chem. Soc. Dalton Trans. **1979** 761/6). – [4] P. W. Alexander (J. Electroanal. Chem. Interfacial Electrochem. **36** [1972] App. 21/App. 24). – [5] P. W. Alexander, L. E. Smythe (J. Electroanal. Chem. Interfacial Electrochem. **81** [1977] 151/9).

2.2.4 Other Diamine Complexes

Complexes Containing 1,3-Diaminopropane
1,3-Diaminopropane = tn

***trans*-[Rh(tn)$_2$Cl$_2$]ClO$_4$.** The complex is prepared by reacting $RhCl_3 \cdot 3H_2O$ with tn·2HCl in water at pH = 7. On addition of $HClO_4$, the compound is isolated as orange-yellow needles [1].

***cis*-[Rh(tn)$_2$Cl$_2$]ClO$_4$** is obtained after the removal of the *trans* isomer described above from the mother liquor. On cooling and recrystallisation from HCl/$LiClO_4$ mixtures, the complex is formed as lemon-yellow crystals. It may also be obtained as a dihydrate by reaction of [Rh(tn)$_2$(C_2O_4)]$ClO_4 \cdot 2.5H_2O$ with HCl, followed by treatment with cold $HClO_4$ [1].

***trans*-[Rh(tn)$_2$Br$_2$]ClO$_4$.** This complex is prepared by refluxing *trans*-[Rh(tn)$_2$Cl$_2$]ClO$_4$ with aqueous NaBr and $LiClO_4$. It forms orange crystals.

[Rh(tn)$_2$(C$_2$O$_4$)]ClO$_4$·2.5H$_2$O is prepared from *trans*-[Rh(tn)$_2$Cl$_2$]ClO$_4$ and $Na_2C_2O_4$ in water containing a trace of $NaBH_4$. On addition of $HClO_4$, the complex forms pale yellow crystals [1].

The electronic spectra of the complexes have been reported [1], and the kinetics of exchange of Cl by Br in, and of base hydrolysis of, *trans*-[Rh(tn)$_2$Cl$_2$]$^+$ giving *trans*-[Rh(tn)$_2$(OH)$_2$]$^+$, were reported [2].

References:

[1] M. P. Hancock (Acta Chem. Scand. A **31** [1977] 678/82). – [2] A. J. Thirst, D. H. Vaughan (J. Inorg. Nucl. Chem. **43** [1981] 2889/92).

Complex Containing 1,3-Diamino-2-methylenepropane, [Rh(dia)$_3$]Cl$_3$
$CH_2{=}C(CH_2NH_2)_2$ = dia

The compound is prepared by boiling a mixture of $RhCl_3 \cdot 3H_2O$ and dia in aqueous ethanol containing HCl. It is isolated as a red solid, whose IR spectrum exhibited ν(CC) at 1580 cm^{-1}. The electronic spectrum of the complex was reported in aqueous solution.

H. Henning, K. Schulze, M. Mühlstädt, E. Hoyer, R. Kirmse, L. S. Emeljanowa (Z. Anorg. Allgem. Chem. **413** [1975] 10/26.

2.3 Saturated Polydentate Nitrogen Donors

Complexes containing three and four nitrogen donor atoms are described in this section.

2.3.1 Triamino Complexes

Complexes Containing Diethylenetriamine
$NH_2CH_2CH_2NHCH_2CH_2NH_2$ = dien

This ligand is occasionally known as 3-aza-pentane-1,5-diamine.

$[Rh(dien)_2]X_3$. The **iodide** (X = I), as a monohydrate, is obtained by heating $RhCl_3 \cdot 3H_2O$ with dien in water followed by addition of NaI [1, 2]. The **chloride** is obtained by treating the iodide with AgCl. The electronic spectrum of $[Rh(dien)_2]^{3+}$ has been measured in the range 25000 to 50000 cm^{-1} (spectra shown in paper) [1]. The luminescence spectra of $[Rh(dien)_2]X_3$ have been measured at 85 K and at room temperature (X = Cl, Br) [3] and at 10 K (X = Cl) [4].

$[Rh(dien)_n(NH_3)_mH]^{2+}$. Unidentified species are formed when $[Rh(NH_3)_5H]^{2+}$ is treated with dien. From ^{1}H NMR studies it is clear that two hydrido complexes were present, with τ(RhH) = 26.1, J(RhH) = 18 Hz and τ = 26.45, J(RhH) = 20.0 Hz, in equal amounts [5].

$Rh(dien)X_3$. The **chloride** (X = Cl) is obtained by refluxing a mixture of $RhCl_3 \cdot 3H_2O$ with dien·3HCl and KOH in water. It is isolated as yellow-orange crystals which is a mixture of 1,2,3-(*fac*) and 1,2,6-(*mer*) isomers. The isomers can be separated by fractional crystallisation. The 1, 2,6-(*mer*) **bromide** (X = Br) is prepared by heating Na_3RhBr_6 with dien·3HBr and NaOH and is isolated as yellow-brown crystals. It is suggested that the 1,2,3-(*fac*) isomer may have been present in solution, but it was not isolated. The 1,2,6-(*mer*) **iodide** (X = I) is prepared by reacting $Rh(dien)Br_3$ with Ag_2O followed by HI, and is isolated as a red-brown solid [6].

The electronic spectra of the complexes, either as aqueous solutions (X = Cl, Br) or as diffuse reflectances (X = I), were measured in the range 20000 to 50000 cm^{-1} [6]. The luminescence spectra of 1,2,6-$Rh(dien)X_3$ (X = Cl or Br) have been reported at 85 K and at room temperature [3].

References:

[1] H.-H. Schmidtke (Z. Physik. Chem. [Frankfurt] **38** [1963] 170/83). – [2] G. W. Watt, B. J. McCormick (Inorg. Chem. **4** [1965] 143/7). – [3] P. E. Hoggard, H.-H. Schmidtke (Ber. Bunsenges. Physik. Chem. **77** [1973] 1052/8). – [4] G. Eyring, H.-H. Schmidtke (Ber. Bunsenges. Physik. Chem. **85** [1981] 597/603). – [5] K. Thomas, G. Wilkinson (J. Chem. Soc. A **1970** 356/60).

[6] H.-H. Schmidtke (Z. Anorg. Allgem. Chem. **339** [1965] 103/12).

Deprotonated Dien Complexes

deprotonated dien = (dien-H) and (dien-2H)

$[Rh(dien\text{-}H)_2]I$ is prepared by treating $[Rh(dien)_2]I_3$ with two mole equivalents of K in liquid NH_3. It is a white, diamagnetic solid which is unstable in air, turning pink in the presence of moisture. X-ray diffraction data (d-spacings) and IR spectra (4000 to 700 cm^{-1}) were measured. The compound reacted with NH_4I in liquid NH_3 regenerating $[Rh(dien)_2]I_3$.

Rh(dien-H)(dien-2H) is obtained by reacting $[Rh(dien)_2]I_3$ with 3.5 mole equivalents of K in liquid NH_3. It is isolated as a tan precipitate, and X-ray powder diffraction data and IR spectral data have been obtained. It dissolved in water giving a basic solution, and reacted with NH_4I in liquid NH_3 reforming $[Rh(dien)_2]I_3$.

F. W. Watt, B. J. McCormick (Inorg. Chem. **4** [1965] 143/7).

Alkylated Dien Complexes

N,N′-tetraethyl-diethylene-triamine = $Et_2NCH_2CH_2NHCH_2CH_2NEt_2$ = Et_4dien

$Rh(Et_4dien)Cl_3$ is formed when $RhCl_3 \cdot 3H_2O$ is heated with Et_4dien in ethanol. It is isolated as a yellow-brown, diamagnetic solid which is a non-conductor in ethanol or dimethylformamide (DMF). The electronic spectrum in DMF was briefly reported [1].

$Rh(Et_4dien)(N_3)_3\cdot(EtOH)_{0.81}$. This complex is obtained from the reaction of $RhCl_3\cdot3H_2O$ with Et_4dien and NaN_3 in methanol/ethanol mixtures. The crystals are monoclinic, space group $P2_1/n$, a = 8.206, b = 22.037, c = 12.490 Å, β = 99.25°, Z = 4; D_{exp} = 1.42, D_{calc} = 1.43 g/cm³. The ethanol molecules occupy voids in the lattice, and the complex has a distorted octahedral geometry, in which the N-C-C-N groupings of the Et_4dien ligands have a gauche conformation, and the azido groups are meridional (1,2,6-isomer) [2].

References:

[1] Z. Dori, H. B. Gray (J. Am. Chem. Soc. **88** [1966] 1394/8). – [2] R. F. Ziolo, R. M. Shelby, R. H. Stanford, H. B. Gray (Cryst. Struct. Commun. **3** [1974] 469/72).

Complexes Containing *cis,cis*-Triaminocyclohexane

$C_6H_9(NH_2)_3$ = *cis, cis*-tach

[Rh(*cis,cis*-tach)$_2$]X_3. The **chloride**, containing some water and HCl of crystallisation, is obtained by heating $RhCl_3\cdot3H_2O$ with *cis,cis*-tach at 100°C followed by evaporation, extraction into ethanol and recrystallisation from conc. HCl. It is formed as white hexagonal plates which have not been further purified. The **perchlorate** ($X = ClO_4$) is prepared by passing an aqueous solution of the chloride down an anion-exchange column (ClO_4^- form). The electronic spectrum of the complex in aqueous solution was reported.

R. A. D. Wentworth, J. J. Felten (J. Am. Chem. Soc. **90** [1968] 621/6).

2.3.2 Tetra-amino Complexes

This section contains complexes derived from $NH_2(CH_2)_nNH(CH_2)_mNH(CH_2)_nNH_2$, and $N(CH_2CH_2NH_2)_3$. Useful brief reviews have been made of the steric constraints imposed on facultative ligands, particularly those of the former type shown above [1, 2]. It has been noted that complexes of the type $[Rh\{NH_2(CH_2)_nNH(CH_2)_mNH(CH_2)_nNH_2\}X_2]^+$ occur exclusively in the *cis* form where n = m = 2, and only in the *trans* form when n = m = 3 or n = 3, m = 2 or 4, but that both *cis* and *trans* isomers are formed when n = 2, m = 3 [2].

References:

[1] B. Bosnich, R. D. Gillard, E. D. McKenzie, G. A. Webb (J. Chem. Soc. A **1966** 1331/9). – [2] R. W. Halliday, R. H. Court (Can. J. Chem. **52** [1974] 3469/73).

Complexes Containing Triethylenetetramine

$\{NH_2CH_2CH_2NHCH_2\}_2$ = trien

This ligand is also occasionally known as 3,6-diazaoctane-1,8-diamine, or 2,2,2-tet. The geometries of the isomers of *cis*-$[Rh(trien)X_2]^+$ are as follows:

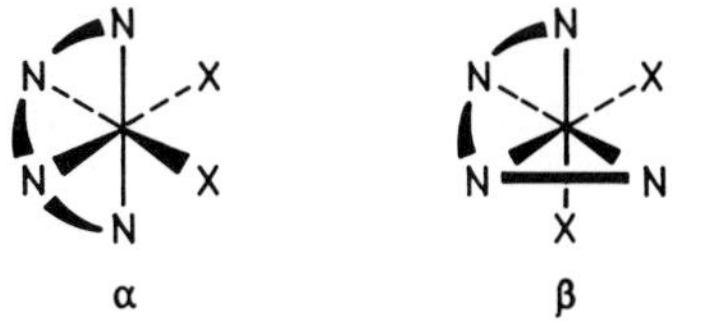

Hydrides

***cis*-[Rh(trien)H_2]BPh_4.** This compound is prepared by reacting *cis*- or *trans*-[Rh(trien)Cl_2]Cl with $NaBH_4$ in ice-cold water, followed by addition of $NaBPh_4$ [1, 2]. It has been isolated but has not been fully characterised because of its apparently explosive nature. The IR spectrum of the compound exhibited ν(RhH) at 1909 cm^{-1}, and the electronic spectrum was briefly reported. The ^{1}H NMR spectrum showed τ(RhH) = 32.5, J(RhH) = 27 Hz [2], although an earlier report suggested that, in a mixture thought to contain *cis*-[Rh(trien)HCl]$^+$ and *cis*-[Rh(trien)H_2]$^+$, a signal at τ = 27.1, J(RhH) = 30 Hz was due to the dihydride [1].

***cis*-[Rh(trien)HCl]$^+$** is thought to be formed as an intermediate during the preparation of *cis*-[Rh(trien)H_2]$^+$ from *cis*-[Rh(trien)Cl_2]$^+$ and $NaBH_4$ in ice-cold water [1, 2]. It has not been fully characterised, but its IR spectrum showed ν(RhH) at 2081 cm^{-1}, and its electronic spectrum has been recorded. The ^{1}H NMR spectrum of solutions containing this species exhibited τ = 28.5, J(RhH) = 27 Hz [1, 2].

References:

[1] R. D. Gillard, G. Wilkinson (J. Chem. Soc. **1963** 3594/9). – [2] J. A. Osborn, R. D. Gillard, G. Wilkinson (J. Chem. Soc. **1964** 3168/73).

Halides

***cis*-[Rh(trien)Cl_2]X.** The **chloride** (X = Cl) is obtained by heating $RhCl_3 \cdot 3H_2O$ with trien·4HCl and KOH in water [1] or refluxing a mixture of $RhCl_3 \cdot 3H_2O$ with trien·4HCl and aqueous NaOH [2]. The latter method affords the *cis*-α isomer as a dihydrate at pH = 8, strict control of the pH of the reaction mixture being essential. By cooling the mother liquor after removal of the *cis*-α form by filtration, the *cis*-β isomer is isolated [2]. Both *cis*-α and *cis*-β forms are obtained as yellow crystals. The **perchlorate** (X = ClO_4) is prepared by refluxing a mixture of Na_3RhCl_6 and trien in water, followed by addition of $HClO_4$. The **tetraphenylborate** (X = BPh_4) is obtained from the perchlorate salt by its treatment with $NaBPh_4$ in water. Both salts from yellow crystals [3].

Partial resolution of the *cis*-α and *cis*-β isomers has been achieved using cation exchange resins [2].

Properties. The IR spectra of *cis*-[Rh(trien)Cl_2]X (X = ClO_4 or BPh_4) have been briefly reported [3] and there has been an IR spectral study, particularly of ν(NH) and ν(ND) (in deuteriated species), of *cis*-α-[Rh(trien)Cl_2]Cl·H_2O (3000 to 3300, 990 to 1100 cm^{-1}) [4].

The electronic spectra of *cis*-[Rh(trien)Cl_2]Cl [1], both the *cis*-α and *cis*-β isomers (in water 250 to 550 nm, shown in paper) [2, 5], and of *cis*-[Rh(trien)Cl_2]ClO_4 [3] have been reported. The optical rotatory dispersion (ORD) and circular dichroism (CD) spectra of *cis*-α and *cis*-β-[Rh(trien)Cl_2]Cl have also been described [2].

The conductivity of *cis*-[Rh(trien)Cl_2]ClO_4 in water was 117 $cm^2 \cdot \Omega^{-1} \cdot mol^{-1}$ (10^{-3} M solution) [3].

The kinetics of reaction of *cis*-α-[Rh(trien)Cl_2]$^+$ with Br^-, I^- and N_3^- have been determined [2]. The photosensitivity of *cis*-α- and *cis*-β-[Rh(trien)Cl_2]$^+$ towards Cl^- loss, aquation and isomerisation in water has been investigated. *Cis*-α-[Rh(trien)Cl_2]$^+$ was converted exclusively to *trans*-[Rh(trien)Cl(H_2O)]$^{2+}$ whereas the *cis*-β isomer afforded a mixture of *cis*-β- (65%) and *trans*-[Rh(trien)Cl(H_2O)]$^{2+}$ (35%) [5]. The nature of the intermediate in the aquation of [Rh(trien)Cl_2]$^+$ has been discussed in general terms [6]. The kinetics and mechanism of chloride ion ($^{36}Cl^-$) exchange with *cis*-[Rh(trien)Cl_2]Cl have been measured [7], and the breakdown products of neutron irradiation of *cis*-[Rh(trien)Cl_2]Cl have been determined [8].

***trans*-[Rh(trien)Cl_2]ClO_4.** This complex is obtained from the mother liquor remaining after the isolation of *cis*-β-[Rh(trien)Cl_2]Cl described on p. 196, by acidification with HCl to pH = 1. On addition of $KClO_4$ the complex forms as yellow crystals. It may also be prepared by evaporation of the mother liquor after isolation of the *cis*-α isomer, filtration and dissolving the filtrate in water to pH = 7 (HCl). Chromatography on an ion exchange column afforded a yellow band which, on evaporation and addition of conc. $HClO_4$, afforded yellow crystals of the pure isomer. The electronic, ORD and CD spectra have been reported in aqueous solution [2].

***cis*-[Rh(trien)Br_2]Br·3H_2O.** This complex is prepared by refluxing $RhBr_3$ and trien·4HBr with either NaOH [2] or KOH [9] in water. The complex, as the *cis*-α isomer, may also be obtained by heating *cis*-α-[Rh(trien)Cl_2]Cl with an excess of NaBr in water [2]. It was isolated as a yellow powder whose IR spectrum in the range 400 to 150 cm^{-1} has been reported [9]. The electronic spectrum of the complex was briefly described [2].

***cis*-α-[Rh(trien)I_2]I** is formed by heating *cis*-α-[Rh(trien)Cl_2]Cl with a large excess of NaI in water. It was isolated as dark red crystals, whose electronic spectrum has been briefly reported [2].

***cis*-[Rh(trien)$(N_3)_2$]N_3** is obtained by reaction of *cis*-α-[Rh(trien)Cl_2]Cl·H_2O with an excess of NaN_3 in hot water. The complex is isolated as orange crystals, and is stable at room temperature in the dark, but is discoloured in daylight within a few hours. The electronic spectrum in water was briefly reported [2].

***cis*-α-[Rh(trien)(NCS)Cl]SCN.** This compound is prepared by heating *cis*-α-[Rh(trien)Cl_2]Cl with a fourfold excess of NaSCN in water. It is formed as a slightly impure yellow solid. The impurity was thought to be [Rh(trien)Cl_2]SCN, whose IR spectrum showed bands at 2120 and 2050 cm^{-1}, consistent with ionic and N-bonded thiocyanate, respectively. The electronic spectrum of the complex was described [2].

[Rh(trien)Cl(H_2O)]$^{2+}$. The *trans* isomer is produced exclusively when *cis*-α-[Rh(trien)Cl_2]$^+$, *cis*-α- or *cis*-β-[Rh(trien)Cl(H_2O)]$^{2+}$ is photolysed in water. The *cis*-α and *cis*-β isomers are prepared by heating the corresponding dichloro cations in water at 60°C. Both *cis* isomers were stable indefinitely in acid solutions, and their electronic spectra were measured in the range 250 to 500 nm (shown in paper) [5].

[Rh(trienH_2)Cl_4]Cl is prepared by heating $RhCl_3$·3H_2O with trien·4HCl in water. It was isolated as brown crystals whose IR and electronic spectra were briefly reported [1].

References:

[1] S. A. Johnson, F. Basolo (Inorg. Chem. **1** [1962] 925/32). – [2] P. M. Gidney, R. D. Gillard, B. T. Heaton, P. S. Sheridan, D. H. Vaughan (J. Chem. Soc. Dalton Trans. **1973** 1462/8). – [3] R. D. Gillard, G. Wilkinson (J. Chem. Soc. **1963** 3193/200). – [4] D. A. Buckingham, D. Jones (Inorg. Chem. **4** [1965] 1387/92). – [5] E. Martins, P. S. Sheridan (Inorg. Chem. **17** [1978] 3631/6).

[6] M. L. Tobe (Inorg. Chem. **7** [1968] 1260/2). – [7] K. W. Bowker, E. R. Gardner, J. Burgess (Trans. Faraday Soc. **67** [1971] 3076/80). – [8] E. R. Gardner, S. Greethong, J. B. Raynor (Radiochim. Acta **14** [1970] 23/7). – [9] K. W. Bowker, E. R. Gardner, J. Burgess (Inorg. Chim. Acta **4** [1970] 626/8).

Complexes Containing 1,4,8,11-Tetraazaundecane

$NH_2(CH_2)_2NH(CH_2)_3NH(CH_2)_2NH_2$ = 2,3,2-tet

The α and β isomers of *cis*-[Rh(2,3,2-tet)$_2$]$^+$ are similar to those described for *cis*-[Rh(trien)X_2]$^+$ (see p. 195).

[Rh(2,3,2-tet)Cl$_2$]Cl·nH$_2$O. The *cis*-α isomer is obtained, as a monohydrate (n = 1) by reaction of RhCl$_3$·3H$_2$O, 2,3,2-tet·4HCl and NaOH in water, and was isolated as pale yellow crystals [1]. The *cis*-β isomer, as a hemihydrate (n = 0.5), is formed similarly, using LiOH instead of NaOH, and was also isolated as yellow crystals [2]. The *trans* isomer, as a hemihydrate (n = 0.5), is prepared by evaporating the mother liquor after removal of the *cis*-α isomer described above, and was isolated as a dark yellow powder. The IR spectra (ν(NH) and δ(NH)) [1] and electronic spectra of these isomers have been reported [1, 2].

Ligand field photolysis of *cis*-β-[Rh(2,3,2-tet)Cl$_2$] in dilute H$_2$SO$_4$ solutions affords *cis*-[Rh(2,3,2-tet)Cl(OH$_2$)]$^{2+}$ and *trans*-[Rh(2,3,2-tet)Cl(OH$_2$)]$^{2+}$ in a ratio of ~1:3. A kinetic study of the chloride equation of *cis*-β-[Rh(2,3,2-tet)Cl$_2$]$^+$ has been made [2].

***trans*-[Rh(2,3,2-tet)Cl$_2$]NO$_3$** is obtained by reaction of *trans*-[Rh(py)$_4$Cl$_2$]Cl·5H$_2$O (py = pyridine) and 2,3,2-tet in water, followed by addition of NaNO$_3$. It was isolated as yellow crystals whose electronic spectrum was reported [3].

***cis*-[Rh(2,3,2-tet)Cl(OH$_2$)]$^{2+}$.** Both the α and β isomers of *cis*-[Rh(2,3,2-tet)Cl$_2$]Cl undergo thermal aquation in dilute H$_2$SO$_4$ solution affording, respectively, the α and β chloro-aquo products. The electronic spectrum of *cis*-β-[Rh(2,3,2-tet)Cl(OH$_2$)]$^{2+}$ was shown (250 to 450 nm in aqueous solution). Ligand field photolysis of the *cis*-β isomer caused its isomerisation to *trans*-[Rh(2,3,2-tet)Cl(OH$_2$)]$^{2+}$ [2].

References:

[1] R. W. Halliday, R. H. Court (Can. J. Chem. **52** [1974] 3469/73). – [2] E. Martins, E. B. Kaplan, P. S. Sheridan (Inorg. Chem. **18** [1979] 2195/9). – [3] B. Bosnich, R. D. Gillard, E. D. McKenzie, G. A. Webb (J. Chem. Soc. A **1966** 1331/9).

Complexes Containing 1,5,8,12-Tetraazadodecane

NH$_2$(CH$_2$)$_3$NH(CH$_2$)$_2$NH(CH$_2$)$_3$NH$_2$ = 3,2,3-tet

***trans*-R,R:S,S-[Rh(3,2,3-tet)Cl$_2$]Cl·H$_2$O** is obtained by refluxing *trans*-[Rh(py)$_4$Cl$_2$]Cl·5H$_2$O with 3,2,3-tet in water, followed by addition of HCl. It was isolated as yellow needles [1].

***trans*-R,R-[Rh(3,2,3-tet)Cl$_2$](BCS)·3H$_2$O** (BCS = C$_{10}$H$_{14}$O$_4$SBr = α-bromocamphor-π-sulphonate). This compound is prepared by treating *trans*-R,R:S,S-[Rh(3,12,3-tet)Cl$_2$]Cl·H$_2$O with NH$_4$[(+)-BCS] in water. The pure isomer is obtained after three fractional crystallisations, and was isolated as fine yellow needles [1].

***trans*-[Rh(3,2,3-tet)Cl$_2$]NO$_3$.** This complex is formed by reaction of RhCl$_3$·3H$_2$O with 3,2,3-tet·4HCl and NaOH in water. It is isolated after recrystallisation from HNO$_3$ as dark yellow prisms [2]. The **(−)-*trans*-R,R** isomer is obtained by dissolving *trans*-R,R-[Rh(3,2,3-tet)Cl$_2$]-(BCS)·3H$_2$O in water and adding HNO$_3$. The complex was isolated as long yellow needles. The **(+)-*trans*-S,S** isomer is formed by treating the filtrate obtained after removal of all *trans*-R,R-[Rh(3,2,3-tet)Cl$_2$](BCS)·3H$_2$O by fractional crystallisation as described above with NH$_4$NO$_3$. Recrystallisation of the product afforded the pure diastereoisomer [1].

The IR spectrum (ν(NH)) of *trans*-[Rh(3,2,3-tet)Cl$_2$]NO$_3$ has been reported [2]. The electronic and CD spectra of *trans*-[Rh(3,2,3-tet)Cl$_2$]NO$_3$ [2] and of *trans*-R,R-[Rh(3,2,3-tet)Cl$_2$]ClO$_4$ (range 250 to 500 nm shown in paper; in H$_2$O, methanol and DMF) [3] have been described, and optical rotations in water reported. The conductivity of *trans*-R,R:S,S-[Rh(3,2,3tet)Cl$_2$]Cl·H$_2$O in water was 74 cm^2·Ω^{-1}·mol^{-1} [1].

References:

[1] B. Bosnich, J. M. Harrowfield, H. Boucher (Inorg. Chem. **14** [1975] 815/28). – [2] R. W. Halliday, R. H. Court (Can. J. Chem. **52** [1974] 3469/73). – [3] B. Bosnich, J. M. Harrowfield (Inorg. Chem. **14** [1975] 828/36).

Complexes Containing $NH_2(CH_2)_3NH(CH_2)_3NH(CH_2)_3NH_2$ (3,3,3-tet) and $NH_2(CH_2)_3NH(CH_2)_4NH(CH_2)_3NH_2$ (3,4,3-tet)

***trans*-[Rh(3,3,3-tet)Cl_2]NO_3·0.5H_2O** is formed by treating $RhCl_3 \cdot 3H_2O$ with 3,3,3-tet·4HCl and NaOH, followed by recrystallisation from HNO_3. The complex was isolated as dark yellow prisms.

***trans*-[Rh(3,4,3-tet)Cl_2]Cl** is obtained by reaction of $RhCl_3 \cdot 3H_2O$ with 3,4,3-tet·4HCl and NaOH, followed by recrystallisation from ethanol. It was isolated as a yellow solid.

The IR spectra (ν(NH)) of these complexes have been tabulated and their electronic spectra briefly reported.

R. W. Halliday, R. H. Court (Can. J. Chem. **52** [1974] 3469/73).

Complexes Containing β,β′,β″-Triaminotriethylamine
$N(CH_2CH_2NH_2)_3$ = tren

For species $[Rh(tren)XY]^+$, the α isomer has X *trans* to the tertiary N atom, while the β isomer has X *cis* to this atom.

$[Rh(tren)(OH_2)_2]^{3+}$. This species is formed by heating β-$[Rh(tren)(OH)Cl]^+$ in aqueous acid. It has not been isolated, but its electronic spectrum has been briefly reported. The α isomer is obtained by thermal aquation of $[Rh(tren)Cl_2]^+$ in aqueous acid [1, 2]. The kinetics of Cl^- anation of this species have been investigated [2].

$[Rh(tren)(OH)_2]^+$ is obtained by heating $[Rh(tren)Cl_2]Cl$ with aqueous NaOH at 80°C [2]. It has not been isolated.

$[Rh(tren)(OH)Cl]^+$. The β isomer is produced by base hydrolysis of $[Rh(tren)Cl_2]Cl$. The α isomer may be formed by careful acidolysis of $[Rh(tren)Cl_2]Cl$, giving α-$[Rh(tren)(OH_2)Cl]^{2+}$, followed by careful neutralisation of the product. It is rapidly converted to $[Rh(tren)(OH)Cl]^+$ [2].

$[Rh(tren)Cl_2]Cl$ is prepared by reaction of $RhCl_3 \cdot 3H_2O$ with tren·3HCl and three mole equivalents of KOH in water [3]. It is isolated as bright yellow crystals which are rather insoluble in water. The kinetics of acid hydrolysis of this species have been investigated [4].

$[Rh(trenH)Cl_3]Cl$ is formed if only two mole equivalents of KOH are used in the above reaction. It is an orange-yellow solid [3].

$[Rh(tren)Br_2]ClO_4 \cdot 0.5H_2O$ is formed as a yellow-orange powder by neutralising an aqueous solution of $[Rh(tren)(OH)_2]^+$ with $HClO_4$, then adding NaBr followed by more $HClO_4$ [2].

$[Rh(tren)I_2]I$ is obtained similarly to the bromide, using an excess of NaI, and is isolated as a rust-brown solid [2].

$[Rh(tren)(N_3)_2]Cl$ is prepared similarly to the bromide, using NaN_3. The red precipitate which is formed is redissolved in water and passed down an anion-exchange column (Cl^- form). It is isolated as yellow crystals [2].

[Rh(tren)(NO_2)$_2$]NO_3 is formed by neutralisation of an aqueous solution of $[Rh(tren)(OH)_2]^+$ with $HClO_4$ and addition of $NaNO_2$ until the solution is colourless. The mixture is partially evaporated, and then treated with HNO_3 giving the complex as white crystals [2].

[Rh(tren)(C_2O_4)]ClO_4·H_2O is prepared by heating $[Rh(tren)Cl_2]Cl$ with aqueous NaOH, neutralisation of the product with $HClO_4$ and refluxing the product mixture with $Na_2C_2O_4$. On addition of $NaClO_4$, the complex is formed as off-white crystals [2].

References:

[1] E. Martins, P. S. Sheridan (Inorg. Chem. **17** [1978] 2822/6). – [2] M. J. Saliby, E. B. Kaplan, P. S. Sheridan, S. K. Madan (Inorg. Chem. **20** [1981] 728/33). – [3] S. A. Johnson, F. Basolo (Inorg. Chem. **1** [1962] 925/32). – [4] S. G. Zipp, S. K. Madan (J. Inorg. Nucl. Chem. **37** [1975] 181/4).

2.3.3 Complexes with Saturated Macrocyclic Nitrogen Donor Ligands

Within this section ligands are treated in order of increasing size of macrocycle.

General Literature:

G. A. Melson, Coordination Chemistry of Macrocyclic Compounds, Plenum, New York 1979.

2.3.3.1 Complexes with 1,4,7,10-Tetra-azacyclododecane
$C_8H_{20}N_4$ = cyclen

***cis*-[RhCl_2(cyclen)]Cl·2H_2O** is prepared by heating a mixture of $RhCl_3 \cdot 3H_2O$, cyclen·4HCl and lithium hydroxide under reflux in water in the presence of a trace of hydrazine, then filtering the mixture and heating the filtrate under reflux overnight. The product is isolated as bright yellow crystals by evaporating the bright yellow filtrate, after removal of metallic rhodium, and crystallising the residue from water-ethanol [1]. Recrystallisation from 5 M HCl-ethanol is also described [2]. The complex has also been prepared by heating a mixture of $RhCl_3 \cdot 3H_2O$ and cyclen in methanol for 2 d and is described as a yellow solid [3]. The electronic spectrum has been recorded; λ_{max} = 365, 302 nm [1, 2], 360, 295 nm [3]. The conductance in aqueous solution is given as 124 $cm^2 \cdot \Omega^{-1} \cdot mol^{-1}$ [3].

The kinetics of base hydrolysis of *cis*-$[RhCl_2(cyclen)]^+$ cation have been investigated, the rate constant is k_{OH} = 37.0 $dm^3 \cdot mol^{-1} \cdot s^{-1}$ at 25°C and I = 0.1 mol/dm^3, activation parameters are $\Delta H^{\ddagger}$ = 7.35 kJ/mol, $\Delta S^{\ddagger}_{298}$ = 32 $J \cdot K^{-1} \cdot mol^{-1}$ and $\Delta G^{\ddagger}_{298}$ = 64.0 kJ/mol. A pH profile for the hydrolysis of *cis*-$[RhCl(OH)(cyclen)]^+$ at 25°C and I = 0.1 mol/dm^3 is reproduced [2].

***cis*-[RhCl_2(cyclen)](PF_6)** is prepared by adding $[NH_4][PF_6]$ to the filtrate left after removal of $[RhCl_2(cyclen)]Cl$ and deposits as a yellow solid. The electronic spectrum shows λ_{max} = 365, 295 nm; conductivity 137 $cm^2 \cdot \Omega^{-1} \cdot mol^{-1}$ [3].

References:

[1] J. P. Collman, P. W. Schneider (Inorg. Chem. **5** [1966] 1380/4). – [2] R. W. Hay, P. R. Norman (J. Chem. Soc. Dalton Trans. **1979** 1441/5). – [3] P. K. Bhattacharya (J. Chem. Soc. Dalton Trans. **1980** 810/2).

2.3.3.2 Complexes with 1,4,7,10-Tetra-azacyclotridecane

***cis*-[RhCl$_2$(C$_9$H$_{22}$N$_4$)]Cl** is obtained as a yellow solid by heating together a mixture of RhCl$_3$·3H$_2$O and free ligand in methanol under reflux for 2 d. The electronic spectrum shows λ_{max} = 350, 300 nm; conductivity 95.6 cm^2·Ω^{-1}·mol^{-1}.

***cis*-[RhCl$_2$(C$_9$H$_{22}$N$_4$)](PF$_6$)** is prepared as a yellow solid by evaporating the filtrate left after removal of [RhCl$_2$(C$_9$H$_{22}$N$_4$)]Cl, dissolving in the minimum volume of water and adding [NH$_4$][PF$_6$]. The electronic spectrum shows λ_{max} = 350, 300 nm; conductivity 128.8 cm^2·Ω^{-1}·mol^{-1}.

P. K. Bhattacharya (J. Chem. Soc. Dalton Trans. **1980** 810/2).

2.3.3.3 Complexes with 1,4,8,11-Tetra-azacyclotetradecane
$C_{10}H_{24}N_4$ = cyclam

***cis*-[RhXY(cyclam)]$^+$, *trans*-[RhXY(cyclam)]$^+$.** Electronic spectra, infrared spectra and kinetic parameters of base hydrolysis for these cations are tabulated at the end of this section.

***cis*-[RhCl$_2$(cyclam)]Cl** is obtained as a yellow precipitate by heating a solution of RhCl$_3$·3H$_2$O and cyclam under reflux in methanol for 5 min. The complex has been resolved using d-ammonium α-bromocamphor-π-sulphonate. Optical rotations are $[\alpha]_{325}$ = +6°, $[\alpha]_{370}$ = −6° for l-*cis*-[RhCl$_2$(cyclam)]Cl [1].

***trans*-[RhCl$_2$(cyclam)]Cl** is obtained by evaporation of the filtrate left after removal of the *cis* isomer and forms yellow crystals from aqueous HCl [1].

***cis*-[RhCl$_2$(cyclam)](ClO$_4$)** is obtained by heating an aqueous mixture of RhCl$_3$·3H$_2$O, cyclam and NaCl under reflux for 30 min and is precipitated as a yellow solid by addition of aqueous HClO$_4$ [1].

***cis*-[RhCl$_2$(cyclam)](PF$_6$)** is precipitated by adding [NH$_4$][PF$_6$] to an aqueous solution of *cis*-[RhCl$_2$(cyclam)]Cl. The electronic spectrum shows λ_{max} = 355, 300 nm [2].

***trans*-[RhCl$_2$(cyclam)](PF$_6$)** is similarly prepared from *trans*-[RhCl$_2$(cyclam)]Cl. The electronic spectrum shows λ_{max} = 400 nm [2].

***cis*-[RhBr$_2$(cyclam)]Br** is prepared by heating *cis*-[RhCl$_2$(cyclam)]Cl with excess NaBr in aqueous solution under reflux for 3 h. It deposits on cooling as orange crystals [1].

***trans*-[RhBr$_2$(cyclam)]Br** is similarly prepared from *trans*-[RhCl$_2$(cyclam)]Cl using a reflux time of 5 h. It deposits as an orange solid [1].

***cis*-[RhI$_2$(cyclam)]I** is prepared by heating a solution of *cis*-[RhCl$_2$(cyclam)]Cl in 0.33 M NaI under reflux for 30 min, and is crystallised from aqueous NaI solution as an orange solid [1].

***trans*-[RhI$_2$(cyclam)]I** is prepared by heating an aqueous mixture of *trans*-[RhCl$_2$(cyclam)]-ClO$_4$ and NaI under reflux for 3 h and deposits as a brown solid which is purified by digestion with aqueous NaI for 3 d [1].

***cis*-[Rh(N$_3$)$_2$(cyclam)](ClO$_4$)** is prepared by heating an aqueous solution of *cis*-[RhCl$_2$-(cyclam)]Cl and excess sodium azide under reflux for 3 h and is precipitated as a yellow solid by addition of aqueous HClO$_4$. It is light-sensitive, decomposing to a brown compound [1].

***trans*-[Rh(N_3)$_2$(cyclam)](ClO_4)** is prepared by a similar procedure using *trans*-[RhCl_2-(cyclam)]Cl and is light-sensitive [1].

***cis*-[Rh(NO_2)$_2$(cyclam)](ClO_4)** is prepared by heating an aqueous solution of *cis*-[RhCl_2-(cyclam)](ClO_4) and $NaNO_2$ under reflux for 3 h, and is precipitated as a pale yellow solid by addition of aqueous $HClO_4$ [1].

***trans*-[Rh(NO_2)$_2$(cyclam)](ClO_4)** is prepared by heating an aqueous mixture of *trans*-[RhCl_2(cyclam)]Cl and $NaNO_2$ under reflux for 20 h, and is precipitated as a white powder by addition of aqueous $HClO_4$ [1].

***cis*-[Rh(NCS)$_2$(cyclam)](ClO_4)** is prepared by heating an aqueous solution of *trans*-[Rh(OH)(H_2O)(cyclam)](ClO_4)$_2$, NaSCN and $NaClO_4 \cdot H_2O$ under reflux for 2 h and precipitates as a yellow-white solid on cooling the solution [1].

***trans*-[Rh(NCS)$_2$(cyclam)](ClO_4)** is prepared by heating an aqueous mixture of *trans*-[Rh(OH)(H_2O)(cyclam)](ClO_4)$_2$, NaSCN and a few drops of $HClO_4$ under reflux for 50 h and is precipitated by addition of solid $NaClO_4 \cdot H_2O$. It is again refluxed (10 h) with aqueous NaSCN and reprecipitated with $NaClO_4 \cdot H_2O$ as a pale yellow solid [1].

***trans*-[RhX(OH)(cyclam)](ClO_4)** (X = Cl, Br, I, N_3). These complexes are prepared by heating an aqueous mixture of the appropriate salt *trans*-[RhX_2(cyclam)]X or *trans*-[Rh(N_3)$_2$(cyclam)]-(ClO_4) and NaOH under reflux for 3 to 10 min. They are precipitated from solution by addition of solid $NaClO_4 \cdot H_2O$ and deposit as white (X = Cl), pale yellow (X = Br, N_3) or orange (X = I) solids [1].

***trans*-[RhX(H_2O)(cyclam)](ClO_4)$_2$** (X = Cl, Br, I, N_3). These complexes are prepared by heating an aqueous mixture of the appropriate salt, *trans*-[RhX_2(cyclam)]X or *trans*-[Rh(N_3)$_2$(cyclam)](ClO_4) and NaOH under reflux for 3 to 10 min, then cooling the solution on ice and neutralising with dilute $HClO_4$. They precipitate from solution on addition of solid $NaClO_4 \cdot H_2O$ as pale yellow (X = Cl), yellow (X = N_3), orange (X = Br) or red (X = I) solids [1].

***cis*-[Rh(OH)(H_2O)(cyclam)](ClO_4)$_2$** is prepared by heating *cis*-[RhCl_2(cyclam)]Cl and excess NaOH under reflux in aqueous solution for 5 min, then adding solid $NaClO_4 \cdot H_2O$ and neutralising the mixture with dilute $HClO_4$. It precipitates as a very pale yellow solid [1].

***trans*-[Rh(OH)(H_2O)(cyclam)](ClO_4)$_2$** is similarly prepared from *trans*-[RhCl_2(cyclam)]Cl but with a reflux time of 100 min, and deposits as a pale yellow solid [1].

***trans*-[RhClBr(cyclam)](ClO_4)** is prepared by heating an aqueous solution of *trans*-[RhCl_2(cyclam)]Cl and NaOH under reflux for 5 min, then adding concentrated HBr and 70% $HClO_4$ and warming the mixture at 55°C for further 20 h. It is recrystallised from water as a yellow solid [1].

***trans*-[RhClI(cyclam)](ClO_4)** is prepared by heating a solution of *trans*-[RhI_2(cyclam)]I in 0.1 M NaCl at 60°C for 2 d and is precipitated as an orange solid by addition of 70% $HClO_4$ [1].

***trans*-[RhCl(N_3)(cyclam)](ClO_4)** is prepared by heating *trans*-[Rh(N_3)$_2$(cyclam)](ClO_4) in 1.0 M HCl at 85°C for 6 h and is precipitated by addition of 70% $HClO_4$. It is recrystallised from aqueous $HClO_4$ as a yellow solid [1].

***trans*-[RhCl(NCS)(cyclam)](ClO_4)** is prepared by heating an aqueous mixture of *trans*-[RhCl(H_2O)(cyclam)](ClO_4)$_2$ and NaSCN under reflux for 15 min, and is precipitated as a yellow solid by addition of $NaClO_4 \cdot H_2O$ [1].

***trans*-[RhBrI(cyclam)](ClO$_4$)** is prepared by heating an aqueous solution of *trans*-[RhI$_2$(cyclam)]I and NaOH under reflux for 15 min, then filtering the solution, adding concentrated HBr to the filtrate and reheating at 70°C for 5 min. It is isolated as orange crystals by dilution of the solution with water followed by addition of $HClO_4$ [1].

***trans*-[RhBr(N$_3$)(cyclam)](ClO$_4$)** is prepared by heating a mixture of *trans*-[Rh(N$_3$)$_2$(cyclam)]-(ClO$_4$) and NaBr at 70°C for 1 h in 1.0 M $HClO_4$. It is precipitated by addition of 70% $HClO_4$ and is crystallised from aqueous $HClO_4$ as orange crystals. It is light sensitive [1].

***trans*-[RhBr(NCS)(cyclam)](ClO$_4$)** is prepared by heating an aqueous solution of *trans*-[RhBr(H$_2$O)(cyclam)](ClO$_4$)$_2$ and NaSCN under reflux for 10 min. It is precipitated as a yellow solid by addition of solid $NaClO_4 \cdot H_2O$ [1].

***trans*-[RhI(N$_3$)(cyclam)](ClO$_4$)** is prepared by heating an aqueous mixture of *trans*-[RhI$_2$(cyclam)]I and NaOH under reflux for 5 min, then filtering the solution and adding NaN_3 to the filtrate. It is isolated as an orange solid by addition of 70% $HClO_4$ [1].

***trans*-[RhI(NCS)(cyclam)](ClO$_4$)** is prepared by heating an aqueous solution of *trans*-[RhI(H$_2$O)(cyclam)](ClO$_4$)$_2$ and NaSCN under reflux for 15 min and is precipitated as a pale orange solid by addition of $NaClO_4 \cdot H_2O$ [1].

***trans*-[Rh(N$_3$)(NCS)(cyclam)](ClO$_4$)** is prepared by heating an aqueous mixture of *trans*-[Rh(N$_3$)(H$_2$O)(cyclam)](ClO$_4$)$_2$ and NaSCN at 70°C for 75 min and is precipitated as a yellow solid by addition of $NaClO_4 \cdot H_2O$ [1].

Details of the infrared spectra of diacidocyclamrhodium(III) complexes [1]:

Complex	Frequency in cm^{-1}				
	ν(NH), s	r(CH$_2$), m	ν(CN or CC), s	ν(Rh-N), m	Other
cis-Cl$_2$	3060	840	1053	505	
	3175	855	1017	459	
			1001		
trans-Cl$_2$	3105	875	1020	493	
cis-Br$_2$	3065	840	1052	477	
	3160	855	1019	450	
			1001		
trans-Br$_2$	3135	890	1020	487	
cis-I$_2$	3065	840	1052	470	
	3140	855	1019	423	
			1000		
trans-I$_2$	3100	865	1018	482	
cis-(N$_3$)$_2$	3080	845	1053	490	351 (Rh-N$_3$)
	3180	860	1029	459	
			1005		
trans-(N$_3$)$_2$	3110	875	1048	501	370 (Rh-N$_3$)
			1025		
cis-(NO$_2$)$_2$	3050	840	1055	479	815, 830 (NO_2^-)
	3215	855	1027	412	
			1002		

Complex	Frequency in cm^{-1}				
	ν(NH), s	r(CH_2), m	ν(CN or CC), s	ν(Rh-N), m	Other
trans-$(NO_2)_2$	3180	880	1055 1030	499	825 (NO_2^-)
cis-$(NCS)_2$	3125 3210	845 860	1056 1034 1010	533 480	2100, 2080; 835, 820 (NCS^-)
trans-$(NCS)_2$	3105	875	1045 1035	500	2090; 830 (NCS^-)
cis-(OH)(H_2O)	3090 3220	830 850	1040 1028	500 450	546 (Rh-OH)
trans-(OH)(H_2O)	3220	870	1040 1025	493	
trans-Cl(OH)	3225	880	1047 1018	501	523, 487 (Rh-OH)
trans-Cl(H_2O)	3225	875	1040	499	
trans-Br(OH)	3220	875	1050 1017	500	521, 482 (Rh-OH)
trans-Br(H_2O)	3225	875	1034	496	
trans-I(OH)	3220	875	1050 1020	501	516, 484 (Rh-OH)
trans-I(H_2O)	3220	870	1035	497	
trans-(N_3)(OH)	3230	880	1028 1010	502	533, 487 (Rh-OH); 343 (Rh-N_3)
trans-(N_3)(H_2O)	3230	875	1050 1030	499	370 (Rh-N_3)
trans-ClBr	3215	875	1040 1018	497	
trans-ClI	3180	875	1022	496	
trans-Cl(N_3)	3200	875	1021	499	362 (Rh-N_3)
trans-Cl(NCS)	3200	875	1047 1020	499	2090; 835 (NCS^-)
trans-BrI	3180	870	1040 1017	490	
trans-Br(N_3)	3190	875	1045	497	352 (Rh-N_3)
trans-Br(NCS)	3200	875	1040 1024	498	2090; 835 (NCS^-)
trans-I(N_3)	3200	870	1040 1021	491	347 (Rh-N_3)
trans-I(NCS)	3200	870	1045	493	835 (NCS^-)
trans-(N_3)(NCS)	3215	870	1040	499	2090; 833 (NCS^-); 357 (Rh-N_3)

r = rocking vibration, s = strong, m = medium

Electronic spectra of $[Rh(cyclam)XY]^{n+}$ (absorption maxima) [1]:

Complex	λ in mμ	Complex	λ in mμ
cis-Cl_2^+	354, 299, 207	*trans*-$N_3(OH)^+$	357, 256
trans-Cl_2^+	406, 310sh, 242sh, 204	*trans*-$N_3(H_2O)^{2+}$	420sh, 363, 261
cis-Br_2^+	367, 309	*cis*-$(OH)(H_2O)^{2+}$	328, 291
trans-Br_2^+	429, 285, 235	*trans*-$(OH)(H_2O)^{2+}$	342, 274
cis-I_2^+	407, 295sh, 260sh, 228	*trans*-$ClBr^+$	418, 312sh, 260sh, 221
trans-I_2^+	515sh, 466, 353, 275, 226	*trans*-ClI^+	493, 445sh, 308, 245
cis-$(N_3)_2^+$	339sh, 262	*trans*-$Cl(N_3)^+$	382, 270, 207
trans-$(N_3)_2^+$	377, 286, 208	*trans*-$Cl(NCS)^+$	368, 252
cis-$(NO_2)_2^+$	293sh	*trans*-BrI^+	497, 459, 321, 256
trans-$(NO_2)_2^+$	320sh, 260sh, 213	*trans*-$Br(N_3)^+$	393, 281, 214
cis-$(NCS)_2^+$	322, 244	*trans*-$Br(NCS)^+$	368, 262
trans-$(NCS)_2^+$	377, 258	*trans*-$I(N_3)^+$	465sh, 417, 300, 280sh, 225
trans-$Cl(OH)^+$	363, 276	*trans*-$I(NCS)^+$	434, 412, 296, 274, 224
trans-$Cl(H_2O)^{2+}$	385, 296sh, 224sh	*trans*-$(N_3)(NCS)^+$	352, 270
trans-$Br(OH)^+$	373, 277sh	*trans*$(OH)_2^+$	341, 277
trans-$Br(H_2O)^{2+}$	468, 403, 310sh, 204	*trans*-$(H_2O)_2^{3+}$	352
trans-$I(OH)^+$	443, 393, 275, 230	*cis*-$(OH)_2^+$	331, 278
trans-$I(H_2O)^{2+}$	494, 341sh, 301sh, 271, 230	*cis*-$(H_2O)_2^{3+}$	296, 251

sh = shoulder

Activation parameters for base hydrolysis of *trans*-$[RhXY(cyclam)]^+$ cations [3]:

Complex X	Y	$\Delta H^{\ddagger}$ in kJ/mol	$\Delta S^{\ddagger}_{298}$ in $J \cdot K^{-1} \cdot mol^{-1}$	10^5 k (55°C) in $mol^{-1} \cdot s^{-1}$
Cl	Cl	147.5±0.8	104.9±2.8	0.750± 0.004
Cl	Br	147.5±0.4	119.1±0.8	3.59 ± 0.04
Cl	I	122.0±1.2	71.0±2.0	121.0 ± 4.0
Br	Br	148.0±0.8	131.7±1.6	13.8 ± 0.4
Br	I	128.7±0.8	97.0±1.6	256.0 ± 4.0
I	I	141.7±0.4	145.9±0.8	753.0 ±13.0

Kinetic data for base hydrolysis of *cis*-$[RhCl_2(cyclam)]^+$ has been reported: k(25°C) = 4.56×10^{-2} $mol^{-1} \cdot s^{-1}$, $\Delta H^{\ddagger} = 82.4$ kJ/mol, $\Delta S^{\ddagger}_{298} = 5.7$ $J \cdot K^{-1} \cdot mol^{-1}$, $\Delta G^{\ddagger}_{298} = 80.7$ kJ/mol [4].

Photolysis (407 nm) of *trans*-$[RhCl_2(cyclam)]^+$ affords *trans*-$[RhCl(H_2O)(cyclam)]^{2+}$ as the major product; the chloride quantum yield (Φ_{Cl}) is 0.0011±0.001 [5].

References:

[1] E. J. Bounsall, S. R. Koprich (Can. J. Chem. **48** [1970] 1481/91). – [2] P. K. Bhattacharya (J. Chem. Soc. Dalton Trans. **1980** 810/2). – [3] H. L. Chung, E. J. Bounsall (Can. J. Chem. **56** [1978] 709/13). – [4] R. W. Hay, P. R. Norman (J. Chem. Soc. Dalton Trans. **1979** 1441/5). – [5] C. Kutal, A. W. Adamson (Inorg. Chem. **12** [1973] 1454/6).

2.3.3.4 Complex with C-*meso*-5,12-Dimethyl-1,4,8,11-tetraazacyclotetradecane, *cis*-[$RhCl_2(C_{12}H_{28}N_4)$]Cl

The compound is prepared by heating a mixture of $RhCl_3 \cdot 3H_2O$ and free ligand under reflux in methanol for 4 h and deposits as fine yellow crystals from the filtered solution on cooling. The electronic spectrum (λ_{max} = 355, 305 nm) has been reported; selected infrared bands in cm^{-1}: ν(NH) = 3160, 3040 (or 3165, 3140 in text); ν(CH_2) = 871, 835; ν(Rh-N) 501; ν(Rh-Cl) = 301.

R. W. Hay, D. P. Piplani (J. Chem. Soc. Dalton Trans. **1977** 1956/62).

2.3.3.5 Complexes with C-*meso*-C-*meso*-5,12-Dimethyl-7,14-diphenyl-1,4,8,11-tetraazacyclotetradecane

***trans*-[$RhCl_2(C_{24}H_{36}N_4)$]Cl·H_2O** is prepared by boiling a mixture of $RhCl_3 \cdot 3H_2O$ and free ligand in ethanol for 10 min and is separated by treating the mixture with activated charcoal then filtering, acidifying the filtrate (HCl) and evaporating to small volume. The second and third crops of crystals are recrystallised from ethanol in the presence of concentrated HCl (1 drop). The electronic spectrum (λ_{max} = 405 nm, CH_3OH solution) has been reported.

***trans*-[$Rh(H_2O)_2(C_{24}H_{36}N_4)$][$ClO_4$]$_3$·2$H_2O$** is prepared by heating a mixture of *trans*-[$RhCl_2(C_{24}H_{36}N_4)$]Cl·H_2O and NaOH under reflux in 1:1 ethanol-water solution for 30 min. It is isolated as very pale yellow crystals by removing the ethanol (rotary evaporator) and adding 3 drops of 70% $HClO_4$, then recrystallising the precipitate from aqueous ethanol acidified with $HClO_4$.

R. W. Hay, P. M. Gidney (J. Chem. Soc. Dalton Trans. **1976** 974/8).

2.3.3.6 Complexes with C-*rac*- and C-*meso*-5,5,7,12,12,14-Hexamethyl-1,4,8,11-tetraazacyclotetradecane

Configurational Isomerism. The cyclic amine ligand 5,5,7,12,12,14-hexamethyl-1,4,8,11-tetraazacyclotetradecane has two chiral centres, the configurations of which distinguish *meso*-$C_{16}H_{36}N_4$ from *rac*-$C_{16}H_{36}N_4$. When coordinated to a metal ion, the four secondary amine nitrogen atoms also become chiral centres and complexes of *meso*- and *rac*-$C_{16}H_{36}N_4$ can each, in principle, exist in 10 non-enantiomorphic configurations. For details of these configurations and the designations used in the following section see

P. O. Whimp, M. F. Bailey, N. F. Curtis (J. Chem. Soc. A **1970** 1956/63).

***trans*-[$RhCl_2$(*meso*-$C_{16}H_{36}N_4$)][ClO_4]** (β and δ isomers). The δ isomer is prepared by heating an aqueous solution of $RhCl_3 \cdot 3H_2O$ (or a $[RhCl_6]^{3-}$ salt) and *meso*-$C_{16}H_{36}N_4 \cdot 2HClO_4$ on a steambath for 4 h, filtering and adding $NaClO_4$. It forms yellow crystals from dilute aqueous HCl. The stable β isomer is obtained when the δ isomer is recrystallised from hot dilute ammonia. The electronic spectra have been recorded: λ_{max} = 410, 313(sh) nm (β isomer); 413, 312(sh) nm (δ isomer).

***trans*-[$RhCl_2$(*meso*-$C_{16}H_{36}N_4$)]Cl·5H_2O** (δ isomer) is prepared by heating an aqueous solution of $RhCl_3 \cdot 3H_2O$ (or a $[RhCl_6]^{3-}$ salt) and *meso*-$C_{16}H_{36}N_4 \cdot 3HCl$ on a steambath for 4 h, then evaporating the filtered solution to dryness and crystallising the residue from ethanol-concentrated HCl. It forms pale yellow crystals.

***trans*-[RhCl$_2$(*rac*-C$_{16}$H$_{36}$N$_4$)](ClO$_4$)·0.5H$_2$O** (β and δ isomers). These are similarly prepared using the *rac* form of the ligand, their formulation as hemihydrates is supported by infrared data: ν(OH) ≈ 3575 to 3590 cm^{-1}. Electronic spectra have been recorded: λ_{max} = 410, 314(sh) nm (β isomer); 415, 312(sh) nm (δ isomer).

***trans*-[RhBr$_2$(*meso*-C$_{16}$H$_{36}$N$_4$)](ClO$_4$)** (β isomer) is prepared by heating an aqueous solution of *trans*-[RhCl$_2$(*meso*-C$_{16}$H$_{36}$N$_4$)](ClO$_4$) and NaBr under reflux for 3 h. It deposits on concentration of the solution and addition of NaClO$_4$ as orange crystals. The electronic spectrum shows λ_{max} = 435, 292(sh) nm.

***trans*-[RhBr$_2$(*rac*-C$_{16}$H$_{36}$N$_4$)](ClO$_4$)** (β isomer) is similarly prepared as orange crystals; λ_{max} = 427, 288(sh) nm.

***trans*-[RhI$_2$(*meso*-C$_{16}$H$_{36}$N$_4$)]I·0.5H$_2$O** (β isomer) is prepared by heating an aqueous solution of *trans*-[RhCl$_2$(*meso*-C$_{16}$H$_{36}$N$_4$)](ClO$_4$) and NaI under reflux for 2 h and deposits from the solution on cooling. It is crystallised from hot water (containing a few drops of HI) as red-brown crystals. The electronic spectrum has been recorded, λ_{max} = 476, 364 nm.

***trans*-[RhI$_2$(*rac*-C$_{16}$H$_{36}$N$_4$)]I·2H$_2$O** (β isomer) is similarly prepared as red-brown crystals; λ_{max} = 474, 364 nm.

***trans*-[Rh(NCS)$_2$(*meso*-C$_{16}$H$_{36}$N$_4$)]SCN·0.5H$_2$O** (β isomer) is prepared by heating an aqueous solution of *trans*-[RhCl$_2$(*meso*-C$_{16}$H$_{36}$N$_4$)](ClO$_4$) and NaSCN under reflux for 4 h and deposits on cooling. It forms yellow crystals from methanol/propan-2-ol. The electronic spectrum shows λ_{max} = 353 and 278 nm, the infrared spectrum ν(CN) = 2122, 2105, 2052; ν(CS) = 836, 750 cm^{-1}.

***trans*-[Rh(NCS)(*rac*-C$_{16}$H$_{36}$N$_4$)]CNS·0.5H$_2$O** (β isomer) is similarly prepared as yellow crystals; λ_{max} = 347, 306, 261 nm; ν(CN) = 2145, 2128, 2023 cm^{-1}; ν(CS) = 828, 760 cm^{-1}.

***trans*-[Rh(O$_2$CCH$_3$)$_2$(*meso*-C$_{16}$H$_{36}$N$_4$)](ClO$_4$)·2H$_2$O** (β isomer) is prepared by heating a mixture of *trans*-[RhCl$_2$(*meso*-C$_{16}$H$_{36}$N$_4$)](ClO$_4$) and AgO$_2$CCH$_3$ (1:2 mole ratio) under reflux in 0.1 M CH$_3$CO$_2$H for 4 d. It is isolated from the concentrated filtrate by addition of NaClO$_4$ and forms pale yellow crystals. The electronic spectrum shows λ_{max} = 362, 284(sh) nm, the infrared spectrum ν_{as}(OCO) = 1595, ν_s(OCO) = 1390, δ(OCO) = 690 cm^{-1}.

RhCl$_3$(*rac*-C$_{16}$H$_{36}$N$_4$) (α isomer) is prepared by heating an aqueous mixture of RhCl$_3$·3H$_2$O, *rac*-C$_{16}$H$_{36}$N$_4$·3HCl·H$_2$O and NaCl on a steambath for several hours. The precipitate is dissolved in boiling aqueous Na$_2$CO$_3$ and reprecipitated as a pale yellow solid by addition of HCl. The electronic spectrum (λ_{max} = 361, 301 nm) has been recorded.

RhBr$_3$(*rac*-C$_{16}$H$_{36}$N$_4$) (α isomer) is prepared by treating [Rh(CO$_3$)(*rac*-C$_{16}$H$_{36}$N$_4$)](ClO$_4$) with hot dilute aqueous HBr and deposits as sparingly soluble orange crystals. The electronic spectrum shows λ_{max} = 333 nm (ν_2 obscured by charge transfer band).

RhI$_3$(*rac*-C$_{16}$H$_{36}$N$_4$) (α isomer) is prepared by treating [Rh(CO$_3$)(*rac*-C$_{16}$H$_{36}$N$_4$)](ClO$_4$) with warm dilute CH$_3$CO$_2$H/NaI and deposits as sparingly soluble dark orange crystals. The electronic spectrum (λ_{max} ≈ 345, 307 nm) has been reported.

***cis*-[Rh(CO$_3$)(*rac*-C$_{16}$H$_{36}$N$_4$)]Cl·2H$_2$O** (α isomer) is prepared by dissolving RhCl$_3$(*rac*-C$_{16}$H$_{36}$N$_4$) in boiling dilute aqueous Na$_2$CO$_3$ and deposits from the filtered solution on cooling and concentration.

The sparingly soluble **perclorate, thiocyanate** or **iodide,** all as monohydrates, crystallised when NaClO$_4$, NaCNS or NaI was added to the filtrate. The electronic spectrum (λ_{max} = 351, 283 nm) and infrared spectrum (1595, 1266, 754, 678 cm^{-1}) of the **perchlorate** salt have been reported.

***cis*-[Rh(C_2O_4)(*rac*-$C_{16}H_{36}N_4$)](ClO_4)** (α isomer) is prepared by dissolving $RhCl_3$(*rac*-$C_{16}H_{36}N_4$) in boiling dilute aqueous Na_2CO_3, then adding oxalic acid and $NaClO_4$. It separates on cooling and is recrystallised from hot water-methanol as pale yellow crystals. The electronic spectrum shows $\lambda_{max}=341$, 279(sh) nm, the infrared spectrum $\nu_{as}(OCO)=1698$, 1670 cm^{-1}, $\nu_s(OCO)=$ 1392, 1257 cm^{-1}, $\delta(OCO)=799$ cm^{-1}.

***cis*-[Rh(OH)(O_2CCH_3)(*rac*-$C_{16}H_{36}N_4$)][B(C_6H_5)$_4$]·3H_2O** (α isomer) is prepared by dissolving [Rh(CO_3)(*rac*-$C_{16}H_{36}N_4$)]ClO_4·H_2O in hot glacial acetic acid, diluting the solution with water, filtering and adding Na[B(C_6H_5)$_4$] in ethanol to the filtrate. The sparingly soluble product deposits and is recrystallised from hot methanol/propan-2-ol as pale yellow needles. The electronic spectrum shows $\lambda_{max}=362$, ~333 nm, the infrared spectrum $\nu(OH)=3560$, $\nu_{as}(OCO)=1570$, $\nu_s(OCO)=1392$, $\delta(OCO)=680$ cm^{-1}.

***cis*-[Rh(NO_3)(*rac*-$C_{16}H_{36}N_4$)][NO_3]$_2$** (α isomer) is prepared by dissolving [Rh(CO_3)(*rac*-$C_{16}H_{36}N_4$)][ClO_4]·H_2O in hot aqueous nitric acid and crystallises on dilution of the solution with water. It is sparingly soluble. The electronic spectrum with $\lambda_{max}=347$, 286 nm and infrared spectrum have been reported; $\nu(NO_3)=1505$, 1350, 1283, 990, 824, 798, 743, 710 cm^{-1}.

***cis*-[Rh(ClO_4)$_2$(*rac*-$C_{16}H_{36}N_4$)](ClO_4)·2H_2O** (α isomer) is prepared by dissolving [Rh(CO_3)(*rac*-$C_{16}H_{36}N_4$)](ClO_4)·H_2O in a small volume of hot aqueous $HClO_4$ and is precipitated by addition of propan-2-ol to the concentrated solution. It forms pale yellow crystals from methanol/propan-2-ol. The infrared spectrum with $\nu(ClO_4)=1135$, 1080, 1004, 852 cm^{-1} has been reported.

***cis*-[Rh(NCS)$_2$(*rac*-$C_{16}H_{36}N_4$)](SCN)·0.5H_2O** (α isomer) is prepared by adding NaSCN (10 fold excess) to a solution of [Rh(CO_3)(*rac*-$C_{16}H_{36}N_4$)](ClO_4)·0.5H_2O in hot 50% acetic acid. It forms yellow crystals from hot water. The electronic spectrum shows $\lambda_{max}=348$, 295, 260 nm, the infrared spectrum $\nu(CN)=2093$, 2046 cm^{-1}, $\nu(CS)=829$, 754 cm^{-1}.

N. F. Curtis, D. F. Cook (J. Chem. Soc. Dalton Trans. **1972** 691/7).

2.3.3.7 Complex with 5,6,12,13-Tetramethyl-1,4,8,11-tetraazacyclotetradeca-4,11-diene, *trans*-[RhCl_2($C_{14}H_{28}N_4$)](ClO_4)·H_2O

This compound is prepared by heating a mixture of $RhCl_3$·3H_2O and the diperchlorate salt of the free ligand under reflux in aqueous methanol for 5 h and is precipitated from solution by addition of 70% $HClO_4$. It is recrystallised from methanol-water as a yellow solid. The electronic spectrum shows $\lambda_{max}=395$ nm (aqueous solution); selected infrared bands (in cm^{-1}) are $\nu(OH)=3450$, $\nu(NH)=3140$, $\nu(C{=}N)=1650$, $\nu(ClO_4)=1098$, 629, $\nu(CH_2)=881$, $\nu(Rh\text{-}N)=500$.

R. W. Hay, D. P. Piplani, B. Jeragh (J. Chem. Soc. Dalton Trans. **1977** 1951/6).

2.3.3.8 Complex with *meso*-5,12-Dimethyl-7,14-diphenyl-1,4,8,11-tetraazacyclotetradeca-4,11-diene, *trans*-[RhCl_2($C_{24}H_{32}N_4$)](ClO_4)

This compound is prepared by boiling a mixture of $RhCl_3$·3H_2O and free ligand in ethanol for 10 min, then diluting the mixture with water, evaporating to half volume, filtering and treating the filtrate with aqueous HCl and $NaClO_4$. It separates in poor yield (6%) as yellow crystals which can be recrystallised from methanol. The electronic spectrum ($\lambda_{max}=402$ nm, CH_3OH solution) has been reported.

R. W. Hay, P. M. Gidney (J. Chem. Soc. Dalton Trans. **1976** 974/8).

2.3.3.9 Complex with 1,4,8,12-Tetraazacyclopentadecane, *trans*-$[RhCl_2(C_{11}H_{26}N_4)][PF_6]$

This compound is obtained in poor yield by heating a mixture of $RhCl_3 \cdot 3H_2O$ and free ligand in methanol under reflux for 3 days. It is isolated by evaporating the mixture to dryness, extracting free ligand ($CHCl_3$) dissolving the residue in water and adding $NH_4[PF_6]$ to precipitate the required salt. The electronic spectrum ($\lambda_{max} = 420$ nm, CH_3CN solution) and conductivity ($150.05\ cm^2 \cdot \Omega^{-1} \cdot mol^{-1}$) have been reported.

P. K. Bhattacharya (J. Chem. Soc. Dalton Trans. **1980** 810/2).

2.3.3.10 Complex with 1,5,9,13-Tetraazacyclohexadecane, *trans*-$[RhCl_2(C_{12}H_{28}N_4)][PF_6]$

This compound is obtained in poor yield by heating a mixture of $RhCl_3 \cdot 3H_2O$ and free ligand in methanol under reflux for 3 d. It is isolated by evaporating the mixture to dryness, extracting free ligand ($CHCl_3$), dissolving the residue in water and adding $[NH_4][PF_6]$ to precipitate the required salt. The electronic spectrum ($\lambda_{max} = 440$ nm, CH_3CN solution) has been reported.

P. K. Bhattacharya (J. Chem. Soc. Dalton Trans. **1980** 810/2).

2.3.3.11 Complex with Tetraazaannulene, $[Rh(C_{28}H_{20}N_4)(CH_3CN)_2](ClO_4)_3 \cdot H_2O$

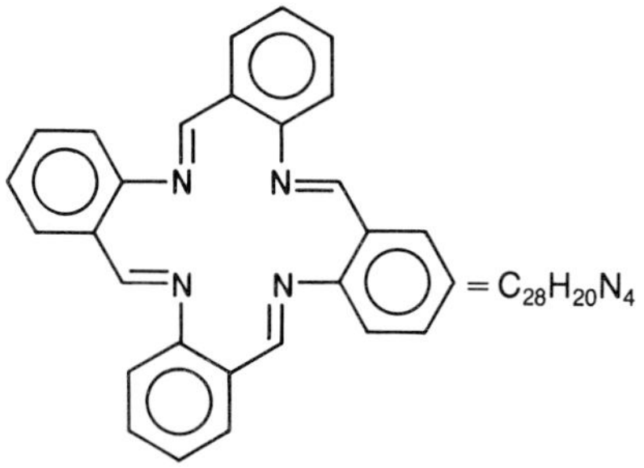

This compound is prepared by heating $RhCl_3 \cdot 3H_2O$ in CH_3CN until a yellow solution forms, then adding the ligand as the perchlorate salt ($C_{28}H_{20}N_4 \cdot 2HClO_4$) and heating the mixture under reflux for 50 h. It is isolated by precipitation with diethyl ether followed by crystallisation from CH_3CN in the presence of a little $HClO_4$.

J. S. Skuratowicz, I. L. Madden, D. H. Busch (Inorg. Chem. **16** [1977] 1721/5, 1722).

2.3.3.12 Complexes with Cyclic Dioxime Ligands

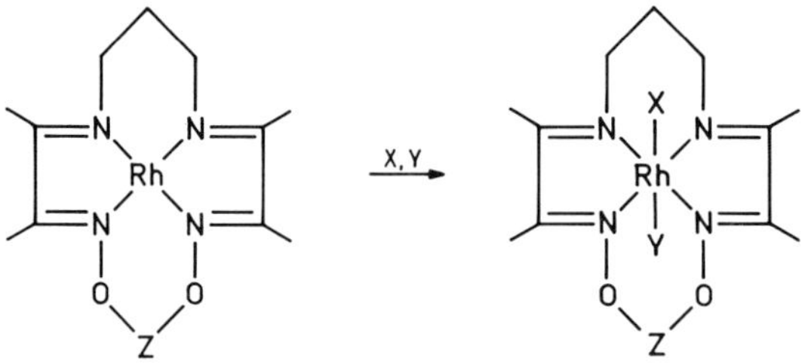

I II

$Rh(C_{11}H_{19}N_4O_2)$ (I, Z = H) has been prepared and the crystal structure has been determined but no details have been published. The complex undergoes "oxidative addition" reactions readily with halogens, alkyl and acyl halides etc.

$Rh(C_{11}H_{18}BF_2N_4O_2)$ (I, Z = BF_2) has been prepared by stirring $RhCl_2(C_{11}H_{19}N_4O_2)$ with $BF_3 \cdot O(C_2H_5)_2$ and reducing the rhodium(III) product to rhodium(I) by unspecified means. This complex undergoes "oxidative addition" reactions readily with halogen, alkyl and aryl halides etc.

***trans*-$RhCl_2(C_{11}H_{19}N_4O_2)$** (II, Z = H, X = Y = Cl) has been prepared by treatment of $Rh(C_{11}H_{19}N_4O_2)$ with CCl_4 or $SnCl(CH_3)_3$, no further details given.

***trans*-$RhHCl(C_{11}H_{19}N_4O_2)$** (II, Z = H, X = H, Y = Cl) has been prepared by addition of 1 mol HCl to $Rh(C_{11}H_{19}N_4O_2)$, no further details given. ν(Rh-H) occurs at 2065 cm^{-1}. Treatment with excess HCl leads to rupture of the oxime bridge.

***trans*-$RhCl(SnCl_3)(C_{11}H_{18}BF_2N_4O_2)$** (II, Z = BF_2, X = Cl, Y = $SnCl_3$) has been prepared by addition of $SnCl_4$ to $Rh(C_{11}H_{18}BF_2N_4O_2)$; no further details given.

***trans*-$RhCl(SnCl_3)(C_{11}H_{18}Cl_3N_4O_2Sn)$** (II, Z = $SnCl_3$, X = $SnCl_3$, Y = Cl) has been prepared by addition of $SnCl_4$ to $Rh(C_{11}H_{19}N_4O_2)$; no further details given.

J. P. Collman, D. W. Murphy, G. Dolcetti (J. Am. Chem. Soc. **95** [1973] 2687/9).

2.4 Unsaturated Bidentate Nitrogen Donors

2.4.1 Dimethylglyoximato Complexes

Dimethylglyoxime, $CH_3(CNOH)_2CH_3$ = DMGH; dimethylglyoximato ligand, $CH_3CNO \cdot CNOH \cdot CH_3^-$ = DMG. $[(DMG)_2\text{-}H]$ = deprotonated dimethylglyoxime, Me = methyl, Et = ethyl, py = pyridine, Ph = phenyl, OAc = acetate.

2.4.1.1 Unsubstituted Complexes

These lack full characterisation.

$[Rh(DMG)_2]^-$ or **Rh(DMG)(DMGH).** This (with the latter formulation) is made as a trihydrated blue-black material from *trans*-$H[Rh(DMG)_2Cl_2]$ and NaOH in boiling ethanol under N_2. It decomposes at 80°C in vacuo. The kinetics of its formation from $[Rh(DMG)_2Cl_2]^-$ were investigated [1]. A species $[Rh(DMG)_2]^-$ is thought to be formed by electrochemical reduction of $[Rh(DMG)_2Cl_2]^-$, $E_{1/2} = -0.87$ V vs. standard calomel electrode [2]; with $[Rh(DMG)_2Cl_2]^-$, $[Rh(DMG)_2]_2$ is said to be formed [2, 3].

A rhodium(I) "rhodoxime", presumably $[Rh(DMG)_2]^-$ or Rh(DMG)(DMGH), has been mentioned but without specific formulation [4]; see also $Rh(DMG)_2H$ below.

$H[Rh(DMG)_3]$ is said to be formed as a black precipitate when $H[Rh(DMG)_2Cl_2]$ in aqueous sodium acetate solution is boiled with DMGH in ethanol and NaCOOH. It is very sensitive to aerial oxidation. It dissolves in NaOH solution to give a dark brown solution. With HCl, and O_2, a mixture of *cis-* and *trans*-$H[Rh(DMG)_2Cl_2]$ is said to be formed [5].

$[Rh(DMG)_2]_2$ is made from $Rh_2(OAc)_4$ and DMGH in a 1:2 mole ratio in acetone [6]. It is also thought to be formed from $[Rh(DMG)_2]^-$ and *trans*-$[Rh(DMG)_2Cl_2]^-$ [2, 3].

It is diamagnetic at 77 K and at room temperature. The infrared spectrum was measured; from this and the diamagnetism a dimeric structure was suggested. The compound is black. The resistance of a polycrystalline sample was measured as $3\times10^{11}\,\Omega\cdot cm$ [6].

It is soluble in water, MeOH, EtOH, CH_3CN, $(CH_3)_2SO$, tetrahydrofuran and pyridine to give yellow or red solutions [3, 6], and with pyridine gives $[Rh(DMG)_2py]_2$ [6]; when electrochemically generated, it is thought to react with H_2 to give $Rh(DMG)_2H$ [2].

$Rh(DMG)_3$ is made from "rhodium sulphate" with DMGH in ethanol under reflux; sodium acetate is then added and then NaOH. The compound is dark brown.

It is insoluble in all organic solvents but dissolves in HCl to give *trans*-$H[Rh(DMG)_2Cl_2]$ [5].

References:

[1] J. D. Miller, F. D. Oliver (J. Chem. Soc. Dalton Trans. **1972** 2469/72). – [2] M. V. Klyuev, M. L. Khidekel', V. V. Strelets (Transition Metal Chem. [Weinheim] **3** [1978] 380/1). – [3] M. V. Klyuev, M. L. Khidekel' (Transition Metal Chem. [Weinheim] **5** [1980] 134/9). – [4] J. H. Weber, G. N. Schrauzer (J. Am. Chem. Soc. **92** [1970] 726/7). – [5] F. P. Dwyer, R. S. Nyholm (J. Proc. Roy. Soc. New South Wales **78** [1944] 266/70).

[6] H. J. Keller, K. Seibold (Z. Naturforsch. **25 b** [1970] 551/2).

2.4.1.2 Substituted Complexes

2.4.1.2.1 Hydrido and Hydroxido Complexes

$Rh(DMG)_2H$ is said to be formed by the action of H_2 on $[Rh(DMG)_2]_2$, generated by electrochemically formed $[Rh(DMG)_2]^-$ and *trans*-$[Rh(DMG)_2Cl_2]^-$. An anodic wave at $E_{1/2} = -0.25$ V may arise from $Rh(DMG)_2H$ [1]. A hydridic species, presumably $Rh(DMG)_2H$ (though not specifically formulated as such in the reference) is made by reaction of $H[Rh(DMG)_2Cl_2]$ with $NaBH_4$ [2].

The rôle of $Rh(DMG)_2H$ in the reductive amination of carbonyl compounds has been investigated [3].

The X-ray photoelectronic spectrum of the compound has been measured: the Rh $3d_{5/2}$ and N1s bonding energies are 309.5 and 400.8 eV, respectively [4]; see also $[Rh(DMG)_2]^-$ and Rh(DMG)(DMGH), p. 210.

$Rh(DMG)_2(OH)$ or **$Rh(DMGH)_2OH$** appears to be formed from $Rh(DMG)_2H$ and OH^- [2].

References:

[1] M. V. Klyuev, M. L. Khidekel', V. V. Strelets (Transition Metal Chem. [Weinheim] **3** [1978] 380/1). – [2] J. H. Weber, G. N. Schrauzer (J. Am. Chem. Soc. **92** [1970] 726/7). – [3] M. V. Klyuev,

M. L. Khidekel' (Transition Metal Chem. [Weinheim] **5** [1980] 134/9). – [4] M. V. Klyuev, B. G. Rogachev, Yu. M. Shul'ga, M. L. Khidekel' (Izv. Akad. Nauk SSSR Ser. Khim. **1979** 1869/70; Bull. Acad. Sci. USSR Div. Chem. Sci. **1979** 1732/3).

2.4.1.2.2 Complexes with N-Donor Ligands

[Rh(DMG)$_2$(NH$_3$)$_2$](NO$_3$). Treatment with 25% NH_4OH of the material obtained by refluxing ethanolic DMGH with the filtrate from the $RhCl_3 \cdot 3H_2O/AgNO_3$ reaction gives yellow-brown crystals of the compound [1].

For the crystal structure of $[Rh(DMG)_2(NH_3)_2]^+$ in $[Rh(DMG)_2Cl_2][Rh(DMG)_2(NH_3)_2]$, see p. 215, for the structure of the cation in $[Rh(DMG)_2(NH_3)_2]I$, see below.

[Rh(DMG)$_2$(NH$_3$)$_2$]Cl·5H$_2$O is isomorphous with its cobalt analogue; the X-ray powder diffraction patterns of both salts are reproduced in the paper [3].

The electronic absorption spectrum of $[Rh(DMG)_2(NH_3)_2]^+$ was measured in aqueous solution from 200 to 350 nm (reproduced in paper) at pH = 3 and (in the presence of KOH) at pH = 12 [4, 11].

With KOH it gives $Rh[(DMG)_2\text{-}H](NH_3)_2 \cdot \frac{1}{2}H_2O$ [4] (see p. 219).

[Rh(DMG)$_2$(NH$_3$)$_2$]I. This is made by addition of NH_4I to $[Rh(DMG)_2(NH_3)_2]Cl$ [2].

The crystals are monoclinic and a single crystal X-ray study shows them to belong to the space group B2/b-C^6_{2h}, a = 23.44(1), b = 11.677(8), c = 11.846(7) Å, γ = 97.87(4)°, Z = 8; density by flotation 2.07, calculated density 2.04 g/cm^3. The crystal structure is shown in **Fig. 13**. There are two non-equivalent $[Rh(DMG)_2(NH_3)_2]^+$ cations. The Rh has slightly distorted octahedral symmetries in both; in one the metal lies at a centre of symmetry while the other metal atom is located on a twofold axis. The Rh-N(DMG) distances vary from 1.94 to 2.04 Å and the Rh-NH$_3$ distances from 2.04 to 2.06 Å; both cations contain O-H···O bonds (2.66 and 2.67 Å) [2].

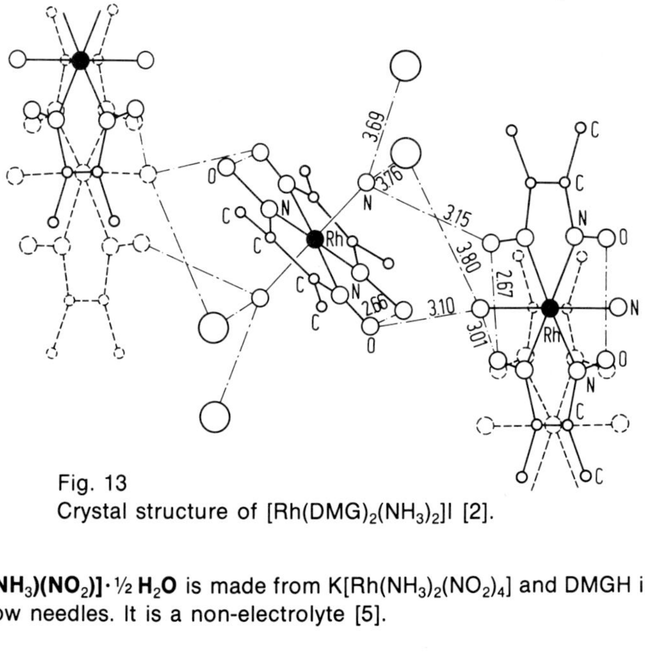

Fig. 13
Crystal structure of $[Rh(DMG)_2(NH_3)_2]I$ [2].

[Rh(DMG)$_2$(NH$_3$)(NO$_2$)]·½H$_2$O is made from $K[Rh(NH_3)_2(NO_2)_4]$ and DMGH in boiling water, and forms yellow needles. It is a non-electrolyte [5].

$[Rh(DMG)_2py]_2$ is made from $[Rh(DMG)_2]_2$ and pyridine in a nitrogen atmosphere. It forms yellow crystals. The structure is likely to involve a metal-metal bonded dimer with pyridine coordinated in the two positions *trans* to the Rh-Rh bond [6].

$[Rh(DMG)_2py_2](NO_3)$. This is made by heating with pyridine the material obtained by refluxing the filtrate from the $RhCl_3 \cdot 3H_2O/AgNO_3$ reaction with ethanolic DMGH. It is yellow [1]. With KOH it gives $Rh[(DMG)_2\text{-}H]py_2 \cdot \frac{1}{2}H_2O$ [4] (see p. 219).

$M[Rh(DMG)_2(NO_2)_2] \cdot 2H_2O$. The **ammonium** salt is made by boiling $Na_3[Rh(NO_2)_6]$ and DMGH together and adding NH_4Cl. The **guanidinium** and **$[Pt(NH_3)_4]^{2+}$** salts were also made by metathesis. It loses its water at 100 to 105°C. The nitro groups are believed to be *trans* to each other [7]. With Na_2SO_3 the sodium salt gives *trans*-$Na_3[Rh(DMG)_2(SO_3)_2]$ [9].

$Rh(DMG)_2(NO_2)(H_2O)$ is made from $(NH_4)[Rh(DMG)_2(NO_2)_2]$ and H_2SO_4 with heating. The compound forms pale yellow crystals. With conc. HCl, $H[Rh(DMG)_2Cl(NO_2)] \cdot H_2O$ is formed [8]. Kinetics of formation of $Rh(DMG)_2(NO_2)(H_2O)$ from $H[Rh(DMG)_2Cl(NO_2)]$ were studied [8]. The acid dissociation constants of the compound were determined; the pK_a is 3.98(7) [10].

References:

[1] O. A. Bologa, A. V. Ablov (Zh. Neorgan. Khim. **23** [1978] 1413/4; Russ. J. Inorg. Chem. **23** [1978] 778/9). – [2] A. A. Dvorkin, Yu. A. Simonov, A. V. Ablov, T. I. Malinovskii, L. A. Nemchinova, O. A. Bologa (Koord. Khim. **4** [1978] 138/42; Soviet J. Coord. Chem. **4** [1978] 111/4). – [3] G. P. Syrtsova, T. S. Bolgar' (Zh. Neorgan. Khim. **15** [1970] 1714/5; Russ. J. Inorg. Chem. **15** [1970] 881). – [4] A. V. Ablov, L. A. Nemchinova, O. A. Bologa (Zh. Neorgan. Khim. **22** [1977] 3384/5; Russ. J. Inorg. Chem. **22** [1977] 1849/50). – [5] V. V. Lebedinskii, I. A. Fedorov (Izv. Sekt. Platiny Drug. Blagorodn. Metal. Inst. Obshch. Neorgan. Khim. Akad. Nauk SSSR No. 22 [1948] 158/67 from C.A. **1951** 58).

[6] H. J. Keller, K. Seibold (Z. Naturforsch. **25b** [1970] 551/2). – [7] V. V. Lebedinskii, I. A. Fedorov (Ann. Sect. Platine Inst. Chim. Gen. [USSR] No. 15 [1938] 19/25 from C.A. **1939** 2060). – [8] G. P. Syrtsova, T. S. Bolgar' (Zh. Neorgan. Khim. **17** [1972] 3015/21; Russ. J. Inorg. Chem. **17** [1972] 1585/9). – [9] G. P. Syrtsova, T. S. Bolgar' (Zh. Neorgan. Khim. **18** [1973] 2706/11; Russ. J. Inorg. Chem. **18** [1973] 1438/40). – [10] G. P. Syrtsova, T. S. Bolgar' (Zh. Neorgan. Khim. **20** [1975] 2767/71; Russ. J. Inorg. Chem. **20** [1975] 1531/4).

[11] G. P. Syrtsova, T. S. Bolgar' (Zh. Neorgan. Khim. **14** [1969] 2425/8; Russ. J. Inorg. Chem. **14** [1969] 1272/4).

2.4.1.2.3 Chloro-Dimethylglyoxime Complexes

***trans*-$H[Rh(DMG)_2Cl_2]$**, sometimes formulated as $[Rh(DMG)(DMGH)Cl_2]$. This is made by reaction of $RhCl_3$ in slightly acid solution with DMGH in ethanol under reflux, and forms golden-yellow rods [1]; yields are reduced if the acidity is too great [2]; from the material obtained by treating $RhCl_3 \cdot 3H_2O$ with $AgNO_3$ and refluxing the filtrate with DMGH in ethanol, then by heating this material to 80°C [3]; by treatment of $RhCl_3$ or "$H_3[RhCl_6]$" with DMGH in 2:1 ethanol/water at an unspecified temperature [4]; from "$Rh(DMG)_3$" and HCl [1]; by reaction of $(NH_4)_2[Rh(NH_3)Cl_5]$ and DMGH in HCl [5].

It forms golden-yellow rods [1] or yellow crystals [2, 3]. It is only slightly soluble in water and in organic solvents, but dissolves with ease in NaOH or NH_4OH solutions and in solutions of sodium acetate or ammonium oxalate. It decomposes violently on heating [1].

The X-ray crystal structure of $H[Rh(DMG)_2Cl_2]$ shows the crystals (obtained from an aqueous solution made acid to pH = 2) to be orthorhombic, space group $Pca2_1$-C_{2v}^5 with Z = 4; a = 8.194(4), b = 14.622(3), c = 11.781(4) Å; the pyknometric density in toluene is 1.91 and the calculated density 1.93 g/cm³. The metal atom, see **Fig. 14**, has a distorted octahedral environment provided by two *trans* chloro ligands ($Rh\text{-}Cl_1$ = 2.330, $Rh\text{-}Cl_2$ = 2.336 Å) with four equatorial nitrogen atoms ($Rh\text{-}N_1$ = 2.021, $Rh\text{-}N_2$ = 2.011, $Rh\text{-}N_3$ = 1.983, $Rh\text{-}N_4$ = 2.028 Å). There are weak hydrogen bonds between the oxygen atoms of the oxime groups ($O_1 \cdots O_4$ 2.79 Å, $O_2 \cdots O_3$ 2.85 Å). The "outer" or acidic hydrogen atom is hydrogen-bonded to one of the oxygen atoms of the oxime groups to give a short bond ($O_1 \cdots O_3$ = 2.54, $O_1\text{-}H_1$ = 1.04, $O_3 \cdots H_1$, 1.69 Å, $O_1H_1O_3$ angle 136°). These bent hydrogen bonds provide the links in the chain structure of the complex [6].

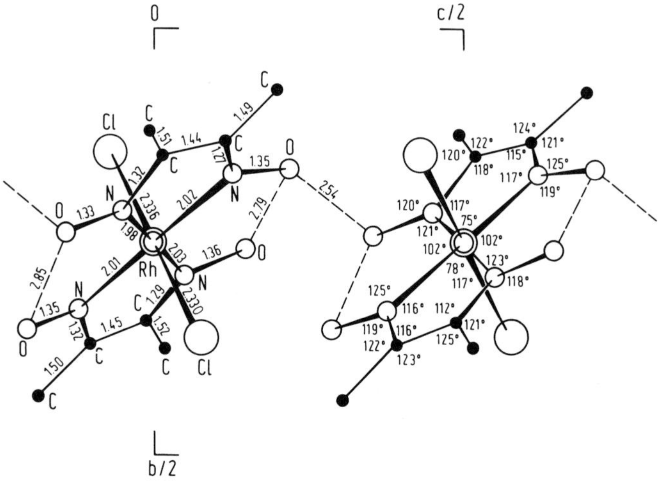

Fig. 14
X-ray crystal structure of $H[Rh(DMG)_2Cl_2]$ [6].

The infrared spectrum was measured from 700 to 4000 cm^{-1} and main absorptions listed [8]. The X-ray photoelectronic spectrum shows the following binding energies: Rh $3d_{5/2}$ = 310.4, N 1s = 401.2 eV [9]. The electronic spectrum was measured (200 to 400 nm, reproduced in paper) [20].

Electrochemical reduction of $[Rh(DMG)_2Cl_2]^-$ in dimethylformamide shows a single cathodic wave at $E_{1/2}$ = −0.87 V (versus standard calomel electrode) due possibly to formation of $[Rh(DMG)_2]^-$ [10]. Kinetics of aquation of $[Rh(DMG)_2Cl_2]^-$ were studied [21].

With PPh_3 and KOH in ethanol-water, $H[Rh(DMG)_2Cl_2]$ gives $Rh(DMG)_2(PPh_3)Cl$ [4]. It is reduced in alkaline aqueous ethanol to Rh(DMG)(DMGH), and the kinetics of this reaction were studied [2]. With $(NH_4)_2SO_3$, salts of $[Rh(DMG)_2(SO_3)_2]^{3-}$ can be obtained [11]. With NH_4X, $NH_4[Rh(DMG)_2ClX] \cdot 2H_2O$ is formed (X = Br, n = 2; X = I, n = 0.5; X = NCS, n = O); with thiourea, $Rh(DMG)_2Cl(CS(NH_2)_2)$ is formed [12]. With NaOH in C_2H_5OH, Rh(DMG)(DMGH) is formed [2].

***cis*-H[Rh(DMG)$_2$Cl$_2$]** is, it is said, obtainable as orange-red prisms mixed with crystals of the yellow *trans* isomer by aerial oxidation of "$H[Rh(DMG)_3]$" in HCl. No analytical data were presented for the complex [1].

Na[Rh(DMG)$_2$Cl$_2$] is made in solution from $Ag[Rh(DMG)_2Cl_2]$ and sodium chloride [1] or by treating, with conc. NaCl solution on a water-bath, the material obtained by refluxing with ethanolic DMGH, the filtrate obtained by treating $RhCl_3 \cdot 3H_2O$ with $AgNO_3$ [3]. The molar conductance of the salt is 65.5 $cm^2 \cdot \Omega^{-1} \cdot mol^{-1}$ [1].

***trans*-K_2[Rh(DMG)(DMGH)Cl_2]·2H_2O** is made from *trans*-H[Rh(DMG)$_2Cl_2$] and KOH at pH = 7 to 9.

The X-ray structure shows the crystals to be monoclinic, space group $P2_1/b\text{-}C^5_{2h}$; a = 7.601(3), b = 15.123(4), c = 15.403(4) Å, γ = 95.04(6)°, Z = 4. The rhodium atom has a distorted octahedral structure with *trans* chloro ligands. Bond lengths (in Å) are: Rh-Cl(1) = 2.355(4), Rh-Cl(2) = 2.364(5), Rh-N(1) = 2.00(1), Rh-N(2) = 2.02(1), Rh-N(3) = 2.05(1), Rh-N(4) = 2.07(1); see also **Fig. 15** according to [13].

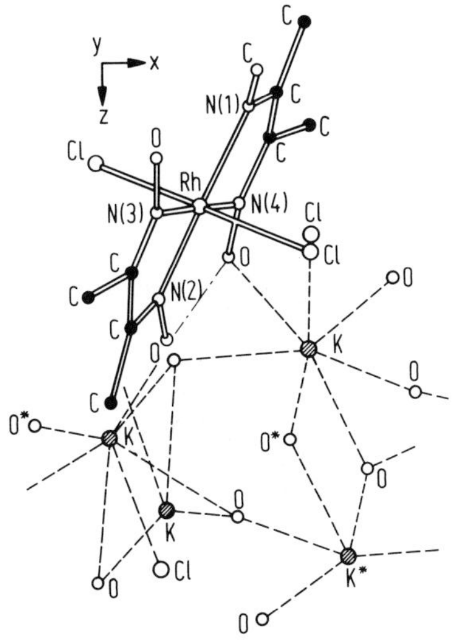

Fig. 15
X-ray crystal structure
of K_2[Rh(DMG)(DMGH)Cl_2]·2H_2O.

(NH_4)[Rh(DMG)$_2Cl_2$]. If this is boiled with NH_4X, NH_4[Rh(DMG)$_2$ClX] is formed (X = Br, I, NCS); thiourea gives Rh(DMG)$_2$Cl[CS(NH_2)$_2$] [12].

Rb[Rh(DMG)$_2Cl_2$]·H_2O. The X-ray crystal structure of the salt is monoclinic, space group $P2_1/b\text{-}C^5_{2h}$ with Z = 4, a = 12.1821(2), b = 9.334(1), c = 13.955(2) Å, γ = 97.10(5)°; measured density 2.03, calculated density 2.03 g/cm^3. The structure, shown in **Fig. 16**, p. 216, contains two centrosymmetric complex ions in which the rhodium atom has a distorted *trans*-octahedral configuration; Rh-Cl 2.329(3) and 2.340(3) Å, Rh-N 1.967 and 2.002 Å for one cation and 1.978, 1.982 Å for the other. There are two intra-complex hydrogen bonds in the equatorial plane O···O 2.67(6) Å. The chelate (NRhN) angle is close to 79° [7].

Ag[Rh(DMG)$_2Cl_2$] is made from H[Rh(DMG)$_2Cl_2$] and $AgNO_3$. It forms yellow crystals and is soluble in dilute HNO_3 [1].

[Rh(DMG)$_2$(NH_3)$_2$][Rh(DMG)$_2Cl_2$]·4H_2O. This is made by slow co-crystallisation of aqueous solutions of H[Rh(DMG)$_2Cl_2$] and [Rh(DMG)$_2$(NH_3)$_2$]Cl. The crystals are monoclinic, space group $P2_1/b\text{-}C^5_{2h}$ with Z = 2; a = 12.515(6), b = 16.894(8), c = 7.846(3) Å with γ = 104.75°; measured and calculated densities 1.74 and 1.75 g/cm^3, respectively. The complex anion is centrosymmetric with a *trans* configuration; Rh-Cl = 2.329(1), Rh-N(1) = 1.983(5) and Rh-N(2) = 1.997(4) Å. There are two equal asymmetric intra-complex hydrogen bonds (O···O = 2.68 Å). The cation is also centrosymmetric and has Rh-NH_3 distances of 2.078 Å with *trans*-ammonia ligands; for the DMG, Rh-N(1) is 2.004(5) and Rh-N(2) = 1.992(4) Å, the intra-complex O···O bond length is 2.680(7) Å [14].

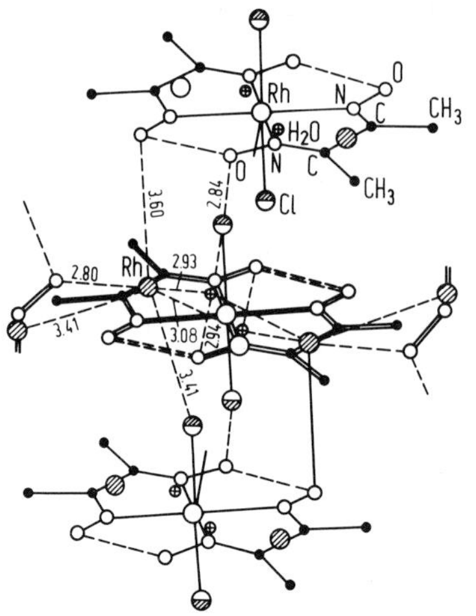

Fig. 16
X-ray crystal structure
of $Rb[Rh(DMG)_2Cl_2]\cdot H_2O$ [7].

$Rh(DMG)_2(NH_3)Cl$ is made by boiling together $Rh(NH_3)_3Cl_3$ and DMGH in aqueous solution. It is a non-electrolyte [5].

$Rh(DMG)_2Cl(H_2O)$ has been mentioned. With KBr it gives $H[Rh(DMG)_2ClBr]$ [16].

$H[Rh(DMG)_2Cl(NO_2)]\cdot H_2O$ is made from $Rh(DMG)_2(NO_2)(H_2O)$ and conc. HCl. It is readily soluble in warm water [15]. The anhydrous complex is made from $H[Rh(DMG)_2Cl_2]$ and KNO_2 in water at 110°C in a sealed ampoule for two hours, and is rust-coloured [16]. Kinetics of aquation of the complex to $Rh(DMG)_2(NO_2)(H_2O)$ were investigated [15].

$Rh(DMG)_2pyCl$ is made from $H[Rh(DMG)_2ClBr]$ and pyridine in a 1:1.5 mol ratio in a sealed ampoule at 110°C. The yellow complex was not obtained in a pure state [16], but a better product was obtained by boiling an aqueous solution of $[Rhpy_4Cl_2]Cl$ and DMGH in water for three days. The product is a non-electrolyte [5]. It is also made from $RhCl_3$ and DMGH in 1:3 boiling water/ethanol with pyridine. It forms yellow crystals [17]. The electronic absorption spectrum was measured from 200 to 350 nm (reproduced in paper) in acid and alkaline media [18].

$Rh(DMG)_2Cl$(aniline) was made in impure form contaminated with $Rh(DMG)_2Br$(aniline) by heating $H[Rh(DMG)_2BrCl]$ and aniline in a 1:1.5 mol mixture at 120°C for two hours. The mixture is moss-green in colour [16].

The electronic absorption spectrum was measured from 200 to 350 nm (reproduced in paper) in acid and alkaline media [18]. With KOH, $Rh[(DMG)_2\text{-}H](aniline)_2\cdot H_2O$ is formed [19]. Evidence for deprotonation to $[Rh(DMG_2\text{-}H)]Cl$-aniline at pH = 9 to 10 was given [18].

$Rh(DMG)_2Cl\cdot L$ (L = *p*-anisidine, *p*-toluidine, *m*-bromoaniline, *o*-chloroaniline). Electronic spectra of these (200 to 350 nm reproduced in paper) were measured in acid and alkaline media [16]. Evidence for deprotonation to $[Rh(DMG_2\text{-}H)]Cl\cdot L$ was given [18].

References:

[1] F. P. Dwyer, R. S. Nyholm (J. Proc. Roy. Soc. N.S. Wales **76** [1944] 266/70). – [2] J. D. Miller, F. D. Oliver (J. Chem. Soc. Dalton Trans. **1972** 2469/72). – [3] O. A. Bologa, A. V. Ablov (Zh. Neorgan. Khim. **23** [1978] 1413/4; Russ. J. Inorg. Chem. **23** [1978] 778/9). – [4] S. A. Shchepinov, E. N. Sal'nikova, M. L. Khidekel' (Izv. Akad. Nauk SSSR Ser. Khim. **1967** 2128/9; Bull. Acad. Sci. USSR Div. Chem. Sci. **1967** 2057). – [5] V. V. Lebedinskii, I. A. Fedorov (Izv. Sekt.

Platiny Drug. Blagorodn. Metal. Inst. Obshch. Neorgan. Khim. Akad. Nauk SSSR No. 22 [1948] 158/67 from C.A. **1951** 58).

[6] A. A. Dvorkin, Yu. A. Simonov, A. V. Ablov, O. A. Bologa, T. I. Malinovskii (Dokl. Akad. Nauk SSSR **217** [1974] 833/5; Doklady Chem. Proc. Acad. Sci. USSR **214/217** [1974] 539/41). – [7] Yu. A. Simonov, L. A. Nemchinova, O. A. Bologa (Kristallografiya **24** [1979] 829/31; Soviet Phys.-Cryst. **24** [1979] 476/7). – [8] J. P. Collman, H. F. Holtzclaw (J. Am. Chem. Soc. **80** [1958] 2054/6). – [9] M. V. Klyuev, B. G. Rogachev, Yu. M. Shul'ga, M. L. Khidekel' (Izv. Akad. Nauk SSSR Ser. Khim. **1979** 1869/70; Bull. Acad. Sci. USSR Div. Chem. Sci. **1979** 1732/3). – [10] M. V. Klyuev, M. L. Khidekel', V. V. Strelets (Transition Metal Chem. [Weinheim] **3** [1978] 380/1).

[11] G. P. Syrtsova, C. T. Tang, T. S. Bolgar' (Zh. Neorgan. Khim. **14** [1969] 2429/33; Russ. J. Inorg. Chem. **14** [1969] 1275/7). – [12] V. V. Lebedinskii, I. A. Fedorov (Izv. Sekt. Platiny Drug. Blagorodn. Metal. Inst. Obshch. Neorgan. Khim. Akad. Nauk SSSR No. 21 [1948] 157/63 from C.A. **1950** 10565). – [13] Yu. A. Simonov, A. A. Dvorkin, L. A. Nemchinova, O. A. Bologa, T. I. Malinovskii (Koord. Khim. **7** [1981] 125/31 from C.A. **94** [1981] No. 112833). – [14] Yu. A. Simonov, L. A. Nemchinova, A. V. Ablov, V. E. Zavodnik, O. A. Bologa (Zh. Strukt. Khim. **17** [1976] 142/6; J. Struct. Chem. [USSR] **17** [1976] 117/21). – [15] G. P. Syrtsova, T. S. Bolgar' (Zh. Neorgan. Khim. **17** [1972] 3015/21; Russ. J. Inorg. Chem. **17** [1972] 1585/9).

[16] A. V. Ablov, L. A. Nemchinova, O. A. Bologa (Zh. Neorgan. Khim. **23** [1978] 2745/9; Russ. J. Inorg. Chem. **23** [1978] 1520/2). – [17] M. E. Vol'pin, A. M. Yurkevich, L. G. Volkova, E. G. Chauser, I. P. Rudakova, I. Ya. Levitin, E. M. Tachkova, T. M. Ushakova (Zh. Obshch. Khim. **45** [1975] 164/9; J. Gen. Chem. [USSR] **45** [1975] 150/4). – [18] A. V. Ablov, L. A. Nemchinova, M. P. Filippov, O. A. Bologa (Zh. Neorgan. Khim. **22** [1977] 425/9; Russ. J. Inorg. Chem. **22** [1977] 231/3). – [19] A. V. Ablov, L. A. Nemchinova, O. A. Bologa (Zh. Neorgan. Khim. **22** [1977] 3384/5; Russ. J. Inorg. Chem. **22** [1977] 1849/50). – [20] G. P. Syrtsova, T. S. Bolgar' (Zh. Neorgan. Khim. **14** [1969] 2425/8; Russ. J. Inorg. Chem. **14** [1969] 1272/4).

[21] G. P. Syrtsova, G. I. Shpakov, V. N. Vakhtina (Zh. Neorgan. Khim. **27** [1982] 1745/9; Russ. J. Inorg. Chem. **27** [1982] 985/7).

For $RhCl_2(DMG)(DMGH)$ see *trans*-$H[RhCl_2(DMG)_2]$, p. 213; for $(NH_4)[Rh(DMG)_2ClBr]$, see below.

2.4.1.2.4 Bromo-Dimethylglyoxime Complexes

$H[Rh(DMG)_2Br_2]$. This is made by heating to 80°C with HBr the product of reaction of $RhCl_3 \cdot 3H_2O$ and $AgNO_3$ with refluxing ethanolic DMGH [1] or from $RhBr_3 \cdot nH_2O$ and DMGH [4] or, as the heptahydrate, from a rhodium sulphato complex and DMGH in KBr solution. It is orange yellow; kinetics of its aquation were studied [5].

$Na[Rh(DMG)_2Br_2]$ is made from NaBr and $H[Rh(DMG)_2Br_2]$ [1].

$Rh(DMG)_2Br(py)$ is made, mixed with the chloro analogue, by reaction of $H[Rh(DMG)_2ClBr]$ and pyridine in a 1:1.5 mol ratio at 110°C for two hours in a sealed ampoule [2].

$Rh(DMG)_2Br(aniline)$ is made, mixed with the chloro analogue, by treatment of $H[Rh(DMG)_2ClBr]$ and aniline in a 1:1.5 mol ratio in a sealed ampoule at 120°C for two hours [2].

$H[Rh(DMG)_2ClBr]$ is made from $Rh(DMG)_2Cl(H_2O)$ and KBr at 110°C in a sealed ampoule. With thiourea it gives $Rh(DMG)_2Br[CS(NH_2)_2]$; aniline or pyridine (L) gives $Rh(DMG)_2L \cdot Cl$ and $Rh(DMG)_2L \cdot Br$ [2].

$(NH_4)[Rh(DMG)_2ClBr] \cdot 2H_2O$ is made by boiling $(NH_4)[Rh(DMG)_2Cl_2]$ with NH_4Br [3].

References:

[1] O. A. Bologa, A. V. Ablov (Zh. Neorgan. Khim. **23** [1978] 1413/4; Russ. J. Inorg. Chem. **23** [1978] 778/9). – [2] A. V. Ablov, L. A. Nemchinova, O. A. Bologa (Zh. Neorgan. Khim. **23** [1978] 2745/9; Russ. J. Inorg. Chem. **23** [1978] 1520/2). – [3] V. V. Lebedinskii, I. A. Fedorov (Izv. Sekt. Platiny Drug. Blagorodn. Metal. Inst. Obshch. Neorgan. Khim. Akad. Nauk SSSR No. 21 **1948** 157/63 from C.A. **1950** 10565). – [4] F. P. Dwyer, R. S. Nyholm (J. Proc. Roy. Soc. N.S. Wales **78** [1944] 266/70). – [5] G. P. Syrtsova, G. I. Shpakov, V. N. Vakhtina (Zh. Neorgan. Khim. **27** [1982] 1745/9; Russ. J. Inorg. Chem. **27** [1982] 985/7).

2.4.1.2.5 Iodo-Dimethylglyoxime Complexes

$H[Rh(DMG)_2I_2]$ is made from an aqueous solution of $RhCl_3$ and KI heated to 80°C with DMGH in ethanol. It forms brown crystals. It is very slightly soluble in water to give an acidic solution [2]. It can also be made from a rhodium sulphato complex with DMGH and KI, and forms a brown heptahydrate. The kinetics of its aquation were studied [4].

$K[Rh(DMG)_2I_2]$ is made by reaction of KI in a water-bath with the material obtained from $RhCl_3 \cdot 3H_2O$, $AgNO_3$ and refluxing ethanolic DMGH. It forms yellow crystals. Its molar conductivity is 110 $cm^2 \cdot \Omega^{-1} \cdot mol^{-1}$ [1].

$Ag[Rh(DMG)_2I_2]$ is a red-brown material made from silver ions and $H[Rh(DMG)_2I_2]$. On heating in water it gives AgI [2].

$(NH_4)[Rh(DMG)_2ICl] \cdot 2H_2O$ is made from $NH_4[Rh(DMG)_2Cl_2]$ and NH_4I in boiling aqueous solution [3].

References:

[1] O. A. Bologa, A. V. Ablov (Zh. Neorgan. Khim. **23** [1978] 1413/4; Russ. J. Inorg. Chem. **23** [1978] 778/9). – [2] F. P. Dwyer, R. S. Nyholm (J. Proc. Roy. Soc. N.S. Wales **78** [1944] 266/70). – [3] V. V. Lebedinskii, I. A. Fedorov (Izv. Sekt. Platiny Drug. Blagorodn. Metal. Inst. Obshch. Neorgan. Khim. Akad. Nauk SSSR No. 21 **1948** 157/63 from C.A. **1950** 10565). – [4] G. P. Syrtsova, G. I. Shpakov, V. N. Vakhtina (Zh. Neorgan. Khim. **27** [1982] 1745/9; Russ. J. Inorg. Chem. **27** [1982] 985/7).

2.4.1.2.6 Sulphido-Dimethylglyoxime Complexes

$H[Rh(DMG)_2S_6]$ is made from "$Rh(DMG)_2Cl_2$" (presumably $H[Rh(DMG)_2Cl_2]$ is meant in the reference) and sodium polysulphide in alcohol, the mixture then being heated on a water-bath. It is yellow.

A polymeric structure $[Rh(DMG)_2S_6H]_n$ was suggested, with bridging $(S_6H)^-$ groups occupying *trans* positions about octahedrally coordinated rhodium.

L. Malatesta, F. Turner (Gazz. Chim. Ital. **72** [1942] 489/91).

For **sulphito** dimethylglyoxime complexes see "Rhodium" Suppl. Vol. B 1, 1982, pp. 150/3; for **phosphine, arsine** and **stibine** dimethylglyoxime complexes see "Rhodium" Suppl. Vol. B 3.

2.4.1.3 Complexes of Deprotonated Dimethylglyoxime

Rh[(DMG)$_2$-H] denotes a Rh(DMG)$_2$ grouping from which one proton has been removed.

Rh[(DMG)$_2$-H](NH$_3$)$_2$·½ H$_2$O is made from [Rh(DMG)$_2$(NH$_3$)$_2$]Cl and KOH in water.

The electronic absorption spectrum (200 to 350 nm, reproduced in paper) in water at pH = 8 was recorded.

Rh[(DMG)$_2$-H]py$_2$·½ H$_2$O is made from [Rh(DMG)$_2$py$_2$]Cl and KOH in water.

The infrared spectrum (200 to 1800 cm^{-1}, reproduced in paper) was measured and compared with that of its cobalt analogue.

Rh[(DMG)$_2$-H](aniline)$_2$·H$_2$O is made as yellow prisms from [Rh(DMG)$_2$(aniline)$_2$]Cl and KOH.

A. V. Ablov, L. A. Nemchinova, O. A. Bologa (Zh. Neorgan. Khim. **22** [1977] 3384/5; Russ. J. Inorg. Chem. **22** [1977] 1849/50).

2.4.1.4 Complexes with Substituted Glyoximes

Methylglyoxime = MGH (MG = [CH$_3$·C(NO)·CH(NOH)]); methylpropylglyoxime = MPGH (MPG = [CH$_3$C(NO)C(NOH)C$_3$H$_7$]$^-$); cyclohexane-1,2-dioneglyoxime and monomethyl-1,2-dioneglyoxime, α-furyldioxime, glyoxime, α-benzylglyoxime = (D-D)H; diphenylglyoxime = DPGM

H[Rh(MG)$_2$X$_2$]. The anhydrous **chloro** complex is made from RhCl$_3$·3H$_2$O and MGH in aqueous C$_2$H$_5$OH; it is dark yellow. The monohydrated **bromo** complex is made from RhBr$_3$·nH$_2$O and MGH in aqueous C$_2$H$_5$OH and is orange, while the dihydrated **iodo** complex is made from a Rh sulphato complex, KI and MGH in aqueous C$_2$H$_5$OH and is brown. The kinetics of aquation of the three complexes were studied [3].

NH$_4$[Rh(MG)$_2$(NO$_2$)$_2$] is made in similar fashion to *trans*-NH$_4$[Rh(DMG)$_2$(NO$_2$)$_2$], and forms fine pale-yellow crystals [1].

H[Rh(MG)$_2$(NO$_2$)Cl]·5H$_2$O forms fine yellow crystals and is made in similar fashion to H[Rh(DMG)$_2$(NO$_2$)Cl] from *trans*-Rh(MG)$_2$(NO$_2$)(H$_2$O) and conc. HCl. The kinetics of aquation and anation for the complex were studied [1].

Rh(MG)$_2$(NO$_2$)(H$_2$O) is made by aquation of H[Rh(MG)$_2$(NO$_2$)Cl] and forms fine bright yellow crystals [1].

NH$_4$[Rh(MPG)$_2$(NO$_2$)$_2$] is made as brown crystals in analogous fashion to NH$_4$[Rh(DMG)$_2$-(NO$_2$)$_2$] [1].

H[Rh(MPG)$_2$(NO$_2$)Cl] forms fine red-yellow crystals and is made in similar fashion to H[Rh(DMG)$_2$(NO$_2$)Cl] from Rh(MPG)$_2$(NO$_2$)(H$_2$O) and conc. HCl. The kinetics of its aquation and anation were studied [1].

Rh(MPG)$_2$(NO$_2$)(H$_2$O) is made by aquation of H[Rh(MPG)$_2$(NO$_2$)Cl]·5H$_2$O, and forms yellow crystals [1].

[Rh(D-D)$_2$]$_n$. Reaction of Rh$_2$(OAc)$_4$ with (D-D)H (cyclohexane-1,2-dioneglyoxime and monomethyl-1,2-dioneglyoxime) gives the black products [2].

$Rh_2(OAc)_4$(D-D). Reaction of $Rh_2(OAc)_4$ with α-furyldioxime or glyoxime (D-D)H gives in acetone red precipitates in which the glyoximates appear to coordinate in *trans* position to the metal-metal bond [2].

$(NH_4)[Rh(D\text{-}D)_2(NO_2)_2] \cdot 2H_2O$ is made from $Na_3[Rh(NO_2)_6]$ and α-benzil dioxime in aqueous solution at 100°C. It forms yellow crystals [4].

$Rh(D\text{-}D)_2(NO_2)(H_2O)$ is made from $NH_4[Rh(D\text{-}D)_2(NO_2)_2]$ and dilute H_2SO_4 (D-D = α-benzil dioxime). It forms pale yellow crystals. The kinetics of formation of this complex from $[Rh(D\text{-}D)_2(NO_2)Cl]^-$ were investigated [4].

$H[Rh(D\text{-}D)_2(NO_2)Cl] \cdot H_2O$ is made from $Rh(D\text{-}D)_2(NO_2)(H_2O)$ and conc. HCl (D-D = α-benzil dioxime). The kinetic and activation parameters of the equilibrium $[Rh(D\text{-}D)_2(NO_2)Cl]^- + H_2O \rightleftharpoons Rh(D\text{-}D)_2(NO_2)(H_2O) + Cl^-$ were studied [4].

$[Rh(D\text{-}D)_2L_2]Cl_2$ with (D-D)H = α-furyldioxime; L = NH_3, py, ½en, thiourea, aniline. These were made from $RhCl_3 \cdot nH_2O$ in ethanol and α-furyldioxime with the ligand L, followed by oxidation at room temperature with NaOCl [5].

$Rh(D\text{-}D)_2(NO_2)_2$. This is made from $RhCl \cdot nH_2O$ and α-furyldioxime with NO_2^- [5].

$Rh(DPGH)_2$ is made from $Rh_2(OAc)_4$ and the oxime in acetone under reflux. It is black. The compound is soluble in alcohol, tetrahydrofuran and pyridine to give intensely yellow solutions, but is insoluble in water. With pyridine it gives $[Rh(DPGH)_2py]_2$ [2].

$Rh(DPGH)_2py_2$ is made from $Rh(DPGH)_2$ and pyridine. It forms yellow crystals [2].

References:

[1] G. P. Syrtsova, T. S. Bolgar', V. C. Khien (Zh. Neorgan. Khim. **21** [1976] 1027/32; Russ. J. Inorg. Chem. **21** [1976] 562/5). – [2] H. J. Keller, K. Seibold (Z. Naturforsch. **25b** [1970] 552/4). – [3] G. P. Syrtsova, G. I. Shpakov, V. N. Vakhtina (Zh. Neorgan. Khim. **27** [1982] 1745/9; Russ. J. Inorg. Chem. **27** [1982] 985/7). – [4] G. P. Syrtsova, T. S. Bolgar' (Zh. Neorgan. Khim. **18** [1973] 2156/62; Russ. J. Inorg. Chem. **18** [1973] 1140/4). – [5] P. Spacu, M. Brezeanu, D. Roman-Vacarescu (Analele Univ. Bucuresti Ser. Stiint Nat. Chim. **13** [1964] 179/83 from C.A. **64** [1964] 12178).

For DMG complex with PPh_3 see "Rhodium" Suppl. Vol. B 3.

2.4.2 Biguanide Complexes

Biguanide (guanylguanide) = $H_2N\underset{\|}{C}NH \cdot \underset{\|}{C} \cdot NH_2$ = big
NH NH

2.4.2.1 Complexes of Unsubstituted Biguanide

$[Rh\,big_3]Y_3$. The heptahydrated **sulphate** is made from $RhCl_3 \cdot 4H_2O$ and biguanide sulphate in NaOH under reflux, and forms silky, light cream-coloured crystals, slightly soluble in cold water and organic solvents. It loses its water of crystallisation at 160°C. It is diamagnetic. The **chloride** is obtained from the sulphate with $BaCl_2$ and forms light cream crystals, soluble in water, diamagnetic, decomposing on boiling with dilute acids or alkalis. The **bromide** and **iodide** are made from the chloride with KBr and KI respectively, and are similar to the chloride. The **nitrate** is made from the sulphate with $Ba(NO_3)_2$ and is light cream in colour, fairly soluble in water, slightly hydrolysed by boiling water. The trihydrated **thiosulphate** is made from the nitrate with $Na_2S_2O_3$ and forms white silky crystals, losing their water of crystallisation at 120°C [1]. The **perchlorate** is made in solution from the chloride with $Ba(ClO_4)_2$ [2].

Optical isomers of $[Rh\,big_3]^{3+}$ have also been made. The (±)-[Rh big_3] chloro(+)-tartrate is made from [Rh big_3]Cl_3 and silver (+) tartrate; this racemic salt on fractional crystallisation gave (−)-[Rh big_3]-chloro(+)-tartrate and the (+)-[Rh big_3]-chloro(+)-tartrate. The former treated with $(NH_4)_2SO_4$ and the resulting sulphate treated with $BaCl_2$ gave (−)-[Rh big_2]Cl_3 while the latter gave (+)-[Rh big_3]Cl_3; the nitrates were similarly made from the corresponding sulphates and $Ba(NO_3)_2$ [3].

The electronic absorption spectrum of the chloride was measured (200 to 400 nm) and assignments proposed [4, 5]; also from 290 to 400 nm (reproduced in paper) [2].

The conductance of the nitrate is 147.9 $cm^2 \cdot \Omega^{-1} \cdot mol^{-1}$ [1].

The chloride decomposes at 320°C; decomposition is complete at 600°C [6]. Kinetics of the acid-catalysed dissociation of $[Rh\,big_3]^{3+}$ to $[Rh\,big_2(H_2O)_2]^{3+}$ have been studied [2].

$[Rh\,big_2Cl_2]Y \cdot nH_2O$. The dihydrated cream-yellow **chloride** is made from $RhCl_3 \cdot 3H_2O$ and biguanide hydrochloride in refluxing water with HCl at pH = 2.5. It is moderately soluble in water and ethanol, and loses its water of crystallisation at 40°C. The tetrahydrated **nitrate** forms fine yellow crystals and is made by metathesis of the chloride with $AgNO_3$; it is less soluble in water than the chloride. The **sulphate** is similarly made metathetically with Ag_2SO_4 and is intermediate in solubility between the chloride and the nitrate [6].

Decomposition of $[Rh\,big_2Cl_2]Cl$ begins at 305°C and is complete at 430°C, giving Rh_2O_3 [6]. Aquation of *cis*-$[Rh\,big_2Cl_2]^+$ to $[Rh\,big_2Cl(H_2O)]^{2+}$ was studied and thermodynamic parameters derived [7, 8].

With excess NaX in refluxing water, $[Rh\,big_2X_2]^+$ is formed (X = Br, I, N_3, NCS). With picolinic acid (pic) or o-phenanthroline (phen), $[Rh\,big_2pic]pic_2 \cdot 3H_2O$ and $[Rh\,big_2phen]Cl_3$ are formed [6].

$[Rh\,big_2Br_2]Y$, $[Rh\,big_2I_2]Y$. The bromide of the cream-coloured **bromo complex** is made from $[Rh\,big_2Cl_2]Cl$ and excess NaBr in water under reflux [6]; the tetrahydrated nitrate is obtained metathetically using $AgNO_3$ [7]. The iodide of the **iodo complex** is similarly made from $[Rh\,big_2Cl_2]Cl$ and NaI [6], and the tetrahydrated nitrate likewise [7].

Aquation of $[Rh\,big_2X_2]^+$ to $[Rh\,big_2X(H_2O)]^{2+}$ (X = Br, I) was studied and thermodynamic parameters evaluated [7].

***cis*-$[Rh\,big_2X_2]Y$** (X = NO_2, N_3, NCS). Reaction of excess $NaNO_2$ with $[Rh\,big_2Cl_2]^+$ in refluxing water followed by addition of HNO_3 gives the white **$[Rh\,big_2(NO_2)_2](NO_3)$**; the white chloride is made by anion exchange. Similar reaction of $[Rh\,big_2Cl_2]^+$ with excess NaN_3 gives the golden-yellow **$[Rh\,big_2(N_3)_2]Cl$**, and NaSCN with $[Rh\,big_2Cl_2]^+$ gives the orange crystalline complex **$[Rh\,big_2(NCS)_2](NCS)$** [6].

The infrared spectrum of the thiocyanato complex indicates that the NCS^- ligand is coordinated via the nitrogen atom; ν(CN) = 2110, 2042 cm^{-1}, ν(CS) = 810, 776 cm^{-1} [6].

$(big\,H_2)[Rh\,big\,Cl_4]_2$. This carmine-red material is made from $RhCl_3 \cdot 3H_2O$ and biguanide dihydrochloride in 10 M HCl under reflux for one hour. The salt is diamagnetic, readily soluble in water but insoluble in organic solvents [6].

$[Rh\,big_2X(H_2O)]^{2+}$ (X = Cl [7, 8], Br [8], I [8]). Kinetics of the formation of these by aquation of $[Rh\,big_2X_2]^+$ and of their aquation to $[Rh\,big_2(H_2O)_2]^{3+}$ have been studied [7, 8].

[Rh big$_2$(H$_2$O)$_2$]$^{3+}$. Kinetics of the formation of this species by aquation of [Rh big$_3$]$^{3+}$ has been studied, and it was made in solution by digesting [Rh big$_2$Cl$_2$]Cl with $AgClO_4$ solution. The electronic spectrum was measured (290 to 370 nm, reproduced in paper) [2].

[Rh big$_2$pic]·pic$_2$·3H$_2$O, [Rh big(phen)$_2$](ClO$_4$)$_3$, [Rh big$_2$phen]Cl$_3$. These are made by reaction of [Rh big$_2$Cl$_2$]Cl·2H$_2$O with picolinic acid (pic) or o-phenanthroline (phen) in aqueous solution at pH = 7; the perchlorate is made in the presence of $HClO_4$. The picolinic acid complex forms pale yellow needles, sparingly soluble in water; the phenanthroline perchlorate is an amorphous material, moderately soluble in water, while the phenanthroline chloride is pale yellow and readily soluble in water [6].

References:

[1] A. I. P. Sinha, S. P. Ghosh (J. Indian. Chem. Soc. **38** [1961] 179/81). – [2] D. Banerjea, B. Chattopadhyay (J. Inorg. Nucl. Chem. **36** [1974] 2351/3). – [3] S. P. Ghosh, A. I. P. Sinha (J. Indian Chem. Soc. **40** [1963] 249/52). – [4] D. Sen (J. Chem. Soc. A **1969** 2900/3). – [5] D. Sen, C. Saha (J. Indian Chem. Soc. **54** [1977] 127/35).

[6] S. P. Ghosh, P. Bhattarcharjee (J. Inorg. Nucl. Chem. **32** [1970] 573/7). – [7] B. Chakravarty, A. K. Sil (Inorg. Chim. Acta **24** [1977] 105/8). – [8] B. Chakravarty, P. K. Das, A. K. Sil (Inorg. Chim. Acta **30** [1978] 149/53).

2.4.2.2 Complexes with Bidentate Substituted Biguanides

R-big = dimethylbiguanide, $N_5C_4H_{11}$; trimorpholinebiguanide = $N_5C_6H_{13}O$;
N^1-phenylbiguanide = $N_5C_8H_{11}$; N-p-chlorophenylbiguanide = $N_5C_8H_{10}Cl$;
N^1-o-tolylbiguanide = $N_5C_9H_{13}$; N^1-p-chlorophenyl-N^5-isopropylbiguanide (paludrine) = $N_5C_{11}H_{16}Cl$

[Rh(R-big)$_3$]Y$_3$. The **chlorides** are made from aqueous solutions of Na$_3$[RhCl$_6$] and the biguanide hydrochloride in a 1:3 molar ratio on a water bath. The **chlorotartrate** of the paludrine complex is made from [Rh(N$_5$C$_{11}$H$_{16}$Cl)$_3$]Cl$_3$ and d-silver tartrate, and the optical isomers are separated by fractional crystallisation. The **[Cr(SCN)$_6$]$^{3-}$** and **nitrate** salts are obtained from the chloro-tartrate and K$_3$[Cr(SCN)$_6$] or $AgNO_3$; the nitrate salt was resolved into its optical isomers. The **[RhCl$_6$]$^{3-}$** salt of the paludrine complex is made directly from Na$_3$[RhCl$_6$] and paludrine in ethanol. Colours are [1]:

[Rh(N$_5$C$_4$H$_{11}$)$_3$]Cl$_3$	light yellow	[Rh(N$_5$C$_{11}$H$_{16}$Cl)$_3$](NO$_3$)$_3$	yellow
[Rh(N$_5$OC$_6$H$_{13}$)$_3$]Cl$_3$	yellow	[Rh(N$_5$C$_{11}$H$_{16}$Cl)$_3$][RhCl$_6$]	pink
[Rh(N$_5$C$_{11}$H$_{16}$Cl)$_3$]Cl$_3$	yellow	[Rh(N$_5$C$_{11}$H$_{16}$Cl)$_3$][Cr(SCN)$_6$]	pink-violet
($\pm$)[Rh(N$_5$C$_{11}$H$_{16}$Cl)$_3$]Cl(C$_4$H$_4$O$_6$)	yellow		

Electronic spectra were measured (200 to 500 nm) for the chlorides in CH_3OH solution (200 to 600 nm, reproduced in paper) and assignments proposed [1].

[Rh(R-big)$_3$]Cl$_4$. The paludrine complex is made from $RhCl_3 \cdot 4H_2O$ and the hydrochloride of the ligand in water-ethanol with $NaClO_2$, the pH being maintained at 7. The complex forms brown crystals [2].

The magnetic moment is μ_{eff} = 1.68 B.M. at 293 K. The electronic absorption spectrum in CH_3OH was measured (240 to 570 nm, reproduced in paper) [2].

***trans*-[Rh(R-big)$_2$Cl$_2$]Y.** The **chloride** is made from $RhCl_3 \cdot 4H_2O$ and the ligand in 1:2 molar proportions in aqueous solution under reflux [3]. For paludrine two forms of [Rh(R-big)$_2$Cl$_2$]Cl were obtained by this method: cooling of the resulting solution gave a yellow complex, thought to be the *cis* form, whereas the mother liquors gave an orange complex, thought to be the *trans* isomer. The yellow form is more soluble in alcohol than the orange [1, 3].

The following compounds were prepared:

$[Rh(N_5C_8H_{10}Cl)_2Cl_2]Cl$	yellow	[3]
$[Rh(N_5C_8H_{10}Cl)_2Cl_2]Cl_2$	brown	[3]
$[Rh(N_5C_8H_{11})_2Cl_2]Cl$	yellow	[3]
$[Rh(N_5C_8H_{11})_2Cl_2]Cl_2$	brown	[3]
$[Rh(N_5C_9H_{13})_2Cl_2]Cl$	yellow	[3]
$[Rh(N_5C_9H_{13})_2Cl_2]Cl_2$	brown	[3]
$[Rh(N_5C_{11}H_{16}Cl)_2Cl_2]$-$[Rh(N_5C_{11}H_{16}Cl)Cl_4]$	orange	[1]
$(N_5C_{11}H_{17}Cl)$-$[Rh(N_5C_{11}H_{16}Cl)Cl_4]$	light yellow	[1]
$[Rh(N_5C_{11}H_{16}Cl)_2Cl_2]Cl$	yellow	[3]; yellow and orange forms [1, 3] (see above)
$[Rh(N_5C_{11}H_{16}Cl)_2Cl_2]Cl_2$	brown	[2, 3]
$[Rh(N_5C_{11}H_{16}Cl)_2Br_2]Br$	dark brown	[3]
$[Rh(N_5C_{11}H_{16}Cl)_2I_2]I$	dark brown	[3]
$[Rh(N_5C_{11}H_{16}Cl)_2(NO_2)_2]NO_2$	yellow-brown	[1, 3]
$[Rh(N_5C_{11}H_{16}Cl)_2$-$(NCO)_2](NCO)$ misprinted as $[Rh(N_5C_{11}H_{12}Cl)(NCO)](NCO)$ in [3]	brown	[3]
$[Rh(N_5C_{11}H_{16}Cl)_2$-$(NCS)_2](NCS)$	brown	[3]
$[Rh(N_5C_{11}H_{16}Cl)_2(N_3)_2](N_3)$	brown	[3]

Infrared spectra of the *cis* and *trans* forms of the paludrine complex $[Rh(N_5C_{11}H_{13}Cl)_2Cl_2]Cl$ (440 to 4000 cm^{-1}) and electronic spectra of the yellow and orange forms of the paludrine complex $[Rh(R\text{-big})_2Cl_2]Cl$ were measured (200 to 550 nm, reproduced in paper) [1].

$[Rh(R\text{-big})_2Cl_2](NO_3)\cdot4H_2O$. Complexes with R-big = methyl, ethyl or phenyl-biguanide were made from $RhCl_3\cdot nH_2O$ and the biguanide chloride in HCl, and the nitrates obtained by metathesis with $AgNO_3$.

Aquation of these species in water and in H_2O-CH_3OH to $[Rh(R\text{-big})_2Cl(H_2O)]^{2+}$ and $[Rh(R\text{-big})_2(H_2O)_2]^{3+}$ was studied; rate and thermodynamic data were obtained [4].

$[Rh(R\text{-big})_2X_2]X$ (X = Br, I, NCO, NCS, N_3). These are made from $RhCl_3\cdot4H_2O$ and the ligand in 1:2 molar ratio in water-ethanol under reflux (X = Cl); for X = Br, I, NCO, NCS and N_3 the chloro complex is refluxed with KX in dimethylformamide. The **nitro** complex of paludrine (X = NO_2) can be similarly made, or directly from $RhCl_3\cdot4H_2O$ and the ligand in 1:2 molar proportions with excess KNO_2. Colours are listed in the table above. The complexes are thought to have a *trans* configuration. Electronic spectra were measured (200 to 600 nm, reproduced in paper) for $[Rh(N_5C_{11}H_{13}Cl)X_2]X$ (X = Br, I, N_3) [3].

$(R\text{-big}\cdot H)[Rh(R\text{-big})]Cl_4$. The paludrine complex is made from $K_2[RhCl_5(H_2O)]$ and paludrine at a 1:2 ratio in water at 35°C. It is orange-yellow [1].

$[Rh(R\text{-big})_2Cl_2][Rh(R\text{-big})]Cl_4$. The paludrine complex is made from $RhCl_3\cdot nH_2O$, and paludrine. It is orange [1].

$[Rh(R\text{-big})_2Cl_2]Cl_2$. These are made from $RhCl_3\cdot4H_2O$ and the ligand in 1:2 molar excess in the presence of $NaClO_2$ and NaOH to maintain the pH at 7 to 8, the reaction being carried out at 70 to 80°C in dimethylformamide, or by oxidation of $[Rh(R\text{-big})_2Cl_2]Cl$ in dimethylformamide with $NaClO_2$ maintained at pH = 7 to 8 with KOH [3]. The chloro complex of paludrine can also be made from $RhCl_3\cdot4H_2O$ and the ligand in water-CH_3OH with $NaClO_2$ at pH = 3 to 4 [3]. Colours are listed in the table above.

The magnetic moment of the paludrine complex (R = $N_5C_{11}H_{16}Cl$) is μ_{eff} = 1.79 B.M. at 293 K [2].

$[Rh(R\text{-}big)_2X_2]Y_n$ ($X=NH_3$, py, NO_2). The **ammine** ($X=NH_3$, $Y=Cl$, $n=4$) of the paludrine complex is made by reaction of $RhCl_3 \cdot 4H_2O$ with paludrine in C_2H_5OH with NH_4OH in the presence of $NaClO_2$. It is brown. The corresponding **pyridine** complex ($X=py$, $Y=Cl$, $n=4$) is similarly made, using paludrine hydrochloride and pyridine. It is brown. The hexahydrated **nitro** complex ($X=Y=NO_2$, $n=1$) is prepared from $RhCl_3 \cdot 4H_2O$ and $NaNO_2$ in H_2O/C_2H_5OH with $NaClO_2$. It is a brown material, sparingly soluble in water [2].

The magnetic moments μ_{eff} at 293 K are 1.65 B.M. for the ammine, and 1.80 B.M. for the nitro complex; at 294 K the moment of the pyridine complex is 1.82 B.M. The electron spin resonance (ESR) spectra of the polycrystalline materials showed signals without hyperfine structure at room temperature. The electronic absorption spectra were measured (240 to 560 nm, that for the pyridine complex reproduced in paper) [2].

References:

[1] P. Spacu, C. Gheorghiu (J. Less-Common Metals **15** [1968] 331/9). – [2] C. Gheorghiu (Rev. Roumaine Chim. **13** [1968] 1181/4). – [3] C. Gheorghiu, L. Burdea (Analele Univ. Bucaresti Chim. **21** [1972] 53/61).

2.4.2.3 Complexes with Tetradentate Substituted Biguanides

R-big = ethylenedibiguanide, $N_{10}C_6H_{16}$ and piperazinedibiguanide, $N_{10}C_8H_{18}$

$[Rh_2(R\text{-}big)_3](SO_4)_3 \cdot nH_2O$. These are made by warming $RhCl_3 \cdot 3H_2O$ and the ligand acid sulphates in aqueous solution in 2:3 molar proportions on a steam bath while gradually increasing the pH to 7 by adding KOH. The ethylenedibiguanide complex (n = 10) is yellow and the piperazinedibiguanide complex (n = 11) is pale yellow [1, 2]

$[Rh(R\text{-}big)X_2]Y \cdot nH_2O$. These (see table) are made from the dichloro complex by refluxing with NaX; for Y = Cl anion exchange was used, and for $Y=ClO_4$ recrystallisation from $HClO_4$ or $NaClO_4$ solutions was used. The **chloro** complexes are made from $RhCl_3 \cdot 3H_2O$ with the ligand hydrochloride in 1:1 molar proportions, the pH being adjusted to ~3 with NH_4OH [1, 2].

Colours of complexes $[RhLX_2]Y \cdot nH_2O$:

$[RhLCl_2]Cl \cdot 2H_2O$	yellow	$[RhL'(NO_2)_2](ClO_4)$	pale yellow
$[RhL'Cl_2]Cl \cdot 3H_2O$	yellow	$[RhL(N_3)_2]Cl$	golden-yellow
$[RhLBr_2](ClO_4)$	yellow	$[RhL'(N_3)_2](N_3)$	orange
$[RhL'Br_2]Br \cdot 6H_2O$	brown-yellow	$[RhL'(N_3)_2]Cl \cdot 2H_2O$	yellow
$[RhLI_2]I$	brown	$[RhL(NCS)_2](SCN)$	orange-yellow
$[RhL'I_2]I$	brown	$[RhL'(NCS)_2](SCN) \cdot 4H_2O$	orange-yellow
$[RhL(NO_2)_2](NO_2)$	pale yellow		

L = ethylenedibiguanide, $N_{10}C_6H_{16}$; L' = piperazinedibiguanide, $N_{10}C_8H_{18}$.

Infrared spectra of $[RhL(NO_2)_2](NO_2)$, $[RhL'(NO_2)_2](ClO_4)$, $[RhL(NCS)_2](SCN)$, $[RhL'(NCS)_2](SCN)$, $[RhL(N_3)_2]Cl$ and $[RhL'(N_3)_2]Cl$ were measured ($\nu(NO_2)$, $\nu(CN)$ and $\nu(N_3)$ frequencies listed). Electronic absorption spectra of $[RhLCl_2]Cl$ and $[RhL'Cl_2]Cl$ were measured and the $^1T_{1g} \leftarrow {}^1A_{1g}$ transitions identified; the conductances were 160.0 and 143.4 $cm^2 \cdot \Omega^{-1} \cdot mol^{-1}$, respectively [1, 2].

References:

[1] B. Chakravarty, P. K. Das, A. K. Sil (Inorg. Chim. Acta **30** [1978] 149/53). – [2] S. P. Ghosh, P. Bhattacharjee, A. Mishra (J. Less-Common Metals **40** [1975] 97/101).

2.5 Unsaturated Cyclic Polydentate Nitrogen Donors

2.5.1 Complexes with Porphyrins

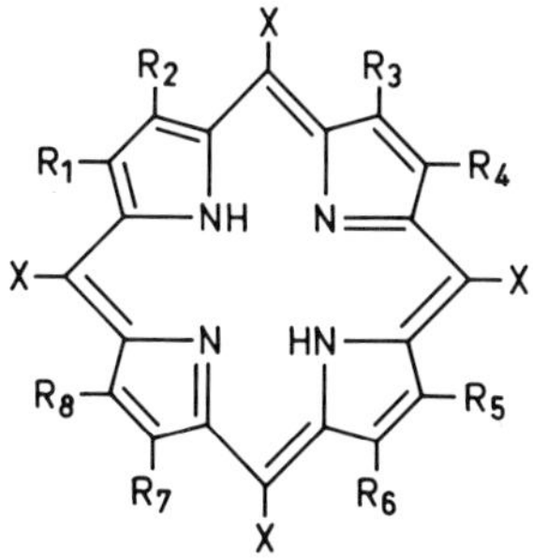

Porphyrin: R_1 to R_8, X = H;
etioporphyrin: $R_1 = R_3 = R_5 = R_7 = CH_3$, $R_2 = R_4 = R_6 = R_8 = C_2H_5$, X = H;
mesoporphyrin: $R_1 = R_3 = R_5 = R_8 = CH_3$, $R_2 = R_4 = C_2H_5$, $R_6 = R_7 = CH_2CH_2CO_2H$, X = H;
mesoporphyrin dimethylester: $R_1 = R_3 = R_5 = R_8 = CH_3$, $R_2 = R_4 = C_2H_5$, $R_6 = R_7 = CH_2CH_2$-CO_2CH_3, X = H; mesoporphyrin diethylester: $R_1 = R_3 = R_5 = R_8 = CH_3$, $R_2 = R_4 = C_2H_5$, $R_6 = R_7 = CH_2CH_2CO_2C_2H_5$, X = H;
hematoporphyrin: $R_1 = R_3 = R_5 = R_8 = CH_3$, $R_2 = R_4 = CH(OH)CH_3$, $R_6 = R_7 = CH_2CH_2CO_2H$, X = H;
octaethylporphyrin: R_1 to $R_8 = C_2H_5$, X = H;
tetraphenylporphyrin: R_1 to R_8 = H, $X = C_6H_5$; tetra-p-sulphonatophenylporphyrin: R_1 to R_8 = H, $X = C_6H_4$-SO_3H;
tetrapyridylporphyrin: R_1 to R_8 = H, X = p-C_5H_4N.

General Literature:

B. D. Berezin, Coordination Compounds of Porphyrins and Phthalocyanines, Wiley, London 1981.

L. J. Boucher, Coordination Chemistry of Porphyrins, in: G. A. Melson, Coordination Chemistry of Macrocyclic Compounds, Plenum, New York 1979.

D. Dolphin, The Porphyrins, Vol. 1/7, Academic Press, New York 1978/79.

K. M. Smith, Porphyrins and Metalloporphyrins, Elsevier, Amsterdam 1975.

J. E. Falk, Porphyrins and Metalloporphyrins, Elsevier, Amsterdam 1964.

Remarks. Much of the published work on rhodium porphyrin complexes involves initial formation of the binuclear dirhodium(I) tetracarbonyl species (structure I, X = $Rh(CO)_2$) from $\{RhCl(CO)_2\}_2$ and free porphyrin, and the subsequent conversion of these products into organorhodium(III) porphyrin species (structure II, below) by oxidative addition of organic halides, carboxylic acid anhydrides and other organic reagents.

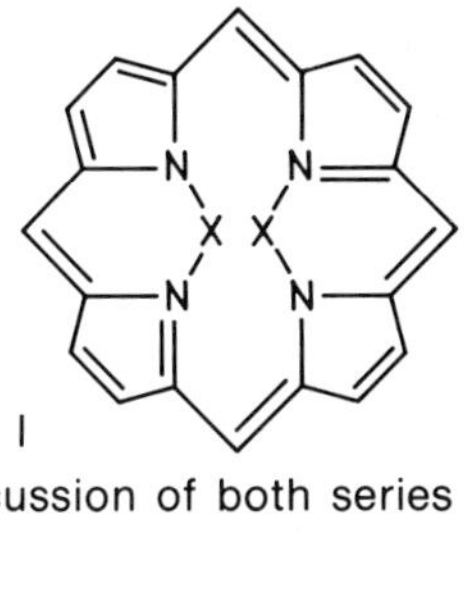

I

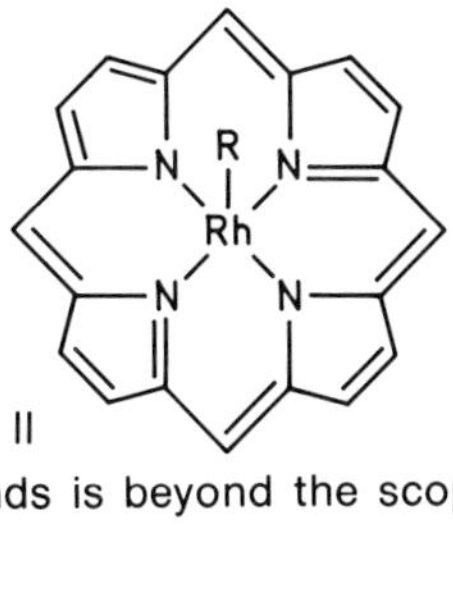

II

A general discussion of both series of compounds is beyond the scope of this volume.

2.5.1.1 Complexes with Etioporphyrin

$C_{32}H_{38}N_4$ = etioporphyrin (1,3,5,7-tetramethyl-2,4,6,8-tetraethylporphyrin) = epH_2

[Rh(ep){NH(CH$_3$)$_2$}$_2$]Cl·2H$_2$O is prepared by heating a mixture of hydrated $RhCl_3$ and free base etioporphyrin under reflux in dimethyl formamide for ~18 h. The mixture is then diluted with water, filtered (to remove excess free ligand), evaporated to dryness and chromatographed (silica gel/chloroform) to yield a dark orange band which is extracted with methanol and crystallised from methanol-chloroform solution as dark brown platelets [1, p. 4822].

The crystals are monoclinic, space group P2/n, lattice parameters a = 14.446(6), b = 10.943(7), c = 12.319(6) Å, β = 105.68(1)°, D_{calc} = 1.32 g/cm^3, Z = 2. The octahedrally coordinated rhodium atom is situated in the plane of the planar porphyrin rings, average bond lengths are Rh-N(ring) = 2.038(6) Å, Rh-N(amine) = 2.090(8) Å [1, pp. 4823/6].

The electronic absorption spectrum with λ_{max} = 546, 514, 397 and 342 nm in CH_2Cl_2 solution at 298 K and emission spectrum with λ_{max} = 733, 657, 600 and 552 nm in 2-methyltetrahydrofuran solution at 298 K and glass at 77 K have been recorded and are reproduced (750 to 300 nm) in the paper [1, pp. 4227/8].

Rh(ep)Cl·2H$_2$O is prepared by heating together free etioporphyrin, {RhCl(CO)$_2$}$_2$ and NaO_2CCH_3 in glacial acetic acid under reflux for 3 h, then pouring the reaction mixture into water. It is extracted from the aqueous solution with chloroform, chromatographed (alumina/chloroform-ethanol) and crystallised (chloroform-ethanol) as red brown needles [2]. The electronic spectrum (λ_{max} = 545, 513, 394, 297, 285 nm) and ^{1}H NMR spectrum have been recorded [2].

Rh(ep)I·H$_2$O is prepared by addition of iodine to a mixture of {RhCl(CO)$_2$}$_2$, free base porphyrin and NaO_2CCH_3 in chloroform [4], or to a solution of preformed {Rh(CO)$_2$}$_2$(ep) in chloroform [3, 4]. The solution which changes from brown to bluish-green then red is chromatographed (silica/chloroform or alumina/chloroform) and the red eluate evaporated then crystallised (chloroform/light petroleum) to yield the complex as purple plates. The compound darkens at 150°C and has a melting point in excess of 300°C. The mass spectrum with m/e = 706 (M^+) and 579 (M − I) and ^{1}H NMR spectrum have been recorded [4].

Rh(ep)(O$_2$CCH$_3$)·H$_2$O is prepared by heating together etioporphyrin, {RhCl(CO)$_2$}$_2$ and sodium acetate in glacial acetic acid under reflux for 5 h. It is isolated as scarlet needles by pouring the reaction mixture into water, extracting the aqueous solution with chloroform, chromatographing the extract (alumina/chloroform) and crystallising (chloroform-ethanol) [2]. The electronic spectrum with λ_{max} = 548, 515, 397, 285 nm, λ_{infl} = 345, 332 nm in $CHCl_3$ solution [2, 5] and NMR spectrum have been recorded: δ = 9.4 (s, 4 × meso-H), 3.9 (q, 4 × C$\underline{H}_2$CH$_3$); 3.4 (s, 4 × CH$_3$); 1.8 (t, 4 × CH$_2$C$\underline{H}_3$); −0.4 (s, CH$_3$CO); −3.2 (m, H$_2$O) [2].

Deuteration at the *meso* sites is achieved by shaking a mixture of Rh(ep)(O$_2$CCH$_3$)(H$_2$O) and CF_3CO_2D in chloroform for 20 min and has been confirmed by ^{1}H NMR [2].

References:

[1] L. K. Hanson, M. Gouterman, J. C. Hanson (J. Am. Chem. Soc. **95** [1973] 4822/9). – [2] R. Grigg, G. Shelton, A. Sweeney, A. W. Johnson (J. Chem. Soc. Perkin Trans. I **1972** 1789/99, 1797). – [3] A. M. Abeysekera, R. Grigg, J. Trocha-Grimshaw, V. Viswanatha, T. J. King (Tetrahedron Letters **1976** 3189/92). – [4] A. M. Abeysekera, R. Grigg, J. Trocha-Grimshaw, V. Viswanatha (J. Chem. Soc. Perkin Trans. I **1977** 1395/403, 1402). – [5] R. Grigg, A. Sweeney, A. W. Johnson (J. Chem. Soc. Chem. Commun. **1970** 1237/8).

2.5.1.2 Complexes with Mesoporphyrin

$C_{34}H_{38}N_4O_4$ = mesoporphyrin (1,3,5,8-tetramethyl-2,4-diethyl-6,7-di(ethyl-2-carboxyl)porphyrin) = mpH_2

Rh(mp). This rhodium(II) complex is observed as a separate band during the chromatographic purification of Rh(mp)Cl (see below). The electronic spectrum with λ_{max} = 551.5, 521 and 406 nm (Soret) and ESR spectrum (g = 2.01) have been reported.

Rh(mp)Cl is prepared by hydrolysis (1% KOH, methanol) of the corresponding dimethylester [Rh(mp-dimethylester)Cl] and can be purified by elution with pyridine/hexane from a polyamide (CC 66) column. The electronic spectrum with λ_{max} = 549, 518, 405.5 nm (Soret) has been reported.

T. S. Srivastava, T. Yonetani (Chromatographia **8** [1975] 124/8).

2.5.1.3 Complex with Mesoporphyrin Dimethylester, Rh(mp-dme)Cl

$C_{36}H_{42}N_4O_4$ = mesoporphyrin dimethylester (1,3,5,8-tetramethyl-2,4-diethyl-6,7-di(ethyl-2-carboxymethyl)porphyrin) = mp-dme·H_2

This compound is prepared by heating a mixture of the free ligand and $\{RhCl(CO)_2\}_2$ under reflux in glacial acetic acid for 3 to 4 h. No other details are given.

T. S. Srivastava, T. Yonetani (Chromatographia **8** [1975] 124/8).

2.5.1.4 Complex with Mesoporphyrin Diethylester, Rh(mp-dee)Cl·2H_2O

$C_{38}H_{46}N_4O_4$ = mesoporphyrin diethylester (1,3,5,8-tetramethyl-2,4-diethyl-6,7-di(ethyl-2-carboxyethyl)porphyrin) = mp-deeH_2

This compound is prepared by mixing together solutions of $\{RhCl(CO)_2\}_2$ or $\{RhCl(C_8H_{14})_2\}_2$ in ethanol and free ligand in boiling glacial acetic acid, then adding NaO_2CCH_3 or Na_2CO_3 and boiling the mixture for 2 h. The solvents are distilled off under reduced pressure and the residue is extracted into chloroform, filtered, evaporated and chromatographed (alumina) to yield dark brown microcrystals [1, 2], which can be recrystallised from concentrated H_2SO_4 [2]. The electronic spectrum has been recorded in chloroform solution (λ_{max} = 544, 512, 397 nm) [1, 2] and in eight other organic solvents [2]. The infrared spectrum, mass spectrum (m/e = 723) and magnetic measurements (μ = 0.4 to 0.5 B.M.) have been reported [1, 2].

Reactions with NaN_3, NaCN and NaSCN have been reported but products were not characterised. Treatment with ethanolic sodium hydroxide gave water soluble rhodium mesoporphyrin products [2].

References:

[1] E. B. Fleischer, N. Sadasivan (J. Chem. Soc. Chem. Commun. **1967** 159/60). – [2] N. Sadasivan, E. B. Fleischer (J. Inorg. Nucl. Chem. **30** [1968] 591/601).

2.5.1.5 Complex with Hematoporphyrin, $[Rh(hp)(H_2O)_2]^+$

$C_{34}H_{38}N_4O_6$ = hematoporphyrin (1,3,5,8-tetramethyl-2,4-di(1-hydroxyethyl)-6,7-di(ethyl-2-carboxyl)porphyrin) = hpH$_2$

Preparation of this species from $\{RhCl(CO)_2\}_2$ and free hematoporphyrin has been mentioned and the value of pK_{OH} for the reaction: $[Rh(hp)(H_2O)_2]^+ + H_2O \overset{K_{OH}}{\rightleftharpoons} [Rh(hp)(OH)(H_2O)] + H_3O^+$ has been given as 5.1 ± 0.1.

G. McLendon, M. Bailey (Inorg. Chem. **18** [1979] 2120/3).

2.5.1.6 Complex with Hematoporphyrin Diethylester, $Rh(hp\text{-}dee)Cl(H_2O)$

$C_{38}H_{46}N_4O_6$ = hematoporphyrin diethylester (1,3,5,8-tetramethyl-2,4-di(1-hydroxyethyl)-6,7-di(ethyl-2-carboxyethyl)porphyrin) = hp-deeH$_2$

A general method of preparation similar to the preparation of mesoporphyrin diethylester complex is mentioned (see 2.5.1.4, p. 227) but no details are given.

N. Sadasivan, E. B. Fleischer (J. Inorg. Nucl. Chem. **30** [1968] 591/601, 592).

2.5.1.7 Complexes with Octaethylporphyrin

$C_{36}H_{46}N_4$ = octaethylporphyrin = oepH$_2$

Na[Rh(oep)] is obtained in solution by addition of NaBH$_4$ in aqueous NaOH to a solution of RhCl(oep)·2H$_2$O in ethanol [1, 2] or tetrahydrofuran [2]. It is extremely air-sensitive and does not appear to have been isolated in solid form, the solutions are brown [1] or brown-red [2] in colour. On acidification (CH$_3$CO$_2$H) the rhodium(III) hydride RhH(oep) is obtained in good yield [3]. Reactions with cyclopropane derivatives [2, 4], alkyl halides [1], olefins and acetylenes [1] afford organorhodium(III) products RhR(oep) (R = organic group).

$\{Rh(oep)\}_2 \cdot C_6H_5CH_3$ is prepared as violet crystals (melting point >310°C) by recrystallising the rhodium(III) hydride RhH(oep) from toluene solution [3] or, more efficiently, by photolysis of solutions of RhH(oep) in toluene under argon [5, 6]. The electronic spectrum (650 to 300 nm) is reproduced [5, 6]. ^{1}H NMR data have been reported, $\delta = 9.23$ ($_1$S, meso protons, 8H), 4.41 (d of quartets, C$\underline{H}_2$, 16H, $J_{gem} = 14$ Hz), 3.92 (d of quartets, C$\underline{H}_2$, 16H) and 1.68 (t, C$\underline{H}_3$, 48H). The complex is diamagnetic, $\chi = -0.65 \times 10^{-6}$ at 25°C [3]. It reacts with carbon monoxide and hydrogen or water to yield a rhodium(III) formyl complex Rh(CHO)(oep) [7]. Reactions with dioxygen and nitric oxide afford the dioxygen complex Rh(O$_2$)(oep) and the nitrosyl Rh(NO)(oep), respectively [5, 6]. Treatment with alkyl halides, olefins and acetylenes yield organorhodium(III) porphyrin products (scheme) [3]:

$$\{Rh(oep)\}_2 \begin{cases} \xrightarrow{C_6H_5CH_2Br} RhBr(oep) + Rh(CH_2C_6H_5)(oep) \\ \xrightarrow{CH_2=CH\text{-}CH_2R} Rh(CH_2\text{-}CH=CHR)(oep) \\ \xrightarrow{HC\equiv CR} (oep)Rh\text{-}CH=CR\text{-}Rh(oep) \end{cases}$$

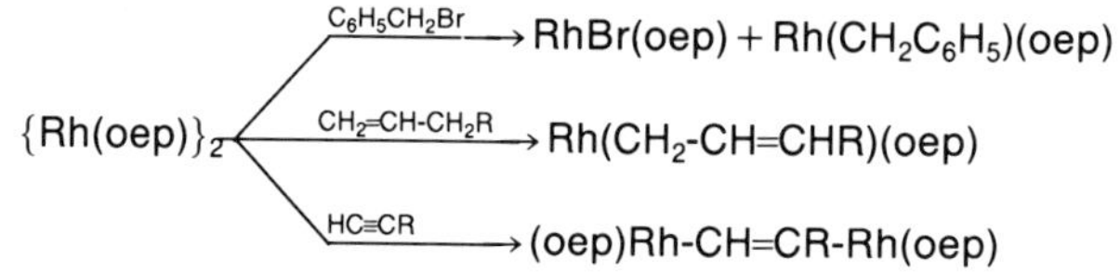

Treatment with trimethyl phosphite, P(OCH$_3$)$_3$, affords Rh(CH$_3$)(oep) and Rh{P(O)(OCH$_3$)$_2$}(oep) [8].

Rh(O$_2$)(oep) is obtained by allowing dry dioxygen to slowly diffuse into a toluene solution of $\{Rh(oep)\}_2$ at $-80°C$, and has been characterised as a Rh^{III}/O$_2^-$ species by electronic and ESR spectra. The electronic spectrum (650 to 350 nm) is reproduced. The ESR spectrum is reproduced; $\langle g \rangle = 2.032$ (toluene, 25°C), $g_1 = 2.100$, $g_2 = 2.010$, $g_3 = 1.988$ (toluene, −160°C). An infrared band at 1075 cm^{-1} is tentatively attributed to $\nu(O_2)$. When solutions of $Rh(O_2)(oep)$ are allowed to warm to 20°C a μ-peroxo complex (oep)Rh-O_2-Rh(oep) is formed [5, 6].

$Rh(O_2)(oep)(C_5H_{11}N)$ is prepared by adding piperidine ($C_5H_{11}N$) to a solution of preformed $Rh(O_2)(oep)$ in toluene at −80°C [5, 6]. The ESR spectrum has been reported and is reproduced; $\langle g \rangle = 2.031$, $g_1 = 2.094$, $g_2 = 2.010$, $g_3 = 1.996$ [5, 6].

$Rh(O_2)(oep)\{P(OC_4H_9)_3\}$ is prepared by adding tributyl phosphite to a solution of preformed $Rh(O_2)(oep)$ in toluene at −80°C [5, 6]. The ESR spectrum has been reported and is reproduced; $\langle g \rangle = 2.032$, $g_1 = 2.084$, $g_2 = 2.004$, $g_3 = 2.000$ [5, 6].

$(oep)RhO_2Rh(oep)$ is postulated as the product formed when solutions of $Rh(O_2)(oep)$ are allowed to warm to 20°C; formation is marked by the disappearance of the ESR signal due to the precursor [5, 6].

Rh(NO)(oep) is prepared by exposing $\{Rh(oep)\}_2$ in the solid state or in degassed toluene solution to nitric oxide [5, 6]. It has also been obtained by treatment of RhH(oep) or RhCl(oep) with nitric oxide in degassed toluene solution. The electronic spectrum (700 to 350 nm) is reproduced. The nitrosyl stretching vibration ν(NO) occurs at 1630 cm^{-1}. $E_{1/2}$ values for first and second oxidations are 0.77 and 1.27 V, respectively. The complex is air-sensitive and diamagnetic [6].

[Rh(NO)(oep)]Cl is a metastable, paramagnetic intermediate formed during the reaction of RhCl(oep) with nitric oxide; it associates to form radical dimers. The electronic spectrum (700 to 350 nm) and ESR spectrum of these solutions are reproduced [6].

$[Rh(NO)(oep)]^+$ is generated by electrolysis of Rh(NO)(oep) in CH_2Cl_2, the electronic spectrum (700 to 350 nm) is reproduced and the ESR spectrum has been recorded; $\langle g \rangle = 2.001$ [6].

$Rh(NO)_2(oep)$ is prepared in solution by reversible addition of nitric oxide to Rh(NO)(oep) and is formulated as $Rh^{III}(oep^-)(NO^-)_2$. The electronic spectrum (700 to 350 nm) and ESR spectrum of these solutions are reproduced [6].

RhH(oep) is prepared by passing dihydrogen gas into a methanolic solution of RhCl(oep) or by the acidification (CH_3CO_2H) of a basic alcoholic solution containing the anion $[Rh(oep)]^-$. It is described as a deep orange precipitate, melting point 280°C, ν(Rh-H) occurs at 2220 cm^{-1} [3]. It reacts in aromatic solvents, spontaneously [3] or on photolysis at 25°C to yield $\{Rh(oep)\}_2$ [6]. Treatment with carbon monoxide affords the formylrhodium(III) complex Rh(CHO)(oep) [9].

$RhCl(oep) \cdot 2H_2O$ is formed, together with $[Rh_2(oepH)Cl(CO)_4]$ by stirring a benzene solution of octaethylporphyrin and $\{RhCl(CO)_2\}_2$ under nitrogen for 2 h, and is separated by chromatography (silica gel/acetone-benzene 1:3) as maroon crystals, melting point 300°C with decomposition. The electronic spectrum with $\lambda_{max} = 554, 520, 403, 339, 285$ nm in $CHCl_3$ solution and the ^1H NMR spectrum have been recorded; $\delta = 1.99$ (t, $CH_2\underline{CH_3}$, 24 H); 4.15 (q, $\underline{CH_2}CH_3$, 16 H); 10.31 (s, meso protons, 4 H) [1, 10]. Oxidation potentials ($E_{1/2}$) for the first and second oxidations are $E_{1/2} = 0.82$ and 1.34 V, respectively [6].

RhBr(oep) is mentioned as one of the products formed on treatment of $\{Rh(oep)\}_2$ with benzyl bromide [3].

$Rh\{P(O)(OCH_3)_2\}(oep)$ is one of the products obtained on treatment of $\{Rh(oep)\}_2$ with trimethyl phosphite in benzene solution. The ^1H NMR spectrum has been recorded; $\delta(CH_3) =$ 0.586 doublet, $^3J(^{31}P-{}^1H) = 12.3$ Hz [8].

References:

[1] H. Ogoshi, J. Setsune, T. Omura, Z. Yoshida (J. Am. Chem. Soc. **97** [1975] 6461/6). – [2] H. Ogoshi, J. Setsune, Z. Yoshida (J. Organometal. Chem. **185** [1980] 95/104). – [3] H. Ogoshi, J. Setsune, Z. Yoshida (J. Am. Chem. Soc. **99** [1977] 3869/70). – [4] H. Ogoshi, J. Setsune, Z. Yoshida (J. Chem. Soc. D **1975** 572/3). – [5] B. B. Wayland, A. R. Newman (J. Am. Chem. Soc. **101** [1979] 6472/3).

[6] B. B. Wayland, A. R. Newman (Inorg. Chem. **20** [1981] 3093/7). – [7] B. B. Wayland, B. A. Woods, R. Pierce (J. Am. Chem. Soc. **104** [1982] 302/3). – [8] B. B. Wayland, B. A. Woods (J. Chem. Soc. D **1981** 475/6). – [9] B. B. Wayland, B. A. Woods (J. Chem. Soc. D **1981** 700/1). – [10] Z. Yoshida, H. Ogoshi, T. Omura, E. Watanabe, T. Kurosaki (Tetrahedron Letters **1972** 1077/80).

2.5.1.8 Complexes with Tetraphenylporphyrin

$C_{44}H_{30}N_4$ = tetraphenylporphyrin = tppH_2

H[Rh(tpp)]·2H$_2$O is prepared by treatment of "Rh(tpp)", subsequently reformulated as Rh(O_2)(tpp) [1] with dihydrogen in propanol or dimethylformamide at 20°C [2]. It contains a titratable proton and has a conductance of 16.6 $cm^2 \cdot \Omega^{-1} \cdot mol^{-1}$ (5×10^{-4} M solution in dimethylformamide). Exposure to oxygen leads to regeneration of the parent complex [2].

{Rh(tpp)}$_n$. A complex of this formulation (n = 1), originally obtained by adding $\{RhCl(CO)_2\}_2$ to a boiling solution of tetraphenylporphyrin and NaO_2CCH_3 in acetic acid under nitrogen [2], has been reformulated as Rh(O_2)(tpp) [1]. Authentic $\{Rh(tpp)\}_2$ is obtained by heating the solid dioxygen adduct at 150°C and sublimes as a diamagnetic solid [1, 3].

Rh(O$_2$)(tpp) is prepared by borohydride reduction of any RhX(tpp) complex followed by addition of dioxygen (no further details given) [3]. Heating the solid product in high vacuum at 150°C removes dioxygen to yield $\{Rh(tpp)\}_2$ [1]. The complex is formulated as a rhodium(III) superoxide (Rh^{III}/O_2^-). The ESR spectrum has been recorded; $\langle g \rangle = 2.033$, $g_1 = 2.084$, $g_2 = 2.025$, $g_3 = 1.993$ [3].

Rh(O$_2$)(tpp){P(OC$_2$H$_5$)$_3$} is prepared by addition of triethyl phosphite to a solution of Rh(O_2)(tpp) at −80°C. The ESR spectrum has been recorded; g-values and A(^{31}P) values (in parentheses) are $\langle g \rangle = 2.032$ (23.7 G), $g_1 = 2.086$ (~0 G), $g_2 = 2.009$ (35.5 G), $g_3 = 2.004$ (35.5 G) [3].

Rh(NO)(tpp) is prepared by treatment of $\{Rh(tpp)\}_2$ with nitric oxide [1] and is the ultimate product when RhCl(tpp) reacts with nitric oxide [3]. The nitrosyl absorption, ν(NO), occurs at 1658 cm^{-1} [1].

RhCl(NO)(tpp) is formed in degassed toluene solution by reversible addition of nitric oxide to RhCl(tpp), and converts to Rh(NO)(tpp) on standing. The ESR spectrum is consistent with a porphyrin cation radical (Rh^{III}tpp)$^{2+}$ formulation [3].

Rh(NO)$_2$(tpp) is formed by reversible binding of nitric oxide to Rh(NO)(tpp). The ESR spectrum is consistent with an $^2A_{1u}$ state [3].

[Rh(tpp){NH(CH$_3$)$_2$}$_2$]Cl·CHCl$_3$ is prepared by heating a mixture of $\{RhCl(CO)_2\}_2$ and tetraphenylporphyrin under reflux in dimethylformamide for 4 h, and is isolated by precipitation with cold water followed by chromatography (alumina/chloroform). It forms red needles from chloroform [4].

The crystals are monoclinic, space group $P2_1/c-C^5_{2h}$ with lattice parameters a = 16.07(3), b = 12.87(2), c = 21.69(4) Å, β = 95.64°, D = 1.41 g/cm³, Z = 4. The cation has axial $NH(CH_3)_2$ ligands (Rh-N = 2.11 Å) and an equatorial porphyrin ligand (Rh-N = 2.03 Å). The molecular configuration is shown in **Fig. 17**. The electronic spectrum (λ_{max} = 535 and 420 nm) has been recorded. The dimethylamine ligands are thought to originate either as impurity in the solvent or via an insertion reaction of the Rh^I(tpp) with the solvent [4].

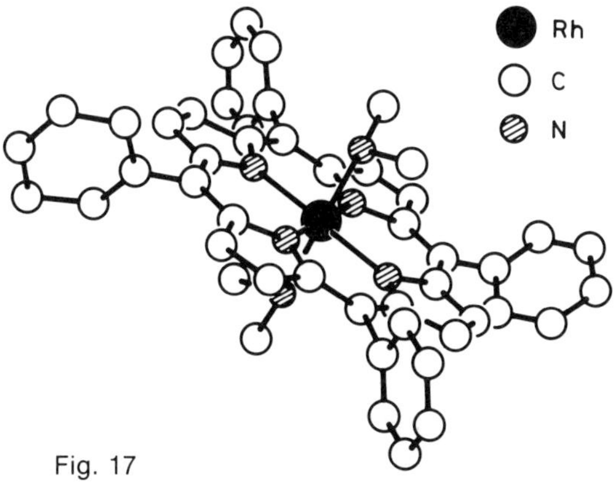

Fig. 17
Structure of rhodium porphyrin.

RhCl(tpp) is prepared by stirring a mixture of $\{RhCl(CO)_2\}_2$ (obtained by carbonylation of $RhCl_3 \cdot 3H_2O$), tetraphenylporphyrin and Na_2CO_3 in CH_2Cl_2 at 20°C for 12 h, then adding N-chlorosuccinimide to the washed (H_2O) and dried (Na_2SO_4) reaction solution. It is purified by chromatography (silica gel: toluene-ethylacetate 90:10) and crystallised from CH_2Cl_2/pentane [5]. An earlier report mentions the complex but gives no preparative details [6].

RhI(tpp) is prepared using the same method as for RhCl(tpp) but replacing the N-chlorosuccinimide by free iodine. It reacts with diazo compounds to yield organorhodium(III) porphyrin complexes [5].

References:

[1] B. B. Wayland, A. R. Newman (J. Am. Chem. Soc. **101** [1979] 6472/3). – [2] B. R. James, D. V. Stynes (J. Am. Chem. Soc. **94** [1972] 6225/6). – [3] B. B. Wayland, A. R. Newman (Inorg. Chem. **20** [1981] 3093/7). – [4] E. B. Fleischer, F. L. Dixon, R. Florian (Inorg. Nucl. Chem. Letters **9** [1973] 1303/5). – [5] H. J. Callot, E. Schaeffer (Nouveau J. Chim. **4** [1980] 311/4).

[6] N. Sadasivan, E. B. Fleischer (J. Inorg. Nucl. Chem. **30** [1968] 591/601).

2.5.1.9 Complexes with Tetra(p-sulphonatophenyl)porphyrin

$C_{44}H_{30}N_4O_{12}S_4$ = tetra(p-sulphonatophenyl)porphyrin = tsppH_6

$Na_3[Rh(tspp)(H_2O)_2] \cdot nH_2O$. Two very similar preparative methods give products with different degrees of hydration, a third gives the potassium complex in solution.

A solution of $RhCl_3 \cdot 3H_2O$ in dimethylformamide is heated until yellow (~20 min) then mixed with Na_4[tsppH_2] in methanol and heated for a further 30 min. The product is isolated and purified by diluting the solution with water then concentrating by evaporation, chromatographing (water-washed alumina/0.02 M NaOH) then passing through a Sephadex G-10 column and evaporating to dryness. The product analyses as $Na_3[Rh(tspp)(H_2O)_2]\cdot 22H_2O$ [1].

A mixture of $\{RhCl(CO)_2\}_2$, Na_4tsppH_2 and sodium acetate in methanol is stirred until all free base is consumed then treated with 3% H_2O_2, heated to decompose excess peroxide and evaporated to dryness. The residue is passed through a cation exchange column (Na^+ form) and the eluate evaporated to yield a crude product which is purified by extraction with methanol and reprecipitation with acetone. The product analyses as $Na_3[Rh(tspp)(H_2O)_2]\cdot 14H_2O$ [2].

A mixture of $\{RhCl(CO)_2\}_2$ and free porphyrin in glacial acetic acid is allowed to react under dinitrogen, the product is then treated with 1 M KOH to remove chloride and carbonyl ligands. Purification is by column chromatography (alumina or polyamide) [3].

Kinetic [1, 2] and thermodynamic [1, 3] studies on the acid dissociation [1, 2, 3] and anation [1, 2] reactions of the $[Rh(tspp)(H_2O)_2]^{3-}$ anion have been reported.

$$[Rh(tspp)(H_2O)_2]^{3-} \underset{+H^+}{\overset{K_1^a,\ -H^+}{\rightleftharpoons}} [Rh(tspp)(OH)(H_2O)]^{4-} \underset{+H^+}{\overset{K_2^a,\ -H^+}{\rightleftharpoons}} [Rh(tspp)(OH)_2]^{5-}$$

$$[Rh(tspp)(H_2O)_2]^{3-} \overset{K_1^L}{\rightleftharpoons} [Rh(tspp)(H_2O)L]^{3-} \overset{K_2^L}{\rightleftharpoons} [Rh(tspp)L_2]^{3-}$$

Values of K_1^a and K_2^a in 1.00 M $NaClO_4$ [1]:

Temp. in °C	K_1^a	pK_1^a	K_2^a	pK_2^a
15	$(2.64 \pm 0.05) \times 10^{-8}$	7.58	$(9.15 \pm 0.04) \times 10^{-9}$	8.04
25	$(9.67 \pm 0.12) \times 10^{-8}$	7.01*)	$(1.59 \pm 0.24) \times 10^{-10}$	9.80
35	$(1.17 \pm 0.04) \times 10^{-7}$	6.93	$(2.52 \pm 0.26) \times 10^{-10}$	9.60

*) Also 7.6 [2] and 6.8 [3].

Values of K_1^L and K_2^L at 25°C in 1.00 M $NaClO_4$ [1]:

Ligand L	K_1^L in M^{-1}	K_2^L in M^{-1}
NCS	$(1.22 \pm 0.04) \times 10^4$	3.68 ± 0.42
I	$(4.12 \pm 0.18) \times 10$	< 0.03
Br	(1.82 ± 0.10)	< 0.03
Cl	(0.74 ± 0.11)	< 0.03

Electronic spectra (Soret bands) have been recorded for species formed in acid dissociation and anation reactions [1]:

λ_{max} in nm	417	417.5	418
Compound	$[Rh(tspp)(H_2O)_2]^{3-}$	$[Rh(tspp)(OH)(H_2O)]^{4-}$	$[Rh(tspp)(OH)_2]^{5-}$
λ_{max} in nm	418	418.5	420
Compound	$[Rh(tspp)Cl(H_2O)]^{4-}$	$[Rh(tspp)Br(H_2O)]^{4-}$	$[Rh(tspp)I(H_2O)]^{4-}$
λ_{max} in nm	420	422	
Compound	$[Rh(tspp)NCS(H_2O)]^{4-}$	$[Rh(tspp)(NCS)_2]^{5-}$	

References:

[1] K. R. Ashley, S.-B. Shyu, J. G. Leipoldt (Inorg. Chem. **19** [1980] 1613/6). – [2] M. Krishnamurthy (Inorg. Chim. Acta **25** [1977] 215/8). – [3] G. McLendon, M. Bailey (Inorg. Chem. **18** [1979] 2120/3).

2.5.1.10 Complex with Tetrapyridylporphyrin, $RhCl(tpyp)(H_2O)$

$C_{40}H_{26}N_8$ = tetrapyridylporphyrin = tpypH_2

A general method of preparation is mentioned but no details are given.

N. Sadasivan, E. G. Fleischer (J. Inorg. Nucl. Chem. **30** [1968] 591/601, 592).

2.5.1.11 Complex with 2,8-Bis(ethoxycarbonyl)-13,17-diethyl-3,7,12,18-tetramethyl-5-thiaporphyrin, $Rh(thia\text{-}p)(O_2CCH_3)_2$

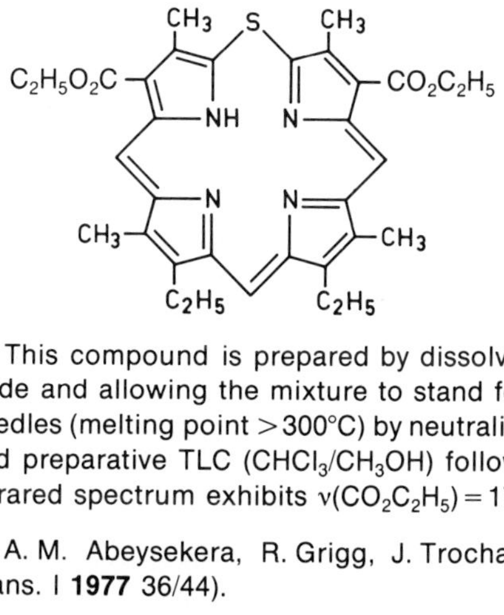

= thia-pH

This compound is prepared by dissolving $Rh(thia\text{-}p)(CO)_2$ in freshly distilled acetic anhydride and allowing the mixture to stand for 2 d at room temperature. It is isolated as brown needles (melting point >300°C) by neutralisation of the solution (K_2CO_3), extraction with $CHCl_3$, and preparative TLC ($CHCl_3/CH_3OH$) followed by crystallisation ($CHCl_3$/light petroleum). The infrared spectrum exhibits $\nu(CO_2C_2H_5) = 1710\ cm^{-1}$ and $\nu(RhO_2CCH_3) = 1630\ cm^{-1}$.

A. M. Abeysekera, R. Grigg, J. Trocha-Grimshaw, V. Viswanatha (J. Chem. Soc. Perkin Trans. I **1977** 36/44).

2.5.2 Complexes with Phthalocyanines

General Literature:

B. D. Berezin, Coordination Compounds of Porphyrins and Phthalocyanines, Wiley, London 1981.

L. J. Boucher, Metal Complexes of Phthalocyanines, in: G. A. Melson, Coordination Chemistry of Macrocyclic Compounds, Plenum, New York 1979, Chapter 7, pp. 461/516.

B. D. Berezin, Koordinatsionnye Soedineniya Porfirinov i Ftalotsiyanina, Moskva 1978.

A. B. P. Lever, The Phthalocyanines, Advan. Inorg. Chem. Radiochem. **7** [1965] 27/114, 62.

I. M. Keen, The Platinum Metal Phthalocyanines, Platinum Metals Rev. **8** [1964] 143/4.

2.5.2.1 Complexes with Phthalocyanine, Rh(pc)

$C_{32}H_{18}N_8$ = phthalocyanine = pcH_2

No preparative details or characterisation have been reported for this apparent rhodium(II) complex. However, fluorescence from upper vibrational levels of the first excited molecular singlet has been recorded for selected metal phthalocyanines including Rh(pc) in solution at 77 K. Vibrational, internal conversion and intersystem crossing rates have been listed [1]. The presence of Rh(pc) as an intermediate in the photolysis of Rh(pc)X (X = Cl, Br, I) has been proposed [2].

References:

[1] E. R. Menzel, K. E. Rieckhoff, E. M. Voigt (Chem. Phys. Letters **13** [1972] 604/7). – [2] K. Schmatz, S. Muralidharan, K. Madden, R. Fessenden, G. Ferraudi (Inorg. Chim. Acta **64** [1982] L23/L24).

2.5.2.2 Miscellaneous Phthalocyanine Complexes

Rh(pc)Cl is prepared by refluxing anhydrous $RhCl_3$ with a five-fold excess of o-$C_6H_4(CN)_2$ for 4 h [1, 2] or by heating together $RhCl_3 \cdot 3H_2O$ and o-$C_6H_4(CONH_2)(CN)$ at 280°C for 4 h [3, 4], and is purified by successive Soxhlet extractions with ethanol, benzene and acetone [3]. It is described as a blue-green solid. The infrared spectrum (3300 to 250 cm^{-1}, Nujol mull) is reproduced and the main bands are tabulated (3300 to 700 cm^{-1}, KBr disc). The electronic spectrum has been reported (λ_{max} = 645, 584 nm, acetone solution) [3].

Rh(pc)X (X = Cl, Br or I). These complexes are prepared by heating together o-C_6H_4-$(CONH_2)(CN)$ and the appropriate hydrated rhodium(III) halide at 280°C for 4 h [5]. Electronic spectra (wavelengths in nm) have been reported for solutions in α-chloronaphthalene (X = Cl), λ_{max} = 656, 633, 593, 573, 555, 420; for X = I, λ_{max} = 656, 632, 593, 573, 556. The complexes display fluorescence (X = Cl, λ_{max} = 662 nm, quantum yield 1×10^{-3}) and phosphorescence (X = Cl, λ_{max} = 987 nm, quantum yield 2×10^{-3}) [5]. The charge transfer and ligand-centred excited state photochemistries of these complexes have been investigated [6].

Rh(pc)(CH_3OH)X (X = Cl, Br or I). These complexes are prepared by treatment of $[Rh(pc)(H_2O)_2](HSO_4)$ with the appropriate halide, NaX, in boiling methanol for 2 h. Electronic spectra (600 to 250 nm) are reproduced. Their photochemistry has been investigated by continuous, flash and laser flash photolysis. Intermediates include rhodium(II) phthalocyanine and rhodium(III) phthalocyanine ligand radical species. ESR data have been recorded for the oxidised ligand radicals, g = 2.0010 to 2.0042 [10]. Visible luminescence ($\lambda_{max} \approx 470$ nm) has been reported for Rh(pc)(CH_3OH)Cl [11].

Rh(pc)(HSO_4)(?). There is some doubt about the stoichiometry of this product: a footnote to one paper [7] refers to the presence of a chlorine atom on the phthalocyanine ligand. A later paper formulates the complex as $[Rh(pc)(H_2O)_2](HSO_4)$ and reports that the original method of synthesis (see p. 235) gives only ~10% yield [10].

This complex is obtained by recrystallisation of Rh(pc)Cl from 96% H_2SO_4 and forms dark blue crystals which are readily soluble in acetone (dark green solution) and absolute alcohol (dark blue solution). The electronic spectrum is reproduced (800 to 200 nm, 16 and 18 M H_2SO_4 solution), λ_{max} = 768, 692, 437, 312, 269 and 224 nm for 18 M H_2SO_4 solution and 768, 698, 439, 308, 269 and 224 nm for 16 M H_2SO_4 solution. Solubility in conc. H_2SO_4 at 25°C ranges from 6.97×10^{-3} mol/l for 14 M H_2SO_4 to 5.48×10^{-2} mol/l for 16.5 M H_2SO_4 [2]. Dissolution involves the equilibrium $Rh(pc)(HSO_4) + H_2SO_4 \rightleftharpoons Rh(pcH)(HSO_4)^+ + HSO_4^-$ for which the equilibrium constant (pK) is reported to be 1.01 ± 0.04 [1] or 0.99 ± 0.04 [8].

Dissociation of protonated $Rh(pc)(HSO_4)$ in concentrated H_2SO_4, $[Rh(pcH)(HSO_4)]^+ + 2H_3O^+ \rightarrow H_2pcH^+ + Rh(HSO_4)^{2+}$, is first order in complex and second order in hydrogen ion concentration [2]. The rate constant (in $l^2 \cdot h^{-1} \cdot mol^{-1}$) for the dissociation has been given as 3.20×10^{-8} at 25°C [9], 0.253×10^{-3} at 120°C [2], 1.02×10^{-3} at 138°C [2, 7] and 3.02×10^{-3} at 155°C. Plots of the rate of dissociation of $Rh(pc)(HSO_4)$ at various temperatures and H_2SO_4 concentrations are reproduced [7].

References:

[1] W. Herr (Z. Naturforsch. **9a** [1954] 180/1). – [2] B. D. Berezin (Dokl. Akad. Nauk SSSR **150** [1963] 1039/42; Dokl. Chem. Proc. Acad. Sci. USSR **148/153** [1963] 478/81). – [3] I. M. Keen, B. W. Malerbi (J. Inorg. Nucl. Chem. **27** [1965] 1311/9). – [4] I. M. Keen (Platinum Metals Rev. **8** [1964] 143/4). – [5] E. R. Menzel, K. E. Rieckhoff, E. M. Voigt (J. Chem. Phys. **58** [1973] 5726/34).

[6] K. Schmatz, S. Muralidharan, K. Madden, R. Fessenden, G. Ferraudi (Inorg. Chim. Acta **64** [1982] L23/L24). – [7] B. D. Berezin (Zh. Fiz. Khim. **38** [1964] 850/7; Russ. J. Phys. Chem. **38** [1964] 462/6). – [8] B. D. Berezin (Zh. Obshch. Khim. **43** [1973] 2718/20; J. Gen. Chem. [USSR] **43** [1973] 2714/7). – [9] B. D. Berezin (Zh. Fiz. Khim. **39** [1965] 1082/6; Russ. J. Phys. Chem. **39** [1965] 576/9). – [10] S. Muralidharan, G. Ferraudi, K. Schmatz (Inorg. Chem. **21** [1982] 2961/7).

[11] S. Muralidharan, G. Ferraudi, L. K. Patterson (Inorg. Chim. Acta **65** [1982] L235/L236).

2.5.2.3 Complex with Tetra-4,4′,4″,4‴t-butylphthalocyanine, Rh(tbpc)Cl

$C_{48}H_{50}N_8$ = tetra-4,4′,4″,4‴-t-butylphthalocyanine = $tbpcH_2$

The compound is prepared by fusing a mixture of $RhCl_3$ and 4-t-butylphthalonitrile in an evacuated glass ampoule at 270°C for 2 h in the presence of ammonium molybdate as catalyst, and is purified by chromatography (aluminium oxide/benzene). The electronic spectrum has been reported: λ_{max} = 658, 628, 592, 344, 328 nm in benzene solution. An infrared band at 314 cm^{-1} is attributed to ν(Rh-Cl).

L. G. Tomilova, E. A. Luk'yanets (Zh. Neorgan. Khim. **22** [1977] 2586/7; Russ. J. Inorg. Chem. **22** [1977] 1403).

2.5.3 Complexes with Corrins

General Literature:

A. Eschenmoser, Roads to Corrins, Quart. Rev. [London] **24** [1970] 366/415, 391.

2.5.3.1 Complexes with 1,2,2,7,7,12,12-Heptamethyl-15-cyanocorrin

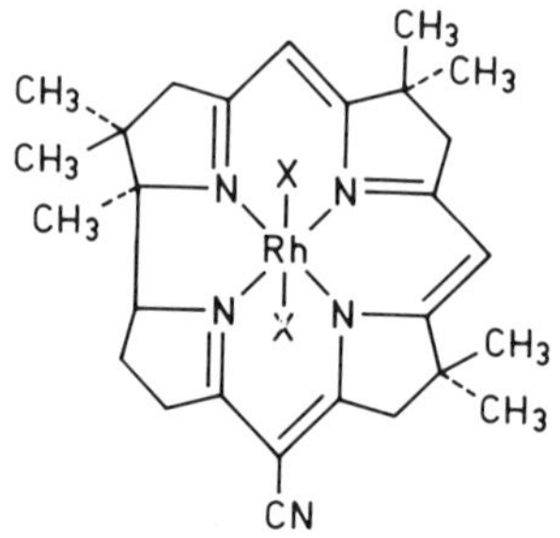

[RhCl(H_2O)(corrin)]Cl is postulated as a product formed during photolysis of $RhCl_2$(corrin) [1].

$RhCl_2$(corrin). No preparative details have been reported. The electronic absorption spectrum has been recorded (λ_{max} = 499, 476, 324 nm) and is reproduced (600 to 300 nm). The electronic emission spectrum has been measured at 77 K. Excitation at 436 and 366 nm led to intense red luminescence either in the solid state or in frozen methanol/water glasses. The emission spectrum (900 to 600 nm, solid sample, excitation wavelength 436 nm, temperature 77 K) is reproduced. Spectral changes (550 to 300 nm) observed during photolysis of $RhCl_2$(corrin) are reproduced [1].

$Rh(CN)_2$(corrin) is prepared from $RhCl_2$(corrin) by an unpublished method. The electronic absorption spectrum has been recorded (λ_{max} = 517, 489, 331 nm) and is reproduced (600 to 300 nm) [1]. The electronic absorption and emission spectra (800 to 400 nm) are reproduced. The complex displays a strong corrin phosphorescence at 77 and 298 K [2].

References:

[1] A. Vogler, R. Hirschmann, H. Otto, H. Kunkely (Ber. Bunsenges. Physik. Chem. **80** [1976] 420/4). – [2] M. Gardiner, A. J. Thomson (J. Chem. Soc. Dalton Trans. **1974** 820/8).

2.5.3.2 Rhodium Analogues of Vitamin B_{12} Co-enzyme and Related Species

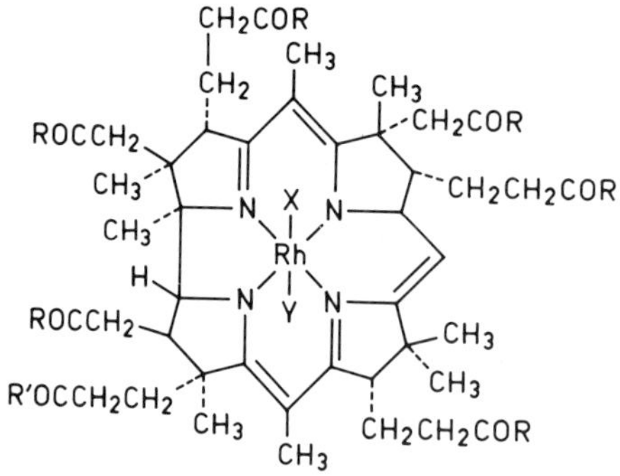

Dicyanorhodibyric Acid (R = NH_2, R′ = OH, X = Y = CN) is prepared by allowing hydrogenobyric acid and sodium acetate to react with $\{RhCl(CO)_2\}_2$ in ethanol-glacial acetic acid solution at 100°C for 1 h, then diluting the mixture with water and adjusting the pH to 9.5 using solid KCN. It is separated and purified by extraction through phenol, and re-extraction into water followed by column (amberlite XAD-2) and paper chromatography to yield a deep red solid. The electronic spectrum (λ_{max} = 528, 497, 350 nm) has been recorded and the Fourier transform proton NMR spectrum is reproduced [1].

Cyanoaquorhodibinamide Cation ($R = NH_2$, $R' = NHCH_2CH(OH)CH_3$, X = CN, $Y = H_2O$) is obtained by treatment of dicyanorhodibinamide with 0.01M HCl for 15 h at room temperature and is stable in solution at pH < 7. It is yellow in colour. The electronic spectrum (600 to 250 nm; λ_{max} = 499, 481, 340 nm) and infrared spectrum (4000 to 800 cm^{-1}; ν(CN) = 2133 cm^{-1}) are reproduced [2].

Cyanohydroxyrhodibinamide ($R = NH_2$, $R' = NHCH_2CH(OH)CH_3$, X = CN, Y = OH) is formed when solutions of cyanoaquorhodibinamide are raised to pH = 11 or above. The electronic spectrum (600 to 250 nm; λ_{max} = 510, 492, 343 nm) is reproduced [2].

Dicyanorhodibinamide ($R = NH_2$, $R' = NHCH_2CH(OH)CH_3$, X = Y = CN) is one of the products obtained by allowing $\{RhCl(CO)_2\}_2$ to react with α-(5,6-dimethyl benzimidazolyl)hydrogenobamide in ethanol-glacial acetic acid solution for 24 h at room temperature, then diluting the mixture with water and adjusting the pH to 9.5 with solid KCN. It is separated by chromatography and crystallised from aqueous acetone as deep red needles. It has also been prepared by hydrolysis of dicyanorhodibalamin using cerous hydroxide. The electronic spectrum (600 to 250 nm; λ_{max} = 528, 497, 350 nm) and infrared spectrum (4000 to 800 cm^{-1}; ν(CN) = 2119 cm^{-1}) are reproduced [2].

Rhodibalamins

$R = NH_2$, R'-Y =

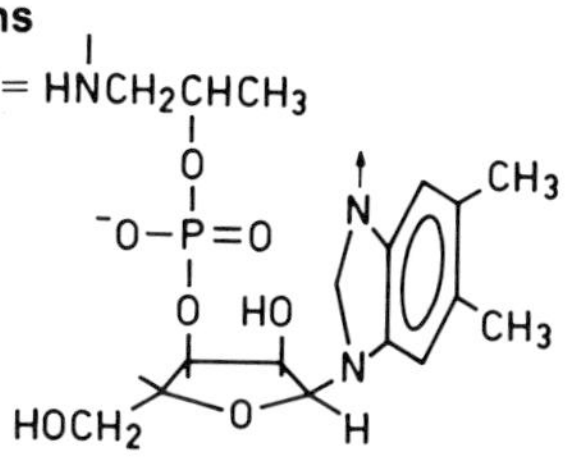

Rhodibalamin(s) - Rhodium(I) Complex (X site vacant). This species is generated in solution by reduction of chlororhodibalamin with sodium borohydride in distilled water under argon. It has been converted in situ into organorhodibalamins by treatment with alkyl halides. The electronic spectrum has been recorded and is reproduced (aqueous solution, pH = 11, 600 to 250 nm), λ_{max} = 385, 298, 274, 260 nm [1].

Aquorhodibalamin ($X = H_2O$) is prepared by treatment of chlororhodibalamin with silver nitrate in distilled water at 100°C for 30 min. It is purified by extraction through phenol followed by paper chromatography. The electronic spectrum is reproduced (aqueous solution, pH = 2.5 and 6.5, 600 to 200 nm), λ_{max} = 503, 477, 340 nm [1].

Chlororhodibalamin (X = Cl) is prepared by allowing a mixture of $\{RhCl(CO)_2\}_2$, α-(5,6-dimethylbenzimidazolyl)hydrogenobamide and sodium acetate in ethanol-glacial acetic acid solution to react for 1 h at 100°C in dim light. It is isolated and purified by extraction through phenol, re-extraction into aqueous solution, paper chromatography and finally crystallisation from aqueous acetone. It forms orange-red rods. The electronic spectrum is reproduced (aqueous solution, pH = 2.5, 7 and 11, 600 to 200 nm), λ_{max} = 513, 485, 342.5 nm [1].

Cyanorhodibalamin (X = CN) is one of the products separated by chromatography from the mixture obtained by allowing $\{RhCl(CO)_2\}_2$ to react with α-(5,6-dimethylbenzimidazolyl)hydrogenobamide in ethanol-glacial acetic acid solution for 24 h, then diluting with water and adjusting the pH to 9.5 with solid KCN. It has also been obtained by treatment of dicyanorhodibalamin with silver nitrate [2]. The electronic spectrum is reproduced (aqueous solution, pH = 2.5, 7 and 11, 600 to 200 nm), λ_{max} = 514, 486(485), 403, 345, 280 nm [1, 2]. The infrared spectrum is reproduced (4000 to 800 cm^{-1}), ν(CN) = 2137 cm^{-1} [2].

Dicyanorhodibalamin (X = Y = CN) is one of the products separated by chromatography from the mixture obtained by allowing $\{RhCl(CO)_2\}_2$ to react with α-(5,6-dimethylbenzimidazolyl)hydrogenobamide in ethanol-glacial acetic acid solution for 24 h, then diluting the solution with water and adjusting the pH to 9.5 with solid KCN [2]. It is also obtained quantitatively by treatment of 5′-deoxyadenosylrhodibalamin with KCN [1]. The electronic spectrum is reproduced (aqueous solution, pH = 7 and 11, 600 to 250 nm), λ_{max} = 528, 497, 403, 350, 289 nm [2].

References:

[1] V. B. Koppenhagen, B. Elsenhans, F. Wagner, J. J. Pfiffner (J. Biol. Chem. **249** [1974] 6532/40). – [2] V. B. Koppenhagen, F. Wagner, J. J. Pfiffner (J. Biol. Chem. **248** [1973] 7999/8002).

2.5.3.3 Complexes with Decobalto-5,6-dioxo-5,6-secocobyrinic Acid Heptamethyl Esters

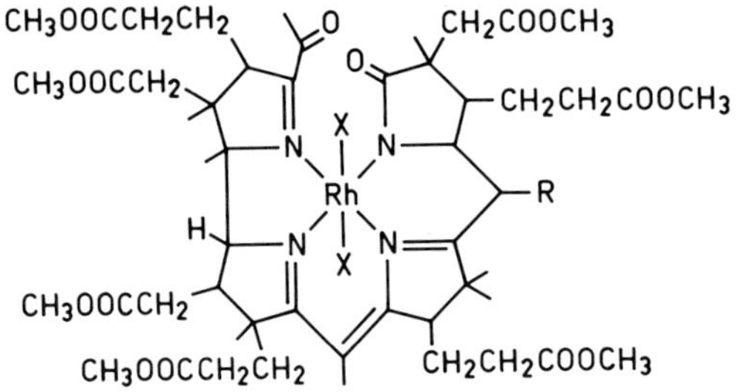

Dichloro-5,6-dioxo-5,6-secorhodibyrinic Acid Heptamethyl Ester (X = Cl, R = H) is prepared by treatment of decobalto-5,6-dioxo-5,6-secocobyrinic heptamethyl ester with $\{RhCl(CO)_2\}_2$ in CH_2Cl_2, and is purified by chromatography. The melting point is given as 102°C. Electronic spectrum: λ_{max} = 479, 324, 286 nm; infrared spectrum: ν_{max} = 2940, 1740, 1560, 1530, 1490, 1430, 1360, 1190, 1170 cm^{-1}; ^1H NMR (100 MHz) spectrum: δ = 0.94, 1.15, 1.29, 1.34, 1.76 (all CH_3); 2.30 (CH_3 on C_{15}), 2.58 (acetyl CH_3), 1.5 to 3.4 (side chain CH_2), 3.56, 3.66, 3.67, 3.69, 3.71, 3.74, 3.79 (all ester OCH_3), 4.24 (d, J = 10 Hz, H on C_{19}), 5.59 (H on C_{10}); mass spectrum: m/e = 1182 (70%) M^+, 1147 (20%) M^+-Cl, 1127 (100%).

Dichloro-10-bromo-5,6-dioxo-5,6-secorhodibyrinic Acid Heptamethyl Ester (X = Cl, R = Br) is similarly prepared using decobalto-10-bromo-5,6-dioxo-5,6-secocobyrinic heptamethyl ester. It melts at 107°C. Electronic spectrum: λ_{max} = 504, 343, 285 nm; infrared spectrum: ν_{max} = 2950, 1730, 1600, 1530, 1500, 1430, 1200, 1180 cm^{-1}; ^1H NMR (90 MHz) spectrum: δ = 1.22, 1.24, 1.29, 1.37, 1.68, 1.80 (all CH_3), 2.27 (CH_3 on C_{15}), 2.56 (acetyl CH_3), 1.9 to 3.3 (side chain CH_2), 3.56, 3.64, 3.67, 3.70, 3.73, 3.78 (all ester OCH_3), 4.06 (d, J = 10 Hz, H on C_{19}); mass spectrum: m/e = 1260 (100%) M^+.

R.-P. Hinze, D. Wullbrandt, H. H. Inhoffen (Liebigs Ann. Chem. **1980** 821/3).

2.6 Complexes of 5- and 6-Membered Heterocyclic Monodentate Nitrogen Donors

2.6.1 Pyrazole Complexes (pyzH = $C_3H_3N_2H$, pyrazole)

***trans*-[Rh(pyzH)$_4$Cl$_2$]Cl·5H$_2$O** is prepared by boiling a solution of $RhCl_3 \cdot 3H_2O$ in aqueous ethanol containing pyrazole. It is recrystallised from water-methanol mixtures and was isolated as yellow-orange crystals [1]. The electronic spectra, electrical conducitivity and polarographic properties (the compound underwent an irreversible two-electron reduction in aqueous solution) are summarised in Table 1, p. 245 [1, 2].

Tris(pyrazolyl)borato Complexes

Tris(pyrazolyl)hydridoborate = $[HB(C_3H_3N_3)_2]^-$ = $[HB(pyz)_3]^-$; tris(3,5-dimethylpyrazolyl)hydridoborate = $[HB(3,5\text{-}Me_2C_3HN_2)_3]^-$ = $[HB(3,5\text{-}Me_2pyz)_3]^-$

The ligands are of the tripodal type, occupying three coordination sites at the metal, and are uninegative. The physical properties of the complexes formed by these ligands (colours, melting points, significant IR and ^{1}H NMR spectral data) are summarised in Table 2, pp. 246/8.

[N(C$_2$H$_5$)$_3$H][Rh{HB(3,5-Me$_2$pyz)$_3$}(O$_2$CCF$_3$)$_2$H]. This complex is formed by reaction of Rh{HB(3,5-Me$_2$pyz)$_3$}(O$_2$CCF$_3$)$_2$(H$_2$O) with hydrogen in toluene to which triethylamine is added [3].

Rh{HB(3,5-Me$_2$pyz)$_3$}(O$_2$CR)$_2$·nH$_2$O (R = CH_3, n = 0; R = CF_3, n = 1). The **acetate** is obtained by refluxing [Rh{HB(3,5-Me$_2$pyz)$_3$}Cl$_2$]$_2$ in toluene with AgO_2CCH_3. The **trifluoroacetate** is obtained in the same way. It was suggested that the acetate contained one mono- and one bidentate carboxylato group [3].

Reaction of the trifluoroacetate with CO and with CH_3CN in dry toluene under nitrogen afforded Rh{HB(3,5-Me$_2$pyz)$_3$}(O$_2$CCF$_3$)$_2$L (L = CO or CH_3CN), respectively [3].

Rh{HB(pyz)$_3$}(O$_2$CCF$_3$)$_2$ is formed by treatment of Rh{HB(pyz)$_3$}Cl$_2$(CH$_3$CN) with AgO_2CCF_3 in dry toluene. On refluxing this complex in a mixture of dry toluene and CH_3CN, Rh{HB(pyz)$_3$}-(O$_2$CCF$_3$)$_2$(CH$_3$CN) is formed [3].

[N(C$_2$H$_5$)$_4$][Rh{HB(3,5-Me$_2$pyz)$_3$}(O$_2$CCF$_3$)$_2$Cl] is prepared by refluxing [N(C$_2$H$_5$)$_4$]-[Rh{HB(3,5-Me$_2$pyz)$_3$}Cl$_3$] with AgO_2CCF_3 in toluene [3].

Rh{HB(pyz)$_3$}(RCOCHCOR)Cl (R = CH_3 or CF_3). The acetylacetonate (R = CH_3) and hexafluoroacetylacetonate (R = CF_3) are obtained by refluxing Rh{HB(pyz)$_3$}Cl$_2$(CH$_3$CN) with Tl(RCOCHCOR) (R = CH_3 or CF_3) in dry toluene under nitrogen. The corresponding complexes containing HB(3,5-Me$_2$pyz)$_3$ have apparently been obtained similarly, but precise details have not been given [3].

Rh{HB(3,5-Me$_2$pyz)$_3$}Cl$_2$(ROH) (R = H or CH_3). The **aquo** complex is formed by refluxing $RhCl_3 \cdot 3H_2O$ with Na[HB(3,5-Me$_2$pyz)$_3$] in 95% ethanol. It may also be produced by refluxing Rh{HB(3,5-Me$_2$pyz)$_3$}Cl$_2$(CH$_3$OH) in toluene which has not been rigorously dried. The **methanol** complex is obtained when $RhCl_3 \cdot 3H_2O$ and Na[HB(3,5-Me$_2$pyz)$_3$] are refluxed in methanol [3].

Rh{HB(3,5-Me$_2$pyz)$_3$}Cl$_2$L (L = CH_3CN, PhCN, C_5H_5N, 4-$CH_3C_5H_4N$, N(C$_2$H$_5$)$_3$, N(C$_3$H$_7$)$_3$, NH_3, 3,5-(CH$_3$)$_2$C$_3$HN$_2$H, bipy, CO, CNBu^n, CNCH$_2$Ph, PPh$_3$, PPh$_2$(CH$_3$), PPh(CH$_3$)$_2$, P(o-CH$_3$-C$_6$H$_4$)$_2$CH$_3$). These complexes are prepared by refluxing Rh{HB(3,5-Me$_2$pyz)$_3$}Cl$_2$(CH$_3$OH) with L in dry toluene under nitrogen. The only exception is the acetonitrile compound, which is formed in neat CH_3CN [3].

The complex containing P(o-CH$_3$C$_6$H$_4$)$_2$(CH$_3$) was isolated with one mole of CH_2Cl_2 of crystallisation [3].

Rh{HB(pyz)$_3$}Cl$_2$L (L = PhCN, C_5H_5N, $N(C_2H_5)_3$, CO, CNBu^n, PPh_3, $PPh_2(CH_3)$, $PPh(CH_3)_2$, $P(C_2H_5)_3$, $AsPh_3$, $AsPh(CH_3)_2$). These complexes are synthesised in the same way as their HB(3,5-Me$_2$pyz)$_3$ analogues (see p. 239) using Rh{HB(pyz)$_3$}Cl$_2$(CH_3CN) [3].

The complex containing $PPh_2(CH_3)$ was isolated containing ½ mole of toluene of crystallisation [3].

Rh{HB(pyz)$_3$}Cl$_2$(CH$_3$CN) is obtained by refluxing [Rh{HB(pyz)$_3$}Cl$_2$]$_2$ in acetonitrile under nitrogen [3].

Rh{HB(3,5-Me$_2$pyz)$_3$}I$_2$(CO) is obtained by iodine oxidation of Rh{HB(3,5-Me$_2$pyz)$_3$}(CO)$_2$ in dichloromethane under nitrogen at room temperature [3].

Rh{HB(3,5-Me$_2$pyz)$_3$}HClL (L = PPh_3, $PPh_2(CH_3)$, $PPh(CH_3)_2$, $P(C_2H_5)_3$, $P(o\text{-}CH_3C_6H_4)_2(CH_3)$, $AsPh(CH_3)_2$). These complexes are formed when [N(C_2H_5)$_3$H][Rh{HB(3,5-Me$_2$pyz)$_3$}HCl$_2$] is refluxed with an equimolar amount of L under nitrogen in toluene. The complexes containing $PPh_2(CH_3)$ and $P(o\text{-}CH_3C_6H_4)_2(CH_3)$ were isolated containing one mole of CH_2Cl_2 of crystallisation [3].

[N(C$_2$H$_5$)$_3$H][Rh{HB(3,5-Me$_2$pyz)$_3$}HCl$_2$] is formed when Rh{HB(3,5-Me$_2$pyz)$_3$}Cl$_2$(CH_3OH) is treated with hydrogen in dry, refluxing toluene containing a small amount of triethylamine [3].

[Rh{HB(pyz)$_3$}Cl$_2$]$_2$ and **[Rh{HB(3,5-Me$_2$pyz)$_3$}Cl$_2$]$_2$** are prepared by refluxing $RhCl_3 \cdot 3H_2O$ with either Na[HB(pyz)$_3$] or Na[HB(3,5-Me$_2$pyz)$_3$] in methanol [3].

[Q][Rh{HB(3,5-Me$_2$pyz)$_3$}Cl$_3$] (Q = $AsPh_4$, $N(C_2H_5)_3H$ or $N(C_2H_5)_4$). The **tetraphenylarsonium** salt is obtained by dissolving Rh{HB(3,5-Me$_2$pyz)$_3$}Cl$_2$(CH_3OH) in the minimum amount of CH_2Cl_2 and adding [$AsPh_4$]Cl. The **triethyl-** and **tetraethyl-ammonium** salts are prepared by refluxing Rh{HB(3,5-Me$_2$pyz)$_3$}Cl$_2$(CH_3OH) in dry toluene containing either [N(C_2H_5)$_3$H]Cl or [N(C_2H_5)$_4$]Cl [3].

[N(CH$_3$)$_4$][Rh{HB(3,5-Me$_2$pyz)$_3$}Cl$_2$X] (X = F or Br) are formed when Rh{HB(3,5-Me$_2$pyz)$_3$}-Cl$_2$(CH_3OH) is refluxed with [N(CH_3)$_4$]X in dry toluene [3].

[N(CH$_3$)$_4$][Rh{HB(pyz)$_3$}Cl$_2$Br] is prepared by refluxing a mixture of Rh{HB(pyz)$_3$}-Cl$_2$(CH_3CN) and [N(CH_3)$_4$]Br in dry toluene [3].

[Rh{HB(pyz)$_3$}$_2$][PF$_6$] is obtained by treatment of $RhCl_3 \cdot 3H_2O$ with two molar equivalents of K[HB(pyz)$_3$], followed by addition of NH_4PF_6 [4].

References:

[1] A. W. Addison, K. Dawson, R. D. Gillard, B. T. Heaton, H. Shaw (J. Chem. Soc. Dalton Trans. **1972** 589/96). – [2] A. W. Addison, R. D. Gillard, D. H. Vaughan (J. Chem. Soc. Dalton Trans. **1973** 1187/93). – [3] S. M. May, P. Reinsalu, J. Powell (Inorg. Chem. **19** [1980] 1582/9). – [4] N. F. Borkett, M. I. Bruce (J. Organometal. Chem. **65** [1964] C51/C52).

2.6.2 Imidazole and Substituted Imidazole Complexes

Imidazole Complexes

Imidazole, 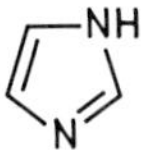 = iz

[Rh(iz)$_4$Cl$_2$]Cl is made from $RhCl_3 \cdot 3H_2O$ in ethanol with imidazole in a 1:5 mole ratio at pH = 4 to 5. The compound forms cream-coloured crystals. The conductance is 115 $cm^2 \cdot \Omega^{-1} \cdot mol^{-1}$. The electronic absorption spectrum was measured.

[Rh(iz)$_2$X$_2$]X (X=Cl, brick-red; Br, light brown; I, dark brown; NO_2, white; NCS, orange-yellow). The **chloro** complex is made by refluxing $RhCl_3 \cdot 3H_2O$ in ethanol with HCl at pH = 2 to 3 and the ligand in a 1:2 metal:ligand ratio. The other species were made by refluxing [Rh(iz)$_4$Cl$_2$]Cl with KX in water for two to three hours. Infrared spectra were measured (X = NCS, NO_2). Molar conductances (in $cm^2 \cdot \Omega^{-1} \cdot mol^{-1}$) are 73 (Cl), 58 (Br), 50 (I).

S. P. Ghosh, P. Bhattarcharjee, L. Dubey, L. K. Mishra (J. Indian Chem. Soc. **54** [1977] 230/8).

For carboxylate imidazole complexes see pp. 26, 34, 37.

N-Methylimidazole Complexes

N-Methylimidazole, 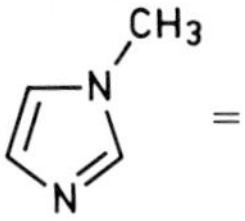 = miz

[Rh(miz)$_5$Cl]Y$_2$, [Rh(miz)$_5$Br](ClO$_4$)$_2$. The **perchlorate** is made by reaction of $RhX_3 \cdot nH_2O$ (X = Cl, Br) in 30% aqueous ethanol with the ligand under reflux, and $NaClO_4$ is then added; the **tetrafluoroborate** of the chloro species is similarly made using HBF_4 in place of $NaClO_4$. The complexes form pale yellow crystals.

^{1}H NMR data were recorded. For other physical data see table below. Electronic spectral band maxima were recorded. With NaX, the chloro species gives *trans*-[Rh(miz)$_4$X$_2$]$^+$ (X = Cl, Br).

***trans*-[Rh(miz)$_4$Cl$_2$]Y, *trans*-[Rh(miz)$_4$Br$_2$]Br·H$_2$O.** The trihydrated **chloride** is made from [Rh(miz)$_5$Cl](ClO$_4$)$_2$ and NaCl in 50% aqueous ethanol under reflux in the absence of O_2. It forms yellow needles. A **perchlorate** was also reported. The orange **bromide** is made in similar fashion to the chloride, NaBr replacing NaCl.

For physical data see table on p. 242. Electronic spectral band maxima were recorded.

A. W. Addison, K. Dawson, R. D. Gillard, B. T. Heaton, H. Shaw (J. Chem. Soc. Dalton Trans. **1972** 589/96).

For carboxylate N-methylimidazole complexes see p. 48.

Substituted N-Methylimidazole Complexes

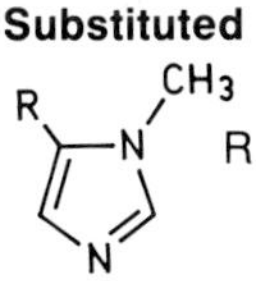 R – miz; substitution in 5-position

***trans*-[Rh(R-miz)$_4$X$_2$]X·nH$_2$O.** These were apparently made by refluxing $RhX_3 \cdot nH_2O$ (X = Cl, Br) with the ligand in 1:4 molar proportion in 30% aqueous ethanol.

Physical properties of N-methylimidazole complexes:

Compound	Conductance in $cm^2 \cdot \Omega^{-1} \cdot mol^{-1}$ in CH_3NO_2	in H_2O
$[Rh(miz)_5Cl](ClO_4)_2$	154	
$[Rh(miz)_5Cl](BF_4)_2$	150	
$[Rh(miz)_5Br](ClO_4)_2$	146	
trans-$[Rh(miz)_4Cl_2]Cl \cdot 3H_2O$	79.5	
trans-$[Rh(miz)_4Br_2]Br \cdot H_2O$	80.2	
trans-$[Rh(NO_2\text{-}miz)_4Cl_2]Cl \cdot 2H_2O$	37.2	105.6
trans-$[Rh(NO_2\text{-}miz)_4Cl_2]Cl \cdot 4H_2O$		
trans-$[Rh(NO_2\text{-}miz)_4Cl_2]ClO_4$	84.3	
trans-$[Rh(NO_2\text{-}miz)_4Br_2]Br \cdot H_2O$	33.4	
trans-$[Rh(Cl\text{-}miz)_4Cl_2]Cl \cdot 2H_2O$		91.8
trans-$[Rh(Br\text{-}miz)_4Br_2]Br \cdot 2H_2O$		112.7

Electronic absorption spectra were measured, and also the ^1H NMR spectrum of *trans*-$[Rh(miz)_4Cl_2]Cl$ in $CDCl_3$-$(CD_3)_2SO$.

A. W. Addison, K. Dawson, R. D. Gillard, B. T. Heaton, H. Shaw (J. Chem. Soc. Dalton Trans. **1972** 589/96).

Benzimidazole Complexes

Benzimidazole, 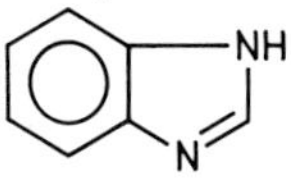 = BzH

"$Rh(Bz)_3 \cdot 3H_2O$" is a light yellow material. No preparative details are given.

$[Rh(BzH)_2(H_2O)_2]Cl_3$ is a light yellow material, made by boiling $[Rh(BzH)_2Cl_2]Cl$ in water. Its conductance is 232 $cm^2 \cdot \Omega^{-1} \cdot mol^{-1}$.

$[Rh(BzH)_2X_2]X$ (X = Cl, brick-red; Br, light brown; I, dark brown; NO_2, white; NCS, yellow-orange). The **chloro** complex is made from $RhCl_3 \cdot 3H_2O$ and the ligand in 1:2 proportion with HCl in ethanol under reflux; the others are made from it and KX in water under reflux. Infrared spectra of the chloro complex shows ν_{RhCl} at 329 cm^{-1}; the ν_{CN} and ν_{CS} frequencies for X = NCS are 2095 and 768 cm^{-1}. Conductances in $cm^2 \cdot \Omega^{-1} \cdot mol^{-1}$ are 82 (Cl), 79 (Br), 75 (I), 73 (NO_2), 68 (NCS).

S. P. Ghosh, P. Bhattarcharjee, L. Dubey, L. K. Mishra (J. Indian Chem. Soc. **54** [1977] 230/8).

Substituted Benzimidazole Complexes

$[Rh(MBzH)_2X_2)]X$ (2-methylbenzimidazole = MBzH). X = Cl, red; Br, light brown; I, deep brown; NCS, orange-yellow; NO_2, white. The **chloro** complex is made from $RhCl_3 \cdot 3H_2O$ and the ligand in 1:2 ratio in refluxing ethanol, and the other complexes by refluxing the chloro complex with KX in water. Infrared spectra of the NCS and NO_2 complexes were measured. Molar conductances (in $cm^2 \cdot \Omega^{-1} \cdot mol^{-1}$) are 50 (Cl) and 40 (I) [1].

$[Rh(GB)_3]Y_3$ (2-guanidinobenzimidazole, GB = $C_8H_9N_5$). The straw-coloured **chloride** is made from $RhCl_3$ and the ligand in 1:3 proportion in water under reflux for three hours; the **perchlorate** is made as pale yellow crystals by addition of $NaClO_4$ [1, 2]. The conductance of the chloride in water is 411 $cm^2 \cdot \Omega^{-1} \cdot mol^{-1}$ [1].

$[Rh(GB)_2Cl_2]NO_3 \cdot 4H_2O$. This is made from $RhCl_3 \cdot 3H_2O$ and the hydrochloride of the ligand in 1:2 molar proportions, followed by addition of conc. HNO_3 at pH = 1 to 2 [1, 2].

$[Rh(GB)_2X_2)]X$ with X = Cl, yellow; Br, bright yellow; I, brown; NO_2, light yellow; NCS, orange-yellow. These were made from $[Rh(GB)_2(H_2O)_2]Cl_3 \cdot H_2O$ and KX in water under reflux. Infrared spectra of the NO_2 and NCS complexes were measured. Conductances in $cm^2 \cdot \Omega^{-1} \cdot mol^{-1}$ are 212 (Cl); 115 (Br); 102 (NO_2) [1, 2]. IR frequencies for X = Cl: ν(RhCl) = 333, 296 cm^{-1} [2].

$[Rh(GB)_2(H_2O)_2]X_3 \cdot nH_2O$. The dull yellow **chloride** (monohydrate) is made from $RhCl_3 \cdot 3H_2O$ and the ligand hydrochloride in a 1:2 ratio, refluxed on a steam bath. The conductance is 368 $cm^2 \cdot \Omega^{-1} \cdot mol^{-1}$. The tetrahydrated **nitrate** (orange-yellow) is made from the chloride with HNO_3 [1, 2].

$[Rh(GB)_2Y_2]Cl_3$, $[Rh(GB)_2LL]Cl_n$ (Y = py, NH_3; LL = 2,2′-bipyridyl, 1,10-phenanthroline (n = 3); α-picolinate (n = 2)). These are made from $[Rh(GB)_2(H_2O)_2]Cl_3 \cdot H_2O$ under reflux with the ligand. They are white or pale yellow. Conductances in $cm^2 \cdot \Omega^{-1} \cdot mol^{-1}$ are 406 (X = NH_3), 271 (LL = α-picolinate); 411 (LL = 2,2′-bipy); 403 (LL = 1,10-phen). Cream-yellow iodides were also made for LL = bipy, phen [1, 2].

$[Rh(PIH)_3](ClO_4)_3 \cdot 5H_2O$ with 2-(2′-pyridyl)imidazole = $C_7H_7N_3$ = PIH. This is made by mixing an aqueous solution of an unspecified rhodium salt with the ligand in ethanol. The yellow product is recrystallised from water-ethanol. It is diamagnetic and has a conductivity of 77.3 $cm^2 \cdot \Omega^{-1} \cdot mol^{-1}$ [3].

$[Rh(PBH)_2X_2]X \cdot nH_2O$ (2-(2′-pyridyl)benzimidazole = $C_{12}H_9N_3$ = PBH). The dihydrated yellow **chloride** is made from $RhCl_3 \cdot 3H_2O$ and the ligand dissolved in dilute HCl at pH = 3.0, and the anhydrous red **iodide** by refluxing with NaI. The X = Br, NO_2^- and N_3^- species were mentioned. The conductance of the chloro complex in water is 112 $cm^2 \cdot \Omega^{-1} \cdot mol^{-1}$ [4].

A *cis* configuration has been assigned to these complexes [5].

References:

[1] S. P. Ghosh, P. Bhattarcharjee, L. Dubey, L. K. Mishra (J. Indian Chem. Soc. **54** [1977] 230/8). – [2] S. P. Ghosh, P. Bhattarcharjee (Indian J. Chem. A **15** [1977] 886/9). – [3] B. Chiswell, F. Lions, B. S. Morris (Inorg. Chem. **3** [1964] 110/4). – [4] S. P. Ghosh, P. Bhattarcharjee, L. K. Mishra (J. Indian Chem. Soc. **51** [1974] 308/14). – [5] B. R. Ramesh, B. S. S. Mallikarjuna, G. K. N. Reddy (Natl. Acad. Sci. Letters [India] **1** [1978] 24/6 from C.A. **89** [1978] No. 122152).

2.6.3 Indazole and Substituted Indazole Complexes

Indazole, 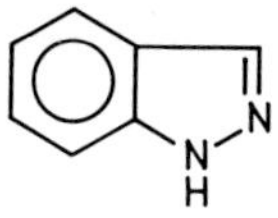 = inz

$Rh(inz)_3Cl_3$. This bright yellow material, melting at 340°C with decomposition, is made by mixing $RhCl_3 \cdot nH_2O$ and the ligand in 1:3 proportion in HCl-ethanol. It is diamagnetic.

The electronic absorption band maxima were recorded; 10 Dq, B and B′ are given as 20109 cm^{-1}, 539.14 cm^{-1} and 0.76, respectively.

The conductance in $(CH_3)_2SO$ and $C_6H_5NO_2$ are, respectively, 9.12 and 9.14 $cm^2 \cdot \Omega^{-1} \cdot mol^{-1}$.

$Rh(NO_2\text{-inz})_3Cl_3$. This is made from 5-nitroindazole and $RhCl_3$ in C_2H_5OH under reflux. It is yellow and melts at 260°C with decomposition.

The complex is said to have a magnetic moment of 2.44 B.M. The infrared spectrum (200 to 4000 cm^{-1}) shows ν(RhCl) at 360, 330 and 285 cm^{-1} with ν(RhN) at 485 cm^{-1}. The electronic absorption spectrum was measured and assigned. The conductances in $(CH_3)_2SO$ and $C_6H_5NO_2$ are 5.70 and 1.14 $cm^2 \cdot \Omega^{-1} \cdot mol^{-1}$, respectively.

$Rh(NH_2\text{-inz})_2Cl_3 \cdot 2H_2O$. This is made from 5-aminoindazole and $RhCl_3$ in refluxing C_2H_5OH. It is brownish-yellow and melts at 360°C with decomposition.

It is diamagnetic. The infrared spectrum (200 to 4000 cm^{-1}) shows ν(RhCl) at 340 cm^{-1} and ν(RhN) at 460 cm^{-1}. The electronic spectrum was measured and assigned. A binuclear structure with two bridging chloro ligands was proposed.

S. A. A. Zaidi, N. Singhal, A. Lal (Transition Metal Chem. [Weinheim] **4** [1979] 133/6).

2.6.4 Thiazole Complexes

***trans*-$[Rh(C_3H_3NS)_4X_2]X \cdot nH_2O$.** The **chloride** (X = Cl), as a pentahydrate (n = 5), and the **bromide** (X = Br), as a dihydrate (n = 2), are prepared by boiling $RhX_3 \cdot 3H_2O$ in ethanol with thiazole [1].

The complexes were isolated as yellow-orange solids. Electronic spectral, conductivity and polarographic data (the complexes underwent an irreversible two-electron reduction in aqueous solution) are shown in Table 1, p. 245 [1, 2].

2.6.5 Triazole Complexes

$Rh(C_6H_4N_3)_3 \cdot 3H_2O$ is obtained by reaction of $Na_3RhCl_6 \cdot 12H_2O$ or rhodium sulphate with benzotriazole in aqueous solution buffered to pH 3.3 to 5.3 using acetate. The corresponding bromobenzotriazolate $Rh(C_6H_3BrN_3)_3 \cdot 3H_2O$ is similarly prepared. Both complexes were isolated as microcrystalline white-yellow solids, whose electronic spectra in n-butanol were obtained in the range 240 to 340 nm (shown in paper) [5].

The compounds lost water at 140 to 350°C, and decomposed at 420 to 470°C [5].

$Rh(C_6H_4N_3)_3 \cdot (C_6H_4N_3H)$, $Rh(C_6H_3BrN_3)_3(C_6H_3BrN_3H)$. The complexes are formed when $Na_3RhCl_6 \cdot 12H_2O$ reacts with an excess of benzotriazole or bromobenzotriazole in aqueous solution as described above. They were isolated as yellow-white solids [5].

2.6.6 Pyrimidine Complexes ($C_4H_4N_2$ = pmd)

***trans*-$[Rh(pmd)_4Cl_2]ClO_4$** is prepared by refluxing $RhCl_3 \cdot 3H_2O$ and pyrimidine in ethanol [1]. It was isolated as yellow-orange crystals. The electronic spectral, electrical conductivity and polarographic data (the complex underwent an irreversible two-electron reduction in aqueous solution) are summarised in Table 1, p. 245 [1, 2].

2.6.7 Pyrazine Complexes ($C_4H_4N_2$ = prz)

$Rh(prz)_3Cl_3$. This complex may be prepared by heating $Na_3RhCl_6 \cdot 12H_2O$ in aqueous solution [4] or in an autoclave at 130°C [5]. It was isolated as yellow-pink [4] or orange-yellow crystals [5].

The IR spectrum of the compound has been correlated with its structure [6], suggested to be in the *mer* configuration [4]. Its electronic spectrum has been described in water and HCl solution in the range 35000 to 48000 cm^{-1} [4].

***trans*-Rh(prz)(przH)Cl$_4$·H$_2$O** is obtained from Rh(prz)$_3$Cl$_3$ in conc. HCl either by refluxing the mixture or by irradiation with UV light for 3 h. It was isolated as orange-red crystals. The electronic spectrum of the compound was reported in HCl solution in the range 35000 to 48000 cm^{-1} [6].

For **[Rh(NH$_3$)$_5$(pz)]$^{3+}$**, see p. 140.

References:

[1] A. W. Addison, K. Dawson, R. D. Gillard, B. T. Heaton, H. Shaw (J. Chem. Soc. Dalton Trans. **1972** 589/96). – [2] A. W. Addison, R. D. Gillard, D. H. Vaughan (J. Chem. Soc. Dalton Trans. **1973** 1187/93). – [3] L. N. Lomakina, I. R. Alimarin (Zh. Neorgan. Khim. **12** [1967] 409/13; Russ. J. Inorg. Chem. **12** [1967] 210/3). – [4] G. Rio, F. Larèze (Bull. Soc. Chim. France **1975** 2393/8). – [5] F. Larèze (Compt. Rend. C **267** [1980] 1119/20).

[6] L. Sebagh, J. Zarembowitch (J. Chim. Phys. **66** [1969] 1974/80).

2.6.8 Tables on Physical Properties

Electronic spectral, electrical conductivity and polarographic properties of rhodium complexes of pyrazole, imidazoles, thiazole and pyrimidine, a) according to [1, 2]:

Table 1

Complex	$\lambda(\varepsilon)$ b)	Λ c) H$_2$O	Λ c) CH$_3$NO$_2$	$E_{1/2}$ d)
trans-[Rh(pyzH)$_4$Cl$_2$]Cl·5H$_2$O	407(88)	77	—	−0.24
[Rh(miz)$_5$Cl](ClO$_4$)$_2$	350(103)		154	−0.57
[Rh(miz)$_5$Cl](BF$_4$)$_2$	350(100)		150	
[Rh(miz)$_5$Br](ClO$_4$)$_2$	370(111)		146	−0.61 e)
trans-[Rh(miz)$_4$Cl$_2$]Cl·3H$_2$O	410(71)		79.5	−0.37
trans-[Rh(miz)$_4$Br$_2$]Br·H$_2$O	439(124)		80.2	−0.34 e)
trans-[Rh(5-NO$_2$-miz)$_4$Cl$_2$]Cl·2H$_2$O	—	105.6	37.2	−0.32 f)
trans-[Rh(5-NO$_2$-miz)$_4$Cl$_2$]ClO$_4$		84.3		
trans-[Rh(5-NO$_2$-miz)$_4$Br$_2$]Br·H$_2$O	439		33.4	
trans-[Rh(5-Cl-miz)$_4$Cl$_2$]Cl·2H$_2$O	409(81)	91.8		−0.50 g)
trans-[Rh(5-Cl-miz)$_4$Br$_2$]Br·2H$_2$O	438(131)	112.7		
trans-[Rh(C$_3$H$_3$NS)$_4$Cl$_2$]Cl·5H$_2$O	409(89)		77	−0.15
trans-[Rh(C$_3$H$_3$NS)$_4$Br$_2$]Br·2H$_2$O	440(144)		73	−0.18
trans-[Rh(pmd)$_4$Cl$_2$]ClO$_4$	409(105)		47	−0.12

a) pyzH = pyrazole; miz-N-methylimidazole, 5-NO$_2$-miz and 5-Cl-miz are its 5-nitro- and 5-chloro-derivatives; C$_3$H$_3$NS = thiazole; pmd = pyrimidine. – b) λ in nm, ε in $cm^2 \cdot l^{-1} \cdot mol^{-1}$. – c) Conductivity in $cm^2 \cdot \Omega^{-1} \cdot mol^{-1}$, in 10^{-3}M solution. – d) In V, using a dropping-mercury electrode, vs. standard calomel electrode. – e) Ill-defined wave. – f) Second wave, $E_{1/2} = -0.54$ V. – g) $E_{1/2}$ varies with time.

Tris(pyrazolyl)borato rhodium complexes, colours, melting points (m.p., dec. = decomposition), IR and ^1H NMR spectral data[a)] [3]:

Table 2

Complex	Colour	m.p. in °C	IR Data in cm^{-1}	^1H NMR τ (J in HZ)
$[N(C_2H_5)_4][RhL(O_2CCF_3)_2H]$	pale yellow plates	209	2050 (RhH)	20.8 (7)
$RhL(O_2CCH_3)$	pale yellow plates	297	—	c)
$RhL(O_2CCF_3)_2 \cdot H_2O$	pale yellow plates	231 to 233	—	c)
$RhL'(O_2CCF_3)_2$	pale yellow plates	178 to 180	—	c)
$RhL'(O_2CCF_3)_2(CH_3CN)$	pale yellow prisms	132 to 137	2340 (CN)	c)
$[N(C_2H_5)_4][RhL(O_2CCF_3)_2Cl]$	pale yellow needles	263 to 265	—	c)
$RhL(O_2CCF_3)_2(CO)$	pale yellow needles	234 to 236	2130 (CO)	c)
$RhL(O_2CCF_3)_2(CH_3CN)$	pale yellow prisms	232 to 233	—	c)
$RhL'(CH_3COCHCOCH_3)Cl$	dull yellow prisms	>350	—	c)
$RhL'(CF_3COCHCOCF_3)Cl$	bright yellow needles	191 to 192	—	c)
$RhL(CH_3COCHCOCH_3)Cl$	pale yellow plates	229 to 231	—	c)
$RhL(CF_3COCHCOCF_3)Cl$	bright yellow plates	255 to 258	—	c)
$RhLCl_2(H_2O)$	pale yellow prisms	315	—	c)
$RhLCl_2(CH_3OH)$	gold prisms	296 to 298	—	c)
$RhLCl_2(CH_3CN)$	yellow plates	259 to 262	2300 (CN)	c)
$RhL'Cl_2(CH_3CN) \cdot CH_3CN$	dull gold prisms	>360	—	c)
$RhLCl_2(PhCN)$	bright yellow needles	298	2260 (CN)	c)
$RhL'Cl_2(PhCN)$	dull gold prisms	206 to 211	—	c)
$RhLCl_2(C_5H_5N)$	pale yellow needles	355	—	c)

$RhL'Cl_2(C_5H_5N)$	bright yellow prisms	331 to 332	—	c)
$RhLCl_2(4\text{-}CH_3C_5H_4N)$	pale yellow prisms	168	—	c)
$RhLCl_2\{N(C_2H_5)_3\}$	gold prisms	280 to 282	—	c)
$RhL'Cl_2\{N(C_2H_5)_3\}$	dull yellow prisms	239 to 241	—	c)
$RhLCl_2\{N(C_3H_7)_3\}$	yellow prisms	226 to 230	—	c)
$RhLCl_2(3,5\text{-}Me_2C_3HN_2H)$	dark gold prisms	165	—	c)
$RhLCl_2(NH_3)$ d)	pale yellow prisms	>350	—	c)
$\{RhLCl_2\}(bipy)$	dark gold prisms	165	—	c)
$RhLCl_2(CO)$	gold microprisms	304 to 306	2100 (CO)	c)
$RhL'Cl_2(CO)\cdot CH_3CN$	dull gold prisms	>350	2120 (CO)	c)
$RhLCl_2(CNBu^n)$	bright gold needles	293	2203 (CN)	c)
$RhL'Cl_2(CNBu^n)$	bright yellow prisms	148 to 150	2240 (CN)	c)
$RhLCl_2(CNCH_2Ph)$	gold tetragonal prisms	>360	2230 (CN)	c)
$RhLI_2(CO)$	dark red crystals	—	2090 (CO)	c)
$RhLClH(PPh_3)$	pale yellow needles	171 to 175	2135 (RhH)	14.25 (8.0; 20)
$RhLClH\{PPh_2(CH_3)\}\cdot CH_2Cl_2$	dark gold prisms	221 to 224	2070	15.2 (10; 24)
$RhLClH\{PPh(CH_3)_2\}$	gold prisms	172 to 175	2175	15.84 (10; 20)
$RhLClH\{P(C_2H_5)_3\}$	bright yellow prisms	230 to 235	—	16.2 (10; 24)
$RhLClH\{P(o\text{-}CH_3C_6H_4)_2(CH_3)\}\cdot CH_2Cl_2$	bright yellow prisms	129 to 133	—	14.66 (7; 23)
$RhLClH\{AsPh(CH_3)_2\}$	bright orange prisms	231 to 235	—	15.48 (6)
$RhLCl_2(PPh_3)$	dark gold needles	199 to 201	—	c)
$RhL'Cl_2(PPh_3)$	pale gold prisms	275 to 278	—	c)
$RhLCl_2\{PPh_2(CH_3)\}$	dark orange needles	222	—	c)

Table 2 (continued)

Complex	Colour	m.p. in °C	IR Data in cm^{-1}	^1H NMR τ (J in HZ)
$RhL'Cl_2\{PPh_2(CH_3)\} \cdot \frac{1}{2} C_6H_5CH_3$	bright gold prisms	273 to 275	—	c)
$RhLCl_2\{PPh(CH_3)_2\}$	orange prisms	232	—	c)
$RhL'Cl_2\{PPh(CH_3)_2\}$	bright gold prisms	291	—	c)
$RhLCl_2\{P(o\text{-}CH_3C_6H_4)_2(CH_3)\} \cdot CH_2Cl_2$	dark yellow prisms	162	—	c)
$RhL'Cl_2\{P(C_2H_5)_3\}$	bright gold prisms	287	—	c)
$RhL'Cl_2(AsPh_3)$	dark gold prisms	110 to 111	—	c)
$RhL'Cl_2\{AsPh(CH_3)_2\}$	pale yellow needles	290	—	c)
$[AsPh_4][RhLCl_3]$	bright gold prisms	339	—	c)
$[N(C_2H_5)_3H][RhLCl_3]$	orange needles	260 to 267	—	c)
$[N(C_2H_5)_4][RhLCl_3]$	bright gold prisms	332 to 342 (dec.)	—	c)
$[N(C_2H_5)_3H][RhLCl_2H]$	dark gold prisms	258	—	c)
$[N(CH_3)_4][RhLCl_2F]$	dark yellow plates	261 to 300 (dec.)	—	c)
$[N(CH_3)_4][RhLCl_2Br]$	dark gold plates	341 (dec.)	—	c)
$[N(CH_3)_4][RhL'Cl_2Br]$	pale yellow prisms	>350	—	c)
$[RhLCl_2]_2$	dark gold plates	228 to 233	—	c)
$[RhL'Cl_2]_2$	dark gold plates	293 to 297	—	c)

a) Ligands: L = HB(3,5-Me_2pyz)$_3$; L' = HB(pyz)$_3$. – b) In $CDCl_3$ and $(CD_3)_2SO$ solutions; τ value $\{^1J(Rh\text{-}H)\}$. – c) ^1H NMR spectra in range τ – 0 – 10 reported in [3]. – d) ^1H NMR data indicate 2 mol NH_3 present, but bonding arrangements are unknown. – e) $(^1J(RhH); ^2J(PH))$.

References:

[1] A. W. Addison, K. Dawson, R. D. Gillard, B. T. Heaton, H. Shaw (J. Chem. Soc. Dalton Trans. **1972** 589/96). – [2] A. W. Addison, R. D. Gillard, D. H. Vaughan (J. Chem. Soc. Dalton Trans. **1973** 1187/93). – [3] S. M. May, P. Reinsalu, J. Powell (Inorg. Chem. **19** [1980] 1582/9).

2.6.9 Pyridine Complexes

py = pyridine, C_5H_5N

This is a large group of compounds, principally of the type *trans*-$[Rh(py)_4X_2]^+$ and $Rh(py)_3X_3$, although there are bis- and mono-pyridine complexes. The large majority of compounds contain Rh^{III}, but Rh^{I} species are implicated in the catalytic formation of the Rh^{III} complexes.

General Literature:

R. D. Gillard, Advances in Rhodium Chemistry, Rec. Chem. Progr. **32** [1971] 17/28.

2.6.9.1 Unsubstituted Pyridine Complexes

Rhodium(I) Complexes

$[Rh\,py_4]^+$. Complexes of this species have not been isolated, but the cation is implicated in the electrochemical reduction of *trans*-$[Rh\,py_4X_2]^+$ (X = Cl or Br) [1], and in the reactions of *trans*-$[Rh\,py_4Cl_2]^+$ with BH_4^-, where *trans*-$[Rh\,py_4HCl]^+$ is apparently formed [2]. Reaction of the latter with base was thought to give $[Rh\,py_4]^+$ which subsequently reacted with water and with halogens (X_2) giving *trans*-$[Rh\,py_4H(OH)]^+$ and *trans*-$[Rh\,py_4X_2]^+$ [2]. It may also be present in the ethanol-catalysed formation of *trans*-$[Rh\,py_4Cl_2]^+$ from $[Rh(H_2O)Cl_5]^{2-}$ and pyridine [3].

$Rh\,py_3Cl$. This compound is formed, together with CO_2, on irradiation of $Rh\,py_3(C_2O_4)Cl$ in vacuo. It is a red solid which dissolved in methanol forming a yellow-black solution. Full characterisation of the species was not possible, but it afforded a ^1H NMR spectrum indicating that it was diamagnetic. It reacted with Cl_2 and with PPh_3 in the presence of CO giving $Rh\,py_3Cl_3$ and $Rh(CO)(PPh_3)_2Cl$, respectively [4].

References:

[1] R. D. Gillard, B. T. Heaton, D. H. Vaughan (J. Chem. Soc. A **1971** 1840/6). – [2] R. D. Gillard, B.T. Heaton (Coord. Chem. Rev. **8** [1972] 149/57). – [3] J. V. Rund, F. Basolo, R. G. Pearson (Inorg. Chem. **3** [1964] 658/61). – [4] A. W. Addison, R. D. Gillard, P. S. Sheridan, L. R. H. Tipping (J. Chem. Soc. Dalton Trans. **1974** 709/16).

Rhodium(II) Complexes

A number of complexes formulated as containing Rh^{II} [1, 2] have been subsequently reformulated as Rh^{III} species. These compounds, and their reformulations, are listed below [3]:

$Rh\,py_6X_2$	X = Cl	now $[Rh\,py_4Cl_2]Cl\cdot5H_2O$
	X = Br	now $[Rh\,py_4Br_2]Br$
	X = I	now $[Rh\,py_5H]I_2$
$[Rh\,py_5X]X$	X = Cl	now $[Rh\,py_4Cl_2](H_5O_2)Cl_2$
	X = Br	now $[Rh\,py_4Br_2]Br$
$[Rh\,py_5Br]I$		now $[Rh\,py_4Br_2]I$
$Rh\,py_4Br_2$		now $[Rh\,py_4Br_2][Rh\,py_2Br_4]$
$[py\,H]_4[Rh_2py_2X_8]$	X = Cl, Br	now $[py\,H][Rh\,py_2X_4]$

References:

[1] M. Delépine (Bull. Soc. Chim. France [4] **45** [1929] 235/8). – [2] F. P. Dwyer, R. S. Nyholm (J. Proc. Roy. Soc. N.S. Wales **76** [1943] 275/80). – [3] B. N. Figgis, R. D. Gillard, R. S. Nyholm, G. Wilkinson (J. Chem. Soc. **1964** 5189/93).

Rhodium(III) Complexes

Hexakis-pyridine Complex, $[Rh\,py_6]^{3+}$

This is said to be formed from $RhCl_3$ in pyridine solutions containing KCl or KBr [1]. The complex was reported to undergo a one-electron polarographic reduction affording, apparently, a Rh^{II} complex. However, species described as $Rh\,py_6X_3$ have been reformulated as containing $Rh\,py_3X_3$ (X = Cl or Br) [2].

References:

[1] J. B. Willis (J. Am. Chem. Soc. **66** [1944] 1067/9). – [2] B. N. Figgis, R. D. Gillard, R. S. Nyholm, G. Wilkinson (J. Chem. Soc. **1964** 5189/93).

2.6.9.2 Substituted Pyridine Complexes

2.6.9.2.1 Pentakis-pyridine Complexes

$[Rh\,py_5H]^{2+}$. This species may be formed when aqueous solutions thought to contain *trans*-$[Rh\,py_4HX]^+$ are boiled with an excess of pyridine. The colourless solutions so formed exhibited a high-field signal at $\tau = 29$.

$[Rh\,py_5H]I_2$ is obtained when *trans*-$[Rh\,py_4I_2]I$ is treated with aqueous $NaBH_4$ or hypophosphorous acid. It was isolated as very pale yellow, air-sensitive crystals. It exhibited ν(RhH) at 1980 or 1990 cm^{-1}.

B. N. Figgis, R. D. Gillard, R. S. Nyholm, G. Wilkinson (J. Chem. Soc. **1964** 5189/93).

2.6.9.2.2 Tetrakis-pyridine Complexes

Hydrido Complexes

Many reactions of *trans*-$[Rh\,py_4X_2]^+$ with hydride-forming reagents, e.g. $NaBH_4$, H_3PO_2, primary and secondary alcohols, etc., may involve the formation of hydrido intermediates, although this has not always been conclusively demonstrated [1]. These species may be important catalytically in substitution reactions of *trans*-$[Rh\,py_4X_2]^+$ [2, 3], although kinetic studies have cast some doubt on this [4].

***trans*-$[Rh\,py_4HX]^+$** (X = Cl or Br). The **chloro** (X = Cl) complex is obtained by treatment of *trans*-$[Rh\,py_4Cl_2]^+$ with aqueous $NaBH_4$ [5 to 8] and is pale brown. The **bromo** (X = Br) compound may be formed either by reaction of *trans*-$[Rh\,py_4Br_2]^+$ with aqueous or ethanolic $NaBH_4$ [6, 8], or by treatment with aqueous H_3PO_2 [8]. It has been isolated, presumably as an aquated bromide, as brown acicular crystals [8].

The IR spectrum of *trans*-$[Rh\,py_4HBr]^+$ exhibited a maximum at 1976 cm^{-1} [6] or ~2000 cm^{-1} [8] thought to be due to ν(RhH). The electronic spectra of *trans*-$[Rh\,py_4HX]^+$ (X = Cl or Br) have been recorded [6]. The complexes exhibited high-field ^1H NMR signals, at $\tau = 28.5$ [6, 8] or 28.6 [5, 6]; no J(RhH) was reported.

trans-$[Rh\,py_4HCl]^+$ reacted with CCl_4 giving *trans*-$[Rh\,py_4Cl_2]^+$ [5].

References:

[1] R. D. Gillard (Coord. Chem. Rev. **8** [1972] 149/57). – [2] R. D. Gillard, J. A. Osborn, G. Wilkinson (J. Chem. Soc. **1965** 1951/65). – [3] R. D. Gillard, B. T. Heaton, D. H. Vaughan (Chem. Commun. **1969** 974/5). – [4] J. V. Rund (Inorg. Chem. **7** [1968] 24/7). – [5] R. D. Gillard, G. Wilkinson (J. Chem. Soc. **1963** 3594/9).

[6] J. A. Osborn, R. D. Gillard, G. Wilkinson (J. Chem. Soc. **1964** 3168/73). – [7] R. D. Gillard, G. Wilkinson (Inorg. Syn. **10** [1967] 64/7). – [8] B. N. Figgis, R. D. Gillard, R. S. Nyholm, G. Wilkinson (J. Chem. Soc. **1964** 5189/93).

Complexes Containing O-Donor Ligands

***trans*-[Rh py$_4$X(H$_2$O)]$^{2+}$** (X = Cl or Br). These ions are obtained either by aquation of *trans*-[Rh py$_4$X$_2$]X · n H$_2$O in the presence of Hg^{2+} [1] or by photolysis of the *trans* dihalo species in water [2, 3].

***trans*-[Rh py$_4$X(H$_2$O)](ClO$_4$)$_2$ · n H$_2$O** (X = Cl or Br). The chloro species (X = Cl) as a monohydrate, and the bromo complex, as a dihydrate, are prepared by heating *trans*-[Rh py$_4$X$_2$]X · 5 H$_2$O (X = Cl or Br) with sodium acetate followed by addition of $NaClO_4$ and recrystallisation from dilute $HClO_4$. The bromide was isolated as orange-red prisms [4].

***trans*-[Rh py$_4$X(OH)]ClO$_4$** (X = Cl or Br). The chloro (X = Cl, monohydrate) and bromo (X = Br) complexes are obtained by neutralisation of *trans*-[Rh py$_4$X(OH$_2$)](ClO$_4$)$_2$ · n H$_2$O with sodium hydroxide. The bromo complex may also be formed by heating an aqueous solution of *trans*-[Rh py$_4$Br$_2$]$^+$ at pH = 6.5 with an excess of KF [4].

The electronic spectra of these salts have been reported briefly [4]. The cations underwent an irreversible two-electron reduction in aqueous solution (0.1 M KNO_3, pH = 8) at the dropping-mercury electrode; $E_{1/2} = -0.34$ V (X = Cl) and -0.26 V (X = Br) vs. standard calomel electrode [5].

References:

[1] A. B. Venediktov, A. V. Belyaev (Koord. Khim. **2** [1976] 1414/21; Soviet J. Coord. Chem. **2** [1976] 1085/91; C. A. **86** [1976] No. 60939). – [2] M. M. Muir, W.-L. Huang (Inorg. Chem. **12** [1973] 1831/5). – [3] M. M. Muir, L. B. Zinner, L. A. Paguaga, L. M. Torres (Inorg. Chem. **21** [1982] 3448/50). – [4] A. W. Addison, K. Dawson, R. D. Gillard, B. T. Heaton, H. Shaw (J. Chem. Soc. Dalton Trans.**1972** 589/96). – [5] A. W. Addison, R. D. Gillard, D. H. Vaughan (J. Chem. Soc. Dalton Trans. **1973** 1187/93).

Complexes Containing N-Donor Ligands

***trans*-[Rh py$_4$(N$_3$)$_2$]X.** The **azide** (X = N$_3$), as a pentahydrate, is obtained by refluxing an aqueous solution of *trans*-[Rh py$_4$Cl$_2$]Cl · 5 H$_2$O with NaN_3. It was isolated as yellow needles which exploded violently when heated. The **tetrafluoroborate** (X = BF$_4$) is prepared from solutions containing the azide by addition of $NaBF_4$ [1].

The IR spectrum of the complex exhibited ν(NN) at 2010 cm^{-1}, and the electronic spectrum of the cation has been briefly reported. The conductivity of the azide salt (in H_2O, 10^{-3} M solution) was 73 $cm^2 \cdot \Omega^{-1} \cdot mol^{-1}$ [1].

trans-[Rh py$_4$(N$_3$)$_2$]$^+$ underwent a two-electron reduction at the dropping-mercury electrode, $E_{1/2} = -0.32$ V vs. standard calomel electrode [2].

References:

[1] A. W. Addison, K. Dawson, R. D. Gillard, B. T. Heaton, H. Shaw (J. Chem. Soc. Dalton Trans. **1972** 589/96). – [2] A. W. Addison, R. D. Gillard, D. H. Vaughan (J. Chem. Soc. Dalton Trans. **1973** 1187/93).

Halo Complexes

Preparation

***trans*-[Rhpy$_4$Cl$_2$]X·nH$_2$O.** The chloride (X = Cl), as a pentahydrate, is made by heating Na_3RhCl_6 with aqueous pyridine [1] or pyridine/methanol [2], or by refluxing $RhCl_3 \cdot 3H_2O$ with pyridine in ethanol [3] or water [4]. It may also be formed by reaction of $RhCl_3 \cdot 3H_2O$ in boiling water with pyridine and sodium hypophosphite [5] or H_3PO_2 [6], reaction of $RhCl_3$ with pyridine and zinc amalgam in methanol, by heating $RhCl_3$ with pyridine either in aqueous solution containing NaCl and hydrazinium hydrochloride, or in a 1:1 mixture of 2-methoxyethanol and ethanol [6]. It can also be obtained by reaction of $[RhCl_5(H_2O)]^{2-}$ with pyridine in aqueous ethanol [7].

The iodide, hydrogen sulphate (as a hexahydrate), perchlorate, nitrate (as a hexahydrate) and thiocyanate are prepared from the chloride by reaction and/or recrystallisation from aqueous KI solution, H_2SO_4, conc. aqueous $KClO_4$ solution, HNO_3, and KNCS in water, respectively [8].

These salts are obtained as yellow or golden-yellow crystals depending on the method of recrystallisation [1 to 8].

***trans*-[Rhpy$_4$Br$_2$]X·nH$_2$O.** The bromide, as a hexahydrate, is obtained by reaction of $RhCl_3 \cdot 3H_2O$ with LiBr in boiling water to which pyridine and some H_3PO_2 are added, the mixture being refluxed [9]. It is also formed by refluxing $Na_3RhBr_6 \cdot 12H_2O$ with pyridine/methanol [10], and by reaction of a refluxing aqueous mixture of $RhBr_3$, pyridine and sodium hypophosphite [5].

The chloride [8], iodide [8, 11], nitrate [8], hydrogen sulphate, all as hexahydrates, and the perchlorate [8, 11], thiocyanate and dithionate [8] are obtained by reaction and/or recrystallisation of *trans*-[Rhpy$_4$Br$_2$]Br with conc. aqueous NaCl, aqueous HI or NaI, conc. HNO_3, H_2SO_4, aqueous $NaClO_4$ or $KClO_4$, KNCS and conc. aqueous $Na_2S_2O_6$, respectively.

These salts are isolated as golden-yellow or orange-yellow crystals [5, 8 to 11].

***trans*-[Rhpy$_4$Br$_2$][Rhpy$_2$Br$_4$]** is formed when [Rhpy$_4$Br$_2$]Br is refluxed with aqueous HBr or KBr and H_3PO_2 [6, 11]. It was isolated as a buff precipitate.

***trans*-[Rhpy$_4$ClBr]BF$_4$.** This complex is prepared by boiling an aqueous solution of *trans*-[Rhpy$_4$Br(H$_2$O)](ClO$_4$)$_2$ with concentrated HCl, followed by addition of HBF_4 [12]. It was isolated as orange needles.

***trans*-[Rhpy$_4$I$_2$]I·5H$_2$O** is obtained by bubbling hydrogen through an ethanolic solution containing RhI_3 and pyridine [12]. It was isolated as golden-yellow crystals.

Adducts of trans-[Rhpy$_4$X$_2$]$^+$

***trans*-[Rhpy$_4$Cl$_2$]X$_3$.** The trichloride is obtained from the reaction between *trans*-[Rh(py)$_4$Cl$_2$]Cl·5H$_2$O and Cl_2, and was isolated as feathery cream crystals. The tribromide is formed by reaction of the monochloride with Br_3^- in chloroform, and was obtained as orange crystals. The tri-iodide is prepared by the Fischer titration of *trans*-[Rhpy$_4$Cl$_2$]Cl·5H$_2$O, and crystallised as brown platelets [13].

***trans*-[Rh py$_4$Br$_2$]Y$_3$.** The tribromide is made by reaction of *trans*-[Rh py$_4$Br$_2$]Br with Br_2 in water, and the tri-iodide is produced by reaction of the monobromide with aqueous KI_3. The former was isolated as a yellow solid and the latter as a yellow-brown solid [11].

***trans*-[Rh py$_4$X$_2$][(H$_5$O$_2$)X$_2$]** (X = Cl or Br). The chloride-containing adducts are prepared by recrystallising *trans*-[Rh py$_4$Cl$_2$]Cl·5H$_2$O from conc. HCl solution, and the corresponding bromide species from conc. HBr solution. They were isolated as yellow (X = Cl) and orange-yellow (X = Br) plates [14, 15].

***trans*-[Rh py$_4$X$_2$][H(ONO$_2$)$_2$]** (X = Cl or Br). The chloride is obtained by treatment of an aqueous solution of *trans*-[Rh py$_4$Cl$_2$]Cl·5H$_2$O with conc. HNO_3. The bromide is obtained similarly from an ethanolic solution of *trans*-[Rh py$_4$Br$_2$]Br·6H$_2$O [15].

***trans*-[Rh py$_4$Cl$_2$]ClO$_4$·L** (L = CH_3NO_2, $C_3H_7NO_2$, NC(CH$_2$)$_4$CN). These solvent adducts are prepared by recrystallisation of *trans*-[Rh py$_4$Cl$_2$]ClO$_4$ from nitromethane or -propane, or from 1,4-dicyanobutane [13].

***trans*-[Rh py$_4$Cl$_2$](HBr$_2$)·3H$_2$O.** This compound is obtained by heating an aqueous ethanolic solution of *trans*-[Rh py$_4$Cl$_2$]Cl·5H$_2$O with NaBr and conc. HBr solution. It was isolated as yellow crystals [12, 13].

References:

[1] S. A. Repin (Zh. Prikl. Khim. **20** [1947] 45/54 from C.A. **1947** 5814). – [2] M. Delépine (Compt. Rend. **236** [1953] 559/62). – [3] J. P. Collman, H. F. Holtzclaw (J. Am. Chem. Soc. **80** [1958] 2054/6). – [4] H.-H. Schmidtke (Z. Physik. Chem. [Frankfurt] **34** [1962] 295/311). – [5] R. D. Gillard, G. Wilkinson (Inorg. Syn. **10** [1967] 64/7).

[6] R. D. Gillard, J. A. Osborn, G. Wilkinson (J. Chem. Soc. **1965** 1951/65). – [7] J. V. Rund, F. Basolo, R. G. Pearson (Inorg. Chem. **3** [1964] 658/61). – [8] J. Meyer, H. Kienitz (Z. Anorg. Allgem. Chem. **242** [1939] 281/301). – [9] E. Pederson, H. Toftlund (Inorg. Chem. **13** [1974] 1603/12). – [10] C. Ouannès, C. Tard (Compt. Rend. **267** [1958] 1202/4).

[11] B. N. Figgis, R. D. Gillard, R. S. Nyholm, G. Wilkinson (J. Chem. Soc. **1964** 5189/93). – [12] A. W. Addison, K. Dawson, R. D. Gillard, B. T. Heaton, H. Shaw (J. Chem. Soc. Dalton Trans. **1972** 589/96). – [13] A. W. Addison, R. D. Gillard (J. Chem. Soc. Dalton Trans. **1973** 2009/12). – [14] R. D. Gillard, G. Wilkinson (J. Chem. Soc. **1964** 1640/6). – [15] D. Dollimore, R. D. Gillard, E. D. McKenzie (J. Chem. Soc. **1965** 4479/82).

[16] R. D. Gillard, R. Ugo (J. Chem. Soc. A **1966** 549/52).

Structures

trans-[Rh py$_4$Cl$_2$][H(ONO$_2$)$_2$] is orthorhombic, space group Pbcn-D_{2h}^{14}, cell dimensions being listed in the table below. The cation is six-coordinate [1, 2].

Cell dimensions of *trans*-[Rh py$_4$Cl$_2$][H(ONO$_2$)]:

a in Å	b in Å	c in Å	β	Z	Ref.
7.54	21.65	14.84	—	—	[1]
7.529(5)	21.717(6)	14.717(7)	—	4	[2]
7.53	21.80	14.82	90°	4	[3]

The *trans* Cl atom has a Rh-Cl distance of 2.34 [1] or 2.331 Å [2], and a Rh-N bond length of 2.09 [1] or 2.060 Å [2], respectively. The pyridine ligands adopt a propellor-like arrangement around the metal. The counter-anion is the hydrogendinitrate ion, with four O atoms lying at the corners of a distorted tetrahedron with an O-O distance of 3.06 Å [1].

trans-[Rh py_4Br_2][H(ONO_2)] forms orthorhombic crystals, space group Pbcn-D_{2h}^{14}, a = 7.64, b = 21.78, c = 14.82 Å, β = 90°, Z = 4 [3].

X-ray powder diffraction data (d spacings) have been obtained from *trans*-[Rh py_4X_2]-[(H_5O_2)X_2] with X = Cl or Br [4].

References:

[1] G. R. Dobinson, R. Mason, D. R. Russell (Chem. Commun. **1967** 62/3). – [2] J. Rozière, M. S. Lehmann, J. Potier (Acta Cryst. B **33** [1979] 1099/102). – [3] R. D. Gillard, R. Ugo (J. Chem. Soc. A **1966** 549/52). – [4] D. Dollimore, R. D. Gillard, E. D. McKenzie (J. Chem. Soc. **1965** 4479/82).

Spectroscopic Studies

Vibrational and Electronic Spectral Studies. The IR spectra of *trans*-[Rh py_4Cl_2]Cl · 5 H_2O in the range 1700 to 600 cm^{-1} [1, 2, 3] and 3200 to 100 cm^{-1} [4] (spectrum shown in paper), and of *trans*-[Rh py_4Br_2]Br in the range 1700 to 600 cm^{-1} [2], have been reported. The IR spectrum of *trans*-[Rh py_4Cl_2](NCS) has been mentioned [5]. The IR spectra of *trans*-[Rh py_4Cl_2]X_3 (X = Cl, Br or I) [6], *trans*-[Rh py_4X_2][HX_2] · 2 H_2O (X = Cl or Br) [6, 7, 8] and *trans*-[Rh py_4X_2][H(ONO_2)$_2$] [9] have been reported in the range 2500 to 700 cm^{-1}.

The electronic spectra of *trans*-[Rh py_4Cl_2]Cl [1] in the range 600 to 200 nm [2], of *trans*-[Rh py_4Cl_2]$^+$ in water [10] and of *trans*-[Rh py_4X_2]X · 5 H_2O (X = Cl, Br or I) [11], have been reported.

References:

[1] R. D. Gillard, G. Wilkinson (J. Chem. Soc. **1964** 1224/8). – [2] H.-H. Schmidtke (Z. Physik. Chem. [Frankfurt] **34** [1962] 295/311). – [3] J. P. Collman, H. F. Holtzclaw (J. Am. Chem. Soc. **80** [1958] 2054/6). – [4] F. Herbelin, J. D. Herbelin, J. P. Mathieu, H. Poulet (Spectrochim. Acta **22** [1966] 1510/22). – [5] R. D. Gillard, J. A. Osborn, G. Wilkinson (J. Chem. Soc. **1965** 1951/65).

[6] A. W. Addison, R. D. Gillard (J. Chem. Soc. Dalton Trans. **1973** 2009/12). – [7] R. D. Gillard, G. Wilkinson (J. Chem. Soc. **1964** 1640/6). – [8] B. N. Figgis, R. D. Gillard, R. S. Nyholm, G. Wilkinson (J. Chem. Soc. **1964** 5189/93). – [9] R. D. Gillard, R. Ugo (J. Chem. Soc. A **1966** 549/52). – [10] C. K. Jørgensen (Acta Chem. Scand. **11** [1957] 151/65).

[11] A. W. Addison, K. Dawson, R. D. Gillard, B. T. Heaton, H. Shaw (J. Chem. Soc. Dalton Trans. **1972** 589/96).

Other Spectral Studies. X-ray photoelectron spectral studies have been made of *trans*-[Rh py_4Cl_2]Cl · n H_2O, where n = 2 [1] or 5 [2]. The Rh $3d_{3/2}$ and $3d_{5/2}$ binding energies were 315.1 and 310.6 eV, respectively [1], the Cl $2p_{1/2}$ and $2p_{3/2}$ binding energies 199.6, 198.2 and 196.5 eV [1] or 199.2 and 197.8 eV [2], respectively. The N1s binding energy was 399.8 eV [2]. X-ray emission and absorption spectroscopic studies have been made of [Rh py_4Cl_2]NO_3 · 6 H_2O [3].

The luminescence spectrum of *trans*-[Rh py_4Cl_2]Cl has been measured at 85 K [4], 77 K [5, 6] and 10 K [7], and the emission bands analysed theoretically [8]. The quantum efficiency in d-d luminescence from *trans*-[Rh py_4Br_2]Br has been measured [9] and a ligand field interpretation of the quantum yield of photosolvation of *trans*-[Rh py_4Cl_2]Cl offered [10].

From ^{1}H NMR spectral studies it was established that there was virtually no exchange of the heterocyclic ligand between $[Rh py_4Cl_2]^+$ and "free" pyridine in solution [11].

References:

[1] A. D. Hamer, D. G. Tisley, R. A. Walton (J. Chem. Soc. Dalton Trans. **1973** 116/20). – [2] J. R. Ebner, D. L. McFadden, D. R. Tyler, R. A. Walton (Inorg. Chem. **15** [1976] 3014/8). – [3] L. N. Mazalov, A. P. Sadovskii, A. V. Belyaev, L. I. Chernyavskii, E. S. Gluskin, L. F. Berkhoer (Izv. Sibirsk. Otd. Akad. Nauk SSSR Ser. Khim. Nauk **1971** No. 2, pp. 51/9 from C.A. **76** [1972] No. 39710). – [4] P. E. Hoggard, H.-H. Schmidtke (Ber. Bunsenges. Physik. Chem. **77** [1973] 1052/8). – [5] M. K. De Armond, J. E. Hillis (J. Chem. Phys. **54** [1971] 2247/53).

[6] J. E. Hillis, M. K. De Armond (J. Lumin. **4** [1971] 273/90). – [7] G. Eyring, H.-H. Schmidtke (Ber. Bunsenges. Physik. Chem. **85** [1981] 597/603). – [8] K. W. Hipps, G. A. Merrell, G. A. Crosby (J. Phys. Chem. **80** [1976] 2232/9). – [9] J. N. Demas, C. A. Crosby (J. Am. Chem. Soc. **92** [1970] 7262/70). – [10] M. J. Incorvia, J. I. Zink (Inorg. Chem. **13** [1974] 2489/94).

[11] Yu. N. Kukushkin, G. M. Khvostik, G. P. Kondratenkov (Koord. Khim. **5** [1979] 1225/8; Soviet J. Coord. Chem. **5** [1979] 966/9; C.A. **91** [1979] No. 184496).

Electrochemical Properties

Conductivities of *trans*-$[Rh(py)_4X_2]Y \cdot nH_2O$ (in solution):

X	Y	n	H_2O	Λ in $cm^2 \cdot \Omega^{-1} \cdot mol^{-1}$ Solvent CH_3NO_2	$PhNO_2$	Other	Ref.
Cl	Cl	5	95.5	77.5			[1]
Br	Br	5	106	69.0			[1]
I	I	5	107				[1]
I	ClO_4	0		88			[1]
Br	ClO_4	0			27.2		[2]
Br	Br_3	0			29.6	164 a)	[2]
Br	I_3	0			27.4	140 a)	[2]
Br	NO_3	0		77		42.7 b)	[3]
Br	$H(NO_3)_2$	0		75.1		81.5 b)	[3]
Cl	NO_3	0	150			44.9 b)	[3]
						132 c)	[3]
Cl	$H(NO_3)_2$	0	510			78.5 b)	[3]
						146 c)	[3]

a) acetone; b) dimethylsulphoxide; c) acetonitrile

Polarographic studies have been made of *trans*-$[Rh(py)_4X_2]^+$ in aqueous solution [4 to 7]. It has been suggested that $[Rh(py)_4Cl_2]^+$ underwent a one-electron reduction [4], although other results indicated a two-electron process [5, 6, 7] (see table on p. 256).

Polarographic reduction of *trans*-$[Rh(py)_4XY]^+$:

X	Y	$E_{1/2}$ [a] in V	No. of electrons	Ref.
Cl	Cl	−0.21	2	[5]
		−0.210		[6]
		−0.39 } −0.95 [b] }	2	[7]
Cl	Br	−0.22	2	[5]
Br	Br	−0.27	2	[5]
		−0.273		[6]
		−0.28 } −0.95 [b] }	2	[7]
I	I	−0.25		[5]

[a] vs. standard calomel electrode; [b] second reduction wave, not clearly defined

References:

[1] A. W. Addison, K. Dawson, R. D. Gillard, B. T. Heaton, H. Shaw (J. Chem. Soc. Dalton Trans. **1972** 589/96). – [2] B. N. Figgis, R. D. Gillard, R. S. Nyholm, G. Wilkinson (J. Chem. Soc. **1964** 5189/93). – [3] R. D. Gillard, R. Ugo (J. Chem. Soc. A **1966** 549/52). – [4] F. Pantani (J. Electroanal. Chem. **5** [1963] 40/7). – [5] A. W. Addison, R. D. Gillard, D. H. Vaughan (J. Chem. Soc. Dalton Trans. **1973** 1187/93).

[6] R. D. Gillard, B. T. Heaton, D. H. Vaughan (J. Chem. Soc. A **1971** 1840/6). – [7] R. D. Gillard, J. A. Osborn, G. Wilkinson (J. Chem. Soc. **1965** 4107/10).

Chemical Reactions of *trans*-$[Rh(py)_4X_2]Y$

trans-$[Rh(py)_4X_2]^+$ is a good, easily accessible starting material for the preparation of other complexes. Thus, with NH_3, ethylenediamine (en), 1,3-diaminopropane (pn) and $(NH_2CH_2CH_2NHCH_2)_2$ (tet), $[Rh(NH_3)_5Cl]^{2+}$, *trans*-$[Rh(en)_2Cl_2]^+$, $[Rh(en)_3]^{3+}$, *trans*-$[Rh(pn)_2Cl_2]^+$ and *trans*-$[Rh(tet)Cl_2]^+$, can be formed, whereas dimethylglyoxime ($DMGH_2$) and CN^- afford *trans*-$[Rh(DMGH)_2Cl_2]^-$ and $Rh(py)_2(CN)_2Cl_2$, respectively [1].

The rates of halogen interchange in aqueous solutions of *trans*-$[Rh(py)_4X_2]X$ (X = Cl or Br) were shown to be markedly inhibited by oxygen, but were catalysed by primary or secondary, but not tertiary alcohols [2].

The aquation of *trans*-$[Rh(py)_4Cl_2]NO_3$ in the presence of Hg^{2+} ions, giving $[Rh(py)_4Cl_2Hg]^{3+}$ and then $[Rh(py)_4Cl(H_2O)]^{2+}$, has been studied kinetically [3]. The photoaquation of *trans*-$[Rh(py)_4Cl_2]^+$ has been investigated and both *trans*-$[Rh(py)_4Cl(H_2O)]^{2+}$ and *trans*-$[Rh(py)_3(H_2O)Cl_2]^+$ were detected [4]. The photochemistry of *trans*-$[Rh(py)_4Cl_2]^+$ was investigated spectrophotometrically in aqueous solutions and in glassy matrices (H_2SO_4, H_3PO_4, ethylene glycol, water mixtures and methanol) at 77 K [5, 6].

The products of γ-radiolysis of $[Rh(py)_4Cl_2]Cl \cdot 6H_2O$ have been studied by electron spin resonance techniques [7], and the characteristics of $[Rh(py)_4Cl_2]^+$ on ion-exchange columns investigated [8]. The surface-active properties and antibacterial activity of *trans*-$[Rh(py)_4Cl_2]Cl \cdot 5H_2O$ have been determined [9].

The thermal decomposition of *trans*-[Rh(py)$_4$Br$_2$]X, where X = [(H_5O_2)Cl_2] [10] and [H(ONO_2)$_2$] [11], have been studied using thermogravimetric analysis.

References:

[1] R. D. Gillard, E. D. McKenzie, M. D. Ross (J. Inorg. Nucl. Chem. **28** [1966] 1429/34). – [2] R. D. Gillard, B. T. Heaton, D. H. Vaughan (Chem. Commun. **1969** 974/5). – [3] A. B. Venediktov, A. V. Belyaev (Koord. Khim. **2** [1976] 1414/21; Soviet J. Coord. Chem. **2** [1976] 1085/91). – [4] M. M. Muir, W.-L. Huang (Inorg. Chem. **12** [1973] 1831/5). – [5] L. A. Il'yukevich, E. N. Leonteva (Prevrashch. Kompleksn. Soedin. Deistviem Sveta Radiats. Temp. **1973** 148/51 from C.A. **82** [1975] No. 24274).

[6] L. N. Neokladovna, L. A. Il'yukevich (Khim. Vysokikh Energ. **12** [1978] 507/9; High Energy Chem. [USSR] **12** [1978] 422/4). – [7] Yu. V. Glazkov, N. I. Zotov, L. A. Il'yukevich, L. N. Neokladovna (Khim. Vysokikh Energ. **9** [1975] 88/90; High Energy Chem. [USSR] **9** [1975] 76/8). – [8] N. K. Pshenitsyn, K. A. Gladyshevskaya, L. M. Rykhova (Anal. Blagorodn. Metal. **1959** 103/14 from C.A. **1960** 16280) – [9] T. R. Thomas (J. Inorg. Biochem. **12** [1980] 187/99). – [10] D. Dollimore, R. D. Gillard, E. D. McKenzie (J. Chem. Soc. **1965** 4479/82).

[11] D. Dollimore, R. D. Gillard, E. D. McKenzie, R. Ugo (J. Inorg. Nucl. Chem. **30** [1968] 2755/8).

2.6.9.2.3 Tris-pyridine Complexes

These species exist as *fac*-(1,2,3) and *mer*-(1,2,6) isomers.

Rh(py)$_3$(N$_3$)$_3$ is formed when *trans*-[Rh(py)$_4$Cl$_2$]Cl·5H_2O in water is treated with NaN_3 and $NaBH_4$. It may also apparently be formed by bubbling N_2 through a cold aqueous ethanolic solution of *trans*-[Rh(py)$_4$Cl$_2$]Cl and an excess of NaN_3. The complex was isolated as orange-yellow crystals. The IR spectrum exhibited ν(NN) at 2008 and 2030 cm^{-1} [1].

Rh(py)$_3$(NCO)$_2$X (X = Cl or Br). These complexes are obtained by heating of *trans*-[Rh(py)$_4$X$_2$]X·5H_2O in water with NaNCO at 50°C for 30 min. The chloride was isolated as yellow platelets, and the bromide as orange-yellow crystals. The IR spectrum of the chloride exhibited ν(CN) at 2235 and 2190 cm^{-1}, and δ(NCO) at 590 cm^{-1}. It was suggested that the NCO group was N-bonded and that the molecule had a *cis* geometry [1].

Rh(py)$_3$(NO$_2$)$_3$ is prepared by warming an aqueous ethanolic solution of *trans*-[Rh(py)$_4$Cl$_2$]Cl·5H_2O with $NaNO_2$ and $NaBH_4$. The complex precipitated as white crystals whose IR spectrum indicated that the NO_2 ligand was probably N-bonded [1].

Rh(py)$_3$(C$_2$O$_4$)X (X = Cl, Br or I). The chloro and bromo complexes are formed by reaction of *trans*-[Rh(py)$_4$X$_2$]X with $K_2C_2O_4$ in boiling water [2] or in the presence of catalytic amounts of $NaBH_4$ in warm water [3]. They are isolated as feathery yellow (X = Cl) or orange (X = Br) crystals. The iodo complex is prepared by reaction of Rh(py)$_3$(C_2O_4)Cl with NaI in the presence of $NaBH_4$ [3]. It was described as being deeply coloured. These compounds are insoluble in common solvents, but dissolve to some extent in aqueous pyridine or dimethylsulphoxide [2, 3].

The IR [2, 3] and Raman [3] spectra of Rh(py)$_3$(C_2O_4)X (X = Cl, Br or I) have been reported and ν(RhX) tentatively assigned [3] as 356 (X = Cl) and 199 (Raman) or 197 (IR) (X = Br) cm^{-1}. The electronic spectra of these complexes have also been briefly reported [2, 3].

Rh(py)$_3$(C_2O_4)X reacted with HX in water or in methanol affording Rh(py)$_3$X$_3$, and can be photolysed in the solid state to afford Rh(py)$_3$Cl [3].

$Rh(py)_3(SCN)_3$. A mixture of *fac*-(1,2,3) and *mer*-(1,2,6) isomers of this compound is obtained by refluxing a mixture of $RhCl_3 \cdot 3H_2O$, KSCN and pyridine in water with or without H_3PO_2. The isomers may be separated by fractional crystallisation [4].

The IR spectra of these species in the range 2100 to 650 cm^{-1} have been measured (shown in paper, in Nujol and CS_2), and the electronic spectra have also been reported [4].

References:

[1] A. W. Addison, K. Dawson, R. D. Gillard, B. T. Heaton, H. Shaw (J. Chem. Soc. Dalton Trans. **1972** 589/96). – [2] R. D. Gillard, E. D. McKenzie, M. D. Ross (J. Inorg. Nucl. Chem. **28** [1966] 1429/34). – [3] A. W. Addison, R. D. Gillard, P. S. Sheridan, L. R. H. Tipping (J. Chem. Soc. Dalton Trans. **1974** 709/16). – [4] R. D. Gillard, J. A. Osborn, G. Wilkinson (J. Chem. Soc. **1965** 1951/65).

Halo Complexes

Preparation

$Rh(py)_3X_3$ (X = Cl, Br or I). The **trichloro** complex (X = Cl) is obtained, as the *fac* isomer, by heating $RhCl_3 \cdot 3H_2O$ carefully in aqueous pyridine, fractionally crystallising the product and collecting the first fraction [1]. If the reaction mixture is heated more strongly or for longer periods in water [2 to 5], or the initial product obtained as in [1] described above (which is a mixture of *fac* and *mer* isomers of $Rh(py)_3Cl_3$, *trans*-$[Rh(py)_4Cl_2]Cl$ and other products [2, 5], is heated in a sealed tube at 150°C for 3 d [5], then the *mer* isomer is formed. The *mer* isomer may also be obtained directly by heating *trans*-$[Rh(py)_4Cl_2]Cl$ in ethanol [1], by heating *trans*-$[Rh(py)_4Br_2]Br \cdot 6H_2O$ with conc. HCl [6], by treatment of $Rh(py)_3(C_2O_4)Cl$ with two equivalents of HCl [7], or by reaction of $[Rh(CO)_2Cl_2]^-$ with pyridine [8]. By recrystallisation of the trichloro complex from $CHCl_3$, $CDCl_3$ or CH_2Cl_2, the crystalline solvates $Rh(py)_3Cl_3 \cdot 2CHCl_3$ [2, 9], $Rh(py)_3Cl_3 \cdot nCDCl_3$ (n = 1 or 2) and $Rh(py)_3Cl_3 \cdot 2CH_2Cl_2$ [9] are formed. The unsolvated and solvated compounds were isolated as orange [1 to 9] or red [2] crystals, although $Rh(py)_3Cl_3$ has been described as orange-brown [2].

The ***mer*-tribromo** compound (X = Br) is formed from *trans*-$[Rh(py)_4Br_2]Br$ either by heating it alone in ethanol [1] or by refluxing it with HBr and H_3PO_2 [10], by treatment of $Rh(py)_3(C_2O_4)Br$ with two equivalents of HBr [7] or by heating $[Rh(CO)_2Br_2]^-$ with pyridine [8]. It may also be formed by reaction of $RhBr_3 \cdot nH_2O$ with pyridine, although details have not been given [3].

The ***mer*-tri-iodo** (X = I) complex is prepared by boiling an ethanolic mixture of *trans*-$[Rh(py)_4Cl_2]Cl$ either with pyridine and I^- [7] or with KI and H_3PO_2 [11], and by reaction of $[Rh(CO)_2Cl_2]^-$ with NaI and pyridine [8]. It was isolated as plum-coloured [7], dark brown [11] or dark red [8] crystals.

$Rh(py)_3XY_2$. The **chloro-dibromo** (X = Cl, Y = Br) compound is obtained by reacting $Rh(py)_3(C_2O_4)Cl$ with two equivalents of HBr, while the **bromo-dichloro** (X = Br, Y = Cl) complex is prepared similarly by reaction of the bromo-oxalate with two equivalents of HCl [7]. These complexes may be recrystallised from chloroform affording the solvates $Rh(py)_3ClBr_2 \cdot nCHCl_3$ (n = 2 or 3) and $Rh(py)_3ClBr_2 \cdot CHCl_3$ [9]. Both the solvated and unsolvated species form deep orange (X = Br, Y = Cl) or deep red (X = Cl, Y = Br) solids [7, 9].

X-ray powder diffraction data (d-spacings) have been reported by $Rh(py)_3Cl_3$ and its bis-chloroform solvate [9].

Spectroscopic Studies

The IR spectrum of *mer*- and *fac*-$Rh(py)_3Cl_3$ have been measured in the ranges 3200 to 100 cm^{-1} (spectrum shown in paper) [12], 3200 to 700 cm^{-1} [7], 1700 to 650 cm^{-1} [1, 5] and 700 to 200 cm^{-1} [13, 14]. The values of the metal-chlorine stretching frequencies were as follows: *mer* isomer 355(s), 332(s), 295(m) cm^{-1}; *fac* isomer 341(s), 325(m) cm^{-1} [13, 14]. IR and Raman spectral data have been obtained in the ranges 1050 to 1000 and 400 to 250 cm^{-1} for $Rh(py)_3Cl_3$, $Rh(py)_3Cl_3 \cdot 2CHCl_3$, $Rh(NC_5D_5)_3Cl_3 \cdot 2CHCl_3$ and $Rh(py)_3Cl_3 \cdot nCDCl_3$ (n=1 or 2), for $Rh(py)_3Cl_2Br \cdot 2CHCl_3$ and $Rh(py)_3ClBr_2 \cdot CHCl_3$ [9].

The electronic spectra of *mer*-$Rh(py)_3Cl_3$ [1, 5, 6, 15, 16] and *fac*-$Rh(py)_3Cl_3$ [1, 5, 16] have also been reported. The electronic spectrum of $Rh(py)_3Br_3$ has been described [1, 10].

The X-ray photoelectron spectrum of $Rh(py)_3Cl_3$ has been measured; the values for Rh $3d_{5/2}$, N 1s and Cl $2p_{3/2}$ binding energies were 310.0, 400.0 and 198.2 eV, respectively [17].

References:

[1] H.-H. Schmidtke (Z. Physik. Chem. [Frankfurt] **34** [1972] 295/311). – [2] J. P. Collman, H. F. Holtzclaw (J. Am. Chem. Soc. **80** [1958] 2054/6). – [3] R. D. Gillard, G. Wilkinson (Inorg. Syn. **10** [1967] 64/7). – [4] J. V. Rund, F. Basolo, R. G. Pearson (Inorg. Chem. **3** [1964] 658/61). – [5] R. D. Gillard, G. Wilkinson (J. Chem. Soc. **1964** 1224/8).

[6] R. D. Gillard, E. D. McKenzie, M. D. Ross (J. Inorg. Nucl. Chem. **28** [1966] 1429/34). – [7] A. W. Addison, R. D. Gillard, P. S. Sheridan, L. R. H. Tipping (J. Chem. Soc. Dalton Trans. **1974** 709/16). – [8] J. V. Kingston, F. T. Mahmoud, G. R. Scollary (J. Inorg. Nucl. Chem. **34** [1972] 3197/201). – [9] A. W. Addison, R. D. Gillard (J. Chem. Soc. Dalton Trans. **1973** 2002/9). – [10] B. N. Figgis, R. D. Gillard, R. S. Nyholm, G. Wilkinson (J. Chem. Soc. **1964** 5189/93).

[11] R. D. Gillard, J. A. Osborn, G. Wilkinson (J. Chem. Soc. **1965** 1951/65). – [12] F. Herbelin, J. D. Herbelin, J. P. Mathieu, H. Poulet (Spectrochim. Acta **22** [1966] 1515/22, 1516). – [13] R. J. H. Clark, C. S. Williams (Inorg. Chem. **4** [1965] 350/7). – [14] R. J. H. Clark (Record Chem. Progr. **26** [1965] 269/82 from C.A. **64** [1966] 7537). – [15] C. K. Jørgensen (Acta Chem. Scand. **11** [1957] 151/65).

[16] E. König, H. L. Schäfer (Z. Physik. Chem. [Frankfurt] **26** [1960] 371/403). – [17] V. I. Nefedov, M. A. Porai-Koshits (Mater. Res. Bull. **7** [1972] 1543/52).

Other Properties and Reactions

The *mer* and *fac* isomers of $Rh(py)_3Cl_3$ could be separated by adsorption chromatography on silica or alumina columns [1]. The behaviour of $Rh(py)_3Cl_3$ in acetone on thin-layer chromatographic plates has been examined [2]. $Rh(py)_3Cl_3$ has been separated from other rhodium-pyridine complexes in aqueous solution using electrophoretic and ion-exchange techniques [3].

Mixtures of $Rh(py)_3Cl_3$ and $NaBH_4$ in dimethylformamide catalytically hydrogenated methyl linoleate [4], various vicinal dihalides, carbonyl, epoxy and nitro groups [5]. Catalytic hydrogenation of folic acid using $Rh(py)_3Cl_3$ with (+)- or (−)-$HCONHCH(CH_3)C_6H_5$ and $NaBH_4$ gave the biologically active form of tetrahydrofolic acid; the (−)-catalyst gave 46% activity and the (+)-catalyst 28% activity [6]. The mixture $Rh(py)_3Cl_3/NaBH_4$ also hydrogenated $C_6H_5C_2C_6H_5$ to *cis*-stilbene and acted as a catalyst for the isomerisation of *cis*- to *trans*-stilbene [7].

Tests have been made for the genetic toxicity and mutagenicity of $Rh(py)_3Cl_3$ [8, 9].

References:

[1] G. B. Kauffman, G. L. Anderson, L. A. Teter (J. Chromatog. **114** [1975] 465/72). – [2] G. B. Kauffman, B. H. Gump, G. L. Anderson, B. J. Stedjee (J. Chromatog. **117** [1976] 455/63). – [3] S. K. Shukla (Ann. Chim. [Paris] [13] **6** [1961] 1383/443). – [4] C. J. Love, F. J. McQuillin (J. Chem. Soc. Perkin Trans. I **1973** 2509/12). – [5] P. Abley, F. J. McQuillin (J. Catal. **24** [1972] 536/40).

[6] B. H. Boyle, M. T. Keating (Chem. Commun. **1974** 375/6). – [7] E. F. Litvin, L. Kh. Freidlin, L. F. Krokhmaleva, L. M. Kozlova, N. M. Nazarova (Izv. Akad. Nauk SSSR Ser. Khim. **1981** 811/5; Bull. Acad. Sci. USSR Div. Chem. Sci. **1981** 592/5). – [8] G. Warren, S. J. Rogers, E. H. Abbott (ACS Symp. Ser. No. 140 [1980] 227/36). – [9] G. Warren, E. H. Abbott, P. Schultz, K. Bennett, S. J. Rogers (Mutat. Res. **88** [1981] 165/73; C.A. **94** [1981] No. 132011).

2.6.9.2.4 Bis-pyridine Complexes

$Rh(py)_2(NH_3)Cl_3 \cdot 0.5H_2O$ is prepared by heating an aqueous solution of $(NH_4)_2[Rh(NH_3)Cl_5]$ containing pyridine and HCl. The complex was isolated as golden-yellow crystals [1].

$[Rh(py)_2(en)Cl_2]Y$ (en = $NH_2CH_2CH_2NH_2$). The chloride (Y = Cl), as a trihydrate, is obtained by heating *trans*-$[Rh(py)_4Cl_2]Cl$ with ethylenediamine hydrate in water containing a little H_2SO_4. It was isolated as yellow crystals. The bromide (Y = Br, mono-hydrate), iodide (Y = I, monohydrate), thiocyanate (Y = NCS, trihydrate), nitrate (Y = NO_3) and perchlorate (Y = ClO_4) are obtained from the chloride by its treatment with conc. aqueous solutions of KBr, KI, KSCN, HNO_3 and $KClO_4$, respectively. The salts were isolated as light or golden-yellow crystals. The compounds $[Rh(py)_2(en)Cl_2]_2Y$, where Y = S_2O_6 (hexahydrate), $PtCl_6$ and Cr_2O_7, are also obtained from the chloride by its treatment with aqueous solutions containing $Na_2S_2O_6$, $H_2PtCl_6 \cdot nH_2O$ and $K_2Cr_2O_7$. They were isolated as light yellow (Y = S_2O_6), dark yellow (Y = Cr_2O_7) or light orange (Y = $PtCl_6$) crystals [2].

$[Rh(py)_2(en)Br_2]Y$. The bromide (Y = Br), as a dihydrate, is obtained by heating *trans*-$[Rh(py)_4Br_2]Br$ with ethylenediamine hydrate in water. It was isolated as orange crystals. The nitrate (Y = NO_3, dihydrate), and dithionate (Y = S_2O_6) are prepared from the bromide by reaction with aqueous solutions containing HNO_3 and $Na_2S_2O_6$, respectively [2].

$Rh(py)_2(CN)_2Cl \cdot 2H_2O$ is obtained by boiling *trans*-$[Rh(py)_4Cl_2]Cl \cdot 5H_2O$ with aqueous KCN [3, 4]. It is insoluble in ethanol, pyridine and dimethylsulphoxide. The IR spectrum in the range 3500 to 650 cm^{-1} has been reported and $\nu(CN)$ = 2152 (terminal mode) and 2021 cm^{-1} (bridging mode) [3].

$[Rh(py)_2Cl_3]_n$. This compound may be obtained by the action of light on aqueous solutions of $K[Rh(py)_2Cl_4]$ [5], or on chloroform, dichloromethane or tetrachloroethane solutions of *mer*-$Rh(py)_3Cl_3$ [6]. It may also be formed by heating *mer*-$Rh(py)_3Cl_3$ in vacuo followed by extraction with CH_2Cl_2, by refluxing *mer*-$Rh(py)_3Cl_3$ in chloroform, or by keeping it in dry CH_2Cl_2 under N_2 for 8 d [7]. The compound is reported to be brick-red [6] or pink [7], and is polymeric, probably having an overall octahedral structure with bridging Cl atoms and *trans*-pyridine ligands [7]. The IR and electronic spectra (reflectance) have been recorded [9].

$Rh(py)_2(H_2O)Cl_3$ is obtained by the action of light on an aqueous solution of $K[Rh(py)_2Cl_4]$ after nine months [5].

$[Rh(py)_2Br_3]_2 \cdot CHCl_3$ is prepared by the action of light on chloroform solutions of $Rh(py)_3Br_3$. It was isolated as a brown precipitate [8].

***trans*-[Rh(py)$_2$Cl$_4$]$^-$.** The pyridinium salt is obtained by refluxing $Rh(py)_6Cl_3$ or $Rh(py)_3Cl_3$ with H_3PO_2, followed by treatment with refluxing HCl [9, 10]. The electronic [7, 11] and luminescence [12] spectra of the anion have been measured in aqueous solutions. The electronic spectra of the *cis* and *trans* isomers of this anion (15000 to 30000 cm^{-1}) have also been tabulated [11].

***trans*-[Rh(py)$_2$Br$_4$]$^-$.** The $[Rh(py)_4Br_2]^+$ salt of this anion is obtained either by boiling an aqueous mixture of *trans*-$[Rh(py)_4Cl_2]Cl \cdot 5H_2O$ with KBr and H_3PO_2 [3] or by refluxing an aqueous solution of *trans*-$[Rh(py)_4Br_2]Br$ containing HBr and H_3PO_2 [10]. The pyridinium salt is obtained by refluxing $Rh(py)_6Br_3$ or $Rh(py)_3Br_3$ with aqueous H_3PO_2, followed by addition of HBr [9, 10]. It was isolated as a buff precipitate, whose IR spectrum has been recorded [3].

$(C_2H_5)_4N[Rh(py)_2(N_3)_4]$ is formed when a hot 60% aqueous ethanolic solution of *trans*-$[Rh(py)_4Cl_2]Cl \cdot 5H_2O$ is treated with NaN_3 and $NaBH_4$. After filtering and standing overnight, addition of $[(C_2H_5)_4N]Cl$ causes precipitation of the complex as a yellow solid. The IR spectrum exhibited ν(NN) at 2015 cm^{-1} [13].

References:

[1] N. A. Vargunin, N. S. Kurnakova, V. V. Lebedinskii (Tr. Krasnoyarsk Gos. Med. Inst. **5** [1958] 101/2 from C.A. **1961** 26818). – [2] J. Meyer, H. Kienitz (Z. Anorg. Allgem. Chem. **242** [1939] 281/301). – [3] R. D. Gillard, J. A. Osborn, G. Wilkinson (J. Chem. Soc. **1965** 1951/65). – [4] R. D. Gillard, G. D. McKenzie, M. D. Ross (J. Inorg. Nucl. Chem. **28** [1966] 1429/34). – [5] M. Delépine (Compt. Rend. **240** [1955] 2468/70).

[6] M. Delépine (Compt. Rend. **236** [1953] 1713/6). – [7] R. D. Gillard, G. Wilkinson (J. Chem. Soc. **1964** 1224/8). – [8] M. Delépine (Compt. Rend. **238** [1954] 27/9). – [9] F. P. Dwyer, R. S. Nyholm (J. Proc. Roy. Soc. N.S. Wales **76** [1943] 275/80). – [10] B. N. Figgis, R. D. Gillard, R. S. Nyholm, G. Wilkinson (J. Chem. Soc. **1964** 5189/93).

[11] C. K. Jørgensen (Acta Chem. Scand. **11** [1957] 151/65). – [12] F. Zuloaga, M. Kasha (Photochem. Photobiol. **7** [1968] 549/55). – [13] A. W. Addison, K. Dawson, R. D. Gillard, B. T. Heaton, H. Shaw (J. Chem. Soc. Dalton Trans. **1972** 589/96).

2.6.9.2.5 Mono-pyridine Complexes

Salts of [Rhpy(NH$_3$)Cl$_3$]$^-$. The **pyridinium** salt of this anion is prepared by heating $[NH_4][Rh(NH_3)Cl_5]$ in water with pyridine and HCl. The **silver** salt is obtained from it by addition of $AgNO_3$ in aqueous HNO_3. The **$[Pt(NH_3)_4]^{2+}$** salt is also obtained from the pyridinium salt by addition of $[Pt(NH_3)_4]Cl_2$ in water. These salts were isolated as red crystals [1].

[Rhpy(NH$_3$)$_3$Cl$_2$]X. The **chloride** (X = Cl) is prepared by reaction of $Rh(NH_3)_3Cl_3$ with an excess of pyridine. The **perchlorate** (X = ClO_4), **tetrachloro-** and **hexachloro-platinum**(II) and (IV) salts (X = $[PtCl_4]^{2-}$ and $[PtCl_6]^{2-}$) were obtained from the chloride by addition of salts of these anions. It was suggested that the Cl atoms in the cation were mutually *cis* [2].

***cis*-[Rhpy(en)$_2$Cl]X$_2$.** The **chloride** (X = Cl, dihydrate) is obtained by heating *trans*-$[Rh(py)_4Cl_2]Cl \cdot 5H_2O$ with ethylenediamine in aqueous solution in a molar ratio of 1:2. It is isolated as light yellow crystals which could be dehydrated by recrystallisation from conc. aqueous NaCl. The **iodide** (X = I, dihydrate), **thiocyanate** (X = SCN), **nitrate** (X = NO_3, dihydrate) and **hexachloroplatinate**(IV) (X = $PtCl_6$) are obtained from the chloride by treatment with conc. aqueous KI, KSCN, HNO_3 and $H_2PtCl_6 \cdot nH_2O$, respectively. They were isolated as light yellow or golden-yellow crystals [3].

[Rhpy(en)$_2$Br]Br$_2$ is formed by heating *trans*-[Rh(py)$_4$Br$_2$]Br with ethylenediamine in aqueous solution in a 1:2 molar ratio. It was isolated as sulphur-yellow crystals [3].

References:

[1] N. A. Vargunin, N. S. Kurnakova, V. V. Lebedinskii (Tr. Krasnoyarsk Gos. Med. Inst. **5** [1958] 97/101 from C.A. **1961** 26818). – [2] N. A. Vargunin (Izv. Vysshikh Uchebn. Zavedenii Khim. Khim. Tekhnol. **5** [1962] 539/43 from C.A. **58** [1963] 5243). – [3] J. Meyer, H. Kienitz (Z. Anorg. Allgem. Chem. **242** [1939] 281/301).

For **[Rh(NH$_3$)$_5$py]$^{3+}$** see p. 140.

2.6.9.2.6 Binuclear Pyridine Complexes

***trans, trans*-[{Rh(py)$_4$(H$_2$O)}$_2$O$_2$](ClO$_4$)$_5$·6H$_2$O** is obtained when *trans*-[Rh(py)$_4$Cl$_2$]Cl·5H$_2$O is treated in aqueous ethanol with NaOH and 10% ozone in oxygen. The solution changes colour from yellow through orange and green, finally becoming deep blue. Addition of and recrystallisation from HClO$_4$ affords the complex as purple rhombs.

The colour of the complex changed as the pH of aqueous solutions was increased, probably due to the equilibrium $[\{Rh(py)_4(H_2O)\}_2O_2]^{5+} \rightleftharpoons [\{Rh(py)_4\}_2(H_2O)(OH)O_2]^{4+} \rightleftharpoons [\{Rh(py)_4(OH)\}_2O_2]^{3+}$. The electronic spectrum of the complex has been reported in dilute H$_2$SO$_4$ solution [1].

***trans, trans*-[{Rh(py)$_4$Cl}$_2$O$_2$](ClO$_4$)$_3$·8H$_2$O** is prepared by reaction of *trans*-[Rh(py)$_4$Cl$_2$]Cl·5H$_2$O in aqueous ethanol with NaOH and 10% ozone in oxygen [1]. The solution undergoes colour changes as described for the aqua species above, and on addition of HClO$_4$, followed by recrystallisation from conc. HCl, affords the complex as deep blue crystals [7].

The compound is paramagnetic, having one unpaired electron per bimetallic cation ($\mu_{eff}=1.67\pm0.05$ B.M.) and exhibits an ESR spectrum in acetone or in 50% H$_2$SO$_4$, giving $g=2.019$ (no Rh hyperfine coupling was observed) [1].

The IR spectrum of the compound was virtually superimposable on that of *trans*-[Rh(py)$_4$Cl$_2$]ClO$_4$, with the exception of additional bands due to water. The electronic spectrum of the complex in dilute H$_2$SO$_4$ was also reported (200 to 700 nm). The complex is a strong oxidant in aqueous solution [1].

***trans, trans*-[{Rh(py)$_4$Cl}$_2$O$_2$](ClO$_4$)$_2$** has not been isolated although it is presumed to be formed by Fe^{2+} reduction in aqueous solution. The electronic spectrum of this species was recorded in the range 200 to 400 nm [1].

***trans, trans*-[{Rh(py)$_4$Cl}$_2$O$_2$](BF$_4$)$_3$.** This complex is prepared in the same way as the tris-perchlorate above, but using HBF$_4$ in place of HClO$_4$. It was isolated as deep blue crystals which behaved as a 1:3 electrolyte in nitromethane ($\Lambda=192$ cm$^2\cdot\Omega^{-1}\cdot$mol^{-1} in 10^{-3}M solution). The electronic spectrum was briefly described [1].

A photographic examination of the complex in water revealed an ill-defined one-electron reduction at zero V vs. standard calomel electrode, and a second unassigned reduction wave at −0.58 V [2].

[pyH]$_4$[Rh$_2$(py)$_2$X$_8$] (X = Cl or Br). These complexes, which are prepared by boiling Rh(py)$_6$Br$_3$ with aqueous H$_3$PO$_2$ followed by treatment with HX [3], have been reformulated as [pyH][Rh(py)$_2$X$_4$] (X = Cl or Br) [4].

References:

[1] A. W. Addison, R. D. Gillard (J. Chem. Soc. A **1970** 2523/6). – [2] A. W. Addison, R. D. Gillard, D. H. Vaughan (J. Chem. Soc. Dalton Trans. **1973** 1187/93). – [3] F. P. Dwyer, R. S. Nyholm (J. Proc. Roy. Soc. N.S. Wales **76** [1943] 275/80). – [4] B. N. Figgis, R. D. Gillard, R. S. Nyholm, G. Wilkinson (J. Chem. Soc. **1964** 5189/93).

2.6.9.3 Complexes of Substituted Pyridines

$4\text{-}CH_3C_5H_4N = \gamma$-pic; $3\text{-}CH_3C_5H_4N = \beta$-pic; $3,5\text{-}(CH_3)C_5H_3N$ = lut; 2,3- or $4\text{-}NH_2C_5H_4N$ = 2,3- or 4-amp; 2,3- or $4\text{-}ClC_5H_4N$ = 2,3- or 4-Clpy; 4-alkyl C_5H_4N = 4-R-py; isoquinoline = iquin

2.6.9.3.1 Rhodium(I) Complexes

Rh(γ-pic)$_3$Cl is obtained when Rh(γ-pic)$_3$(C_2O_4)Cl is irradiated by UV light in dichloromethane. It has apparently not been isolated, but forms a brown solution which reacted with Cl_2 giving *mer*-Rh(γ-pic)$_3$Cl$_3$ [1].

Rh(2-amp)$_2$X (X = Cl or Br). These compounds are obtained when $[Rh(CO)_2X_2]^-$ (X = Cl or Br) react with 2-aminopyridine in ethanol. They form yellow solids [2].

References:

[1] A. W. Addison, R. D. Gillard, P. S. Sheridan, L. R. H. Tipping (J. Chem. Soc. Dalton Trans. **1974** 709/16). – [2] J. V. Kingston, F. T. Mahmoud, G. R. Scollary (J. Inorg. Nucl. Chem. **34** [1972] 3197/201).

2.6.9.3.2 Rhodium(III) Complexes

Tetrakis-substituted Complexes

***trans*-[RhL$_4$X$_2$]Y · nH$_2$O** (L = β- or γ-$CH_3C_5H_4N$, 3- or 4-$C_2H_5C_5H_4N$, 4-n- or i-$C_3H_7C_5H_4N$, 3,5-$(CH_3)_2C_5H_3N$, 3- or 4-$NH_2C_5H_4N$, 3-ClC_5H_4N, 4-$(HO)_2CHC_5H_4N$, 4-$CH_3COC_5H_4N$, 4-$HO_2CC_5H_4N$, 4-$OHCC_5H_4N$; X = Cl, Br, NO_3, ClO_4, BF_4 or NO_3). The **chlorides** (X = Y = Cl) and **bromides** (X = Y = Br) are generally prepared by boiling an aqueous ethanolic solution of $RhX_3 \cdot nH_2O$ (X = Cl or Br) with the substituted pyridine. The red-brown precipitates which form initially dissolve giving a yellow-orange solution from which yellow-orange crystals are obtained on cooling. The **nitrate** (Y = NO_3), **tetrafluoroborate** (Y = BF_4) and **perchlorate** (Y = ClO_4) salts are obtained by addition of HNO_3, HBF_4 or $HClO_4$ to aqueous solutions containing $[RhL_4X_2]^+$ [1]. The compounds prepared in this way are summarised, together with conductivity and polarographic data, in the table on p. 264.

trans-[Rh(γ-pic)$_4$X$_2$]X (X = Cl or Br) may also be prepared by heating $Na_3RhX_6 \cdot nH_2O$, where X = Cl [2, 3] and Br [3], with γ-picoline in aqueous ethanol.

trans-[Rh(3- or 4-amp)$_4$Cl$_2$]Cl may also be prepared from $RhCl_3$ and 3- or 4-aminopyridine in ethanol or ethanol-butanol mixtures [4].

The IR spectra of *trans*-[Rh(3- or 4-amp)$_4$Cl$_2$]Cl have been described [4]. The electronic spectra of these species have been reported [5], and these are characterised by the $^1E_g \leftarrow {}^1A_{1g}$ transition at 409 ± 2 (X = Cl) and 439 ± 2 nm (X = Br), respectively [1]. The solid-state emission spectrum of *trans*-[Rh(γ-pic)$_4$Br$_2$]Br at 2 K has been measured and analysed [6].

Electrical conductivities, indicating that these species are 1:1 electrolytes in solution, have been determined [1, 4]; see table on p. 264 for data. It has been reported that *trans*-

$[Rh(\gamma\text{-pic})_4Cl_2]^+$ underwent a one-electron reduction [2], and this has been substantiated by a more extensive study of a series of substituted pyridine complexes (table below) [7].

Photolysis of *trans*-$[RhL_4Cl_2]^+$ (L = ligands listed in table) resulted in replacement of one L group, giving $[RhL_3(H_2O)Cl_2]^+$, and of Cl^-, giving $[RhL_4(H_2O)Cl]^{2+}$ [5].

Conductivities Λ [1, 4, 8] and polarographic data [1, 7] for complexes of the type *trans*-$[RhL_4X_2]Y \cdot nH_2O$:

L	X	Y	n	Λ[a]		$E_{1/2}$[b]
				H_2O	CH_3NO_2	in V
3-methylpyridine	Cl	Cl	1		73.0	−0.24
	Br	Br	1		77.5	−0.26
4-methylpyridine	Cl	Cl	3		77.5	−0.39
	Cl	ClO_4	0		75	
	Cl	BF_4	0		90	
	Br	Br	2		77.5	−0.31
3-ethylpyridine	Cl	Cl	2		73.0	−0.27
4-ethylpyridine	Cl	ClO_4	0		75.0	−0.36
	Br	Br	1		77.5	
4-n-propylpyridine	Cl	NO_3	0		82.5	
	Br	NO_3	0		69.0	
4-i-propylpyridine	Br	Br	4		77.5	
4-n-butylpyridine	Cl	ClO_4	0		69.0	
3,5-dimethylpyridine	Cl	Cl	0		73.0	−0.11[c]
3-aminopyridine	Cl	Cl	0	83		−0.20
	Br	Br	0	85		
4-aminopyridine	Cl	Cl	2	95.5		−0.47
				93.4		
	Br	Br	2	98		
3-chloropyridine	Cl	Cl	5		52	−0.14
4-(1-hydroxypropyl)pyridine	Cl	ClO_4	0	103		
pyridine-4-carboxylic acid	Cl	Cl	3			
3-acetylpyridine	Cl	ClO_4	0		68	
isoquinoline	Cl	Cl	3		61	−0.22
3-formylpyridine	Cl	Cl	0	88		

[a] conductivity in $cm^2 \cdot \Omega^{-1} \cdot mol^{-1}$, ~$10^{-3}$M solution; [b] using a dropping mercury electrode in water (10^{-4}M in 0.1M KCl), vs. standard calomel electrode; [c] ill-defined wave

$[Rh(\gamma\text{-pic})_4(H_2O)Cl]Cl \cdot 4H_2O$ is obtained either from the reaction of $Na_3RhCl_6 \cdot nH_2O$ in aqueous ethanol with a deficiency of γ-picoline, or recrystallising $Rh(\gamma\text{-pic})_3Cl_3 \cdot 2CHCl_3$ from wet chloroform [3].

$[Rh\{4\text{-}(HO)_2CH\text{-}py\}_4Cl_2]Cl$. This complex is obtained by passing hydrogen through an aqueous solution containing $RhCl_3$ and 4-formylpyridine. It was isolated as yellow-orange rhombic crystals which behaved as a 1:1 electrolyte in water (Λ = 79.8 $cm^2 \cdot \Omega^{-1} \cdot mol^{-1}$ in 10^{-3}M solution). The IR and electronic spectra of the complex were briefly reported [8].

$[Rh(4\text{-}CHO\text{-}py)_4Cl_2]Cl$ is prepared by dehydration of the above complex at 80°C and 0.005 Torr for 18 h. The IR spectrum exhibited ν(CO) at 1715 cm^{-1}, and the electronic spectrum was briefly described [8].

$[Rh(3\text{-}CHO\text{-}py)_4Cl_2]Cl$. This complex is formed when hydrogen is passed through an aqueous solution containing $RhCl_3$ and 3-formylpyridine. It was observed as yellow needles whose conductivity has been determined (see table) [8].

References:

[1] A. W. Addison, K. Dawson, R. D. Gillard, B. T. Heaton, H. Shaw (J. Chem. Soc. Dalton Trans. **1972** 589/96). – [2] F. Pantani (J. Electroanal. Chem. **5** [1963] 40/7). – [3] C. Ouannès, C. Tard (Compt. Rend. **247** [1958] 1202/4). – [4] C. McRobbie, H. Frye (Australian J. Chem. **25** [1972] 893/6). – [5] M. M. Muir, L. B. Zinner, L. A. Paguaga, L. M. Torres (Inorg. Chem. **21** [1982] 3448/50).

[6] K. W. Hipps, G. A. Merrell, G. A. Crosby (J. Phys. Chem. **80** [1976] 2232/9). – [7] A. W. Addison, R. D. Gillard, D. H. Vaughan (J. Chem. Soc. Dalton Trans. **1973** 1187/93). – [8] R. D. Gillard, B. T. Heaton (J. Chem. Soc. A **1968** 1405/6).

Tris-substituted Complexes

$Rh(\gamma\text{-pic})_3(N_3)_3$ is prepared by reaction of *trans*-$[Rh(\gamma\text{-pic})_4Cl_2]Cl$ with NaN_3 and $NaBH_4$ in aqueous solution. The complex was formed as explosive yellow-orange crystals, whose ^{1}H NMR spectrum has been reported [1].

$Rh(\gamma\text{-pic})_3X_3$ (X = Cl, Br or I). The **chloride** (X = Cl) is prepared by refluxing $RhCl_3 \cdot 3H_2O$ with γ-picoline in aqueous ethanol. It is isolated as yellow crystals [2], and if recrystallised from chloroform is obtained as the orange chloroformate $Rh(\gamma\text{-pic})_3Cl_3 \cdot 1.5CHCl_3$ [3]. The chloride may also be obtained by addition of Cl_2 to $Rh(\gamma\text{-pic})_3Cl$, itself produced by photolysis of $Rh(\gamma\text{-pic})_3(C_2O_4)Cl$ [4]. The **bromide** (X = Br) and **iodide** (X = I) are prepared like the chloride, from RhX_3 (X = Br or I) and appear to dimerise in solution [2].

The far IR spectra of the chloride and bromide have been measured (400 to 60 cm^{-1}), and tentative assignments of ν(RhX) made [2]. The IR spectrum of the chloride as chloroformate has been briefly discussed [3].

The ^{1}H NMR spectrum of $Rh(\gamma\text{-pic})_3Cl_3$ has been reported [4]. X-ray photoelectron spectra of this compound afforded the following binding energies (in eV): Rh $3d_{3/2}$ = 314.0, Rh $3d_{5/2}$ = 309.6, Cl $2p_{1/2}$ = 198.9, Cl $2p_{3/2}$ = 197.3 [5].

$Rh(3\text{- or }4\text{-Et-py})_3Cl_3$. These complexes are obtained by refluxing $RhCl_3$ with 3- or 4-ethylpyridine in methanol. They were isolated as orange crystals, whose IR spectra have been recorded (3500 to 650 cm^{-1}) [6].

$Rh(3\text{- or }4\text{-CN-py})_3Cl_3$. These compounds are prepared by refluxing $RhCl_3$ with 3- or 4-cyanopyridine in ethanol. The 3-cyano species was yellow-orange and its 4-cyano analogue yellow. Conductivity studies in dimethylformamide established that they were non-electrolytes [6].

$Rh(lut)_3Cl_3$ is prepared by reaction of $Rh(lut)_3(C_2O_4)Cl$ with aqueous methanolic HCl [4]. The ^{1}H NMR spectrum of the complex has been reported [1].

$Rh(4\text{-Ph-pyr})_3X_3$ (X = Cl, Br or I). These complexes are prepared by reaction of $RhX_3 \cdot nH_2O$ with 4-phenylpyridine in refluxing ethanol. The bromide and iodide appear to dimerise in solution. The far IR spectrum (400 to 60 cm^{-1}) of the chloride and bromide have been reported and tentative assignments made of ν(RhX) [2].

Rh(DAP)Cl$_3$·2H$_2$O, Rh(MEDP)Cl$_3$·2H$_2$O, Rh(EDAP)Cl$_3$·2H$_2$O (DAP = 2,6-diaminopyridine, MEDP = 4-methoxy-2,6-diaminopyridine, EDAP = 4-ethoxy-2,6-diaminopyridine). These complexes are prepared by heating RhCl$_3$ with the ligands in ethanol. They were isolated as dark brown precipitates which exhibited antitumour activity [7].

Rh(γ-pic)$_3$(C$_2$O$_4$)Cl·nH$_2$O. The **tetrahydrate** (n = 4) is obtained by treatment of *trans*-[Rh(γ-pic)$_4$Cl$_2$]Cl with K$_2$C$_2$O$_4$ in boiling aqueous ethanol followed by recrystallisation from water. It is isolated as yellow crystals, and the **anhydrate** (n = 0) is obtained from it by drying in a vacuum desiccator [4]. The electronic spectrum of the compound has been briefly reported, and the compound underwent a one-electron polarographic reduction at −0.41 V vs. standard calomel electrode [8]. The complex was photolysed (UV light) affording Rh(γ-pic)$_3$Cl [4].

Rh(lut)$_3$(C$_2$O$_4$)Cl is prepared in the same way as its γ-picoline analogue using 3,5-lutidine, and isolated as yellow anhydrous platelets [4].

References:

[1] A. W. Addison, K. Dawson, R. D. Gillard, B. T. Heaton, H. Shaw (J. Chem. Soc. Dalton Trans. **1972** 589/96). – [2] I. I. Bhayat, W. R. McWhinnie (Spectrochim. Acta A **28** [1972] 743/51). – [3] A. W. Addison, R. D. Gillard (J. Chem. Soc. Dalton Trans. **1973** 2002/9). – [4] A. W. Addison, R. D. Gillard, P. S. Sheridan, L. R. H. Tipping (J. Chem. Soc. Dalton Trans. **1974** 709/16). – [5] A. D. Hamer, D. G. Tisley, R. A. Walton (J. Chem. Soc. Dalton Trans. **1973** 116/20).

[6] C. McRobbie, H. Frye (Australian J. Chem. **25** [1972] 893/6). – [7] A. Vassilian, A. B. Bikhazi, H. A. Tayim (J. Inorg. Nucl. Chem. **41** [1979] 775/8). – [8] A. W. Addison, R. D. Gillard, D. H. Vaughan (J. Chem. Soc. Dalton Trans. **1973** 1187/93).

Binuclear Complexes

[{Rh(γ-pic)$_4$Cl}$_2$O$_2$]X$_3$ (X = ClO$_4$ or BF$_4$). The **perchlorate** (X = ClO$_4$), as a dihydrate, is prepared by treating *trans*-[Rh(γ-pic)$_4$Cl$_2$]Cl·3H$_2$O in an aqueous ethanolic solution containing NaOH with 10% ozone in oxygen. After the colour of the solution becomes blue, it is chlorinated briefly and HClO$_4$ is added. The blue precipitate is recrystallised from conc. HCl. The **tetrafluoroborate** (X = BF$_4$) is obtained similarly, but using HBF$_4$ in place of HClO$_4$ [1].

The magnetic moments of the perchlorate and tetrafluoroborate were 1.63 ± 0.02 and 1.64 ± 0.02 B.M., respectively, consistent with one unpaired electron per bimetallic cation. The complexes exhibited a single-line ESR signal (g = 2.019) at room temperature in acetone or 50% H$_2$SO$_4$ solution [1].

The IR spectra of the complexes have been discussed, but few significant details published. The electronic spectra have been reported. The electrical conductivities of the perchlorate and tetrafluoroborate in nitromethane (10^{-3} M solutions) were 193 and 210 $cm^2 \cdot \Omega^{-1} \cdot mol^{-1}$, respectively, consistent with their formulations as 1:3 electrolytes [1]. The polarographic behaviour of the species in aqueous solution has been examined, and an ill-defined one-electron process at zero V (vs. standard calomel electrode) and an irreversible 6-electron reduction at −0.79 V detected [2].

[{Rh(γ-pic)$_4$Cl}$_2$O$_2$](ClO$_4$)$_2$·2H$_2$O. Treatment of [{Rh(γ-pic)$_4$Cl}$_2$O$_2$](ClO$_4$)$_3$ in dilute H$_2$SO$_4$ with FeSO$_4$·7H$_2$O affords this complex which is precipitated as yellow crystals on addition of NaClO$_4$. The ^{1}H NMR spectrum of the compound has been reported, indicating that it is diamagnetic, and the electronic spectrum has been described. The conductivity of the salt in nitromethane (10^{-3}M solution) was 154 $cm^2 \cdot \Omega^{-1} \cdot mol^{-1}$, consistent with the formulation as a 1:2 electrolyte [1]. A polarographic study of the complex revealed an irreversible 6-electron reduction in water at −0.79 V vs. standard calomel electrode [2].

References:

[1] A. W. Addison, R. D. Gillard (J. Chem. Soc. A **1970** 2523/6). – [2] A. W. Addison, R. D. Gillard, D. H. Vaughan (J. Chem. Soc. Dalton Trans. **1973** 1187/93).

For other complexes of substituted pyridines see pp. 306/12.

2.6.10 Bipyridyl Complexes (bipyridyl = bipy, $C_{10}H_8N_2$)

While rhodium(III) forms the largest number of 2,2'-bipyridyl complexes, of the type $\{Rh(bipy)_3]^{3+}$, $[Rh(bipy)_2XY]^+$ and $[Rh(bipy)X_4]^-$, compounds containing rhodium(II), (I), 0 and (-I) are also known. Of this latter group $[Rh(bipy)_2]^+$ is significant.

Almost all of the complexes of the type $[Rh(bipy)_2XY]^+$ adopt the *cis* geometry. There is considerable interest in the electronic spectral properties of $[Rh(bipy)_3]^{3+}$ and $[Rh(bipy)_2XY]^+$, especially in relation to the photochemical generation of hydrogen from water.

2.6.10.1 Unsubstituted Complexes

Rhodium(-I) Complexes

The species $[Rh(bipy)_2]^-$ has apparently been detected in a cyclic voltammetric examination of $[Rh(bipy)_2]^+$ in aqueous solution, but it has not been isolated or characterised.

H. Caldararu, M. K. DeArmond, K. W. Hanck, V. E. Sahini (J. Am. Chem. Soc. **98** [1976] 4455/7).

Rhodium(0) Complexes

$[Rh(bipy)_2]_n$. Monomeric and dimeric complexes have apparently been identified by cyclic voltammetric studies of $[Rh(bipy)_3]^{3+}$ and $[Rh(bipy)_2]^+$ in aqueous solution. Electrolytic reduction in aqueous solution of $[Rh(bipy)_3]^{3+}$ at −1.2 V affords dark purple $[Rh(bipy)_2]^+$ in solution, and continued electrolysis at −1.5 V gives a black solid which is thought to be $Rh(bipy)_2$. The ESR spectrum of the compound in acetonitrile revealed two signals, in the region g = 4 and g = 2. The latter signal at 77 K afforded $g_{\parallel} = 1.98 \pm 0.003$ and $g_{\perp} = 2.01 \pm 0.003$.

H. Caldararu, M. K. DeArmond, K. W. Hanck, V. E. Sahini (J. Am. Chem. Soc. **98** [1976] 4455/7).

Rhodium(I) Complexes

Rhodium(I) complexes, possibly $[Rh(bipy)_2]^+$, are implicated in the photochemical generation of hydrogen from water using $[Rh(bipy)_3]^{3+}$ in the presence of $[Ru(bipy)_3]^{2+}$ and triethanolamine [1, 2]. They are also detected during cyclic voltammetric studies of $[Rh(bipy)_3]^{3+}$ when, upon reduction, two irreversible one-electron reductions are observed [3, 4]. The first is due to the formation of $[Rh(bipy)_3]^{2+}$ (see "Rhodium(II) Complexes", p. 269) which is then reduced to $[Rh(bipy)_3]^+$, itself undergoing ligand dissociation (apparently irreversible) to give $[Rh(bipy)_2]^+$ [2, 5, 6]. A rhodium(I) species is also apparently formed in the reactions of $[Rh(bipy)_3]^{3+}$ with $e^-(aq.)$, $(CH_3)_2\dot{C}OH$, $CO_2^{\dot{-}}$, etc. in aqueous solution, possibly via $[Rh(bipy)_3]^{2+}$ which disproportionated into $[Rh(bipy)_3]^{3+}$ and $[Rh(bipy)_2]^+$ [7, 8].

A spectrophotometric study of the behaviour of $[Rh(bipy)_2]^+$ in aqueous solution as a function of pH revealed the existence of purple $[Rh(bipy)_2]_2^{2+}$, colourless $[Rh(bipy)_2H(H_2O)]^{2+}$ (formally containing Rh^{III}) and brown $[\{Rh(bipy)_2\}_2H]^{3+}$ (electronic spectra of these species were briefly reported) [9]. From other studies, it appeared that the following species related to $[Rh(bipy)_2]^+$ could be identified using a combination of synthesis and electronic spectral studies: (a) a red-violet species which may be $[Rh(bipy)_2(OH)_n]^{1-n}$, possibly dimeric with OH bridges, which was very sensitive to pH; (b) violet, relatively insoluble $[Rh(bipy)_2]X$ (X = Cl, ClO_4, etc.); (c) a transient green species produced when $[Rh(bipy)_2(OH)_n]^{1-n}$ was acidified, possibly $[Rh(bipy)_2(OH_2)_n]^+$; and (d) a colourless species present in acid solutions, possibly $[Rh(bipy)_2H]^{2+}$, which formally contains Rh^{III} [7].

$[Rb(bipy)_2]X$ (X = ClO_4 or Cl). The **perchlorate**, as a trihydrate, can be prepared by reduction of $[Rh(bipy)_3](ClO_4)_3$ in aqueous solution with BH_4^-, sodium in zinc amalgam [10], reduction of an aqueous mixture of $RhCl_3$ and bipyridyl with sodium amalgam, followed by addition of $NaClO_4$ [3], reaction of $[Rh(bipy)_3](ClO_4)_3$ in aqueous ethanol with either solid $NaBH_4$ or fresh sodium amalgam, addition of solid $[Rh(bipy)_2Cl_2](ClO_4)\cdot 2H_2O$ to NaOH in ethanol [12], or reaction of $[Rh(H_2O)_6](ClO_4)_3$ in water with Na_2CO_3 (adjusting pH to 8), and adding bipyridyl in warm ethanol, the mixture being refluxed [13]. The anhydrous **chloride** is obtained by treating the $[Rh(C_8H_{14})Cl]_2$ (C_8H_{14} = cyclooctene) in benzene with an aqueous alkaline solution of bipyridyl [9]. The cation may also be formed when $[Rh(bipy)_2Cl_2]Cl\cdot 2H_2O$ in aqueous NaOH or ethanolic solution is treated with hydrogen [12, 14].

The perchlorate was isolated as red-violet [3, 10, 11], violet [12] or purple-red [13] crystals, and the chloride as a purple solid [9]. Salts of the cation are very air-sensitive as solids or in solution [3, 9 to 13].

The IR spectrum of $[Rh(bipy)_2]ClO_4\cdot 3H_2O$ has been tabulated in the range 3000 to 600 cm^{-1} [11], 400 to 60 cm^{-1} (Nujol mull) [11] and 800 to 700 cm^{-1} [15]. It was suggested on the basis of IR spectral data, that $[Rh(bipy)_2]ClO_4\cdot 3H_2O$ actually contained the tetragonal cation $[Rh(bipy)_2(H_2O)_2]^+$ [11]. The electronic spectrum of the perchlorate has been determined in the ranges 220 to 320 nm (shown in paper) [11], and of the chloride in ethanol, ethanol/water, methanol, dimethylacetamide or dichloromethane solution in the range 250 to 560 nm [9]. The electronic spectrum of $[Rh(bipy)_2]^+$ in aqueous NaOH was measured between 300 and 570 nm (shown in paper) [9]. The extinction curves and absorption spectra of $[Rh(bipy)_2]^+$ in aqueous solution were reported over the range 300 to 2000 nm [16, 17].

$[Rh(bipy)_2]^+$ reacted with aqueous bromine giving $[Rh(bipy)_2Br_2]^+$ with chemiluminescence [18]. Species derived from $[Rh(bipy)_2]^+$ (but not identified) act, in the presence of hydrogen, as homogeneous catalysts for the reduction of organic carbonyl groups in the presence of CC double bonds [14, 19].

$Rh(bipy)_2NO_3\cdot 3H_2O$ is prepared by treatment of an ethanolic alkaline solution of $[Rh(bipy)_2Cl_2]NO_3$ with $NaBH_4$. On addition of $NaNO_3$ the complex precipitated as a red-violet solid. The complex was paramagnetic (μ = 1.83 B.M.), and its IR and electronic spectra were reported in the ranges 3000 to 600 cm^{-1} and 400 to 600 nm, respectively. It was suggested that the compound could be either a mixture of Rh^{II} and Rh^0 species, or $[Rh(bipy)_2(H_2O)_2]NO_3$ [11].

References:

[1] G. M. Brown, S.-F. Chan, C. Creutz, H. A. Schwarz, N. Sutin (J. Am. Chem. Soc. **101** [1979] 7638/40). – [2] S.-F. Chan, M. Chou, C. Creutz, T. Matsubara, N. Sutin (J. Am. Chem. Soc. **103** [1981] 369/79). – [3] G. M. Waind, B. Martin (J. Inorg. Nucl. Chem. **8** [1958] 551/6). – [4] G. M. Waind, B. Martin (Chem. Coord. Compds. Symp., Rome 1957, pp. 551/6). – [5] G. Kew, M. K. DeArmond, K. Hanck (J. Phys. Chem. **78** [1974] 727/34).

[6] C. Creutz, A. D. Keller, N. Sutin, A. P. Zipp (J. Am. Chem. Soc. **104** [1982] 3618/27). – [7] Q. G. Mulazzani, S. Emmi, M. Z. Hoffman, M. Venturi (J. Am. Chem. Soc. **103** [1981] 3362/70). – [8] Q. G. Mulazzani, M. Venturi, M. Z. Hoffman (J. Phys. Chem. **86** [1982] 242/7). – [9] M. Chou, C. Creutz, D. Mahajan, N. Sutin, A. P. Zipp (Inorg. Chem. **21** [1982] 3989/97). – [10] B. Martin, G. M. Waind (Proc. Chem. Soc. [London] **1958** 169).

[11] B. Martin, W. R. McWhinnie, G. M. Waind (J. Inorg. Nucl. Chem. **23** [1961] 207/23). – [12] J. D. Miller, F. D. Oliver (J. Chem. Soc. Dalton Trans. **1972** 2473/7). – [13] G. S. Kulasingam, W. R. McWhinnie, J. D. Miller (J. Chem. Soc. A **1969** 521/4). – [14] G. Mestroni, R. Spogliarich, A. Camus, F. Martinelli, G. Zassinovich (J. Organometal. Chem. **157** [1958] 345/52). – [15] W. R. McWhinnie (J. Inorg. Nucl. Chem. **26** [1964] 15/9).

[16] J. Csaszar (Magy. Kem. Folyoirat **66** [1960] 267/71; C.A. **1961** 7035). – [17] J. Csaszar (Acta Chim. Acad. Sci. Hung. **24** [1960] 55/65; C.A. **1961** 9038). – [18] A. Vogler, L. El-Sayed, R. G. Jones, J. Namnath, A. W. Adamson (Inorg. Chim. Acta **53** [1981] L35/L37). – [19] G. Zassinovich, A. Camus, G. Mestroni (Inorg. Nucl. Chem. Letters **12** [1976] 865/7).

Rhodium(II) Complexes

$[Rh(bipy)_3]^{2+}$. This species is detected by polarographic reduction of $[Rh(bipy)_3]^{3+}$ in aqueous solution [1, 2]. It is apparently labile, readily losing one ligand [3, 4]. $[Rh(bipy)_3]^{2+}$ is also implicated in the photochemical production of hydrogen from water using $[Rh(bipy)_3]^{3+}$ in the presence of $[Ru(bipy)_3]^{2+}$ and triethanolamine [5, 6], and in the reduction of $[Rh(bipy)_3]^{3+}$ by $e^-(aq)$, $CO_2^{-\cdot}$ or $(CH_3)_2\dot{C}OH$ [7, 8].

References:

[1] G. M. Waind, B. Martin (J. Inorg. Nucl. Chem. **8** [1958] 551/6). – [2] G. M. Waind (Chem. Coord. Compds. Symp., Rome 1957, pp. 551/6). – [3] G. Kew, M. K. DeArmond, K. Hanck (J. Phys. Chem. **78** [1974] 727/34). – [4] C. Creutz, A. D. Keller, N. Sutin, A. P. Zipp (J. Am. Chem. Soc. **104** [1982] 3618/27). – [5] G. M. Brown, S.-F. Chan, C. Creutz, H. A. Schwarz, N. Sutin (J. Am. Chem. Soc. **101** [1979] 7638/40).

[6] S.-F. Chan, M. Chou, C. Creutz, T. Matsubara, N. Sutin (J. Am. Chem. Soc. **103** [1981] 369/79). – [7] Q. G. Mulazzani, S. Emmi, M. Z. Hoffman, M. Venturi (J. Am. Chem. Soc. **103** [1981] 3362/70). – [8] Q. G. Mulazzani, M. Venturi, M. Z. Hoffman (J. Phys. Chem. **86** [1982] 242/7).

Rhodium(III) Complexes, $[Rh(bipy)_3]X_3$

The **hexafluorophosphate** is made by refluxing $[Rh(bipy)_2Cl_2]Cl \cdot 2H_2O$ and bipyridyl in 75% ethanol, followed by evaporation to dryness. The residue is dissolved in aqueous $NaCl/PO_4^{3-}$ buffer, chromatographed using aqueous NaCl as the eluant, the salt being precipitated by addition of $NaPF_6$ [1]. It may also be obtained by treatment of $[Rh(bipy)_3]Cl_3$ in aqueous solution with PF_6^- [2] and is isolated as a white solid [1, 2]. The **chloride,** as a 4.5-hydrate, is prepared by fusing $RhCl_3 \cdot 3H_2O$ and bipyridyl at 270 to 273°C (the boiling point of bipyridyl) for 10 to 15 min, then dissolving the cooled mixture in aqueous ethanol and either precipitating the product with acetone [2], or treating the solution with animal charcoal before addition of acetone [3]. If the charcoal is omitted, the salt may be isolated as pink crystals, but when it was included, a white solid was obtained [2, 3]. These preceding routes may not give very good yields. $[Rh(bipy)_3]Cl_3$ as the pentahydrate, is better obtained by refluxing a mixture of $RhCl_3 \cdot 3H_2O$ and bipyridyl in 50% aqueous ethanol containing a small amount of N_2H_5Cl, the crude product being recrystallised from hot ethanol after treatment with activated charcoal [6]. The chloride may also be obtained using N-ethylmorpholine as reducing agent/catalyst [7].

The **bromide,** as a 4.5-hydrate, is obtained in the same way as the chloride, using $RhBr_3 \cdot 3H_2O$, and is also obtained as a white crystalline solid. The **thiocyanate** and **tri-iodide**, as trihydrates, are formed by treatment of an ethanolic solution of the trichloride with aqueous LiSCN or LiI, respectively, and were isolated as yellow solids. The **perchlorate,** as a 3.5-hydrate, is prepared by treating an aqueous solution of the trichloride with $NaClO_4$, being isolated as white crystals [3]. It may be obtained in the anhydrous form, by fusing $RhCl_3 \cdot 3H_2O$ with bipyridiyl at ~270°C for 20 min, cooling and dissolving the residue in water to which $NaClO_4$ is added. In this procedure, $[Rh(bipy)_2Cl_2]ClO_4$ precipitates first and after filtration, $[Rh(bipy)_3](ClO_4)_3$ crystallises from the filtrate as pink crystals [4]. The perchlorate can also be made by refluxing $RhCl_3 \cdot 3H_2O$ and bipyridyl in ethanol containing a small amount of $N_2H_4 \cdot H_2O$ for 24 h. On addition of $HClO_4$ the complex was again isolated as pink crystals, and addition of charcoal did not always result in the production of colourless material [5]. The **nitrate** can be obtained in the same way as the perchlorate, but by replacing $NaClO_4$ by KNO_3 [4].

The **sulphate,** as a heptahydrate, is obtained from the chloride by passage of the salt over an anion exchange resin in the sulphate form. It was isolated as hygroscopic yellow crystals [7].

$[Rb(bipy)_3](H_2PO_3) \cdot xH_2O$ is prepared similarly [7]. D-(+)$_{350}$-$[Rh(bipy)_3]^{3+}$ is resolved using (+)-tris{L-cysteine-sulphinato(2)-S,N} cobaltate(III), (+)-$[CoL_3]^{3-}$, by converting the tris(bipyridyl) cation to the Cl^- form via anion exchange resins, and adding K_3[(+)-CoL_3] to the racemic product. On cooling, the less soluble diastereoisomer precipitated [8].

References:

[1] G. A. Crosby, W. H. Elfring (J. Phys. Chem. **80** [1976] 2206/11). – [2] R. E. DeSimone, R. S. Drago (J. Am. Chem. Soc. **92** [1970] 2343/52). – [3] C. M. Harris, E. D. McKenzie (J. Inorg. Nucl. Chem. **25** [1963] 171/4). – [4] B. M. Martin, E. M. Waind (J. Chem. Soc. **1958** 4284/8). – [5] C. Creutz, A. D. Keller, N. Sutin, A. P. Zipp (J. Am. Chem. Soc. **104** [1982] 3618/27).

[6] J. E. Hillis, M. K. DeArmond (J. Lumin. **4** [1971] 273/90). – [7] M. Kirch, J.-M. Lehn, J.-P. Sauvage (Helv. Chim. Acta **62** [1979] 1345/84). – [8] L. S. Dollimore, R. D. Gillard (J. Chem. Soc. Dalton Trans. **1973** 933/40).

Properties of $[Rh(bipy)_3]^{3+}$ Salts

Spectroscopic Properties. The electronic spectrum of $[Rh(bipy)_3]^{3+}$ has been measured in the ranges 300 to 2000 nm [1], 260 to 340 nm in water (circular dichroism, CD, spectrum also reported) [2], 200 to 350 nm (CD spectrum included) [3] and 250 to 370 nm at 83 K [4]. The electronic spectrum of $[Rh(bipy)_3]Cl_3$ was tabulated within the ranges 200 to 750 nm at 80 K [5] and 230 to 360 nm [6]. The visible and UV spectrum of $[Rh(bipy)_3](PF_6)_3$ between 25000 and 50000 cm^{-1} in aqueous solution has been reported [7] as has that of $[Rh(bipy)_3](ClO_4)_3$ in the ranges 200 to 800 nm (shown in paper) [8], in 98% ethanol [9, 10, 11], and 200 to 600 nm (spectrum from 200 to 360 nm shown in paper) [12].

Photoluminescence has been observed from $[Rh(bipy)_3](ClO_4)_3$ [7, 8], and lifetime and quantum yields for non-radiative processes reported [13]. Emission spectra at 77 K in glassy solutions and at room temperature have been reported for $[Rh(bipy)_3]Cl_3$ [5], and emission photoselection studies have been made of the cation [4]. Phosphorescence decay studies have been made of $[Rh(bipy)_3]Cl_3$ at liquid helium temperatures [14].

The ^1H NMR spectrum of $[Rh(bipy)_3](PF_6)_3$ has been obtained in D_2SO_4 and CH_3CN solutions [15]. The natural abundance ^{15}N NMR spectrum of $[Rh(bipy)_3]Cl_3$ in water gave $\delta(N) = 186.1$ ppm (relative to $^{15}NH_4Cl$ as internal reference), $J(RhN) = 18.1$ Hz [16].

References:

[1] J. Csaszar (Acta Chim. Acad. Sci. Hung. **24** [1960] 55/65; C.A. **1961** 9038). – [2] S. F. Mason, B. J. Peart, R. E. Waddell (J. Chem. Soc. Dalton Trans. **1973** 944/9). – [3] S. F. Mason, B. J. Peart (J. Chem. Soc. Dalton Trans. **1973** 949/55). – [4] M. K. DeArmond, C. M. Carlin, W. L. Huang (Inorg. Chem. **19** [1980] 62/71). – [5] M. K. DeArmond, J. E. Hillis (J. Chem. Phys. **54** [1971] 2247/53).

[6] C. M. Harris, E. D. McKenzie (J. Inorg. Nucl. Chem. **25** [1973] 171/4). – [7] G. A. Crosby, W. H. Elfring (J. Phys. Chem. **80** [1976] 2206/11). – [8] D. H. W. Carstens, G. A. Crosby (J. Mol. Spectrosc. **34** [1970] 113/35, 117). – [8] J. Csaszar, E. Horvath (Magy. Tudoman. Akad. Kem. Tudoman. Osztal. Kozlem. **14** [1960] 377/84 from C.A. **1961** 16138). – [10] J. Csaszar (Magy. Kem. Folyoirat **66** [1960] 267/71; C.A. **1961** 7035).

[11] A. Kiss, J. Csaszar (Acta Chim. Acad. Sci. Hung. **38** [1963] 421/34; C.A. **60** [1964] 2436). – [12] B. Martin, G. M. Waind (J. Chem. Soc. **1958** 4284/8). – [13] J. E. Hillis, M. K. DeArmond (J. Lumin. **4** [1971] 273/90). – [14] W. Halper, M. K. DeArmond (Chem. Phys. Letters **24** [1974] 114/6). – [15] R. E. DeSimone, R. S. Drago (J. Am. Chem. Soc. **92** [1970] 2343/52).

[16] K. S. Bose, E. H. Abbott (Inorg. Chem. **16** [1977] 3190/3).

Electrochemical Properties. The electrical conductivities of $[Rh(bipy)_3]Cl_3 \cdot 4.5\,H_2O$, $[Rh(bipy)_3]Br_3 \cdot 4.5\,H_2O$, $[Rh(bipy)_3](SCN)_3 \cdot 3\,H_2O$ and $[Rh(bipy)_3]I_3 \cdot 3\,H_2O$ in nitromethane ($\sim 10^{-3}$M) were 230, 240, 248 and 251 $cm^2 \cdot \Omega^{-1} \cdot mol^{-1}$ [1].

The electrochemical behaviour of $[Rh(bipy)_3]^{3+}$ was established by cyclic voltammetry which revealed the following relationships [2, 3, 4]:

$$[Rh(bipy)_3]^{3+} + e^- \rightleftharpoons [Rh(bipy)_3]^{2+}$$
$$[Rh(bipy)_3]^{2+} + e^- \rightleftharpoons [Rh(bipy)_3]^{+}$$
$$[Rh(bipy)_3]^{+} \rightarrow [Rh(bipy)_2]^{+} + bipy$$

E_{pc} for the first reduction was −0.75 V (versus standard hydrogen electrode; sweep rate 20 mV/s, pyrolytic graphite electrode) and −0.83 V (versus standard calomel electrode) [4].

References:

[1] C. M. Harris, E. D. McKenzie (J. Inorg. Nucl. Chem. **25** [1963] 171/4). – [2] G. Kew, M. K. DeArmond, K. Hanck (J. Phys. Chem. **78** [1974] 727/34). – [3] S.-F. Chan, M. Chou, C. Creutz, T. Matsubara, N. Sutin (J. Am. Chem. Soc. **103** [1981] 369/79). – [4] C. Creutz, A. D. Keller, N. Sutin, A. P. Zipp (J. Am. Chem. Soc. **104** [1982] 3618/27).

Other Properties. $[Rh(bipy)_3]^{3+}$ assisted in the photochemical generation of hydrogen from water in the presence of $[Rh(bipy)_3]^{2+}$ and triethanolamine [1, 2, 3] or 1,2,4,5-tetramethyl-3,6-dipyridyl-1,2,4,5-tetrazene [4], in the presence of Pt catalysts [5, 6, 7] and platinised TiO_2 supports [8].

$[Rh(bipy)_3]^{3+}$, on irradiation, underwent addition of α-hydroxyalkyl radicals at the ligands [9] and this was substantiated by ESR spectral studies of reactions involving $CH_3CHOH^{\cdot}$, $CH_3CH_2^{\cdot}$, $CH_3^{\cdot}$ or $CH_3O^{\cdot}$ radicals [10].

References:

[1] G. M. Brown, S.-F. Chan, C. Creutz, H. A. Schwarz, N. Sutin (J. Am. Chem. Soc. **101** [1979] 7638/40). – [2] S.-F. Chan, M. Chou, C. Creutz, T. Matsubara, N. Sutin (J. Am. Chem. Soc. **103** [1981] 369/79). – [3] M. Kirch, J.-M. Lehn, J.-P. Sauvage (Helv. Chim. Acta **62** [1979] 1345/84). – [4] X. Wang, F. Wang, W. Gu, C. Gu (Taiyangneng Xuebao **3** [1982] 233/4; C.A. **98** [1983] No. 19384). – [5] J.-M. Lehn, J.-P. Sauvage (Nouv. J. Chim. **1** [1977] 449/51).

[6] Z. Shi, H. Tang, X. Zhu (Cuihua Xuebao **2** [1981] 201/8; C.A. **96** [1982] No. 172011). – [7] K. Kalyanasundaram (Nouv. J. Chim. **3** [1979] 511/5). – [8] T. Li, C. Yu, H. Tang, Y. Chen (Cuihua Xuebao **2** [1981] 156/8; C.A. **95** [1981] No. 195079). – [9] M. Venturi, S. Emmi, P. G. Fuochi, Q. G. Mulazzani (J. Phys. Chem. **84** [1980] 2160/6). – [10] M. K. DeArmond, W. Halper (J. Phys. Chem. **75** [1971] 320/4).

2.6.10.2 Substituted Complexes

2.6.10.2.1 Rhodium(II) Complexes

[Rh(bipy)$_2$Cl]ClO$_4$·2H$_2$O is claimed to be formed by BH_4^-, sodium or zinc amalgam reduction of [Rh(bipy)$_2$Cl$_2$]ClO$_4$ in aqueous solution. It is described as a reddish-violet solid which is diamagnetic and may be a Cl-bridged dimer [1].

{Rh(bipy)$_2$Cl(NO$_3$)·2H$_2$O}$_n$. This complex is formed by sodium amalgam reduction of [Rh(bipy)$_2$Cl$_2$]NO$_3$·2H$_2$O in ethanol under N_2. It is a red-violet solid and was essentially diamagnetic; μ_{eff} = 0.16 B.M. at 10°C [2]. It was suggested that the complex was a Cl-bridged dimer. The IR spectrum was tabulated between 3000 and 600 cm^{-1} (Nujol mull), and the electronic spectrum reported in the range 400 to 600 nm [2].

{Rh(bipy)$_2$Cl(ClO$_4$)·2H$_2$O}$_n$ is obtained by sodium amalgam reduction of [Rh(bipy)$_2$Cl$_2$]ClO$_4$ in ethanol under N_2. It was isolated as an essentially diamagnetic red-violet solid; μ_{eff} = −0.04 B.M. at 14°C. The IR spectrum of the compound was tabulated in the range 3000 to 600 cm^{-1}, and it was also reported to be a 1:1 electrolyte in ethanol [2].

[Rh$_2$(bipy)$_4$(CH$_3$CN)$_2$](ClO$_4$)$_4$. This complex is prepared by exposing acetonitrile solutions of [Rh(bipy)$_2$]ClO$_4$·3H$_2$O in dry air for 24 h. The initially purple solution turns yellow-brown and on evaporation the complex is formed as brown crystals. The complex was diamagnetic and was thought to contain a Rh-Rh bond [3].

References:

[1] B. Martin, G. M. Waind (Proc. Chem. Soc. **1958** 169). – [2] B. Martin, W. R. McWhinnie, G. M. Waind (J. Inorg. Nucl. Chem. **23** [1961] 207/23). – [3] H. Caldararu, M. K. DeArmond, K. Hanck (Inorg. Chem. **17** [1978] 2030/2).

2.6.10.2.2 Rhodium(III) Complexes

Complexes of the type [Rh(bipy)$_2$XY]$^+$ have the *cis* geometry.

Hydrido, Alkyl and Cyano Complexes

[Rh(bipy)$_2$H]$^{2+}$ or **[Rh(bipy)$_2$H(H$_2$O)]$^{2+}$.** This species has not been isolated but is thought to be present in aqueous acidic solutions containing [Rh(bipy)$_2$]$^+$. It is colourless [1, 2].

[Rh(bipy)$_2$HCl$_2$]$_2$ is obtained by treatment of [Rh(bipy)$_2$]Cl with 12 M HCl. A green solution is formed from which the complex was obtained as a green solid. The IR spectrum exhibited ν(RhH) = 2140 cm^{-1} and ν(RhCl) or ν(RhN) = 325 or 305 cm^{-1}. The compound dissolved in alkali giving a gray solution, and reacted with bipyridyl giving [Rh(bipy)$_2$]$^+$ [2].

[Rh(bipy)$_2$HCl]Cl is obtained from the filtrate after removal of the green [Rh(bipy)$_2$HCl$_2$]$_2$ described above. It is obtained as yellow crystals whose IR spectrum (Nujol mull) exhibited ν(RhH) = 2070 cm^{-1}. The electronic spectrum of the compound was recorded in HCl, methanol and in aqueous NaOH in the range 240 to 340 nm [2].

[Rh(bipy)$_2$DCl]Cl is prepared by treating a suspension of [Rh(bipy)$_2$]Cl in C$_2$H$_5$OD with DCl in D$_2$O. It was isolated as a green solid [2].

[Rh(bipy)$_2$RX]ClO$_4$ (R = CH_3, C_2H_5, Ph, X = I; R = CH_2Ph, X = Cl or Br). These complexes are prepared by dissolving $RhCl_3 \cdot 3H_2O$ in methanol under N_2, adding an excess of RX and bipyridyl in methanol, and then adding $NaBH_4$ slowly until a brown solution is formed. After refluxing this mixture for 2 h, the complexes are isolated on cooling and addition of $NaClO_4$. Alternatively, [Rh(bipy)$_2$Cl$_2$]Cl·2H_2O can be treated with an excess of RX and $NaBH_4$ in methanol, and the $NaBH_4$ can be replaced by sodium amalgam. The compounds were isolated as yellow (R = Ph, X = I), brown-yellow (R = CH_2Ph, X = Cl or Br), orange-yellow (R = CH_3, X = I) or orange-brown (R = C_2H_5, X = I) crystals [3].

The IR spectra of the complexes were measured in Nujol. Electrical conductivity measurements (~10^{-3} M CH_3NO_2) established that the complexes were 1:1 electrolytes in solution: Λ (in $cm^2 \cdot \Omega^{-1} \cdot mol^{-1}$) = 123 (R = CH_3), 105 (R = C_2H_5), 59 (R = CH_2Ph, X = Cl), 78 (R = CH_2Ph, X = Br) and 84 (R = Ph) [3].

[Rh(bipy)$_2$(CN)$_2$]Cl·4H$_2$O is prepared by boiling an aqueous solution of [Rh(bipy)$_2$Cl$_2$]Cl·2H_2O and NaCN until it becomes colourless, and then adding HCl carefully and boiling for 2 to 3 min. The complex was isolated as colourless crystals whose IR spectrum exhibited ν(CN) = 2153 and 2141 cm^{-1}. The electronic spectrum was obtained in the range 200 to 500 nm, and the electrical conductivity of the complex was 75 $cm^2 \cdot \Omega^{-1} \cdot mol^{-1}$ (~10^{-3} M in H_2O) [4].

References:

[1] Q. G. Mulazzani, S. Emmi, M. Z. Hoffman, M. Venturi (J. Am. Chem. Soc. **103** [1981] 3362/70). – [2] M. Chou, C. Creutz, D. Mahajan, N. Sutin, A. P. Zipp (Inorg. Chem. **21** [1982] 3989/97). – [3] I. I. Bhayat, W. R. McWhinnie (J. Organometal. Chem. **46** [1972] 159/65). – [4] P. M. Gidney, R. D. Gillard, B. T. Heaton (J. Chem. Soc. Dalton Trans. **1972** 2621/8).

Complexes Containing O-Donor Ligands

[Rh(bipy)$_2$(H$_2$O)$_2$](ClO$_4$)$_3$·2H$_2$O. This complex is prepared by heating [Rh(bipy)$_3$]$_2$(SO$_4$)$_3$·7H_2O in ethanol/water mixtures containing NaOH, followed by addition of H_2SO_4 and $LiClO_4$ [1], or by reaction of [Rh(bipy)$_2$Cl$_2$]Cl·2H_2O with N_2H_5Cl in aqueous NaOH followed by neutralisation of the mixture with conc. $HClO_4$ [2, 3]. The complex is obtained as colourless crystals [1, 2, 3]. The electronic spectrum of the species has been determined in aqueous solution in the range 200 to 400 nm [1, 3] and 200 to 500 nm [2], and the ^{1}H NMR spectrum has been reported [2]. The electrical conductivity of the complex was 225 $cm^2 \cdot \Omega^{-1} \cdot mol^{-1}$ (~10^{-3}M in CH_3NO_2) [2].

[Rh(bipy)$_2$(OH)$_2$]$^+$. This species, as a hydrated hydroxide can be obtained by passing the bis-aquo cation described above down an anion-exchange resin in the OH^- form, and can be isolated, by evaporation of the aqueous solution, as a yellow solid [2]. It is also formed by passage of the bis-aquo cation through an anion-exchange resin in the SO_4^{2-} form and adjusting the pH of the solution to ~7 using NaOH [1]. The electronic spectrum of the complex has been measured between 200 and 400 nm [1, 3] and 200 and 500 nm [2], and the ^{1}H NMR spectrum has been measured [2]. The pK_a values for this species were 4.4 (pK_a1) and 6.4 (pK_a2) [1]. The cation underwent an irreversible two-electron polarographic reduction at ca. −0.95 V (using a pyrolitic graphite electrode in aqueous NaOH solutions, vs. standard calomel electrode) [3].

[Rh(bipy)$_2$Cl(H$_2$O)](ClO$_4$)$_2$·H$_2$O is obtained by vigorously boiling [Rh(bipy)$_2$Cl$_2$]Cl·2H_2O with aqueous NaOH followed by addition of $HClO_4$ to pH ≈ 1.0. It was isolated as a yellow solid whose electronic spectrum (aqueous solution) was tabulated between 200 and 500 nm. The electrical conductivity of the species was 154 $cm^2 \cdot \Omega^{-1} \cdot mol^{-1}$ (~10^{-3}M in H_2O) [2].

$[Rh_2(bipy)_4Cl_2(O_2)](ClO_4)_3$. This complex is prepared by reacting $[Rh_2(bipy)_4(CH_3CN)_2](ClO_4)_4$ with Cl_2 in air, or by electrochemical reduction of $[Rh(bipy)_3](ClO_4)_3 \cdot 2H_2O$ in acetonitrile at −1.2 V (versus standard calomel electrode) and leaving the reduced solution in air. The complex was isolated as red crystals, and is paramagnetic ($\mu_{eff} = 1.80$ B.M.) with one unpaired electron per bimetallic cation. The complex exhibited an ESR spectrum in acetonitrile: g(iso) = 2.040(3), and at low temperatures $g_1 = 2.088(8)$, $g_2 = 2.020(9)$, $g_3 = 2.004(5)$. On exposure to air, solutions containing this paramagnetic species afforded additional ESR signals: g(iso) = 2.032(8); $g_1 = 2.088(8)$, $g_2 = 2.020(9)$, $g_3 = 1.992(6)$ [4]. It was suggested that this complex was a dimeric superoxo (O_2^-) adduct of Rh^{III}.

On heating the compound at 135°C, the ESR signal disappeared, but reappeared on addition of Cl_2 in aerated CH_3CN [4].

References:

[1] M. Kirch, J.-M. Lehn, J.-P. Sauvage (Helv. Chim. Acta **62** [1979] 1345/84). – [2] P. M. Gidney, R. D. Gillard, B. T. Heaton (J. Chem. Soc. Dalton Trans. **1972** 2621/8). – [3] S.-F. Chan, M. Chou, C. Creutz, T. Matsubara, N. Sutin (J. Am. Chem. Soc. **103** [1981] 369/79). – [4] H. Caldararu, M. K. DeArmond, K. Hanck (Inorg. Chem. **17** [1978] 2030/2).

Complexes Containing N-Donor Ligands

$[Rh(bipy)_2Cl(NO_2)]X \cdot nH_2O$. The **chloride,** as a 1.5-hydrate, is prepared by vigorous boiling of a mixture of $[Rh(bipy)_2Cl_2Cl_2]Cl \cdot 2H_2O$ and $NaNO_2$, followed by addition of HCl. The **perchlorate,** as a dihydrate, is obtained from the chloride by treatment with ClO_4^-. The complexes were isolated as yellow crystals whose IR spectra exhibited $\nu(NO_2) = 1410$, 1315 and 870 cm^{-1}, consistent with the presence of a $Rh\text{-}NO_2$ bond. The electronic spectrum of the compound in water was determined in the range 200 to 500 nm and the electrical conductivity was 85 $cm^2 \cdot \Omega^{-1} \cdot mol^{-1}$ (~10^{-3}M in H_2O) [1].

$[Rh(bipy)_2(phen)]X_3$ (phen = o-phenanthroline). The **tris-hexafluorophosphate** is obtained by refluxing a mixture of $[Rh(bipy)_2Cl_2]Cl \cdot 2H_2O$ and phenanthroline in 75% ethanol. The mixture, after evaporation, is dissolved in an aqueous $NaCl/PO_4^{3-}$ buffer solution, chromatographed using NaCl as eluant, and on addition of $NaPF_6$, the complex is formed as white crystals. The **tris-chloride,** as a pentahydrate, is prepared from the hexafluorophosphate by dissolving it in acetone and adding aqueous LiCl [2].

The electronic spectrum of this species has been measured in aqueous media in the range 25000 to 50000 cm^{-1} [2]. The emission spectrum of the complex has been measured in methanol and water at 77 K and at room temperature, in the range 400 to 700 nm [2, 3, 4]. Lifetime and quantum yields of the emission were determined [2, 3].

$[Rh(bipy)_2Cl(NH_3)](ClO_4)_2 \cdot 2H_2O$ is formed by heating $[Rh(bipy)_2Cl_2]Cl \cdot 2H_2O$ with conc. aqueous ammonia. Its electronic spectrum in aqueous solution was reported in the range 200 to 500 nm, and its electrical conductivity was 196 $cm^2 \cdot \Omega^{-1} \cdot mol^{-1}$ (~10^{-3}M in H_2O) [1].

$[Rh(bipy)_2Cl(py)](ClO_4)_2$ (py = C_5H_5N, pyridine) is produced by boiling $[Rh(bipy)_2Cl_2]Cl \cdot 2H_2O$ with pyridine in water, followed by addition of $NaClO_4$. The electronic spectrum was recorded (water, 200 to 500 nm) and the electrical conductivity was 194 $cm^2 \cdot \Omega^{-1} \cdot mol^{-1}$ (~10^{-3}M in H_2O) [1].

References:

[1] P. M. Gidney, R. D. Gillard, B. T. Heaton (J. Chem. Soc. Dalton Trans. **1972** 2621/8). – [2] G. A. Crosby, W. H. Elfring (J. Phys. Chem. **80** [1976] 2206/11). – [3] W. Halper, M. K. DeArmond

(J. Lumin. **5** [1972] 225/37). – [4] R. J. Watts, J. van Houten (J. Am. Chem. Soc. **100** [1978] 1718/21).

Halide Complexes

$[Rh(bipy)_2Cl_2]X \cdot nH_2O$. The **chloride,** as a dihydrate, is prepared by heating $RhCl_3 \cdot 3H_2O$ and bipyridyl in aqueous ethanol [1, 2, 3], ethanol/2-methoxyethanol mixtures [2], or by treatment of $RhCl_3 \cdot 3H_2O$ and bipyridyl in aqueous ethanol with N_2H_5Cl [2]. It is also formed by treatment of $[H_3O][Rh(bipy)Cl_4]$ with bipyridyl and N_2H_5Cl in water [4]. The **nitrate,** as a dihydrate, and the anhydrous **perchlorate** are obtained by dissolving the chloride in water and adding $NaNO_3$ and $NaClO_4$, respectively. These species were isolated as yellow crystals [2]. The **tetrachloro (bipyridyl)rhodate(III)** (X = $Rh(bipy)Cl_4$) is formed if the reaction between $RhCl_3 \cdot 3H_2O$ and bypyridyl in aqueous ethanol is stopped at the point when the mixture becomes orange, and was isolated as orange crystals [3].

$[Rh(bipy)_2Cl_2][(H_5O_2)Cl_2]$ is obtained by treatment of $[Rh(bipy)_2Cl_2]Cl \cdot 2H_2O$ with conc. HCl at 90°C and is isolated as yellow crystals [5, 6].

$[Rh(bipy)_2Cl_2][H(ONO_2)_2]$ is prepared by suspending $[Rh(bipy)_2Cl_2]NO_3$ in 8 M HNO_3 [7].

$[Rh(bipy)_2Br_2]X \cdot nH_2O$. The **bromide,** as a dihydrate, is prepared by heating an aqueous solution of $RhCl_3 \cdot 3H_2O$ with an excess of KBr, and then adding bipyridyl in ethanol containing a small amount of N_2H_5Cl [2]. The **tetrabromo(bipyridyl)rhodate(III)** (X = $Rh(bipy)Br_4$) is formed when $RhCl_3 \cdot 3H_2O$, bipyridyl and an excess of KBr are refluxed in aqueous ethanol [3].

$Rh_2(bipy)_3Br_6$ is prepared by heating $[RhBr_6]^{3-}$ with bipyridyl in ethanol, and was isolated as a fawn precipitate [2].

$[Rh(bipy)_2I_2]I$ is formed by heating an aqueous solution of $[Rh(bipy)_2Cl_2]Cl \cdot 2H_2O$ with KI and H_3PO_2 [2], or with NaI with or without N_2H_5Cl [4]. It was isolated as crude red crystals [4].

References:

[1] S. A. Johnson, F. Basolo, R. G. Pearson (J. Am. Chem. Soc. **85** [1963] 1741/7). – [2] R. D. Gillard, J. A. Osborn, G. Wilkinson (J. Chem. Soc. **1965** 1951/65). – [3] I. I. Bhayat, W. R. McWhinnie (Spectrochim. Acta A **28** [1972] 743/51). – [4] P. M. Gidney, R. D. Gillard, B. T. Heaton (J. Chem. Soc. Dalton Trans. **1972** 2621/8). – [5] R. D. Gillard, G. Wilkinson (J. Chem. Soc. **1964** 1640/6).

[6] D. Dollimore, R. D. Gillard, E. D. McKenzie (J. Chem. Soc. **1965** 4479/82). – [7] R. D. Gillard, R. Ugo (J. Chem. Soc. A **1966** 549/52).

Structures

From X-ray diffraction measurements, d-spacings have been obtained for $[Rh(bipy)_2Cl_2]Cl \cdot 2H_2O$ [1 to 4], $[Rh(bipy)_2Br_2]Br \cdot 2H_2O$ [1] and $[Rh(bipy)_2Cl_2][(H_5O_2)Cl_2]$ [4, 5].

References:

[1] R. D. Gillard, B. T. Heaton (J. Chem. Soc. A **1969** 451/4). – [2] M. M. Muir, W.-L. Huang (Inorg. Chem. **12** [1973] 1831/5). – [3] P. Andersen, J. Josephsen (Acta Chem. Scand. **25** [1971] 3255/60). – [4] E. D. McKenzie, R. A. Plowman (J. Inorg. Nucl. Chem. **32** [1970] 199/212). – [5] D. Dollimore, R. D. Gillard, E. D. McKenzie (J. Chem. Soc. **1965** 4479/82).

Spectroscopic Properties

The IR spectra of $[Rh(bipy)_2X_2]X \cdot 2H_2O$ (X = Cl or Br) have been tabulated in the range 3500 to 700 cm^{-1} [1], and 500 to 180 cm^{-1} (no assignments made) [2], of $[Rh(bipy)_2Cl_2]Cl \cdot 2H_2O$ in the range 400 to 60 cm^{-1}, $\nu(RhCl) = 357$ and 350.5 cm^{-1} [3]. The far IR and Raman spectra of $[Rh(bipy)_2Cl_2]Cl \cdot 2H_2O$ permitted assignment of $\nu(RhCl)$ at 352 and 348 cm^{-1} (IR) and 352 and 347 cm^{-1} (Raman) [4]. The IR spectrum of $[Rh(bipy)_2Cl_2][(H_5O_2)Cl_2]$ has been reported in the range 4000 to 700 cm^{-1} [5]. The IR spectra of $[Rh(bipy)_2X_2][Rh(bipy)X_4]$ (X = Cl and Br) have been reported in the range 400 to 60 cm^{-1}; $\nu(RhCl) = 357$, 350, 336, 328, 321(sh) and 316(sh) cm^{-1}; $\nu(RhBr) = 338$, 333, 327 and 316(sh) cm^{-1} [3].

The electronic spectra of $[Rh(bipy)_2X_2]X \cdot 2H_2O$ have been measured in aqueous solution in the ranges 200 to 400 nm (X = Cl, Br) [2], 200 to 450 nm (X = Cl, shown in paper together with circular dichroism spectrum) [6] and 200 to 750 nm (X = Cl, shown in paper) [7]. The visible and UV spectra of $[Rh(bipy)_2X_2]Y$ (X = Y = Cl or I; X = Br, Y = NO_3) was measured in the range 350 to 600 nm in ethanol/methanol glasses at 77 K [8, 9].

The luminescence spectra, quantum yields and lifetimes have been obtained as rigid glasses and as solids for $[Rh(bipy)_2X_2]Y \cdot nH_2O$ (X = Cl, Y = Cl, n = 2; X = Br, Y = NO_3; X = Y = I, n = 0) [8, 9, 10], and for $[Rh(bipy)_2Cl_2]Cl \cdot 2H_2O$ [7, 11]. The emission spectra of $[Rh(bipy)_2X_2]X \cdot 2H_2O$ (X = Cl or Br) have been assigned as phosphorescence and were measured at room temperature, 77 K and liquid helium temperatures [12, 13].

References:

[1] R. D. Gillard, J. A. Osborn, G. Wilkinson (J. Chem. Soc. **1965** 1951/65). – [2] R. D. Gillard, B. T. Heaton (J. Chem. Soc. A **1969** 451/4). – [3] I. I. Bhayat, W. R. McWhinnie (Spectrochim. Acta A **28** [1972] 743/51). – [4] G. C. Kulasingam, W. R. McWhinnie, J. D. Miller (J. Chem. Soc. A **1969** 521/4). – [5] R. D. Gillard, G. Wilkinson (J. Chem. Soc. **1964** 1640/6).

[6] P. M. Gidney, R. D. Gillard, B. T. Heaton (J. Chem. Soc. Dalton Trans. **1972** 2621/8). – [7] M. K. DeArmond, J. E. Hillis (J. Chem. Phys. **54** [1971] 2247/53). – [8] D. H. W. Carstens, G. A. Crosby (J. Mol. Spectrosc. **34** [1970] 113/35). – [9] G. A. Crosby, J. N. Demas (J. Am. Chem. Soc. **92** [1970] 7262/70). – [10] S. H. Peterson, J. N. Demas, J. Kennelly, H. Gafney, D. P. Novak (J. Phys. Chem. **83** [1979] 2991/6).

[11] J. E. Hillis, M. K. DeArmond (J. Lumin. **4** [1971] 273/90). – [12] Y. Ohashi, K. Yoshihara, S. Nagakura (J. Mol. Spectrosc. **38** [1971] 43/52). – [13] W. Halper, M. K. DeArmond (Chem. Phys. Letters **24** [1974] 114/6).

Electrochemical and Chemical Reactions

Electrical conductivity data have been obtained for $[Rh(bipy)_2Cl_2]Cl \cdot 2H_2O$ in $(CH_3)_2SO$ (83 $cm^2 \cdot \Omega^{-1} \cdot mol^{-1}$) [1] and in water (102 $cm^2 \cdot \Omega^{-1} \cdot mol^{-1}$) [2], for $[Rh(bipy)_2Cl_2]NO_3$ and $[Rh(bipy)_2Cl_2][H(ONO_2)_2]$ in water (162 to 89 and 510 to 458 $cm^2 \cdot \Omega^{-1} \cdot mol^{-1}$, respectively, depending on concentration) [3]. The conductivity of $[Rh(bipy)_2X_2][Rh(bipy)X_4]$ in $(CH_3)_2SO$ was 20 (X = Cl) and 45 (X = Br) $cm^2 \cdot \Omega^{-1} \cdot mol^{-1}$ [1], of $[Rh(bipy)_2Br_2]Br \cdot 2H_2O$ in water 102 $cm^2 \cdot \Omega^{-1} \cdot mol^{-1}$ [2] and of $[Rh(bipy)_2I_2]I$ in water 104 $cm^2 \cdot \Omega^{-1} \cdot mol^{-1}$ [4].

The electrochemical properties of $[Rh(bipy)_2Cl_2]^+$ in aqueous solution have been investigated by cyclic voltammetry, and it was shown that Cl^- was eliminated following the first reduction step [5].

The thermal decomposition of $[Rh(bipy)_2Cl_2][(H_5O_2)Cl_2]$ has been studied using thermogravimetric analysis and differential thermal analysis, and it was suggested that intermediate stages in the weight loss were probably due to the formation of $[Rh(bipy)_2Cl_2]HCl_2$ [6, 7].

The sensitised photolytic decomposition of $[Rh(bipy)_2Cl_2]Cl$ in ethanol glasses was investigated at 77 K using ESR spectroscopy, and it was established that radical addition (e.g., $CH_3CHOH^{\cdot}$, $CH_3CH_2^{\cdot}$ $CH_3^{\cdot}$ or $CH_3O^{\cdot}$) occurred at the heterocyclic ligand [8]. The kinetics of the autocatalytic reduction of $[Rh(bipy)_2Cl_2]ClO_4 \cdot 2H_2O$ in aqueous alkaline ethanol under hydrogen were determined [9].

References:

[1] I. I. Bhayat, W. R. McWhinnie (Spectrochim. Acta A **28** [1972] 743/51). – [2] M. M. Muir, H.-L. Huang (Inorg. Chem. **12** [1973] 1831/5). – [3] R. D. Gillard, R. Ugo (J. Chem. Soc. A **1966** 549/52). – [4] P. M. Gidney, R. D. Gillard, B. T. Heaton (J. Chem. Soc. Dalton Trans. **1972** 2621/8). – [5] G. Kew, M. K. DeArmond (J. Phys. Chem. **78** [1974] 727/34).

[6] D. Dollimore, R. D. Gillard, E. D. McKenzie (J. Chem. Soc. **1965** 4479/82). – [7] D. Dollimore, R. D. Gillard, E. D. McKenzie, R. Ugo (J. Inorg. Nucl. Chem. **30** [1968] 2755/8). – [8] M. K. DeArmond, W. Halper (J. Phys. Chem. **75** [1971] 3230/4). – [9] J. D. Miller, F. D. Oliver (J. Chem. Soc. Dalton Trans. **1972** 2473/7).

$[Rh(bipy)(PPh_3)_2H_2]ClO_4$. This complex is obtained by reaction of $Rh(bipy)_2Cl_2]Cl \cdot 2H_2O$ in methanol under N_2 with a small amount of $NaBH_4$ and a large excess of PPh_3. After stirring, addition of $NaClO_4$ affords the complex [1]. Alternatively, it is formed by addition of an ether solution of PPh_3 to $[Rh(bipy)(C_8H_{12}]ClO_4$ (C_8H_{12} = cycloocta-1,5-diene) in methanol under H_2, followed by an aqueous solution of $NaClO_4$ [2]. The compound was isolated as white crystals [1, 2], whose IR spectrum exhibited ν(RhH) at 2120 and 2050 cm^{-1} (Nujol mull) [1] or 2041 to 2058 cm^{-1} [2]. The ^1H NMR spectrum of the complex showed τ(RhH) = 25.7, $^1J(RhH)$ = $^2J(PH)$ = 15 Hz. The electrical conductivity of the complex was 78 $cm^2 \cdot \Omega^{-1} \cdot mol^{-1}$ (~10^{-3} M CH_3NO_2 solution) [1].

$[bipyH][Rh(bipy)Cl_4]$. This complex is prepared by treating $RhCl_3 \cdot 3H_2O$ in hot HCl solution with bipyridyl, and was isolated as red-orange crystals [3].

$[Rh(bipy)_2X_2][Rh(bipy)X_4]$ (X = Cl or Br) has been described on p. 275.

References:

[1] I. I. Bhayat, W. R. McWhinnie (J. Organometal. Chem. **46** [1972] 159/65). – [2] C. Cocevar, G. Mestroni, A. Camus (J. Organometal. Chem. **35** [1972] 389/95). – [3] E. D. McKenzie, R. A. Plowman (J. Inorg. Nucl. Chem. **32** [1970] 199/212).

2.6.10.3 Heterocyclic Ring-Substituted Complexes

3,3′- and 4,4′-dimethyl- and -diphenylbipyridyls abbreviated as 3,3′-, 4,4′-Me_2bipy and 4,4′-Ph_2bipy.

Unsubstituted Complexes

$[Rh(4,4'\text{-}Me_2bipy)_3]X_3$. The **perchlorate** is made by refluxing an ethanolic solution of $RhCl_3 \cdot 3H_2O$ containing 4,4′-Me_2bipy and a small amount of $N_2H_4 \cdot H_2O$ followed by addition of $HClO_4$. It was isolated as pink crystals [1]. The **chloride,** as a heptahydrate, is prepared by heating $RhCl_3 \cdot 3H_2O$ with N-ethylmorpholine and 4,4′-Me_2bipy in aqueous ethanol for 25 to 30 min and was obtained as yellow crystals. The **sulphate,** $[Rh(4,4'\text{-}Me_2bipy)_3]_2(SO_4)_3 \cdot 5H_2O$, is obtained from the chloride by passage over an anion-exchange resin in the SO_4^{2-} form, and was isolated as yellow crystals [2].

A cyclic voltammetric study of $[Rh(4,4'\text{-}Me_2bipy)_3]^{3+}$ revealed that it underwent ligand dissociation with reduction [1]:

$$[Rh(4,4'\text{-}Me_2bipy)_3]^{3+} + e^- \rightleftharpoons [Rh(4,4'\text{-}Me_2bipy)_3]^{2+}$$
$$[Rh(4,4'\text{-}Me_2bipy)_3]^{2+} + e^- \rightleftharpoons [Rh(4,4'\text{-}Me_2bipy)_3]^{+}$$
$$[Rh(4,4'\text{-}Me_2bipy)_3]^{+} \rightarrow [Rh(4,4'\text{-}Me_2bipy)_2]^{+} + 4,4'\text{-}Me_2bipy$$

E_{pc} for the first reduction wave was −0.86 V (vs. standard hydrogen electrode) or −0.93 V (vs. standard calomel electrode), in aqueous solutions containing NaOH and Na_2SO_4. The waves were scanrate-dependent, and a second reduction wave could be detected at −1.10 V (vs. standard calomel electrode) [1].

The complex, in the presence of $[Ru(bipy)_3]^{2+}$ and triethanolamine, caused hydrogen generation from water on irradiation with visible light [2].

$[Rh(3,3'\text{-}Me_2bipy)_3]^{3+}$. The cation has been optically resolved using (+)- or (−)-tris-{L-cysteinesulphinato(2-)-S, N}cobaltate(III) [3, 4].

Crystals of $[Rh(3,3'\text{-}Me_2bipy)_3](ClO_4)_3 \cdot 3H_2O$ were orthorhombic, space group $P2_12_12_1\text{-}D_2^4$, a = 17.675(2), b = 18.151(2), c = 12.637(4) Å, Z = 4, and the X-ray structure determination showed that Rh-N = 2.039(5) Å and that the absolute configuration of the cation was $\Lambda(\delta\lambda\lambda)$. The pyridine rings of the heterocyclic ligand were twisted relative to each other because of the steric interaction between the methyl groups [5].

The electronic and circular dichroism spectra of (+)- and (−)-$[Rh(3,3'\text{-}Me_2bipy)_3]^{3+}$ have been measured in aqueous solution in the ranges 25000 to 50000 cm^{-1} (shown in paper) [3] and 500 to 700 nm (shown in paper) [4].

Substituted Complexes

$[Rh(4,4'\text{-}Me_2bipy)_2X_2]X$, $[Rh(4,4'\text{-}Ph_2bipy)_2X_2]X$ (X = Cl or Br). The **chlorides,** as 1,5-hydrates, are prepared by refluxing $RhCl_3 \cdot 3H_2O$ with 4,4'-Me_2bipy or 4,4'-Ph_2bipy in ethanol and were isolated as yellow crystals. The **bromide** (4,4'-Ph_2bipy complex only) is obtained from solutions containing the chloride by addition of an excess of KBr [6].

$[4,4'\text{-}Me_2bipyH][Rh(4,4'\text{-}Me_2bipy)Br_4]$ is obtained by heating $RhCl_3 \cdot 3H_2O$ with 4,4'-Me_2bipy in aqueous ethanol followed by addition of KBr [6].

The IR spectra of these complexes have been measured in the range 400 to 60 cm^{-1} and assignments made of ν(RhX) (see table below). The electrical conductivities of the complexes were determined in CH_3NO_2 (see table below) [6].

IR spectral and conductivity data [6]:

Complex	ν(RhX) in cm^{-1}	Λ in $cm^2 \cdot \Omega^{-1} \cdot mol^{-1}$ [a]
$[Rh(4,4'\text{-}Me_2bipy)_2Cl_2]Cl \cdot 1.5H_2O$	349, 341	66
$[4,4'\text{-}Me_2bipyH][Rh(4,4'\text{-}Me_2bipy)Br_4]$	192	79
$[Rh(4,4'\text{-}Ph_2bipy)_2Cl_2]Cl$	358, 348	69
$[Rh(4,4'\text{-}Ph_2bipy)_2Br_2]Br$	294, 245	68
$[Rh(4,4'\text{-}Ph_2bipy)_2I_2]I$	—	79

[a] in ~10^{-3}M CH_3NO_2.

References:

[1] C. Creutz, A. D. Keller, N. Sutin, A. P. Zipp (Inorg. Chem. **21** [1982] 2477/82). – [2] M. Kirch, J.-M. Lehn, J.-P. Sauvage (Helv. Chim. Acta **62** [1979] 1345/84). – [3] T. M. Suzuki (Bull. Chem. Soc. Japan **52** [1979] 433/6, 436). – [4] L. S. Dollimore, R. D. Gillard (J. Chem. Soc. Dalton Trans. **1973** 933/40, 938). – [5] S. Oha, M. Miyamae, S. Sato, Y. Saito (Acta Cryst. B **35** [1979] 1470/2).

[6] I. I. Bhayat, W. R. McWhinnie (Spectrochim. Acta A **28** [1972] 743/51).

2.6.11 o-Phenanthroline Complexes (o-phenanthroline = phen)

Rhodium(III) forms complexes mainly of three types with o- or 9,10-phenanthroline: $[Rh(phen)_3]^{3+}$, $[Rh(phen)_2XY]^+$ and $[Rh(phen)X_4]^-$. There are a few examples of Rh^I complexes and a species containing Rh^0 has been observed.

2.6.11.1 Unsubstituted Complexes

Rhodium(0) Complexes

$[Rh(phen)_2]_n$ (n = 1 or 2). Electrochemical reduction of $[Rh(phen)_3]^{3+}$ at −1.2 V (versus standard calomel electrode) in aqueous solution gave purple $[Rh(phen)_2]^+$ which, on continued electrolysis at −1.5 V, afforded the compound as a black solid. The species exhibited two ESR signals at $g \approx 4$ and $g \approx 2$ in acetonitrile, and at 77 to 200 K, the signal at $g \approx 2$ was resolved into $g_{\parallel} = 1.97 \pm 0.003$ and $g_{\perp} = 2.01 \pm 0.003$. It was thought that the compound had a distorted square-planar structure.

H. Caldararu, M. K. DeArmond, K. W. Hanck, V. E. Sahini (J. Am. Chem. Soc. **98** [1976] 4455/7).

Rhodium(I) Complexes

$[Rh(phen)_2]X$ (X = Cl or PF_6). The **chloride** is obtained by treating $[Rh(C_8H_{14})Cl]_2$ (C_8H_{14} = cyclooct-1-ene) in benzene with an aqueous alkaline solution of o-phenanthroline [1]. The **hexafluorophosphate** was formed by reaction of an acetonitrile solution of $[Rh(phen)(ED)]PF_6$ (ED = hexa-1,5-diene) with o-phen [2]. The chloride was isolated as a purple solid [1] and the hexafluorophosphate as violet crystals [2]; both are sensitive to oxygen.

The electronic spectrum of the chloride has been measured in ethanolic NaOH solution in the range 260 to 570 nm [1, 3].

References:

[1] M. Chou, C. Creutz, D. Mahajan, N. Sutin, A. P. Zipp (Inorg. Chem. **21** [1982] 3989/97). – [2] G. Mestroni, G. Zassinovich, A. Camus (J. Organometal. Chem. **140** [1977] 63/72). – [3] J. Cszaszar (Magy. Kem. Folyoirat **66** [1960] 267/71; C.A. **1961** 7035).

Rhodium(II) Complexes

$[Rh(phen)_3]^{2+}$ has been detected by cyclic voltammetric studies of $[Rh(phen)_3]^{3+}$ in aqueous solution [1, 2]. It is formed by one-electron reduction of the Rh^{III} trication, $E_{1/2} = -0.67$ V in aqueous NaOH solution [1], but readily loses a ligand by dissociation [1, 2].

The electronic spectrum of this transient dication was detected when $[Rh(phen)_3]^{3+}$ was irradiated in the presence of $NHPh_2$ [3].

References:

[1] S.-F. Chan, M. Chou, C. Creutz, T. Matsubara, N. Sutin (J. Am. Chem. Soc. **103** [1981] 369/79). – [2] C. Creutz, A. D. Keller, N. Sutin, A. P. Zipp (J. Am. Chem. Soc. **104** [1982] 3618/27). – [3] R. Ballardini, G. Varani, V. Balzani (J. Am. Chem. Soc. **102** [1980] 1719/20).

Rhodium(III) Complexes

$[Rh(phen)_3]X_3 \cdot nH_2O$. The **perchlorate,** as a tetrahydrate, is obtained either by refluxing an ethanolic solution of $RhCl_3 \cdot 3H_2O$ containing o-phenanthroline and a small amount of $N_2H_4 \cdot H_2O$ followed, on cooling, by addition of $HClO_4$ [1], or by addition of $NaClO_4$ to an aqueous solution of $[Rh(phen)_3]Cl_3 \cdot 4H_2O$ [2]. It was isolated as pink crystals [1, 2]. The **hexafluorophosphate** is prepared by fusing $RhCl_3 \cdot 3H_2O$ with o-phenanthroline in a test-tube, and heating at ~270°C for 10 to 15 min. On cooling, dissolving in aqueous ethanol, and precipitating with acetone, $[Rh(phen)_3]Cl_3$ is obtained which is dissolved in water and treated with PF_6^-. The complex was obtained as a white solid [3]. The **chloride,** as a tetrahydrate, may be prepared by boiling an aqueous ethanolic solution of $RhCl_3 \cdot 3H_2O$ and o-phenanthroline hydrate, followed by recrystallisation from water and acetone to which charcoal is added [2], or by heating $RhCl_3 \cdot 3H_2O$ with o-phenanthroline hydrate and N-ethylmorpholine in aqueous ethanol [4]. It was isolated as white crystals [2, 4]. The **bromide,** as a pentahydrate, is formed by boiling an aqueous ethanolic mixture of $RhBr_3$ and o-phenanthroline hydrate, and recrystallising from aqueous acetone. It was obtained as colourless crystals which became light yellow when anhydrous. The **iodide,** as a 5,5-hydrate, and **thiocyanate,** as a 3,5-hydrate, are prepared by adding KI and KSCN, respectively, to $[Rh(phen)_3]Cl_3$ in water, and were both obtained as yellow crystals. The **oxalate,** as a 22-hydrate, is prepared by boiling $Rh_2O_3 \cdot nH_2O$ with KHC_2O_4 in water, adding o-phenanthroline hydrate, refluxing for two days, and then adding ethanol. It was isolated as brown-rose microcrystals [2].

$[Rh(phen)_3]^{3+}$ is optically resolved by conversion to typical salts of the chloride form by use of anion exchange resins in the Cl^- form, and addition of the K^+ salt of (+)-tris{L-cysteinesulphinato-(2-S,N}cobaltate(III). The diastereoisomers are separated by fractional crystallisation [5].

References:

[1] R. E. DeSimone, R. S. Drago (J. Am. Chem. Soc. **92** [1970] 2343/52). – [2] E. D. McKenzie, R. A. Plowman (J. Inorg. Nucl. Chem. **32** [1970] 199/212). – [3] C. Creutz, A. D. Keller, N. Sutin, A. P. Zipp (J. Am. Chem. Soc. **104** [1982] 3618/27). – [4] M. Kirch, J.-M. Lehn, J.-P. Sauvage (Helv. Chim. Acta **62** [1979] 1345/84). – [5] L. S. Dollimore, R. D. Gillard (J. Chem. Soc. Dalton Trans. **1973** 933/40).

Properties of Rhodium(III) Complexes

Spectroscopic Properties. The electronic spectrum of $[Rh(phen)_3]^{3+}$ has been measured in the range 300 to 2000 nm [1, 2], 200 to 400 nm, together with its circular dichroism spectrum [3], 250 to 500 nm, with CD and optical rotatory dispersion data [4], and 200 to 350 nm [5]. The absorption spectrum of $[Rh(phen)_3](ClO_4)_3$ in water has been reported in the ranges 200 to 800 nm (shown in paper) [6], and of $[Rh(phen)_3]Cl_3$ in glassy solutions at 77 K [7].

Emission spectra have been obtained from $[Rh(phen)_3]^{3+}$ at 77 K [6 to 9] and it was established that irradiation of $[Rh(phen)_3]^{3+}$ in the presence of $NHPh_2$ gave $[Rh(phen)_3]^{2+}$ transiently [10]. Lifetime and quantum yields have been determined for non-radiative processes in $[Rh(phen)_3]^{3+}$ [11]. The phosphorescence decay of $[Rh(phen)_3]Cl_3$ has been measured at liquid helium temperatures [12, 13].

The ^1H NMR spectrum of $[Rh(phen)_3](PF_6)_3$ has been determined in D_2SO_4 and CH_3CN solutions [14] and the natural abundance ^{15}N NMR spectrum of $[Rh(phen)_3]Cl_3$ revealed $\delta(^{15}N) = 180.8$ ppm, $^1J(RhN) = 18.4$ Hz [15].

References:

[1] J. Csaszar (Acta Chim. Acad. Sci. Hung. **24** [1960] 55/65; C.A. **1961** 9038). – [2] J. Csaszar (Magy. Kem. Folyoirat **66** [1960] 267/71; C.A. **1961** 7035). – [3] S. F. Mason, B. J. Peart (J. Chem. Soc. Dalton Trans. **1973** 949/55). – [4] L. S. Dollimore, R. D. Gillard (J. Chem. Soc. Dalton Trans. **1973** 933/40). – [5] E. D. McKenzie, R. A. Plowman (J. Inorg. Nucl. Chem. **32** [1970] 199/212).

[6] D. H. W. Carstens, G. A. Crosby (J. Mol. Spectrosc. **34** [1970] 113/35). – [7] M. K. DeArmond, J. E. Hillis (J. Chem. Phys. **54** [1971] 2247/53). – [8] M. K. DeArmond, C. M. Carlin, W. L. Huang (Inorg. Chem. **19** [1980] 62/71). – [9] G. A. Crosby, D. H. W. Carstens (Mol. Lumin. Intern. Conf., Chicago 1968 [1969], pp. 309/20; C.A. **71** [1969] No. 26353). – [10] R. Ballardini, G. Varani, V. Balzani (J. Am. Chem. Soc. **102** [1980] 1719/20).

[11] J. E. Hillis, M. K. DeArmond (J. Lumin. **4** [1971] 273/90). – [12] W. Halper, M. K. DeArmond (Chem. Phys. Letters **24** [1974] 114/6). – [13] F. Bolletta, A. Rossi, F. Barigelletti, S. Dellonte, V. Balzani (Gazz. Chim. Ital. **111** [1981] 155/8). – [14] R. E. DeSimone, R. S. Drago (J. Am. Chem. Soc. **92** [1970] 2343/52). – [15] K. S. Bose, E. H. Abbott (Inorg. Chem. **16** [1977] 3190/3).

Other Properties. The cyclic voltammograms of $[Rh(phen)_3]^{3+}$ in aqueous solutions containing NaOH [1], or NaOH and Na_2SO_4 [2], exhibited two irreversible electron transfers, and ligand dissociation:

$$[Rh(phen)_3]^{3+} + e^- \rightleftharpoons [Rh(phen)_3]^{2+}$$
$$[Rh(phen)_3]^{2+} + e^- \rightleftharpoons [Rh(phen)_3]^{+}$$
$$[Rh(phen)_3]^{+} \rightarrow [Rh(phen)_2]^{+} + phen$$

E_{pc} for the first reduction wave was −0.67 V (vs. standard calomel electrode, SCE) [1], or −0.71 V (versus standard hydrogen electrode) [2], and −0.80 V (vs. SCE). A second wave, at −1.06 V (versus SCE) could also be detected, and the potentials depended on the cyclic voltammogram scan rate [2].

The stability constants for the formation of outer-sphere complexes between $[Rh(phen)_3]^{3+}$ (as $[Fe(CN)_6]^{3-}$ salt) and NO_3^-, HCO_2^-, $CH_3CO_2^-$ and $C_2H_5CO_2^-$ have been determined [3].

Addition of radicals, e.g., $CH_3CHOH^{\cdot}$, $CH_3CH_2^{\cdot}$, $CH_3^{\cdot}$ or $CH_3O^{\cdot}$, to $[Rh(phen)_3]^{3+}$ occurred when $[Rh(phen)_3]^{3+}$ was irradiated in ethanol solution [4, 5] and an ESR spectral study established that the radical attached to the 9-position of the o-phenanthroline ring system [5].

References:

[1] S.-F. Chan, M. Chou, C. Creutz, T. Matsubana, N. Sutin (J. Am. Chem. Soc. **103** [1981] 369/79). – [2] C. Creutz, A. D. Keller, N. Sutin, A. P. Zipp (J. Am. Chem. Soc. **104** [1982] 3618/27). – [3] A. K. Pyartman, M. V. Sof'in, V. E. Mironov (Izv. Vysshikh Uchebn. Zavedenii Khim. Khim. Tekhnol. **24** [1981] 1465/8; C.A. **96** [1982] No. 111102). – [4] M. Venturi, S. Emmi, D. G. Fuochi, Q. G. Mulazzani (J. Phys. Chem. **84** [1980] 2160/6). – [5] M. K. DeArmond, W. Halper (J. Phys. Chem. **75** [1971] 3230/4).

2.6.11.2 Substituted o-Phenanthroline Complexes, $[Rh\,phen_xY_y]^{n+}$

Complexes of the type $[Rh(phen)_2XY]^+$ have a *cis*-geometry.

2.6.11.2.1 BIs-phenanthroline Complexes of Rhodium(III)

2.6.11.2.1.1 Hydrido Complexes

$[Rh(phen)_2HCl]Cl$ is prepared by treating $[Rh(phen)_2]Cl$ with 12 M HCl. It may be recrystallised from dichloromethane/diethylether mixtures, and was isolated as yellow crystals.

The IR spectrum exhibited $\nu(RhH) = 2090\ cm^{-1}$ and the electronic spectrum in acid solution exhibited $\lambda_{max} = 273$ nm, and in NaOH solution, $\lambda_{max} = 535$, 273 nm.

M. Chou, C. Creutz, D. Mahajan, N. Sutin, A. P. Zipp (Inorg. Chem. **21** [1982] 3989/97).

2.6.11.2.1.2 Cyano Complexes

$[Rh(phen)_2(CN)_2]X$. The **chloride,** as a 2.5-hydrate, is obtained by boiling $[Rh(phen)_2Cl_2]Cl \cdot 3H_2O$ with aqueous sodium cyanide followed by careful addition of conc. HCl. It was isolated as colourless crystals. The **perchlorate** is prepared by addition of $NaClO_4$ to an aqueous solution of the chloride salt. The IR spectrum of $[Rh(phen)_2(CN)_2]^+$ exhibited $\nu(CN) = 2147$ and $2141\ cm^{-1}$, and the electronic spectrum was recorded in the range 200 to 500 nm. The electrical conductivity of the chloride salt ($\sim 10^{-3}$ M solution in water) was $81\ cm^2 \cdot \Omega^{-1} \cdot mol^{-1}$.

$[Rh(phen)_2Cl(CN)]Cl \cdot H_2O$ is formed by heating an aqueous solution of $[Rh(phen)_2Cl_2]Cl \cdot 3H_2O$ with NaCN at 90°C for 1 h. Treatment with conc. HCl followed by concentration of the solution afforded the compound as pale yellow crystals.

P. M. Gidney, R. D. Gillard, B. T. Heaton (J. Chem. Soc. Dalton Trans. **1972** 2621/8).

2.6.11.2.1.3 Complexes Containing Aquo Ligands

$[Rh(phen)_2(H_2O)_2]X_3 \cdot nH_2O$. The **perchlorate,** as a dihydrate, is prepared by boiling $[Rh(phen)_2Cl_2]Cl \cdot 3H_2O$ in aqueous NaOH containing N_2H_5Cl for 10 min followed by evaporation and neutralisation of the mixture with conc. $HClO_4$. On cooling solutions containing the perchlorate, the compound formed as pale yellow crystals [1].

X-ray diffraction studies of the **nitrate** ($X = NO_3$), as a dihydrate, gave the d-spacings [2]. The electronic spectrum of the **tris-perchlorate** was recorded in the range 200 to 500 nm, and its electrical conductivity ($\sim 10^{-3}$ M aqueous solution) was $201\ cm^2 \cdot \Omega^{-1} \cdot mol^{-1}$ [1].

$[Rh(phen)_2Cl(H_2O)](ClO_4)_2 \cdot H_2O$. This is prepared from $[Rh(phen)_2Cl_2]Cl \cdot 3H_2O$ with boiling aqueous NaOH, followed by addition of $HClO_4$ to pH ≈ 1.0. On cooling it was isolated as yellow crystals. The electronic spectrum was recorded in the range 200 to 500 nm, and the electrical conductivity of the compound ($\sim 10^{-3}$ M solution in water) was $153\ cm^2 \cdot \Omega^{-1} \cdot mol^{-1}$ [1].

References:

[1] P. M. Gidney, R. D. Gillard, B. T. Heaton (J. Chem. Soc. Dalton Trans. **1972** 2621/8). – [2] P. Andersen, J. Josephsen (Acta Chem. Scand. **25** [1971] 3255/60).

2.6.11.2.1.4 Complexes with N-Donor Ligands

[Rh(phen)$_2$(NH$_3$)$_2$](ClO$_4$)$_3$·3H$_2$O is prepared by boiling an aqueous solution of [Rh(phen)$_2$Cl$_2$]Cl·3H$_2$O with conc. aqueous ammonia and N_2H_5Cl. On addition of $NaClO_4$, the complex was precipitated initially as light brown crystals which were further purified by recrystallisation from water in the presence of charcoal, which afforded colourless crystals [1]. The electronic spectrum of the complex was reported in the range 200 to 500 nm, and its electrical conductivity (~10^{-3} M solution in water) was 293 $cm^2 \cdot \Omega^{-1} \cdot mol^{-1}$ [1].

The complex has been optically resolved using (+)-tris[L-cysteinesulphinato(2-)-S,N]cobaltate(III), and the absorption, optical rotatory dispersion and circular dichroism spectra of the compound measured in the range 500 to 750 nm [2] (spectra shown in paper).

[Rh(phen)$_2$Cl(NH$_3$)](ClO$_4$)$_2$·H$_2$O is prepared by boiling [Rh(phen)$_2$Cl$_2$]Cl·3H$_2$O in conc. aqueous ammonia. On addition of $NaClO_4$, the complex was formed as yellow crystals. The electronic spectrum was measured in the range 200 to 500 nm, and its electrical conductivity was 230 $cm^2 \cdot \Omega^{-1} \cdot mol^{-1}$ (~10^{-3}M solution in water) [1].

[Rh(phen)$_2$Cl(C$_5$H$_5$N)](ClO$_4$)$_2$·H$_2$O is obtained by boiling a mixture of [Rh(phen)$_2$Cl$_2$]Cl·3H$_2$O and pyridine in water, followed by addition of $NaClO_4$. The electronic spectrum of the complex has been reported in the range 200 to 500 nm, and its electrical conductivity (~10^{-3} M solution in water) was 230 $cm^2 \cdot \Omega^{-1} \cdot mol^{-1}$ [1].

[Rh(phen)$_2$(bipy)]X$_3$. The **chloride** is obtained by refluxing a mixture of [Rh(phen)$_2$Cl$_2$]Cl and bipyridyl (mole ratio 1:1) in aqueous ethanol [3]. The **hexafluorophosphate** is prepared similarly, the [Rh(phen)$_2$(bipy)]Cl$_3$ formed initially being dissolved in aqueous NaCl/PO_4^{3-} buffer, chromatographed with aqueous NaCl as eluant, and the solution treated with $NaPF_6$ [4]. The salts were isolated as colourless crystals [3, 4].

The electronic spectrum of the tri-cation has been measured in aqueous solution in the range 25000 to 50000 cm^{-1} [4]. Emission spectra for the tris-chloride have been analysed, and quantum yields and lifetimes determined [5]. The luminescence spectrum of [Rh(phen)$_2$-(bipy)]$^{3+}$ was measured in the range 400 to 700 nm in methanol/water glasses at 77 K [6].

Cyclic voltammetric studies of [Rh(phen)$_2$(bipy)]$^{3+}$ established that the species underwent four one-electron transfers, the first two being followed by chemical reactions. Addition of the first electron appeared to preferentially involve the o-phenanthroline ligands [7].

The sensitised photolytic decomposition of glassy ethanol solutions of [Rh(phen)$_2$(bipy)]$^{3+}$ at 77 K appeared to involve, on the basis of ESR spectral studies, the addition of radicals, e.g. $CH_3CHOH^\cdot$, $CH_3CH_2^\cdot$, $CH_3^\cdot$ or $CH_3O^\cdot$, to the o-phenanthroline ligands [3].

[Rh(phen)$_2$Cl(NO$_2$)]Cl·1.5H$_2$O is prepared by boiling [Rh(phen)$_2$Cl$_2$]Cl·3H$_2$O rigorously with aqueous $NaNO_2$, followed by addition of conc. HCl [1]. The complex was isolated as yellow crystals whose IR spectrum exhibited $\nu(NO_2)$ = 1410, 1315 and 820 cm^{-1}, indicating that the complex contained the Rh-NO$_2$ group. The electronic spectrum was measured in the range 200 to 500 nm and the electrical conductivity (~10^{-3} M solution in water) was 99 $cm^2 \cdot \Omega^{-1} \cdot mol^{-1}$ [1].

References:

[1] P. M. Gidney, R. D. Gillard, B. T. Heaton (J. Chem. Soc. Dalton Trans. **1972** 2621/8). – [2] L. S. Dollimore, R. D. Gillard (J. Chem. Soc. Dalton Trans. **1973** 933/40). – [3] M. K. DeArmond,

W. Halper (J. Phys. Chem. **75** [1971] 3230/4). – [4] G. A. Crosby, W. H. Elfring (J. Phys. Chem. **80** [1976] 2206/11). – [5] W. Halper, M. K. DeArmond (J. Lumin. **5** [1972] 225/37).

[6] J. E. Hillis, M. K. DeArmond (J. Lumin. **4** [1971] 273/90). – [7] G. Kew, K. Hanck, M. K. DeArmond (J. Phys. Chem. **79** [1975] 1828/35).

2.6.11.2.1.5 Complexes with Halides

Synthetic Studies

[Rh(phen)$_2$Cl$_2$]Cl·3H$_2$O. This complex may be prepared by the following methods:

(I) boiling a mixture of $RhCl_3 \cdot 3H_2O$ and o-phenanthroline monohydrate in water followed by addition of hydrazine [1];

(II) refluxing a mixture of $RhCl_3 \cdot 3H_2O$ and o-phenanthroline monohydrate in water for two days, adding 10 M HCl, refluxing for a further two days, removal of [phen H][Rh(phen)Cl$_4$] by filtration, and crystallisation of the dichloro complex from the filtrate [2];

(III) heating finely ground $RhCl_3 \cdot 3H_2O$ with o-phenanthroline at 180°C for 5 min and extraction of the residue with water, followed by treatment with charcoal and concentration of the solution [3];

(IV) fusion of a mixture of $Na_3RhCl_6 \cdot 12H_2O$ with o-phenanthroline monohydrate for 10 min, adding ethanol and heating for ~5 h, extraction of the light pink precipitate so formed with HCl [2];

(V) formation of a saturated solution of $RhCl_3 \cdot 3H_2O$ in acetonitrile (with refluxing), addition of anhydrous o-phenanthroline, refluxing followed by concentration in vacuo and recrystallisation of the product from methanol [2];

(VI) refluxing $RhCl_3$ with o-phenanthroline in methanol (molar ratio 1:2) for several days [2];

(VII) boiling a solution of [H$_3$O][Rh(phen)Cl$_4$], o-phenanthroline and N_2H_5Cl in water or in aqueous ethanol [1];

(VIII) refluxing [phen H][Rh(phen)Cl$_4$] with a four-fold excess of o-phenanthroline in water followed by evaporation to dryness and recrystallisation from water [1];

(IX) boiling [phen H][Rh(phen)Cl$_4$] in ethyleneglycol, followed by passing through an ion-exchange column. (H^+ form) and addition of HCl [4];

(X) heating [Rh(phen)$_3$]Cl$_3$·nH$_2$O slowly to 200°C in vacuo and maintaining this for 500 h, followed by addition of methanol to the dark green compound produced by loss of H_2O and phenanthroline on cooling, yellow crystals were formed which were recrystallised aqueous LiCl solutions [2];

(XI) heating a mixture of *trans*-[Rh(py)$_4$Cl$_2$]Cl·5H$_2$O with o-phenanthroline at 110°C until there was no smell of pyridine, and extraction of the product with water followed by work-up as in (III), above [3].

[Rh(phen)$_2$Cl$_2$]Cl·3H$_2$O is isolated as three isomorphous forms. The α form is prepared by methods (V), (VI) and (VIII) described above, or by recrystallisation of any of the products obtained from routes (I) to (XI) from "laboratory reagent grade" methanol. The ε form may be obtained by methods (II), (IV) and (X), or by recrystallisation of any form of the cation from water containing dilute HCl or a low concentration of LiCl. The δ form, together with the α isomorph, can be prepared by recrystallisation from more concentrated aqueous LiCl solutions. Thus, when [Rh(phen)$_2$Cl$_2$][(H$_5$O$_2$)Cl$_2$] is heated to 190°C to remove HCl, and the residue is dissolved in hot water containing HCl, yellow needles and chunky yellow prisms are formed. The needles can be dissolved in ethanol leaving some of the chunky prisms, which are the δ isomorph. On evaporating and cooling the filtrate after removal of the δ form, yellow needles of the α form are obtained [2].

[Rh(phen)$_2$Cl$_2$][(H$_5$O$_2$)Cl$_2$] is prepared from the chloride by dissolving it in, and recrystallising from, conc. HCl. It is obtained as yellow crystals [2, 5].

[Rh(phen)$_2$Cl$_2$]X (X = NO$_3$ or ClO$_4$). The **nitrate,** as a tetrahydrate, is obtained by dissolving any soluble salt of $[Rh(phen)_2Cl_2]^+$ in dilute HNO_3, or adding a soluble nitrate to aqueous solutions of the cation. It forms an insoluble pale yellow precipitate [2] which may be recrystallised from nitrate solutions incorporating nitrate salts, e.g. $[Rh(phen)_2Cl_2]NO_3 \cdot 1.5 NaNO_3$ [6].

The **perchlorate**, as a trihydrate (referred to as the α form) is obtained from the chloride by addition of $LiClO_4$ in methanol, and the anhydrous form (known as the β form) is precipitated from aqueous solutions of the chloride by adding $LiClO_4$. Both are yellow solids [2].

[Rh(phen)$_2$Cl$_2$][Rh(phen)Cl$_4$] is prepared by treating aqueous $RhCl_3$ with o-phenanthroline in warm ethanol, followed by addition of N_2H_5Cl to the red precipitate, and boiling the mixture for 10 min. The complex is formed as a buff precipitate [1].

The optical isomers of $[Rh(phen)_2Cl_2]^+$ have been partially resolved by chromatography on a column packed with Cellex CM cation exchange cellulose, which contains the Na^+ salts of carbomethoxy groups bound to an optically active matrix [7].

[Rh(phen)$_2$Br$_2$]X. The **bromide** is prepared by refluxing an aqueous ethanolic solution of $RhCl_3 \cdot 3H_2O$, o-phenanthroline monohydrate and KBr. After removal of the red-brown precipitate of [phen H][Rh(phen)Br$_4$], the yellow compound is crystallised from the filtrate [8]. The **perchlorate** is obtained from [phen H][Rh(phen)Br$_4$] by boiling in ethyleneglycol, passage through an ion exchange resin in the H^+ form, and addition of $NaClO_4$ [4].

[Rh(phen)$_2$I$_2$]I is formed by reaction of $[Rh(phen)_2Cl_2]Cl \cdot 3H_2O$ with NaI in refluxing aqueous ethanol. It was isolated as a light brown solid [1].

Rh$_2$(phen)$_3$X$_6$. The **chloride** is obtained from $RhCl_3 \cdot 3H_2O$ and o-phenanthroline monohydrate in water after 4 d at room temperature. It was formed as orange-rose crystals which were isolated after 6 months. The **bromide** is obtained by boiling $RhBr_3$ and o-phenanthroline in water, and was isolated as an orange-brown precipitate [2].

References:

[1] P. M. Gidney, R. D. Gillard, B. T. Heaton (J. Chem. Soc. Dalton Trans. **1972** 2621/8). – [2] E. D. McKenzie, R. A. Plowman (J. Inorg. Nucl. Chem. **32** [1970] 199/212). – [3] R. D. Gillard, B. T. Heaton (J. Chem. Soc. A **1969** 451/4). – [4] J. A. Broomhead, W. Crumley (Inorg. Chem. **10** [1971] 2002/9). – [5] D. Dollimore, R. D. Gillard, E. D. McKenzie (J. Chem. Soc. **1965** 4479/82).

[6] J. D. Miller, F. D. Oliver (J. Chem. Soc. Dalton Trans. **1972** 2473/7). – [7] D. E. Schwab, J. V. Rund (J. Inorg. Nucl. Chem. **32** [1970] 3949/50). – [8] G. C. Kulasingam, W. R. McWhinnie, J. D. Miller (J. Chem. Soc. A **1969** 521/4).

Crystallographic, Spectroscopic and Other Properties

Unit cell data have been obtained from $[Rh(phen)_2Cl_2]NO_3 \cdot 3H_2O$: a = 12.66, b = 13.41, c = 15.53 Å, α = γ = 90°, β = 98.9° [1]. The d-spacings for $[Rh(phen)_2]Cl_2]Cl \cdot 3H_2O$ [2, 3, 4], its α, δ and ε isomorphs [5], for $[Rh(phen)_2Cl_2][(H_5O_2)Cl_2]$ [6], $[Rh(phen)_2Cl_2]ClO_4 \cdot 3H_2O$ [5] and $[Rh(phen)_2Cl_2]ClO_4$ [2, 5], $[Rh(phen)_2Cl_2]NO_3 \cdot 3H_2O$ [1, 5], $[Rh(phen)_2Br_2]Br \cdot 2H_2O$ [3, 4] and $[Rh(phen)_2I_2]I$ [4] have been measured.

The IR spectra of $[Rh(phen)_2Cl_2]^+$ and $[Rh(phen)_2Br_2]^+$ (various salts) were obtained in the range 4000 to 600 cm^{-1} [7] and 500 to 180 cm^{-1} [3], and assignments of ν(RhCl) and ν(RhBr) were made [2, 7, 8]; see table below.

The electronic spectrum of $[Rh(phen)_2Cl_2]^+$ was measured in water in the ranges 200 to 350 nm (shown in paper) [5], 200 to 400 nm [2], 200 to 450 nm (shown in paper, together with circular dichroism spectrum) [9] and 16000 to 30000 cm^{-1} (solution and diffuse reflectance, shown in paper) [8]. The electronic spectrum of $[Rh(phen)_2Cl_2]Cl \cdot 3H_2O$ was measured in the range 350 to 600 nm at 77 K in ethanol/methanol glasses [10, 11, 12].

The visible and ultraviolet spectra of $[Rh(phen)_2Br_2]^+$ have been reported in aqueous solution in the ranges 200 to 450 nm [10] and 16000 to 30000 cm^{-1} (solution and diffuse reflectance spectrum, shown in paper) [8]. The spectra in water and methanol were also obtained in the range 200 to 350 nm [5]. The absorption spectra of $[Rh(phen)_2Br_2]Br$ and $[Rh(phen)_2I_2]I$ were recorded in the range 200 to 800 nm at 77 K in rigid glasses [12].

The optical rotatory dispersion spectrum has been obtained from $[Rh(phen)_2Cl_2]^+$ in the range 450 to 650 nm [2].

The luminescence spectrum of $[Rh(phen)_2Cl_2]^+$ has been investigated at 77 K and room temperature [11, 12, 13], and lifetime and quantum yields determined [10, 14, 15]. Similar data have been obtained from $[Rh(phen)_2Br_2]^+$ [10, 16] and $[Rh(phen)_2I_2]^+$ [10]. The phosphorescence of $[Rh(phen)_2Cl_2]Cl$ has been measured at liquid helium temperatures and compared with data obtained at 77 K [10].

The electrical conductivities of $[Rh(phen)_2Cl_2]Cl \cdot 3H_2O$ and $[Rh(phen)_2Br_2]Br$ in $(CH_3)_2SO$ [8] and in water [4] (both ~10^{-3} M solutions) were 29.4 and 98 (Cl) and 41.3 and 95.5 (Br) $cm^2 \cdot \Omega^{-1} \cdot mol^{-1}$, respectively.

Thermogravimetric studies have been made of $[Rh(phen)_2Cl_2][(H_5O_2)Cl_2]$ and the data shown to be consistent with loss of HCl and water [6].

Photoaquation of $[Rh(phen)_2X_2]^+$ (X = Cl or Br) afforded $[Rh(phen)_2(H_2O)X]^{2+}$ [4]. Photolysis of $[Rh(phen)_2Cl_2]Cl$ in ethanol glasses at 77 K resulted in radical attack (by $CH_3CHOH^\bullet$, $CH_3CH_2^\bullet$, $CH_3^\bullet$ or $CH_3O^\bullet$) on the phenanthroline ligand, as established by ESR spectroscopy [17]. The rates of nitration of the o-phenanthroline ligand in $[Rh(phen)_2Cl_2]^+$ have been measured and shown to be significantly greater than that in the uncomplexed ligand [18].

Far IR and Raman spectral data from $[Rh(phen)_2X_2]^+$ (X = Cl, Br):

Complex	IR ν(RhX) in cm^{-1}	Raman ν(RhX) in cm^{-1}	Ref.
$[Rh(phen)_2Cl_2]Cl \cdot 2H_2O$	352	354	[8]
	355, 347		[7]
$[Rh(phen)_2Cl_2]Cl \cdot 3H_2O$	352, 331		[2]
$[Rh(phen)_2Cl_2]NO_3$ (RT)	353, 349		[8]

Complex		IR ν(RhX) in cm^{-1}	Raman ν(RhX)	Ref.
	(80 K)	357, 351		[8]
$[Rh(phen)_2Br_2]Br$		204, 109		[7]
		239		[2]

References:

[1] P. Andersen, J. Josephsen (Acta Chem. Scand. **25** [1971] 3255/60). – [2] J. A. Broomhead, W. Grumley (Inorg. Chem. **10** [1971] 2002/9). – [3] R. D. Gillard, B. T. Heaton (J. Chem. Soc. A **1969** 451/4). – [4] M. M. Muir, W.-L. Huong (Inorg. Chem. **12** [1973] 1831/5). – [5] E. D. McKenzie, R. A. Plowman (J. Inorg. Nucl. Chem. **32** [1970] 199/212).

[6] D. Dollimore, R. D. Gillard, E. D. McKenzie (J. Chem. Soc. **1965** 4479/82). – [7] I. I. Bhayat, W. R. McWhinnie (Spectrochim. Acta A **28** [1972] 743/51). – [8] G. C. Kulasingam, W. R. McWhinnie, J. D. Miller (J. Chem. Soc. A **1969** 521/4). – [9] P. M. Gidney, R. D. Gillard, B. T. Heaton (J. Chem. Soc. Dalton Trans. **1972** 2621/8). – [10] G. A. Crosby, J. N. Demas (J. Am. Chem. Soc. **92** [1970] 7262/70).

[11] M. K. DeArmond, J. E. Hillis (J. Chem. Phys. **54** [1971] 2247/53). – [12] D. H. W. Carstens, G. A. Crosby (J. Mol. Spectrosc. **34** [1970] 113/35). – [13] G. A. Crosby, D. H. W. Carstens (Mol. Lumin. Intern. Conf., Chicago 1968 [1969], pp. 309/20 from C. A. **71** [1969] No. 26353). – [14] J. E. Hillis, M. K. DeArmond (J. Lumin. **4** [1971] 273/90). – [15] S. H. Petersen, J. N. Demas, T. Kennelly, H. Gafney, D. P. Novak (J. Phys. Chem. **83** [1979] 2991/6).

[16] W. Halper, M. K. DeArmond (Chem. Phys. Letters **24** [1974] 144/6). – [17] M. K. DeArmond, W. Halper (J. Phys. Chem. **75** [1971] 3230/4). – [18] R. D. Gillard, R. P. Houghton, J. N. Tucker (Transition Metal Chem. [Weinheim] **1** [1976] 67/9).

2.6.11.2.2 Mono-phenanthroline Complexes

$[phen\,H][Rh(phen)Cl_4]$. This complex may be prepared either by heating $RhCl_3 \cdot 3H_2O$ and o-phenanthroline monohydrate in 2M HCl solution [1, 2, 3] or warming the two compounds in acetone [4]. It was isolated as orange crystals [1 to 4].

$[H_3O][Rh(phen)Cl_4]$ is obtained from the $[phen\,H]^+$ salt by covering it with 10M HCl and heating the mixture [1], or by shaking the $[phen\,H]^+$ salt with a cation exchange resin (H^+ form) in water for 2 h, filtering and adding conc. HCl, the mixture being refluxed for 20 min, cooled and evaporated [3]. It was isolated as orange-red crystals [1, 3].

Other Salts of $[Rh(phen)Cl_4]^-$. The **potassium** salt, as a monohydrate, is prepared by heating the $[phen\,H]^+$ salt with KCl in 5M HCl solution, and was obtained as orange crystals [1]. The **ammonium** salt, also as a monohydrate, is formed by dissolving the $[phen\,H]^+$ salt in a saturated aqueous solution of NH_4Cl, adjusting the pH to 10 to 12 by adding aqueous ammonia, shaking and filtering off the orange solid [3]. The **$[Rh(phen)_2Cl_2]^+$** salt is formed by boiling a mixture of $RhCl_3 \cdot 3H_2O$, o-phenanthroline and N_2H_5Cl in aqueous ethanol, and removing the salt, after crystallisation, by filtration as a buff precipitate [2].

The d-spacings of [phen H][Rh(phen)Cl_4] have been measured from X-ray powder diffraction data [1]. The far IR and Raman spectra of [phen H][Rh(phen)Cl_4] showed ν(RhCl) = 342 (IR) and 348 cm^{-1} (Raman) [4], 343 (s), 332 (sh) and 319 (s) cm^{-1} (IR) [3], or 344 (m), 329 (sh) and 324 (ms) cm^{-1} (IR) [5]. The electronic spectrum of [phen H][Rh(phen)Cl_4] was recorded in the range 16000 to 30000 cm^{-1} in water, and as diffuse reflectance [4], and of $[Rh(phen)Cl_4]^-$ (various salts) in water in the range 200 to 400 nm [3]. The absorption spectrum of [H_3O][Rh(phen)Cl_4] was recorded in methanol in the range 200 to 350 nm [1]. The emission spectrum of K[Rh(phen)Cl_4] was measured over the range 11500 to 13550 cm^{-1}, and luminescence lifetimes reported [6]. The electrical conductivity of the $[phen H]^+$ salt in $(CH_3)_2SO$ solution (~10^{-3} M) was 42.1 $cm^2 \cdot \Omega^{-1} \cdot mol^{-1}$ [4].

[phen H][Rh(phen)Br_4] is prepared by heating a mixture of $RhBr_3 \cdot nH_2O$ and o-phenanthroline in conc. HBr [1, 3] and, after removal of an olive-green precipitate [1], could be isolated as mauve [1] or maroon [3] crystals. It is also obtained by refluxing an aqueous ethanolic solution containing $RhCl_3 \cdot 3H_2O$, o-phenanthroline monohydrate and KBr, and was isolated as a dark red-brown solid [4].

[H_3O][Rh(phen)Br_4] is obtained from the $[phen H]^+$ salt by covering it with conc. HBr solution and heating. On cooling, the complex was formed as orange-brown crystals [3].

The IR spectrum of the $[phen H]^+$ salt was recorded in the range 4000 to 60 cm^{-1}; ν(RhBr) = 205, 192 and 182 cm^{-1} [5]. The electronic spectrum of the $[phen H]^+$ salt was measured in the range 16000 to 30000 cm^{-1} (aqueous solution and diffuse reflectance). The electrical conductivity of [phen H][Rh(phen)Br_4] in $(CH_3)_2SO$ (~10^{-3} M) was 42.1 $cm^2 \cdot \Omega^{-1} \cdot mol^{-1}$ [4].

[phen H][Rh(phen)I_4] is prepared by refluxing a mixture of $RhCl_3 \cdot 3H_2O$, o-phenanthroline and KI in aqueous ethanol, and was isolated as chocolate-brown crystals. The electrical conductivity of the salt in $(CH_3)_2SO$ (~10^{-3} M) was 45.2 $cm^2 \cdot \Omega^{-1} \cdot mol^{-1}$ [4].

$[Rh(phen)Cl_3]_n$ is obtained by dissolving [H_3O][Rh(phen)Cl_4] in water, evaporating to dryness and repeating the process until an insoluble orange solid is formed. The compound is also insoluble in ethanol and diethylether [1].

References:

[1] E. D. McKenzie, R. A. Plowman (J. Inorg. Nucl. Chem. **32** [1970] 199/212). – [2] P. M. Gidney, R. D. Gillard, B. T. Heaton (J. Chem. Soc. Dalton Trans. **1972** 2621/8). – [3] J. A. Broomhead, W. Grumley (Inorg. Chem. **10** [1971] 2002/9). – [4] G. C. Kulasingam, W. R. McWhinnie, J. D. Miller (J. Chem. Soc. A **1969** 521/4). – [5] I. I. Bhayat, W. R. McWhinnie (Spectrochim. Acta A **28** [1972] 743/51).

[6] R. J. Watts, J. van Houten (J. Am. Chem. Soc. **96** [1974] 4334/5).

2.6.12 Complexes of Substituted Phenanthrolines

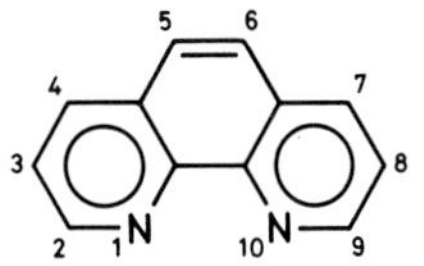

Ligands: 5-Rphen, R = Cl, Br, NO_2, CH_3, Ph; 5,6- and 4,7-$(CH_3)_2$phen, 4,7-Ph_2phen; 3,4,7,8-$(CH_3)_4$phen, 2,4,7,8-$(CH_3)_4$phen

Tris(ligand) Complexes

[Rh(5-Rphen)$_3$](ClO_4)$_3$ (R = Cl, Br, Ph and CH_3), **[Rh{5,6-$(CH_3)_2$phen}$_3$](ClO_4)$_3$, [Rh{4,7-$(CH_3)_2$phen}$_3$](ClO_4)$_3$.** These complexes are obtained by refluxing an aqueous ethanolic solution of $RhCl_3 \cdot 3H_2O$, a small amount of hydrazine hydrate, and the appropriate phenanthroline. On cooling and concentration the complexes were isolated as pink crystals [1].

The cyclic voltammograms of these compounds have been investigated in aqueous solution. All complexes underwent two irreversible one-electron reductions, and dissociation of a ligand according to the scheme below [1]:

$$[Rh(ligand)_3]^{3+} + e^- \rightleftharpoons [Rh(ligand)_3]^{2+}$$
$$[Rh(ligand)_3]^{2+} + e^- \rightleftharpoons [Rh(ligand)_3]^{+}$$
$$[Rh(ligand)_3]^{+} \rightarrow [Rh(ligand)_2]^{+} + ligand.$$

The cathodic peak potentials in aqueous solution containing NaOH and Na_2SO_4 were measured. The rate constants for the quenching of $[Ru(phen)_3]^{2+*}$ in solution by these complexes have been determined [1].

Bis(ligand) Complexes

[Rh(5-NO_2phen)$_2$$Cl_2$]$NO_3$ · 3.5 $NaNO_3$ · 5 H_2O has been reported only briefly and is precipitated from solutions containing $[Rh(5\text{-}NO_2phen)_2Cl_2]^+$ by addition of $NaNO_3$. The kinetics of the reduction of this complex by aqueous ethanolic alkali to Rh^I species have been determined [2].

[Rh{3,4,7,8-$(CH_3)_4$phen}$_2$$Cl_2$]$NO_3$ · 0.5 $NaNO_3$ · 3 H_2O has been obtained similarly, and the kinetics of its reduction determined [2].

[Rh{2,4,7,8-$(CH_3)_4$phen}$_2$$Cl_2$]Cl · 3 H_2O is prepared by refluxing an aqueous ethanolic solution of $RhCl_3 \cdot 3H_2O$ and the phenanthroline. It was isolated as yellow crystals [3]. The far IR and Raman spectra of the complex afforded ν(RhCl) = 327 (IR) and 326 cm^{-1} (Raman) at room temperature, and 334 (IR) and 330 (IR) cm^{-1} at 80 K [2]. The electronic spectrum of the complex in aqueous solution was measured in the range 16000 to 30000 cm^{-1}, and the electrical conductivity of the complex in $(CH_3)_2SO$ (~10^{-3} M solution) was 37.1 $cm^2 \cdot \Omega^{-1} \cdot mol^{-1}$ [3].

[Rh{5,6-$(CH_3)_2$phen}$_2$(bipy)]$^{3+}$ and **[Rh{5,6-$(CH_3)_2$phen}(bipy)$_2$]$^{3+}$** have been briefly described, and their emission spectra and luminescence decay behaviour recorded [4].

Mono(ligand) Complexes

Salts of **[Rh(5-Rphen)Cl_4]$^-$** (R = Cl, NO_2 and CH_3), **[Rh{4,7-$(CH_3)_2$phen}Cl_4]$^-$** and **[Rh{5,6-$(CH_3)_2$phen}Cl_4]$^-$**. The $[R\text{-}phenH]^+$, $[R_2\text{-}phenH]^+$ and K^+ salts of these adducts and the effect of the ligand substituents on the d-d luminescence of the compounds have been briefly reported [5].

References:

[1] C. Creutz, A. D. Keller, N. Sutin, A. P. Zipp (J. Am. Chem. Soc. **104** [1982] 3618/27). – [2] J. D. Miller, F. D. Oliver (J. Chem. Soc. Dalton Trans. **1972** 2473/7). – [3] G. C. Kulasingam, W. R. McWhinnie, J. D. Miller (J. Chem. Soc. A **1969** 521/4). – [4] R. J. Watts, J. van Houten (J. Am. Chem. Soc. **100** [1978] 1718/21). – [5] R. J. Watts, J. van Houten (J. Am. Chem. Soc. **96** [1974] 4334/5).

2.6.13 Terpyridyl Complexes

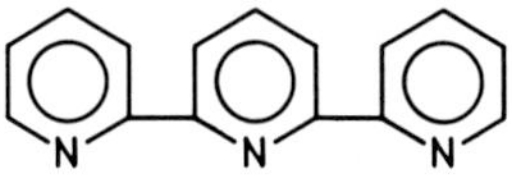

= terpy

[Rh(terpy)$_2$]X$_3$. The **chloride**, as a heptahydrate, may be obtained by fusing $RhCl_3$ with terpyridyl and boiling the melt briefly before extracting with aqueous ethanol, treating with animal charcoal, and precipitation with acetone [1]. It may also be prepared, as a pentahydrate, by treating an ethanolic solution of $RhCl_3 \cdot 3H_2O$ and terpyridyl with N-ethyl-morpholine [2]. It is obtained as colourless crystals (yellow when anhydrous) [1, 2]. The **bromide**, as a hexahydrate, may also be obtained by fusion of $RhBr_3$ with terpyridyl, and is isolated as pale cream crystals (yellow when anhydrous). The **thiocyanate**, as a tetrahydrate, the **iodide** and **perchlorate**, as dihydrates, are prepared by treatment of aqueous or aqueous ethanolic solutions of the chloride with NaSCN, LiI and $NaClO_4$, respectively. They were obtained as yellow (bright yellow when anhydrous), bright yellow (orange when anhydrous) and colourless crystals, respectively [1]. The perchlorate is also prepared by refluxing an aqueous ethanolic solution containing $RhCl_3 \cdot 3H_2O$, a small amount of hydrazine hydrate, and terpyridyl. It was isolated as pale pink crystals which could be decolourised by recrystallisation in the presence of charcoal [3].

The electronic spectrum of $[Rh(terpy)_2]^{3+}$ has been recorded in methanol in the range 230 to 360 nm [1] (shown in paper). The cyclic voltammogram of $[Rh(terpy)_2]^{3+}$ in aqueous solution containing NaOH and Na_2SO_4 exhibited $E_{pc} = -0.55$ V (vs. standard hydrogen electrode, cyclic sweep rate 20 mV/s) and -0.66 and -0.85 V (vs. standard calomel electrode, sweep rate 1 mV/s. These processes represented irreversible one-electron reductions corresponding to the formation of $[Rh(terpy)_2]^{2+}$ and $[Rh(terpy)_2]^{+}$; ligand dissociation was also observed [3]. Molar conductivities (in $cm^2 \cdot \Omega^{-1} \cdot mol^{-1}$) of $[Rh(terpy)_2]X_3 \cdot nH_2O$ in $\sim 10^{-3}$ M nitromethane solution: 254 (X = Cl), 238 (X = Br), 251 (X = I), 252 (X = ClO_4) [1].

The role of $[Rh(terpy)_2]^{3+}$ in the generation of hydrogen from water using light in the presence of $[Ru(bipy)_3]^{2+}$ and triethanolamine has been investigated [2].

Rh(terpy)X$_3$ (X = Cl, Br or I). These compounds are obtained by refluxing RhX_3 (X = Cl, Br or I) with terpyridyl in aqueous ethanol. Their IR spectra have been measured in the range 400 to 60 cm^{-1}, and for $Rh(terpy)Cl_3$, ν(RhCl) = 357, 350, 336, 328, 321 and 316 cm^{-1}. The electrical conductivities (in $cm^2 \cdot \Omega^{-1} \cdot mol^{-1}$) of the compounds in $(CH_3)_2SO$ ($\sim 10^{-3}$ M solution) were 8 (Cl), 11 (Br) and 21 (I), respectively, consistent with their formulation as nonelectrolytes [4].

References:

[1] C. M. Harris, E D. McKenzie (J. Inorg. Nucl. Chem. **25** [1963] 171/4). – [2] M. Kirch, J.-M. Lehn, J.-P. Sauvage (Helv. Chim. Acta **62** [1979] 1345/84). – [3] C. Creutz, A. D. Keller, N. Sutin, A. P. Zipp (J. Am. Chem. Soc. **104** [1982] 3618/27). – [4] I. I. Bhayat, W. R. McWhinnie (Spectrochim. Acta A **28** [1972] 743/51).

2.7 Complexes with Miscellaneous Nitrogen Donor Ligands

2.7.1 Complexes with Di-(2-pyridyl)amine

$C_{10}H_9N_3$ = di(2-pyridyl)amine = dipyam

***trans*-[RhCl$_2$(dipyam)$_2$]Cl** is prepared by heating a mixture of $RhCl_3 \cdot 3H_2O$ in ethanol and dipyam in acetone under reflux for 15 min, it separates from the cooled concentrated reaction solution and forms yellow crystals from aqueous ethanol. The same complex can be isolated from the product mixture obtained on heating an intimate mixture of $RhCl_3 \cdot 3H_2O$ and dipyam at 212°C for 14 h (see below). The electronic spectrum shows λ_{max} = 403, 336, 298 nm, DMF solution; 329, 293 nm, H_2O solution, the far infrared spectrum (667 to 222 cm^{-1}, assignments given) has been reported; conductivity 45 $cm^2 \cdot \Omega^{-1} \cdot mol^{-1}$ in 10^{-3} M DMF solution; magnetic moment μ_{eff} = 0. With aqueous hydrochloric acid an adduct $[RhCl_2(dipyam)_2]Cl \cdot 2H_2O \cdot HCl$, thought to contain the $H_5O_2^+$ cation, is formed.

***trans*-[RhBr$_2$(dipyam)$_2$]Br** is prepared by heating an aqueous ethanolic mixture of $RhCl_3 \cdot 3H_2O$, NaBr (10-fold excess) and dipyam under reflux for 30 min and forms orange-brown crystals from DMF/CCl_4 solution. The electronic spectrum shows λ_{max} = 407, 330, 292 nm DMF in solution; the far infrared spectrum (667 to 222 cm^{-1}, assignments given) has been reported; conductivity 33 $cm^2 \cdot \Omega^{-1} \cdot mol^{-1}$ in 10^{-3} M DMF solution; magnetic moment μ_{eff} = 0.

***trans*-[RhI$_2$(dipyam)$_2$]I** is similarly prepared using excess NaI and forms brown crystals. The electronic spectrum shows λ_{max} = 403, 340, 289 nm in DMF solution; the far infrared spectrum (667 to 222 nm, assignments given) has been reported; conductivity 92 $cm^2 \cdot \Omega^{-1} \cdot mol^{-1}$ in 10^{-3} M DMF solution; magnetic moment μ_{eff} = 0.65 B.M.

[Rh(dipyam)$_3$]Cl$_3$·3H$_2$O is prepared by heating an intimate mixture of $RhCl_3 \cdot 3H_2O$ and dipyam at 212°C for 14 h and is separated from the product mixture by extraction with water followed by fractional crystallisation of the extract. The pale yellow complex loses water on heating at 137°C and 0.1 Torr for 6 h to yield the orange-yellow **anhydrous** form. The electronic spectrum shows λ_{max} = 418, 338, 308 nm in DMF solution; the far infrared spectrum (667 to 222 cm^{-1}, assignments given) has been reported; conductivity 287 $cm^2 \cdot \Omega^{-1} \cdot mol^{-1}$ in 10^{-4} M DMF solution; magnetic moment μ_{eff} = 0.

G. C. Kulasingham, W. R. McWhinnie (J. Chem. Soc. **1965** 7145/8).

2.7.2 Complexes with Arylazooximes

ArN=N-C(R)=NOH = arylazooximes

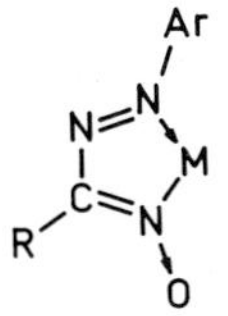

These ligands were originally formulated as N,O donor chelates [1] but have subsequently been characterised as N,N′ donor chelates on the basis of infrared evidence [2].

***cis/trans*-Rh[ArN=NC(R)=NO]** (Ar = C_6H_5, R = CH_3, n-C_3H_7, p-$C_6H_4CH_3$; Ar = o-$C_6H_4CH_3$, R = H). These complexes are obtained as *cis/trans* isomer mixtures by heating an aqueous ethanolic solution of $Rh(NO_3)_3 \cdot 3H_2O$ and the appropriate arylazooxime under reflux for 1 h and adjusting to pH ≈ 4 by addition of $NaHCO_3$ at 15 min intervals. The products are precipitated by addition of water, washed with 5% ammonia solution, then extracted into chloroform and isolated by evaporation of the extract. Separation into *cis* and *trans* isomers is achieved by chromatography (alumina/benzene); the *cis* isomers are more strongly held on the column. The complexes which are described as red to black crystalline solids dissolve in organic solvents to form red solutions. Isomeric purity is preserved during prolonged reflux in benzene. Physical and spectroscopic data are tabulated below. Proton NMR data are also recorded in the original paper and a spectrum (Ar = o-$C_6H_4CH_3$, R = H) is reproduced. The electronic spectrum (700 to 300 nm) is reproduced for one complex with Ar = C_6H_5, R = CH_3 [2].

R	Ar	Isomer	Relative Yield	M. P. in °C	λ_{max} in nm
H	o-$C_6H_4CH_3$	*cis*	1	175	461 321(sh)
		trans	3	252	461 321(sh)
CH_3	C_6H_5	*cis*	1	220	481 310(sh)
		trans	1	233	476 321(sh)
n-C_3H_7	C_6H_5	*cis*	1	106	481 292
		trans	1	126	476 310(sh)
p-$C_6H_4CH_3$	C_6H_5	*cis*	1	287	515 315
		trans	1	247	505 304

M.P. = melting point, sh = shoulder.

References:

[1] A. Chakravorty, K. C. Kalia (Inorg. Nucl. Chem. Letters **3** [1967] 319/21). – [2] K. C. Kalia, A. Chakravorty (Inorg. Chem. **8** [1969] 2586/90).

2.7.3 Complex with β-2-Furaldoxime (?)

$RhCl_3(C_5H_5NO_2)_3$ has been reported but no details are given in the abstract.

N. K. Pshenitsyn, G. A. Nekrasova (Izv. Sekt. Platiny Drug. Blagorodn. Metal. Inst. Obshch. Neorgan. Khim. Akad. Nauk SSSR **30** [1955] 142/58 from C.A. **1956** 9926).

2.7.4 Complexes with Dimethylglyoxime Monomethylether

$CH_3O \cdot N=C(CH_3)\text{-}C(CH_3)=NOH$ = dmgmeH

Rh(dmgme)(dmgmeH)Cl_2 is mentioned as a precursor, no other details given in abstract [1]. The crystals are monoclinic, space group B2/b-C^6_{2h}, unit cell dimensions are a = 14.17(2), b = 14.48(1), c = 8.86(2) Å, γ = 111.97(2)°; Z = 4; D_{calc} = 1.72 g/cm³. The rhodium is octahedrally coordinated with axial chloride ligands [2].

Rh(dmgme)(dmgmeH)X_2 (X = Br, I, SCN). These are prepared by heating an ethanolic solution of the corresponding dichloride and excess HX on a water bath for 30 to 40 min [1].

Rh(dmgme)(dmgmeH)X(SCN) (X = Cl, Br, I). These are prepared by heating Rh(dmgme)(dmgmeH)$(SCN)_2$ with KX in 1:1 molar ratio (no other details given in the abstract) [1].

References:

[1] O. A. Bologa, Yu. A. Simonov (Izv. Akad. Nauk Mold. SSR Ser. Biol. Khim. Nauk **1981** 83/4 from C.A. **96** [1982] No. 134766). – [2] A. A. Dvorkin, Yu. A. Simonov, O. A. Bologa, T. I. Malinovskii (Izv. Akad. Nauk Mold. SSR Ser. Biol. Khim. Nauk **1980** 67/71 from C.A. **93** [1980] No. 17256).

2.7.5 Complexes with Amine-oxime Ligands

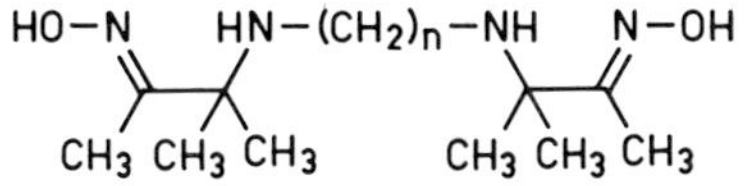

n = 2; 3,3′-(1,2-ethanediyldiamino)bis(3-methyl-2-butanone)dioxime = LH_2
n = 4; 3,3′-(1,4-butanediyldiamino)bis(3-methyl-2-butanone)dioxime = $L'H_2$

$[Rh_2(LH)(L)Cl]Cl_2 \cdot 11H_2O$ is prepared by heating under reflux an aqueous mixture of $RhCl_3 \cdot 3H_2O$, the ligand hydrochloride ($LH_2 \cdot HCl$) and KOH. It deposits as yellow crystals on evaporation. The crystals are monoclinic, space group C2/c-C^6_{2h}, lattice parameters a = 20.813(8), b = 17.527(8), c = 13.484(8) Å, β = 118.36(5)°; D_{exp} = 1.437 g/cm³, Z = 4 (dimer units). The binuclear structure and associated bond lengths are shown in **Fig. 18**. Infrared data have been recorded.

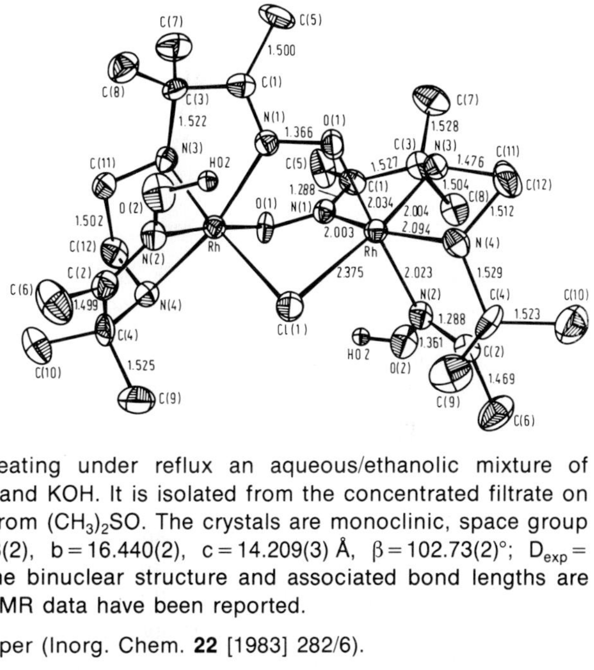

Fig. 18
Structure and bond lengths of $[Rh_2(LH)(L)Cl]Cl_2 \cdot 11H_2O$.

$Rh_2(L'H)_2Cl_4$ is prepared by heating under reflux an aqueous/ethanolic mixture of $RhCl_3 \cdot 3H_2O$, the free ligand ($L'H_2$) and KOH. It is isolated from the concentrated filtrate on cooling and forms yellow crystals from $(CH_3)_2SO$. The crystals are monoclinic, space group $P2_1/n$, lattice parameters a = 8.223(2), b = 16.440(2), c = 14.209(3) Å, β = 102.73(2)°; D_{exp} = 1.628 g/cm³, Z = 2 (dimer units). The binuclear structure and associated bond lengths are shown in **Fig. 19**. Infrared and ^1H NMR data have been reported.

S. Siripaisarnpipat, E. O. Schlemper (Inorg. Chem. **22** [1983] 282/6).

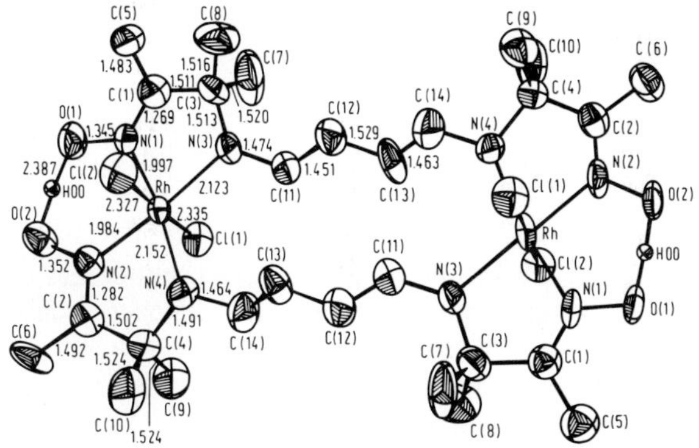

Fig. 19
Structure and bond lengths of $Rh_2(L'H)_2Cl_4$.

2.7.6 Complex with p-Nitrosodimethylaniline

This reagent ($C_8H_{10}N_2O$) is reported to form a cherry-red complex with rhodium(III) on warming in aqueous ethanol solution at pH = 2.2 [1]. The reaction has been developed as a sensitive colorimetric method for determination of rhodium. The electronic spectrum (560 to 490 nm) is reproduced [2], λ_{max} occurs at ~510 nm [2, 3]. Extraction into $CHCl_3$ [3] and the effects of changes in pH, buffers, temperature and concentration have been reported [2]. Procedures for colorimetric determination of rhodium are described [2, 3].

References:

[1] J. J. Kirkland, J. H. Yoe (Anal. Chem. **26** [1954] 1340/4, 1343). – [2] R. B. Wilson, W. D. Jacobs (Anal. Chem. **33** [1961] 1652/4). – [3] A. I. Ivankova, D. P. Shcherbov, I. D. Vvedenskaya (Issled. Tsvet. Fluorestsentn. Reakts. Opred. Blagorod. Met. **1969** 25/30; C.A. **74** [1971] No. 150754).

2.7.7 Complexes with Bis(pyrazyl)alkanes

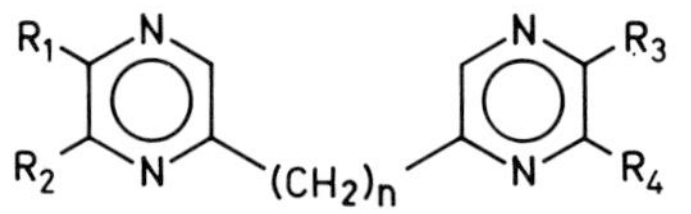

= bpa

R_1, R_2, R_3, R_4 = H, C_5H_{11}, C_7H_{15} or C_9H_{19}, n = 2,5 or 7

{**RhCl(bpa)**}$_2$. These complexes, which are formulated as binuclear, chloride-bridged derivatives of square-planar rhodium(I), are obtained by heating a mixture of $RhCl_3 \cdot 3H_2O$ and the appropriate free bpa ligand under reflux in ethanol for 15 to 60 min and separate from solution as yellow-brown solids. With one exception (see table) they are too insoluble in organic solvents to permit recrystallisation. The complexes are diamagnetic. Electronic spectra are given in the following table.

Complex R_1/R_3	R_2/R_4	n	Electronic Spectrum[a] λ_{max} in nm	
n-C_9H_{19}	H	2[b]	634 (sh)	446
				421 ($CHCl_3$)
				426 (C_6H_6)
			422	273 (C_2H_5OH)
H	n-C_7H_{15}	2		401
H	n-C_9H_{19}	2		410
n-C_5H_{11}	H	5		403
H	n-C_5H_{11}	7		398

[a] All spectra diffuse reflectance unless solvent specified; [b] recrystallised ($CHCl_3$), molecular weight (ebullioscopic $CHCl_3$) 1132, dimer requires 1153.

RhClI$_2$(bpa) (R_1, R_3 = n-C_9H_{19}, R_2, R_4 = H, n = 2) is prepared by addition of a slight excess of iodine to a chloroform solution of the corresponding rhodium(I) complex, and deposits as dark brown crystals. The complex is soluble in organic solvents and has a molecular weight corresponding to a monomer.

B. J. Edmondson, A. B. P. Lever (J. Inorg. Nucl. Chem. **29** [1967] 2777/86).

2.7.8 Complex with Di(2-pyridyl)ketone, *trans*-$[RhCl_2(dpk)_2][RhCl_4(dpk)]$

$(2\text{-}C_5H_4N)_2CO$ = dpk

This compound is prepared by heating an ethanolic solution of $RhCl_3 \cdot 3H_2O$ and free ligand under reflux for 30 min and deposits from the cooled solution as a yellow solid. The electronic spectrum with $\lambda_{max} = 455$ nm (diffuse reflectance) and infrared spectrum with $\nu(CO) = 1705$, 1698(sh), 1694, 1685(sh); $\nu(RhCl) = 355$, 335, 325 cm^{-1} have been recorded; the conductivity (10^{-3}M DMF solution) is 51 $cm^2 \cdot \Omega^{-1} \cdot mol^{-1}$.

R. R. Osborne, W. R. McWhinnie (J. Less-Common Metals **17** [1969] 53/7).

2.7.9 Complex with 2,6-Dipicolinic Acid Hydrazide, $[RhCl(dph)_2]Cl_2$

$2,6(H_2N\text{-}NH\text{-}CO\text{-})_2C_5H_3N$ = dph

This compound is prepared by heating an ethanolic solution of $RhCl_3 \cdot 3H_2O$ and free ligand under reflux for an unspecified time, then concentrating the solution (pH = 3) to half volume. It separates as a yellow solid. The electronic spectrum with $\lambda_{max} = 800$, 572, 488, 370, 303 nm has been recorded and assignments given, values for 10 Dq and Racah parameters B and C are 24500, 406 and 4000 cm^{-1}, respectively. Infrared data (1700 to 200 cm^{-1}) are reported and assignments given; the complex is diamagnetic.

V. B. Rana, S. K. Sahni, S. K. Sangal (Acta Chim. [Budapest] **101** [1979] 405/16).

2.7.10 Complex with N,N′-Dibenzylidene Dipicolinic Acid Hydrazide, $[RhCl(dbpa)]Cl_2$

$2,6\text{-}(C_6H_5 \cdot CH{=}N\text{-}NH\text{-}CO\text{-})_2C_5H_3N$ = dbpa

This compound is prepared by heating an ethanolic solution of $RhCl_3 \cdot 3H_2O$ and free ligand under reflux for an unspecified time, then concentrating the solution (pH = 3) to half volume. It separates as a yellow solid. The electronic spectrum with $\lambda_{max} = 800$, 549, 489, 373, 298 nm has been recorded and assignments given, values for 10 Dq and Racah parameters B and C are 24425, 397 and 3975 cm^{-1}, respectively. Infrared data (1700 to 200 cm^{-1}) are reported and assignments given; the complex is diamagnetic.

V. B. Rana, S. K. Sahni, S. K. Sangal (Acta Chim. [Budapest] **101** [1979] 405/16).

2.7.11 Complexes with 2,2′-Diaminobiphenyls

=dabp ($R = CH_3$, OCH_3)

$[RhCl_2(dabp)_2]Cl$. These complexes are obtained as dirty yellow ($R = CH_3$) or dirty green ($R = OCH_3$) precipitates by warming an ethanolic solution of $RhCl_3 \cdot 3H_2O$ and the free ligand. Infrared data have been reported; conductivities in 10^{-3} M $(CH_3)_2SO$ solution are 40.28 ($R = CH_3$) and 43.01 ($R = OCH_3$) $cm^2 \cdot \Omega^{-1} \cdot mol^{-1}$.

V. Kumar, M. M. Haq, N. Ahmad (Indian J. Chem. A **17** [1979] 305/7).

2.7.12 Complexes with 8-Aminoquinoline
8-aminoquinoline = $C_9H_8N_2$ = 8-aq

[Rh(8-aq)$_2$Cl$_2$]Cl is prepared by heating a mixture of $RhCl_3 \cdot 3H_2O$ and free ligand in aqueous acetone, and separates as a bright yellow precipitate. Electronic (λ_{max} = 282 nm, DMF solution) and far infrared spectra (RhCl = 335 and 310 cm^{-1}) have been reported.

Rh(8-aq)$_2$(H$_2$O)Cl$_3$ is prepared by heating a mixture of $[RhCl(CO)_2]_2$ and free ligand under reflux in ethanol for 2 h. It is a diamagnetic yellow solid and is a non-electrolyte in DMF. Electronic (λ_{max} = 318, 278 nm, DMF solution) and far infrared spectra (Rh-Cl = 335, 308 cm^{-1}) have been recorded.

H. A. Hudali, J. V. Kingston, H. A. Tayim (Inorg. Chem. **18** [1979] 1391/4).

2.7.13 Complexes with Hydrotripyrazolylborate Anion
$HB(C_3H_3N_2)_3$ = hydrotripyrazolylborate = $HBpz_3$

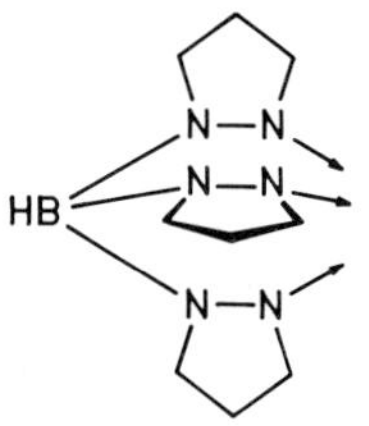

{**RhCl$_2$(HBpz$_3$)**}$_2$ is prepared by heating a methanolic mixture of $RhCl_3 \cdot 3H_2O$ and Na[HBpz$_3$] under reflux for 4 h and separates as dull gold plates, melting point 293 to 297°C.

RhCl$_2$(HBpz$_3$)(CH$_3$CN) is prepared by heating {RhCl$_2$(HBpz$_3$)}$_2$ under reflux in CH_3CN for 24 h and separates as bright yellow crystals or dull gold prisms, melting point >360°C.

RhCl$_2$(HBpz$_3$)(C$_6$H$_5$CN) is prepared by heating a mixture of RhCl$_2$(HBpz$_3$)(CH$_3$CN) and C_6H_5CN under reflux in dry toluene for 24 h and is crystallised from dry toluene as dark gold prisms, melting point 206 to 211°C.

RhCl$_2$(HBpz$_3$)L. Compounds of this stoichiometry were prepared from RhCl$_2$(HBpz$_3$)(CH$_3$CN) and free ligand (L) by the above method and crystallised from dry toluene; for colours and melting points see below.

RhCl$_2$(HBpz$_3$)(C$_5$H$_5$N): bright yellow prisms, melting point 331 to 332°C.

RhCl$_2$(HBpz$_3$){N(C$_2$H$_5$)$_3$}: dull yellow prisms, melting point 239 to 241°C.

RhCl$_2$(HBpz$_3$)(CO)(CH$_3$CN): dull gold prisms, melting point >350°C.

RhCl$_2$(HBpz$_3$)(n-CNC$_4$H$_9$): bright yellow prisms, melting point 148 to 150°C.

RhCl$_2$(HBpz$_3$){P(C$_6$H$_5$)$_3$}: pale gold prisms, melting point 275 to 278°C.

RhCl$_2$(HBpz$_3$){P(CH$_3$)(C$_6$H$_5$)$_2$}·0.5C$_6$H$_5$CH$_3$: bright gold prisms, melting point 273 to 275°C.

RhCl$_2$(HBpz$_3$){P(CH$_3$)$_2$(C$_6$H$_5$)}: bright gold prisms, melting point 291°C.

RhCl$_2$(HBpz$_3$){P(C$_2$H$_5$)$_3$}: bright gold prisms, melting point 287°C.

RhCl$_2$(HBpz$_3$){As(C$_6$H$_5$)$_3$}: dark gold prisms, melting point 110 to 111°C.

RhCl$_2$(HBpz$_3$){As(CH$_3$)$_2$(C$_6$H$_5$)}: pale yellow needles, melting point 290°C.

[N(CH$_3$)$_4$][RhCl$_2$Br(HBpz$_3$)] is prepared by refluxing a mixture of [N(CH$_3$)$_4$]Br and RhCl$_2$(HBpz$_3$)(CH$_3$CN) in dry toluene for 24 h and is crystallised from dry toluene as pale yellow prisms, melting point >350°C.

RhCl(CH$_3$COCHCOCH$_3$)(HBpz$_3$) is prepared by heating a mixture of RhCl$_2$(HBpz$_3$)(CH$_3$CN) and Tl(CH$_3$COCHCOCH$_3$) in dry toluene under reflux for 24 h and is crystallised from dry toluene as dull yellow prisms, melting point >350°C.

RhCl(CF$_3$COCHCOCF$_3$)(HBpz$_3$) is similarly prepared using Tl(CF$_3$COCHCOCF$_3$) and is crystallised from dry toluene as bright yellow needles, melting point 191 to 192°C.

Rh(O$_2$CCF$_3$)$_2$(HBpz$_3$) is prepared by heating RhCl$_2$(HBpz$_3$)(CH$_3$CN) and AgO$_2$CCF$_3$ under reflux in toluene for 4 h and is isolated from the filtered reaction mixture as pale yellow plates, melting point 178 to 180°C.

Rh(O$_2$CCF$_3$)$_2$(HBpz$_3$)(CH$_3$CN) is prepared by heating the preceding compound with CH$_3$CN in boiling dry toluene for 8 h and is isolated as pale yellow prisms, melting point 132 to 137°C.

Proton NMR data and selected infrared bands have been reported for all compounds mentioned in this section.

S. May, P. Reinsalu, J. Powell (Inorg. Chem. **19** [1980] 1582/9).

2.7.14 Complexes with Hydrotris(3,5-dimethylpyrazolyl)borate Anion

HB(C$_5$H$_7$N$_2$)$_3$ = hydrotris(3,5-dimethylpyrazolyl)borate = {HB(3,5-Me$_2$pz)$_3$}

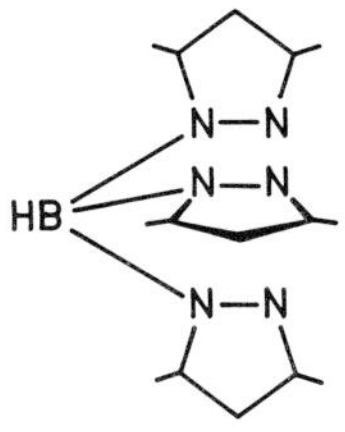

[RhCl$_2${HB(3,5-Me$_2$pz)$_3$}]$_2$ is prepared by heating a mixture of RhCl$_3$·3H$_2$O and Na[HB(3,5-Me$_2$pz)$_3$] under reflux in methanol for 1$^1/_2$ h. It deposits from the cooled concentrated reaction mixture and forms dark gold plates, melting point 228 to 233°C, from methanol.

RhCl$_2${HB(3,5-Me$_2$pz)$_3$}(CH$_3$OH) is prepared by heating a mixture of RhCl$_3$·3H$_2$O and Na[HB(3,5-Me$_2$pz)$_3$] under reflux in methanol for a total of 3$^1/_2$ h. It separates from the reaction mixture and is crystallised from methanol as gold prisms, melting point 296 to 298°C.

RhCl$_2${HB(3,5-Me$_2$pz)$_3$}(H$_2$O) is prepared by heating a mixture of Na[HB(3,5-Me$_2$pz)$_3$] and RhCl$_3$·3H$_2$O in 95% ethanol for 2 h or by heating RhCl$_2${HB(3,5-Me$_2$pz)$_3$}(CH$_3$OH) in wet toluene. It forms pale yellow prisms, melting point 315°C.

RhCl$_2${HB(3,5-Me$_2$pz)$_3$}(CH$_3$CN) is prepared by heating RhCl$_2${HB(3,5-Me$_2$pz)$_3$}(CH$_3$OH) under reflux in neat CH$_3$CN and forms pale yellow plates, melting point 259 to 262°C.

RhCl$_2${HB(3,5-Me$_2$pz)$_3$}L. Compounds of this stoichiometry are prepared by heating a mixture of RhCl$_2${HB(3,5-Me$_2$pz)$_3$}(CH$_3$OH) and the appropriate ligand (L) under reflux in dry toluene for 3 h and are crystallised from CH$_2$Cl$_2$/hexanes; for colours and melting points see p. 298.

$RhCl_2\{HB(3,5\text{-}Me_2pz)_3\}(C_6H_5CN)$: bright yellow needles, melting point 298°C.

$RhCl_2\{HB(3,5\text{-}Me_2pz)_3\}(C_5H_5N)$: pale yellow needles, melting point 355°C.

$RhCl_2\{HB(3,5\text{-}Me_2pz)_3\}(4\text{-}CH_3C_5H_4N)$: pale yellow prisms, melting point 168°C.

$RhCl_2\{HB(3,5\text{-}Me_2pz)_3\}(NH_3)$: pale yellow prisms, melting point 350°C.

$RhCl_2\{HB(3,5\text{-}Me_2pz)_3\}\{N(C_2H_5)_3\}$: gold prisms, melting point 280 to 282°C.

$RhCl_2\{HB(3,5\text{-}Me_2pz)_3\}\{N(C_3H_7)_3\}$: yellow prisms, melting point 226 to 230°C.

$RhCl_2\{HB(3,5\text{-}Me_2pz)_3\}$(3,5-dimethylpyrazole): dark gold prisms, melting point 165°C.

$[RhCl_2\{HB(3,5\text{-}Me_2pz)_3\}]_2$(bipyridyl): dark gold needles, melting point 352 to 355°C (decomposition).

$RhCl_2\{HB(3,5\text{-}Me_2pz)_3\}(CO)$: gold microprisms, melting point 304 to 306°C.

$RhCl_2\{HB(3,5\text{-}Me_2pz)_3\}(n\text{-}CNC_4H_9)$: bright gold needles, melting point 293°C.

$RhCl_2\{HB(3,5\text{-}Me_2pz)_3\}(CNCH_2C_6H_5)$: gold tetragonal prisms, melting point >360°C.

$RhCl_2\{HB(3,5\text{-}Me_2pz)_3\}\{P(C_6H_5)_3\}$: dark gold needles, melting point 199 to 201°C.

$RhCl_2\{HB(3,5\text{-}Me_2pz)_3\}\{P(CH_3)(C_6H_5)_2\}$: dark orange needles, melting point 222°C.

$RhCl_2\{HB(3,5\text{-}Me_2pz)_3\}\{P(CH_3)_2(C_6H_5)\}$: orange prisms, melting point 232°C.

$RhCl_2\{HB(3,5\text{-}Me_2pz)_3\}\{P(CH_3)(o\text{-}C_6H_4CH_3)_2\}CH_2Cl_2$: dark yellow prisms, melting point 162°C.

$RhCl(CH_3COCHCOCH_3)\{HB(3,5\text{-}Me_2pz)_3\}$: pale yellow plates, melting point 229 to 231°C.

$RhCl(CF_3COCHCOCF_3)\{HB(3,5\text{-}Me_2pz)_3\}$: bright yellow plates, melting point 255 to 258°C.

$RhCl\{S_2CN(C_2H_5)_2\}\{HB(3,5\text{-}Me_2pz)_3\}$: pale gold needles, melting point 350 to 353°C.

$[As(C_6H_5)_4][RhCl_3\{HB(3,5\text{-}Me_2pz)_3\}]$ is prepared by mixing stoichiometric amounts of $[As(C_6H_5)_4]Cl$ and $RhCl_2\{HB(3,5\text{-}Me_2pz)_3\}$ in CH_2Cl_2 and crystallises from solution on addition of hexane as bright gold prisms, melting point 339°C.

$[N(CH_3)_4][RhCl_2Br\{HB(3,5\text{-}Me_2pz)_3\}]$ is prepared by heating equimolar quantities of $[N(CH_3)_4]Br$ and $RhCl_2\{HB(3,5\text{-}Me_2pz)_3\}(CH_3OH)$ under reflux in dry toluene, and crystallising the precipitated solid from CH_2Cl_2/hexane. It forms dark gold plates, melting point 341°C (decomposition).

$[N(CH_3)_4][RhCl_2F\{HB(3,5\text{-}Me_2pz)_3\}]$ is similarly prepared using $[N(CH_3)_4]F$ and forms dark yellow plates, melting point 261 to 300°C (decomposition).

$[NH(C_2H_5)_3][RhCl_3\{HB(3,5\text{-}Me_2pz)_3\}]$ is similarly prepared using $[NH(C_2H_5)_3]Cl$ and forms orange needles, melting point 266 to 267°C.

$[N(C_2H_5)_4][RhCl_3\{HB(3,5\text{-}Me_2pz)_3\}]$ is similarly prepared using $[N(C_2H_5)_4]Cl$ and forms bright orange prisms, melting point 332 to 342°C (decomposition).

$[NH(C_2H_5)_3][RhHCl_2\{HB(3,5\text{-}Me_2pz)_3\}]$ is prepared by bubbling hydrogen through a refluxing mixture of $RhCl_2\{HB(3,5\text{-}Me_2pz)_3\}(CH_3OH)$ and $[NH(C_2H_5)_3]Cl$ in dry toluene for 5 h and separates as dull gold prisms, melting point 258°C.

$RhHCl\{HB(3,5\text{-}Me_2pz)_3\}(L)$. These complexes are prepared by heating equimolar portions of $[NH(C_2H_5)_3][RhHCl_2\{HB(3,5\text{-}Me_2pz)_3\}]$ and the appropriate free phosphine or arsine in dry toluene for 2 h and are crystallised from CH_2Cl_2/hexane; for colours and melting points see p. 299.

$RhHCl\{HB(3,5\text{-}Me_2pz)_3\}\{P(C_6H_5)_3\}$: pale yellow needles, melting point 171 to 175°C.

$RhHCl\{HB(3,5\text{-}Me_2pz)_3\}\{P(CH_3)(C_6H_5)_2\}$: dark gold prisms, melting point 221 to 224°C.

$RhHCl\{HB(3,5\text{-}Me_2pz)_3\}\{P(CH_3)_2(C_6H_5)\}$: gold prisms, melting point 172 to 175°C.

$RhHCl\{HB(3,5\text{-}Me_2pz)_3\}\{P(C_2H_5)_3\}$: bright yellow prisms, melting point 230 to 235°C.

$RhHCl\{HB(3,5\text{-}Me_2pz)_3\}\{P(CH_3)(o\text{-}C_6H_4CH_3)\}$: bright yellow prisms, melting point 129 to 133°C.

$RhHCl\{HB(3,5\text{-}Me_2pz)_3\}\{As(CH_3)_2(C_6H_5)\}$: bright orange prisms, melting point 231 to 235°C.

$Rh(O_2CCH_3)_2\{HB(3,5\text{-}Me_2pz)_3\}$ is prepared by heating a mixture of $RhCl_2\{HB(3,5\text{-}Me_2pz)_3\}(CH_3OH)$ and $Ag(O_2CCH_3)$ under reflux in dry toluene. It forms pale yellow plates (melting point 297°C) from CH_2Cl_2/hexane.

$Rh(O_2CCF_3)_2\{HB(3,5\text{-}Me_2pz)_3\}(H_2O)$ is similarly prepared using $Ag(O_2CCF_3)$ and forms pale yellow plates (melting point 231 to 233°C) from CH_2Cl_2/hexane.

$Rh(O_2CCF_3)_2\{HB(3,5\text{-}Me_2pz)_3\}(CH_3CN)$ is prepared by heating the preceding complex under reflux with CH_3CN in dry toluene. It forms pale yellow prisms (melting point 232 to 233°C) from CH_2Cl_2/hexane.

$[N(C_2H_5)_4][Rh(O_2CCF_3)_2Cl\{HB(3,5\text{-}Me_2pz)_3\}]$ is prepared by heating a mixture of $[N(C_2H_5)_4]$-$[RhCl_3\{HB(3,5\text{-}Me_2pz)_3\}]$ and $Ag(O_2CCF_3)$ under reflux in dry toluene. It forms pale yellow needles (melting point 263 to 265°C) from acetone.

$[NH(C_2H_5)_3][Rh(O_2CCF_3)H\{HB(3,5\text{-}Me_2pz)_3\}]$ is prepared by bubbling H_2 through a boiling solution of $Rh(O_2CCF_3)_2\{HB(3,5\text{-}Me_2pz)_3\}(H_2O)$ in dry toluene for 4 h and adding $N(C_2H_5)_3$ after 30 min. It forms pale yellow plates, melting point 209°C.

Proton NMR data and selected infrared bands have been reported for all compounds mentioned in this section.

S. May, P. Reinsalu, J. Powell (Inorg. Chem. **19** [1980] 1582/9).

2.7.15 Complexes with 2,2′-Biquinoline ($C_{18}H_{12}N_2$)

$RhCl_3(C_{18}H_{12}N_2)$ is described as a mononuclear five-coordinate rhodium(III) complex but is probably binuclear with chloride bridges. It is prepared by warming an ethanolic solution of $RhCl_3 \cdot 3H_2O$ and free ligand for a few minutes and deposits as a dirty yellow solid. Infrared data are given; the conductivity is 4.00 $cm^2 \cdot \Omega^{-1} \cdot mol^{-1}$ for a 10^{-3} M solution in $(CH_3)_2SO$ [1].

***cis*-$[RhCl_2(C_{18}H_{12}N_2)]Cl \cdot nH_2O$(?).** This has been briefly reported and given this formulation on the basis of infrared and electronic spectral data [2].

References:

[1] V. Kumar, M. M. Haq, N. Ahmad (Indian J. Chem. A **17** [1979] 305/7). – [2] B. R. Ramesh, B. S. S. Mallikarjuna, G. K. N. Reddy (Natl. Acad. Sci. Letters [India] **1** [1978] 24/6 from C.A. **89** [1978] No. 122152).

2.7.16 Complex with Phthalimide Dithiosemicarbazone

This reagent reacts with rhodium(III) in a slightly acid medium to form an orange-red complex, $\lambda_{max} = 410$ to 420 nm. The application of this reaction to the photometric determination of rhodium is described.

M. Guzman, D. Perez Bendito, F. Pino (Anales Quim. **72** [1976] 651/6).

2.8 Complexes with Nucleotides and Nucleosides

For nucleotide and nucleoside complexes with rhodium carboxylates see pp. 25, 48.

2.8.1 Nucleotide Complexes

$Rh(ade)_2Cl_3 \cdot CH_3OH$ (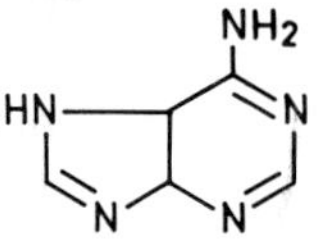 = adenine = ade) is made from $RhCl_3 \cdot nH_2O$ in CH_3OH with adenine in a 1:3 molar ratio and refluxed for 12 h. The complex was precipitated with ether. It melts at 280°C with decomposition. The infrared spectrum was measured (200 to 4000 cm^{-1}); ν(RhCl) lies at 290 and 275 cm^{-1} and ν(RhN) at 575 cm^{-1} [1].

For adenine complexes with rhodium carboxylates see pp. 25, 48.

$Rh(cyt)_2Cl_3(CH_3OH)_2$ (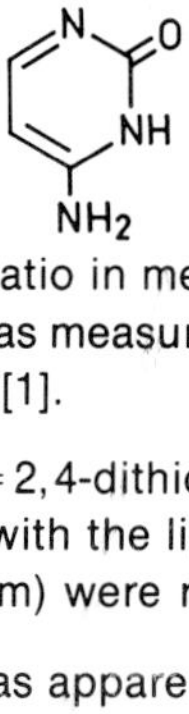 = cytosine = cyt). This is made by refluxing $RhCl_3 \cdot nH_2O$ and cytosine in a 1:3 molar ratio in methanol for 10 to 12 h. It melts with decomposition at 240°C. The infrared spectrum was measured (200 to 4000 cm^{-1}); ν(RhCl) is given as 320 and 310 cm^{-1} and ν(RhN) at 525 cm^{-1} [1].

$Rh(tu)_2Cl$ ($S_2N_2C_4H_3$ = 2,4-dithiouracil = tu). This is made by reaction of an unspecified rhodium salt in ethanol with the ligand. It is brown. The maxima in the electronic absorption spectrum (300 to 1000 nm) were recorded [2].

$[Rh(tu)_2H_2O]_2$. This was apparently made from $[Rh(CCl_3COO)_2(C_2H_5OH)]_2$ and the ligand in *n*-butanol at 70°C for 3 h. It is brown and diamagnetic. The infrared spectrum was measured (300 to 4000 cm^{-1}). The electronic spectrum has a maximum at 17700 cm^{-1} [3].

$Rh(SN_2C_5H_4O)_3Cl_3$ ($SN_2C_3H_6$ = 6-methyl-2-uracil). This orange material is made from $RhCl_3 \cdot 3H_2O$ in ethanol acidified with HCl together with the ligand under reflux.

The infrared spectrum (200 to 2000 cm^{-1}) was measured (1050 to 1820 cm^{-1} portion reproduced in paper); ν(RhS) and ν(RhCl) bands are assigned to 260, 243 and 220 cm^{-1}. Coordination through sulphur and the N(3) nitrogen atom was tentatively proposed. The X-ray photoelectronic spectrum was measured (160 to 288, 307 to 403 and 525 to 538 eV portions reproduced in paper) [4].

References:

[1] B. T. Khan, A. Mehmood (J. Inorg. Nucl. Chem. **40** [1978] 1938/9). – [2] J. S. Dwivedi, U. Agarwala (Indian J. Chem. **10** [1972] 657/9). – [3] J. S. Dwivedi, U. Agarwala (Z. Anorg. Allgem.

Chem. **397** [1973] 74/82, 80). – [4] J. R. Lusty, J. Peeling, M. A. Abdel-Aal (Inorg. Chim. Acta **56** [1981] 21/6).

2.8.2 Nucleoside Complexes

$Rh(ad)Cl_3(CH_3OH)_2$ ($N_5C_{10}H_{13}O_4$ = adenosine = ad) is a light orange material, melting at 278 to 280°C with decomposition, made by refluxing $RhCl_3 \cdot nH_2O$ and adenosine in acidified methanol for 4h.

The infrared spectrum was measured (200 to 4000 cm^{-1}); ν(RhCl) was assigned to a band at 340 cm^{-1} and ν(RhN) to one at 490 cm^{-1}. The conductance of the complex in dimethylformamide (DMF) is 45 $cm^2 \cdot \Omega^{-1} \cdot mol^{-1}$ [1].

$[Rh(ad)(NH_3)_4Cl]Cl_2 \cdot H_2O$. This is light yellow, melting at 218 to 220°C with decomposition. It is made by refluxing adenosine with $[Rh(NH_3)_5Cl]Cl_2$ in water for 38h. The infrared spectrum was measured (200 to 4000 cm^{-1}); ν(RhCl) is at 280 cm^{-1} and ν(RhN) at 490 cm^{-1}. The conductance in dimethylformamide (DMF) is 197 $cm^2 \cdot \Omega^{-1} \cdot mol^{-1}$ [1].

$Rh(ad)Cl_3[(CH_3)_2SO]_2$. This is a pale yellow material made by reaction of equimolar quantities of $RhCl_3[(CH_3)_2SO]_3$ with adenosine in water. The ^1H NMR spectrum was measured, and it appears that coordination occurs via the N(7) binding site [2].

For rhodium **acetate adenosine** complexes see p. 25.

$[Rh(xanth)Cl_2(H_2O)_3]Cl$ ($N_4C_9H_{10}O_6$ = xanthosine = xanth). This is a yellowish-green material melting with decomposition from 238 to 240°C, made from $RhCl_3 \cdot nH_2O$ and adenosine in refluxing aqueous acidified methanol. The infrared spectrum was measured (200 to 4000 cm^{-1}); ν(RhCl) was assigned to a band at 350 cm^{-1} and ν(RhN) to one at 500 cm^{-1}. The molar conductance in dimethylformamide (DMF) is 74 $cm^2 \cdot \Omega^{-1} \cdot mol^{-1}$ [1].

$[Rh(xanth)_2(NH_3)_2Cl_2]Cl \cdot 2H_2O$. This green material, which melts with decomposition at 226 to 228°C, is made from $RhCl_3 \cdot nH_2O$ and xanthosine in refluxing acidified aqueous methanol. The infrared spectrum was measured (200 to 4000 cm^{-1}); ν(RhCl) is at 360 cm^{-1} and ν(RhN) at 500 cm^{-1}. The molar conductance in dimethylformamide (DMF) is 75 $cm^2 \cdot \Omega^{-1} \cdot mol^{-1}$ [1].

$[Rh(H_3ATP)(H_2O)_4]Cl_2$ (adenosine triphosphate, $N_5C_9H_{16}O_{13}P_3$ = H_4ATP). This is made by reaction of H_4ATP in acidified aqueous methanol and $RhCl_3 \cdot nH_2O$ at room temperature for 60h. The complex (colour not specified) melts above 300°C and is soluble in water. ^1H NMR and infrared data suggest coordination through N(7) and a phosphate oxygen atom. The conductance in water is 520 $cm^2 \cdot \Omega^{-1} \cdot mol^{-1}$ [3].

$[Rh(H_4ATP)(NH_3)_5]Cl_3$. This is made from $[Rh(NH_3)_5Cl]Cl_2$ in water with H_4ATP at room temperature for 50h. The complex (colour not specified) is soluble in water and melts at 59 to 60°C. ^1H NMR and infrared data seem to suggest that the purine N(7) is the donor site. The conductance in water is 464 $cm^2 \cdot \Omega^{-1} \cdot mol^{-1}$ [3].

References:

[1] B. T. Khan, M. R. Somayajulu, M. M. Taqui Khan (J. Inorg. Nucl. Chem. **40** [1978] 1251/3). – [2] N. Farrell (J. Chem. Soc. Chem. Commun. **1980** 1014/6). – [3] B. T. Khan, M. R. Somayajulu, M. M. Taqui Khan (Indian J. Chem. A **17** [1979] 359/60).

For other H_4ATP rhodium carboxylate complexes see pp. 37, 43/4.

2.9 Organic Chelates Containing Nitrogen and Oxygen Donor Atoms

2.9.1 Complexes with Schiff Bases

2.9.1.1 Complexes with N,N′-Ethylenebis(salicylaldimine)

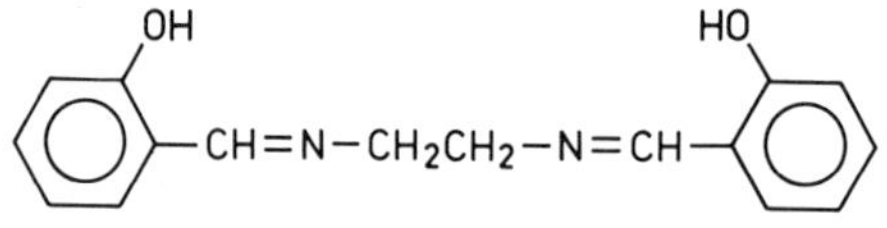

$= H_2(salen)$

$Na_2[Rh(salen)]$(?). This formal rhodium(0) complex is thought to be present in the dark brown solutions obtained on reducing RhCl(salen)py (see below) with $NaBH_4$ and $PdCl_2$ in methanolic base (NaOH) or with sodium amalgam in THF. These solutions react with organic halides (RX) to form organo-rhodium complexes RhR(salen)py [1].

Na[Rh(salen)]THF is prepared by addition of RhCl(salen)py to an equimolar amount of $Na_2[Rh(salen)]$ (prepared from RhCl(salen)py and Na/Hg) and is precipitated from solution as a green solid by addition of n-hexane. It is very oxygen-sensitive, oxidising to a bright orange powder [1].

$\{Rh(salen)py\}_2$ is prepared by dissolving anhydrous $RhCl_3$, $RhCl_3 \cdot 3H_2O$ or $Na_3RhCl_6 \cdot 12H_2O$ in methanolic pyridine solution in the presence of Zn amalgam, adding methanolic H_2(salen) solution and stirring the mixture for 1 h. It separates as a thick creamy yellow precipitate. The infrared spectrum has been recorded (3600 to 400 cm^{-1}) and is reproduced (1800 to 500 cm^{-1}). A binuclear Rh-Rh bonded structure is proposed [1].

RhCl(salen)py is prepared by adding zinc powder and $RhCl_3 \cdot nH_2O$ (n = 0 or 3) or $Na_3[RhCl_6] \cdot 12H_2O$ to a hot (80 to 90°C) solution of H_2(salen) in neat pyridine, and deposits from the orange reaction mixture as a pale yellow powder. Infrared data have been reported (3600 to 400 cm^{-1}) and the spectrum (1800 to 500 cm^{-1}) is reproduced [1].

RhX(salen)py (X = Br, I). These complexes are mentioned as products obtained by treatment of $Na_2[Rh(salen)]$ with perfluoroalkyl bromides and iodides [1].

$[C_5H_5NH][RhCl_2(salen)]$ is prepared by adding $Na_3[RhCl_6] \cdot 12H_2O$ or $RhCl_3 \cdot 3H_2O$ followed by zinc powder to a boiling solution of H_2(salen) in neat pyridine and is isolated from the cooled, filtered reaction mixture by concentration and addition of methanol. It forms a yellow precipitate. The infrared spectrum has been recorded (3600 to 400 cm^{-1}) and is reproduced (1800 to 500 cm^{-1}) [1].

$[Rh(salen)py_2][PF_6]$. The cationic complex is prepared by adding H_2(salen), zinc powder and $[RhCl_2py_4]Cl \cdot 5H_2O$ successively to boiling pyridine. The salt is obtained by evaporating the cooled, filtered reaction solution to dryness, extracting with boiling water and adding aqueous $K(PF_6)$ to the extract. The salt is an orange solid, soluble in acetone, dimethyl sulphoxide and halocarbons, less soluble in alcohols and insoluble in hydrocarbon solvents. Proton NMR data have been recorded [2].

***trans*-$RhCl_2$(H-salen)** is reported to form as a yellow powder when an aqueous alcoholic solution of $H[RhCl_2(dmg)_2]$, NaO_2CCH_3 and H_2(salen) is boiled for 1½ to 2 min, cooled, filtered and treated with concentrated HCl [3]. However, recent attempts to repeat this synthesis were unsuccessful [1]. The original authors reported that the complex was soluble in alcohol but sparingly soluble in water and formed violet solutions with aqueous bases (NH_3, Na_2CO_3, NaO_2CCH_3).

***trans*-$RhBr_2$(H-salen)** was reported to be similarly prepared from $H[RhBr_2(dmg)_2]$ as a yellow microcrystalline solid [3].

References:

[1] R. J. Cozens, K. S. Murray, B. O. West (J. Organometal. Chem. **38** [1972] 391/402). – [2] C. A. Rogers, B. O. West (J. Organometal. Chem. **70** [1974] 445/53). – [3] F. P. Dwyer, R. S. Nyholm (J. Proc. Roy. Soc. N. S. Wales **79** [1945] 126/8).

2.9.1.2 Complex with N,N′-Ethylenebis-5-chlorosalicylaldimine, RhCl(5-Cl-salen)py

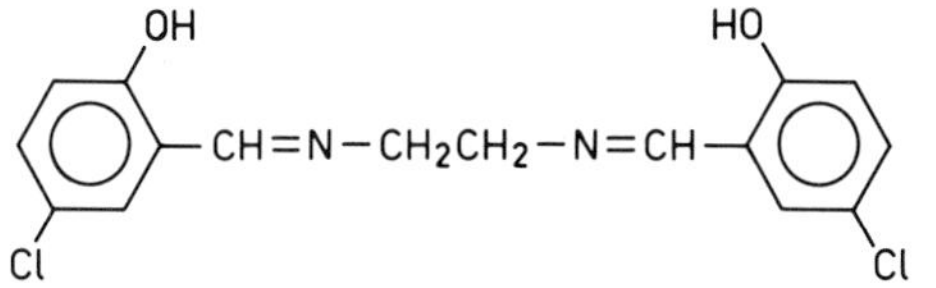

$= H_2$(5-Cl-salen)

This compound is prepared by adding zinc powder and $RhCl_3 \cdot nH_2O$ (n = 0 or 3) or $Na_3[RhCl_6] \cdot 12H_2O$ to a hot solution of H_2(5-Cl-salen) in neat pyridine and deposits as an extremely insoluble solid.

R. J. Cozens, K. S. Murray, B. O. West (J. Organometal. Chem. **38** [1972] 391/402, 399).

2.9.1.3 Complexes with N,N′-Trimethylenebis(salicylaldimine)

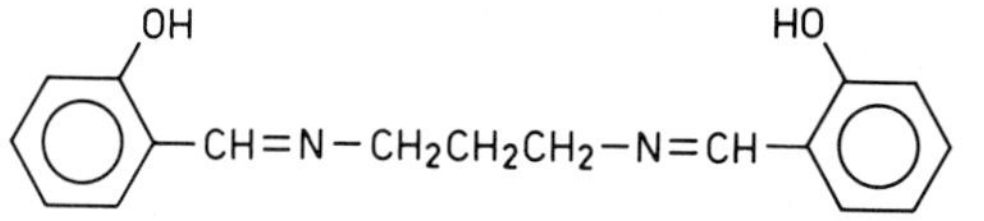

$= H_2$sal-1,3-pn

[Rh(sal-1,3-pn)py$_2$][PF$_6$] is prepared by boiling a solution of H_2(sal-1,3-pn), zinc dust and $[RhCl_2py_4]Cl \cdot 5H_2O$ in pyridine. The cooled solution is filtered, evaporated to dryness and the residue extracted with boiling water. Addition of $K(PF_6)$ to the aqueous extract leads to precipitation of the required product as a yellow solid. It is soluble in acetone, dimethylformamide, and halocarbons, less soluble in alcohols and insoluble in hydrocarbon solvents. Proton NMR data have been reported. The conductivity (nitrobenzene) is in the range expected for a 1:1 electrolyte. A *trans* arrangement has been proposed for the pair of pyridine ligands.

RhCl(sal-1,3-pn)py is prepared by boiling the residue from the aqueous extraction (preceding preparation) with methanol and filtering off the insoluble yellow complex.

C. A. Rogers, B. O. West (J. Organometal. Chem. **70** [1974] 445/53).

2.9.1.4 Complexes with N,N′-Tetramethylenebis(salicylaldimine)

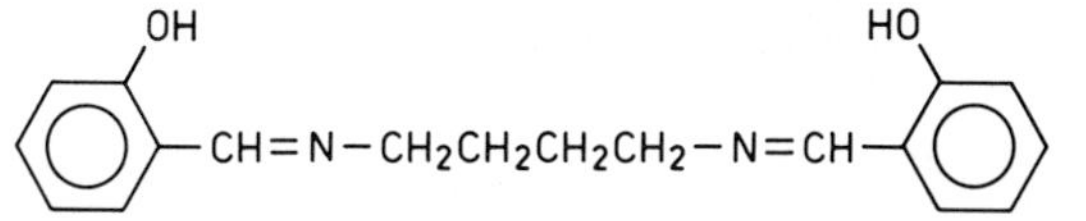

$= H_2$(sal-1,4-bn)

[Rh(sal-1,4-bn)py$_2$][PF$_6$] is prepared by boiling a solution of H_2(sal-1,4-bn), zinc dust and $[RhCl_2py_4]Cl \cdot 5H_2O$ in pyridine. The cooled solution is filtered, evaporated to dryness and the residue extracted with boiling water. Addition of $K(PF_6)$ to the aqueous extract leads to precipitation of the required product as a yellow solid. The complex is soluble in acetone, dimethylsulphoxide and halocarbon solvents, less soluble in alcohols and insoluble in hydrocarbon solvents. Proton NMR data have been reported. Conductivity data are consistent with formulation as a 1:1 electrolyte [1].

RhCl(sal-1,4-bn)py is prepared by boiling the residue from the aqueous extraction (see preceding preparation) with methanol and filtering off the insoluble yellow powder [1].

The crystals are monoclinic, space group $P2_1/c-C^5_{2h}$, lattice parameters a = 8.893(4), b = 11.519(6), c = 22.087(11) Å, β = 104.17°, D_{exp} = 1.55, D_{calc} = 1.55 g/cm³, Z = 4. The complex has the *cis-β* configuration, see **Fig. 20;** principal bond distances in Å: Rh-Cl = 2.347(4), Rh – O = 2.021(10), 2.024(9), Rh – N = 2.026(13), 2.030(12) [2].

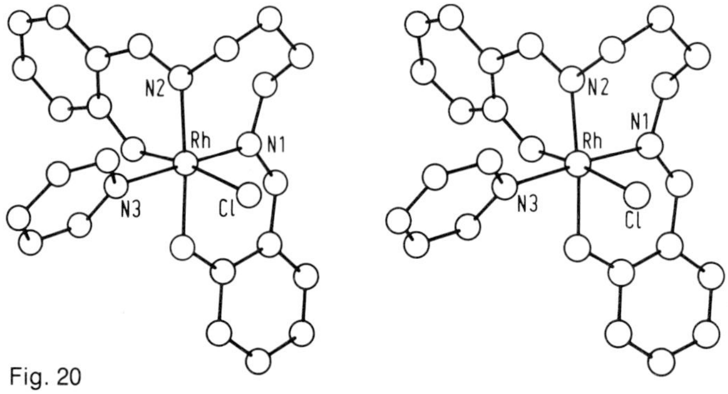

Fig. 20
Molecular structure of RhCl(sal-1,4-bn)py (stereoscopic view).

References:

[1] C. A. Rogers, B. O. West (J. Organometal. Chem. **70** [1974] 445/53). – [2] B. M. Gatehouse, B. E. Reichert, B. O. West (Acta Cryst. B **32** [1976] 30/34).

2.9.1.5 Complex with N-Salicylidene S-Alaninate, $K[RhL_2]$

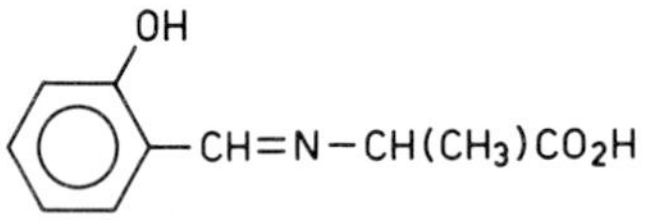

= H_2L

This compound is prepared by adding KOH and $RhCl_3$ to a boiling solution of L-alanine and salicylaldehyde and then heating the whole under reflux for 30 min. The product is precipitated with ether and chromatographed (anhydrous Al_2O_3, 96% ethanol) to give three fractions:

I {Λ(SS)}$K[RhL_2]\cdot 2H_2O$; II {Λ(SR) + Δ(SR)}$K[RhL_2]$; III {Δ(SS)}$K[RhL_2]\cdot 4H_2O$.

Electronic spectra and ORD curves (440 to 220 nm) are reproduced for Λ(SS) and Δ(SS) diastereomers. Proton NMR data have been recorded for fractions I → III. Activation parameters have been measured for the OH^- catalysed epimerisation of Λ(SS) and Δ(SS) diastereomers: E_A = 21.8 ± 1.0 kcal/mol (91.8 ± 4.0 kJ/mol), $\Delta S^{\neq} = -65.5 \pm 3\ J\cdot mol^{-1}\cdot K^{-1}$. Kinetics indicate that the α-proton of the amino-acid fragment is more mobile in the rhodium complex than in the analogous cobalt species.

Yu. N. Belokon, V. M. Belikov, V. A. Karginov, P. V. Petrovskii (Izv. Akad. Nauk SSSR Ser. Khim. **23** [1974] 2303/7; Bull. Acad. Sci. USSR Div. Chem. Sci. **23** [1974] 2218/21).

2.9.1.6 Complexes with Salicylaldoxime, $[RhL_3]$
$C_6H_4(OH)CH = NOH = HL$

Two products of this stoichiometry differing in colour, crystallo-optic properties and solubility in organic solvents are thought to be geometric isomers. No other details given in abstract.

N. K. Pshenitsyn, G. A. Nekrasova (Izv. Sekt. Platiny Drug. Blagorodn. Metal Inst. Obshch. Neorgan. Khim. Akad. Nauk SSSR No. 30 [1955] 59/70 from C.A. **1956** 9926).

2.9.1.7 Complex with N-(4-Methyl-7-hydroxy-8-aceto-coumarinylidene)-o-aminophenol, $H[Rh(C_{18}H_{13}NO_4)_2]$

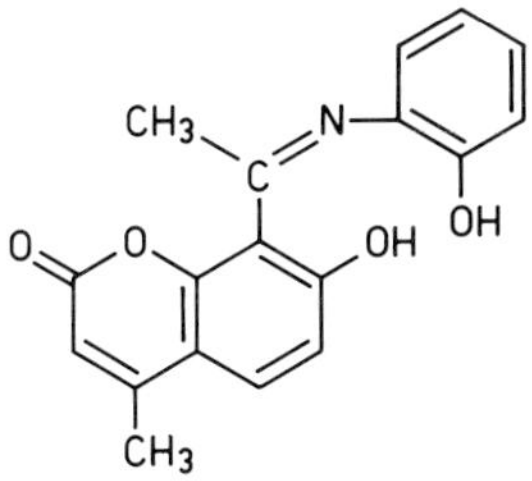

$= C_{18}H_{15}NO_4$

This compound is prepared by heating a mixture of $RhCl_3$ and the free ligand in ethanol at pH = 3.5 on a water bath for 2 to 5 h and deposits from solution as blackish-brown crystals. It is insoluble in water but soluble in organic solvents such as acetone, dimethyl sulphoxide and dimethyl formamide. The conductivity (in nitromethane) is given as 110 $cm^2 \cdot \Omega^{-1} \cdot mol^{-1}$. The electronic spectrum is reproduced (1000 to 280 nm) and assignments given, $\lambda_{max} = 685$ ($^1A_{1g} \rightarrow {}^3T_{1g}$), 518 ($^1A_{1g} \rightarrow {}^1T_{1g}$), 388 nm ($^1A_{1g} \rightarrow {}^1T_{2g}$); other data given are 10 Dq = 21650 cm^{-1}, $\beta = 422$ cm^{-1}; C = 2350 cm^{-1}, $\beta = 0.58$, C/B = 5.6, LFSE = 148.45 kcal/mol (620.5 kJ/mol).

P. Singh, G. P. Pokhariyal, V. Singh, S. C. Singh, G. K. Agrawal (Acta Chim. [Budapest] **104** [1980] 63/7).

2.9.1.8 Complex with N-(4-Methyl-7-hydroxy-8-aceto-coumarinylidene)-anthranilic Acid, $H[Rh(C_{19}H_{13}NO_5)_2]$

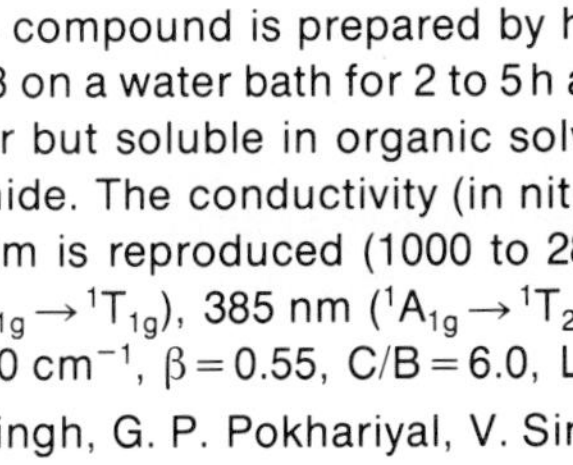

$= C_{19}H_{15}NO_5$

This compound is prepared by heating a mixture of $RhCl_3$ and the free ligand in ethanol at pH = 4.8 on a water bath for 2 to 5 h and deposits from solution as brown crystals. It is insoluble in water but soluble in organic solvents such as acetone, dimethyl sulphoxide and dimethyl formamide. The conductivity (in nitromethane) is given as 100 $cm^2 \cdot \Omega^{-1} \cdot mol^{-1}$. The electronic spectrum is reproduced (1000 to 280 nm) and assignments are given, $\lambda_{max} = 667$ ($^1A_{1g} \rightarrow {}^3T_{1g}$), 510 ($^1A_{1g} \rightarrow {}^1T_{1g}$), 385 nm ($^1A_{1g} \rightarrow {}^1T_{2g}$); other data given are 10 Dq = 21900 cm^{-1}, B = 400 cm^{-1}, C = 2300 cm^{-1}, $\beta = 0.55$, C/B = 6.0, LFSE = 150.17 kcal/mol (627.7 kJ/mol).

P. Singh, G. P. Pokhariyal, V. Singh, S. C. Singh, G. K. Agrawal (Acta Chim. [Budapest] **104** [1980] 63/7).

2.9.1.9 Complexes with N-4-Methylphenacylidene)anthranilic Acid and N-(4-Methylphenacylidene-o-aminophenol

p-$CH_3 \cdot C_6H_4 \cdot CO \cdot CH = N$-$C_6H_4CO_2H$-o $= C_{16}H_{13}NO_3$
p-$CH_3C_6H_4COCH = N \cdot C_6H_4OH$-o $= C_{15}H_{13}NO_2$

$Rh(C_{16}H_{12}NO_3)_2Cl$ is obtained as blackish-brown solid by heating a stoichiometric mixture of $RhCl_3 \cdot 3H_2O$ and the free ligand under reflux in ethanol for 3 to 4.5 h. Infrared and electronic spectra have been reported.

RhL_2Cl are obtained as dark solids by heating a stoichiometric mixture of $RhCl_3 \cdot 3H_2O$ and the free ligand under reflux in ethanol for 3 to 4.5 h. Infrared and electronic spectra have been reported.

G. P. Pokhariyal, S. K. Sharma (J. Ind. Chem. Soc. **58** [1981] 1199/201).

2.9.1.10 Complexes with Miscellaneous Schiff Bases

H_2SB = Schiff base = N,N'-(1,2-phenylene)bis(salicylaldimine) or N,N'-(4-methyl-1,2-phenylene)bis(salicylaldimine) or N,N'-(4,5-dimethyl-1,2-phenylene)bis(salicylaldimine); HSB' = Schiff base = N-methylsalicylaldimine or N-p-tolylsalicylaldimine

$[Rh(SB)py_2][PF_6]$. These complexes are all prepared by boiling a solution of the appropriate Schiff base, zinc dust und $[RhCl_2py_4]Cl$ in neat pyridine. The solution is cooled, filtered, evaporated to dryness and the residue extracted with boiling water. Addition of KPF_6 to the aqueous extract affords the required products as orange or red solids. They are soluble in acetone, dimethyl sulphoxide and halocarbon solvents, less soluble in alcohols and insoluble in hydrocarbon solvents. Proton NMR data have been reported. Conductivity measurements (nitrobenzene) confirm that the complexes are 1:1 electrolytes.

$[Rh(SB')_2py_2][PF_6]$. These complexes are similarly prepared using two molecules of the bidentate Schiff base per mole of rhodium and precipitate as yellow solids. They are soluble in acetone, dimethyl sulphoxide and halocarbon solvents, less soluble in alcohols and insoluble in hydrocarbon solvents. Proton NMR data have been reported. Conductivity measurements (nitrobenzene) confirm that the complexes are 1:1 electrolytes.

C. A. Rogers, B. O. West (J. Organometal. Chem. **70** [1974] 445/53).

2.9.2 Complexes with Acyl and Aroylpyridines

Complex with 2-Acetylpyridine, *trans*-$[RhCl_2(CH_3COpy)_2][RhCl_4(CH_3COpy)]$

$CH_3COC_5H_4N$ = 2-acetylpyridine = CH_3COpy; DMF = dimethylformamide

This compound is prepared by heating an ethanolic solution of $RhCl_3 \cdot 3H_2O$ and free ligand under reflux for 30 min, and deposits from the cooled reaction mixture as an orange solid. The electronic (reflectance) spectrum ($\lambda_{max} = 461$ nm) and selected infrared bands have been reported; $\nu(CO) = 1572$(sh), 1567, 1563 cm^{-1}(sh), $\nu(RhCl) = 370$, 332 cm^{-1}. The molar conductivity has been given as 51 (no units specified) for a 10^{-3}M solution in DMF.

R. R. Osborne, W. R. McWhinnie (J. Less-Common Metals **17** [1969] 53/7).

Complexes with 2-Benzoylpyridine

$C_6H_5COC_5H_4N$ = 2-benzoylpyridine = PhCOpy

***trans*-[RhCl$_2$(PhCOpy)$_2$][RhCl$_4$(PhCOpy)]** is prepared by heating an ethanolic solution of $RhCl_3 \cdot 3H_2O$ and free ligand under reflux for 30 min, and deposits from the cooled reaction solution as an orange solid. The electronic (reflectance) spectrum with λ_{max} = 485, 440(sh) nm and selected infrared bands have been reported; ν(CO) = 1545, 1520 to 1510 cm^{-1}, ν(RhCl) = 372, 341, 300(sh), 326 cm^{-1}. The molar conductivity has been given as 62 (no units specified) for 10^{-3}M solution in DMF.

***trans*-[RhBr$_2$(PhCOpy)$_2$][RhBr$_4$(PhCOpy)]** is obtained as a dark orange solid by heating a mixture of $RhCl_3 \cdot 3H_2O$, dry LiBr and free ligand under reflux in ethanol for 30 min. (The original paper refers to *trans*-[RhBr$_2$(PhCOpy)$_2$][RhCl$_4$(PhCOpy)] in the experimental section.) The electronic (reflectance) spectrum with λ_{max} = 481 (sh), 442 nm and selected infrared bands have been reported; ν(CO) = 1547, 1522 to 1515 cm^{-1}; ν(RhBr) = 273, 234 cm^{-1}. Molar conductivity data has been given as 42 (units not specified) for 10^{-3}M solution in DMF.

***trans-trans*-[RhCl$_2$(PhCOpy)$_2$][ClO$_4$] (yellow isomer)** is prepared by heating an aqueous ethanolic solution of $RhCl_3 \cdot 3H_2O$, free ligand and sodium perchlorate under reflux for 30 min or by metathesis of [RhCl$_2$(PhCOpy)$_2$][RhCl$_4$(PhCOpy)] and $NaClO_4$ in aqueous ethanol. The electronic (reflectance) spectrum with λ_{max} = 465, 420 nm and selected infrared bands have been reported; ν(CO) = 1530 to 1520, ν(RhCl) = 377 cm^{-1}. The molar conductivity has been given as 71 for 10^{-3}M DMF solution and 88 (no units specified) for 10^{-3}M CH_3NO_2 solution. An all *trans* stereochemistry is proposed for the cation.

***trans-cis*-[RhCl$_2$(PhCOpy)$_2$][ClO$_4$] (orange isomer)** is prepared as an orange solid by warming the *trans-trans* isomer in distilled water for a few minutes. The electronic (reflectance) spectrum with λ_{max} = 549(sh), 465(sh), 425(sh) nm and selected infrared bands have been reported; ν(CO) = 1530 to 1515, ν(RhCl) = 373 cm^{-1}. The molar conductivity has been given as 75 for 10^{-3} DMF solution and 94 (units not specified) for 10^{-3}M CH_3NO_2 solution.

***trans-trans*-[RhBr$_2$(PhCOpy)$_2$][ClO$_4$]** is prepared by metathesis of [RhBr$_2$(PhCOpy)$_2$]-[RhBr$_4$(PhCOpy)] and $NaClO_4$ in aqueous suspension and separates as a pale orange solid. The electronic (reflectance) spectrum with λ_{max} = 400(sh) nm and selected infrared bands have been reported; ν(CO) = 1527 to 1520 cm^{-1}. The molar conductivity has been given as 69 for 10^{-3}M DMF solution and 93 (no units specified) for 10^{-3}M CH_3NO_2 solution.

RhCl$_2$(OH)(PhCOpy)$_2$ is obtained as a red solid by refluxing *trans-trans*-[RhCl$_2$(PhCOpy)$_2$]-[RhCl$_4$(PhCOpy)] or [RhCl$_2$(PhCOpy)$_2$][ClO$_4$] in distilled water for 15 min. The electronic (reflectance) spectrum with λ_{max} = 523, 450 nm and selected infrared bands have been reported; ν(OH) = 3395, ν(CO) = 1554, ν(RhCl) = 349 cm^{-1}. The molar conductivity has been given as 7.1 for 10^{-3}M DMF solution (units not specified). The deuteriated complex RhCl$_2$(OD)(PhCOpy)$_2$ has also been mentioned; ν(OD) = 2520 cm^{-1}. A structure with *trans* chloride ligands and one monodentate (O-bonded) PhCOpy ligand has been proposed.

RhBr$_2$(OH)(PhCOpy)$_2$ is obtained as a red solid by refluxing *trans-trans*-[RhBr$_2$(PhCOpy)$_2$]-[RhBr$_4$(PhCOpy)] or *trans-trans*-[RhBr$_2$(PhCOpy)$_2$][ClO$_4$] in distilled water for 15 min. The electronic (reflectance) spectrum with λ_{max} = 515, 444 nm and selected infrared bands have been reported; ν(OH) = 3410, ν(CO) = 1554, ν(RhBr) = 263 cm^{-1}. The molar conductivity has been given as 17 (units not specified) for 10^{-3}M DMF solution.

***mer*-RhCl$_3$(PhCOpy)$_2$** is obtained as a yellow solid by heating RhCl$_2$(OH)(PhCOpy)$_2$ in 75% aqueous ethanol for 3h. The electronic (reflectance) spectrum with λ_{max} = 465(sh), 400 nm and selected infrared bands have been recorded; ν(CO) = 1669, 1565, ν(RhCl) = 365, 342, 314 cm^{-1}.

The chloro ligands are arranged in a *mer* configuration and one PhCOpy ligand is bonded through the pyridine nitrogen alone. $\Lambda_{mol} = 6$ (units not specified) for 10^{-3} M DMF solution.

***mer*-RhBr$_3$(PhCOpy)$_2$** is similarly obtained as a yellow solid by heating RhBr$_2$(OH)(PhCOpy)$_2$ in aqueous ethanol for 3h. The electronic (reflectance) spectrum with $\lambda_{max} > 333$ nm and selected infrared bands have been reported; ν(CO) = 1668, 1565, ν(RhBr) = 275, 261, 234 cm^{-1}. The complex is a non-electrolyte, $\Lambda_{mol} = 4$ (units not specified) for 10^{-3}M DMF solution.

RhCl$_3$(PhCOpy)H$_2$O·H$_2$O is obtained by boiling an aqueous suspension of *trans-trans*-[RhCl$_2$(PhCOpy)$_2$][RhCl$_4$(PhCOpy)], filtering off the insoluble RhCl$_2$(OH)(PhCOpy)$_2$ and allowing the filtrate to stand for several days. It deposits as orange crystals. The electronic (reflectance) spectrum with $\lambda_{max} = 440$ nm and selected infrared bands have been reported; ν(OH) = 3520 to 3440, ν(CO) = 1537 to 1534, ν(RhCl) = 354, 328 cm^{-1}. The complex is a non-electrolyte, $\Lambda_{mol} = 2.5$ (units not specified) for a 10^{-3}M solution in DMF.

RhBr$_3$(PhCOpy)H$_2$O·H$_2$O is obtained in a similar manner from *trans-trans*-[RhBr$_2$-(PhCOpy)$_2$][RhBr$_4$(PhCOpy)] and is isolated as deep orange crystals. The electronic (reflectance) spectrum contains no resolved maximum below 333 nm. Selected infrared bands have been reported; ν(OH) = 3500, ν(CO) = 1540 to 1530, ν(RhBr) = 269, 232 cm^{-1}. The complex is a non-electrolyte, $\Lambda_{mol} = 7.2$ (units not specified) for 10^{-3}M DMF solution.

[Ph$_4$As][RhCl$_4$(PhCOpy)] is prepared as an orange solid by heating an ethanolic mixture of RhCl$_3$·3H$_2$O, [Ph$_4$As]Cl and free ligand under reflux for 30 min. The electronic (reflectance) spectrum ($\lambda_{max} = 440$ nm) and selected infrared bands have been reported; ν(RhCl) = 347, 336 cm^{-1}.

R. R. Osborne, W. R. McWhinnie (J. Chem. Soc. A **1968** 2153/5).

Complexes with 2-(m-Aminobenzoyl)pyridine

m-H$_2$NC$_6$H$_4$COC$_5$H$_4$N = m-H$_2$NC$_6$H$_4$COpy = 2-(m-aminobenzoyl)pyridine

***trans*-[RhCl$_2$(m-H$_2$NC$_6$H$_4$COpy)$_2$][RhCl$_4$(m-H$_2$NC$_6$H$_4$COpy)]·3H$_2$O** is prepared by heating an ethanolic solution of RhCl$_3$·3H$_2$O and free ligand under reflux for 30 min, and deposits as a fawn solid. The electronic (reflectance) spectrum with $\lambda_{max} = 500$(sh), 405(sh) nm and selected infrared bands have been reported; ν(CO) = 1545, 1535, 1530(sh); ν(RhCl) = 332 cm^{-1}. The molar conductivity has been given as 20 (units not specified) for 10^{-3}M solution in DMF.

***trans*-RhCl$_2$(OH)(m-H$_2$NC$_6$H$_4$COpy)$_2$·2H$_2$O** separates as a red solid from the mother liquor in the preceding preparation on addition of light petroleum (boiling point 100 to 120°C). Selected infrared bands have been recorded; ν(CO) = 1545, 1532, 1515 cm^{-1}. The complex is a non-electrolyte ($\Lambda_{mol} = 0$ for a 10^{-3}M DMF solution).

***trans*-[RhCl$_2$(m-H$_2$NC$_6$H$_4$COpy)$_2$][ClO$_4$]·6H$_2$O.** A general method of preparation involving heating RhCl$_3$·3H$_2$O and the free ligand under reflux in aqueous ethanolic solution for 30 min has been reported; the origin of the perchlorate anions is not stated. The complex deposits as an orange solid. The electronic (reflectance) spectrum with $\lambda_{max} = 469$(sh), 380 nm and selected infrared bands have been reported; ν(CO) = 1550, 1545, 1540 and 1535(sh) cm^{-1}. Molar conductance data have been given as 31, 38 and 45 for 10^{-3}, 5×10^{-4} and 2×10^{-4}M DMF solutions, respectively (units not specified).

R. R. Osborne, W. R. McWhinnie (J. Less-Common Metals **17** [1969] 53/7).

Complexes with 2-(m-Nitrobenzoyl)pyridine

m-$O_2NC_6H_4COC_5H_4N$ = m-$O_2NC_6H_4COpy$ = 2-(m-nitrobenzoyl)pyridine

***trans-trans*-[$RhCl_2$(m-$O_2NC_6H_4COpy)_2$][$RhCl_4$(m-$O_2NC_6H_4$COpy)]·6H_2O** is prepared by heating an ethanolic solution of $RhCl_3 \cdot 3H_2O$ and the free ligand under reflux for 30 min, and deposits from the cooled solution as an orange solid. The electronic (reflectance) spectrum shows λ_{max} = 425 nm; selected infrared bands: ν(OH) = 3480, ν(CO) = 1545 (sh), 1538 (sh), 1534 and 1525 (sh), ν(RhCl) = 358, 333 cm^{-1}; conductivity Λ_{mol} = 36, 42, 53 (units not specified) for 10^{-3}, 5×10^{-4} and 2.5×10^{-4} M DMF solutions, respectively.

***trans*-[$RhCl_2$(m-$O_2NC_6H_4COpy)_2$](ClO_4)** is prepared by heating an aqueous ethanolic solution of $RhCl_3 \cdot 3H_2O$ and free ligand under reflux for 30 min and, after removal of solid $RhCl_2$(OH)(m-$O_2NC_6H_4COpy)_2$, is precipitated from the filtrate as pale orange crystals by addition of water. It is also obtained by metathesis using $NaClO_4$. The electronic (reflectance) spectrum shows λ_{max} = 510, 418 nm; selected infrared bands: ν(CO) = 1532, 1527 (sh) cm^{-1}; conductivity Λ_{mol} = 36, 42, 54 (units not specified) for 10^{-3}, 5×10^{-4} and 2.5×10^{-4} M solutions in DMF.

***trans*-$RhCl_2$(OH)(m-$O_2NC_6H_4COpy)_2$** is obtained as a red solid during the synthesis of the preceding complex. The electronic (reflectance) spectrum shows λ_{max} = 523, 450 nm; selected infrared bands: ν(OH) = 3390, ν(CO) = 1530, ν(RhCl) = 343 cm^{-1}; molar conductivity 4 (units not specified) for a 10^{-3} M solution in DMF.

R. R. Osborne, W. R. McWhinnie (J. Less-Common Metals **17** [1969] 53/7).

2.9.3 Complexes with 2-Hydroxypyridines

Complexes with 6-Methyl-2-hydroxypyridine

C_6H_7NO = 6-methyl-2-hydroxypyridine = Hmhp

$Rh_2(mhp)_4$ is one of three products obtained by treating $RhCl_3 \cdot 3H_2O$ or $Rh_2(O_2CCH_3)_4$-$(CH_3OH)_2$ with Na[mhp] in methanol. It is separated from the product mixture by chromatography and crystallises as air stable yellow-brown pyramids [1].

The X-ray crystal structure has been reported [1, 2]. The crystals are orthorhombic, space group Pbca-D_{2h}^{15}. Lattice parameters are a = 15.643(3), b = 16.083(3), c = 18.666(4) Å; D(X-ray) = 1.805 g/cm^3, Z = 8. The molecules have a "lantern" structure with a pair of rhodium atoms bridged by four mhp ligands orientated head-to-tail in 2:2 ratio as shown in **Fig. 21** [1].

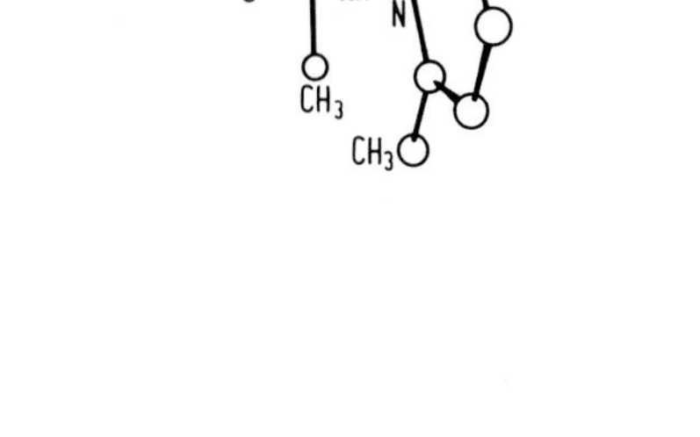

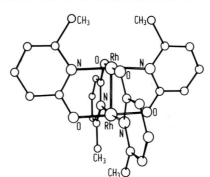

Fig. 21
Molecular structure of $Rh_2(mhp)_4$.

Principal bond distances in Å: $Rh_I-Rh_{II}=2.359(1)$, $Rh_I-N=2.035(2)$, 2.043(2), $Rh_{II}-N=2.043(2)$, 2.050(2), $Rh_I-O=2.016(2)$, 2.016(2), $Rh_{II}-O=2.024(2)$, 2.012(2) [2].

The He^I and He^{II} photoelectron spectra of $Rh_2(mhp)_4$ are reproduced (9 to 6 eV) and the following assignments have been made [1]:

Peak	Ionization Energy in eV	Assignment	Orbital*) Character
A	6.49(2)	Rh – Rh δ^*	a_2
B	7.25(2)	Rh – Rh π^*	e
C	7.64(2)	mhp π	b_1, e
D	8.00(2)	Rh – Rh δ	b_1
E	8.37(2)	mhp π	a_2
F	8.53(3)	Rh – Rh $\pi(\sigma)$	$e(a_1)$

*) in D_{2d} symmetry

$Rh_2(mhp)_4(Hmhp)\cdot0.5C_6H_5CH_3$ is isolated by chromatography from the product mixture obtained by treatment of $RhCl_3\cdot3H_2O$ or $Rh_2(O_2CCH_3)_4(CH_3OH)_2$ with Na[mhp] in methanol. It forms green crystals from $CH_2Cl_2/C_6H_5CH_3$ [3].

The crystals are monoclinic, space group $P2_1/n$, lattice parameters are a = 12.025(2), b = 21.487(4), c = 13.086(3) Å, β = 104.18(1)°, Z = 4. The molecules have the familiar "lantern" structure but with the bridging mhp ligands arranged head-to-tail in 3:1 fashion. The Hmhp ligand occupies the least hindered axial site (see figure in original paper). Principal bond lengths are $Rh_I-Rh_{II}=2.383(1)$, $Rh-O_{eq}=2.024(4)$, $Rh-O_{ax}=2.195(4)$ Å [3].

$\{Rh_2(mhp)_4\}_2\cdot2CH_2Cl_2$ is separated by chromatography from the product mixture obtained by treatment of $RhCl_3\cdot3H_2O$ or $Rh_2(O_2CCH_3)_4(CH_3OH)_2$ with Na[mhp] in methanol. It forms green crystals from CH_2Cl_2/petroleum ether.

The crystals are tetragonal, space group $I4_1cd$-C_{4v}^{12}. Lattice parameters are a = 21.126(4), c = 24.593(5) Å, Z = 8 (for formula given). The molecules consist of $Rh_2(mhp)_4$ lantern units (3:1 arrangement of mhp ligands within each lantern) linked in pairs by bridging oxygen atoms from mhp ligands, see **Fig. 22.** Principal bond distances are Rh – Rh = 2.369(1), $Rh-O_{eq}$ = 2.020(3) and 2.045(3) for O of Rh-O-Rh-O bridge, $Rh-O_{ax}=2.236(3)$ Å [3].

Fusion of $Rh_2(O_2CCH_3)_4$ with 6-methyl-2-hydroxypyridine (Hmhp) for 15 min at 160°C gives a dark green melt. After removal of excess Hmhp by vacuum sublimation, recrystallisation of the residue from CH_3CN or CH_2Cl_2 yielded the following five products [4].

$Rh_2(mhp)_4\cdot H_2O$ is obtained from solution in CH_3CN by slow evaporation and deposits as large orange-brown prisms [4].

The crystals are triclinic, space group $P\bar{1}$-C_i^1, lattice parameters a = 12.330(3), b = 19.360(4), c = 10.754(2) Å, α = 94.42(2)°, β = 96.36(2)°, γ = 81.89(2)°, D_{calc} = 1.76 g/cm³, Z = 4. The structure contains two independent "lantern" units (see figure in original paper). In each the bridging mhp ligands are arranged head-to-tail in 2:2 fashion. Principal bond distances in Å: $Rh_I-Rh_{II}=2.370(1)$, $Rh_I-O=2.018(5)$, 2.017(5), $Rh_{II}-O=2.019(5)$, 2.029(5), $Rh_I-N=2.038(5)$, 2.052(5), $Rh_{II}-N=2.053(8)$, 2.038(7) (for unit A); $Rh_I-Rh_{II}=2.365(1)$, $Rh_I-O=2.018(5)$, 2.014(5), $Rh_{II}-O=2.017(5)$, 2.029(5), $Rh_I-N=2.039(9)$, 2.041(10), $Rh_{II}-N=2.030(7)$, 2.037(7) (for unit B). The molecules of H_2O occupy general positions in the unit cell and do not interact with the vacant axial sites [4].

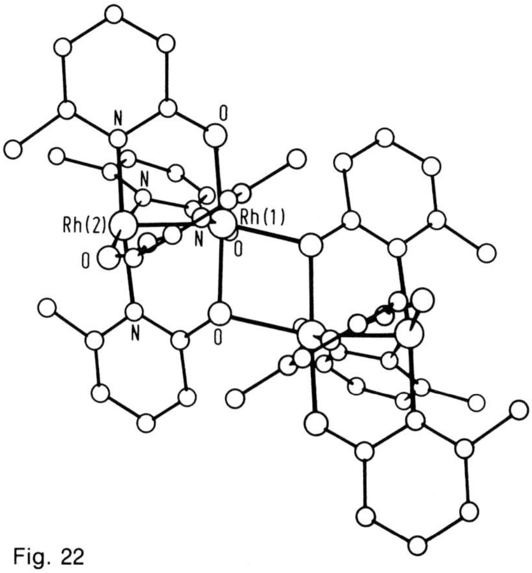

Fig. 22
Molecular structure of $\{Rh_2(mhp)_4\}_2 \cdot 2CH_2Cl_2$.

$Rh_2(mhp)_4(CH_3CN)$ is obtained from solution in CH_3CN and deposits on slow evaporation as very irregular dark green crystals [4].

The crystals are monoclinic, space group $P2_1/n$, lattice parameters a = 9.159(3), b = 20.185(2), c = 13.906(2) Å, β = 92.69(2)°, D_{calc} = 1.76 g/cm³, Z = 4. The molecule has a lantern structure with the four bridging mhp ligands arranged in 3:1 fashion and the CH_3CN ligand occupying the least hindered axial site (see figure in original paper). Principal bond distances in Å: Rh_I-Rh_{II} = 2.372(1), Rh_I-O = 2.015(5), $Rh_{II}-O$ = 2.015(5), 2.018(5), 2.027(4), Rh_I-N = 2.067(5), 2.073(5), 2.031(6), $Rh_{II}-N$ = 2.080, $Rh_I-N_{CH_3CN}$ = 2.152(7) [4].

$Rh_2(mhp)_4(C_3H_4N_2) \cdot 0.5\,CH_3CN$ ($C_3H_4N_2$ = imidazole) is obtained from solution in CH_3CN by slow crystallisation over a period of 4 weeks and deposits as irregular pink prisms or plates. The origin of the imidazole is unknown [4].

The crystals are monoclinic, space group $C2/c\text{-}C^6_{2h}$, lattice parameters a = 11.827(3), b = 15.714(2), c = 31.228(5) Å, β = 95.95(2)°, D_{calc} = 1.67 g/cm³, Z = 8. The molecule has a "lantern" structure with the four bridging mhp ligands arranged in 3:1 fashion. The imidazole ligand occupies the least hindered axial site, the CH_3CN molecules are not coordinated to the rhodium (see figure in original paper). Principal bond distances in Å: Rh_I-Rh_{II} = 2.384(1), Rh_I-O = 2.014(3), $Rh_{II}-O$ = 2.025(3), 2.018(3), 2.054(3), Rh_I-N = 2.062(4), 2.043(4), 2.059(4), $Rh_{II}-N$ = 2.094(4), $Rh_{II}-N$(imidazole) = 2.144(4) Å [4].

References:

[1] M. Berry, C. D. Garner, I. H. Hillier, A. A. MacDowell, W. Clegg (J. Chem. Soc. Chem. Commun. **1980** 494/5). – [2] W. Clegg (Acta Cryst. B **36** [1980] 2437/9). – [3] M. Berry, C. D. Garner, I. H. Hillier, W. Clegg (Inorg. Chim. Acta **45** [1980] L209/L210). – [4] F. A. Cotton, T. R. Felthouse (Inorg. Chem. **20** [1981] 584/600).

$Rh_2(O_2CCH_3)_2(mhp)_2(C_3H_4N_2)$ ($C_3H_4N_2$ = imidazole). For discussion of these complexes see rhodium(II) acetates, p. 34.

Complexes with 6-Chloro-2-hydroxypyridine

C_5H_4ClNO = 6-chloro-2-hydroxypyridine = Hchp

Preparation. Fusion of $Rh_2(O_2CCH_3)_4$ with 6-chloro-2-hydroxypyridine (Hchp) under argon at 140°C for 30 min followed by removal of excess Hchp under vacuum and recrystallisation of the residue from CH_2Cl_2/hexane gives the following two products.

$Rh_2(chp)_4$ is obtained as small octahedral golden crystals.

The crystals are orthorhombic, space group Pbca-D_{2h}^{15}, lattice parameters a = 15.992(2), b = 18.740(3), c = 15.490(3) Å, D_{calc} = 2.06 g/cm³, Z = 8. The molecule has the "lantern" structure with the chp ligands arranged in 2:2 fashion (see figure in original paper). Principal bond distances in Å: Rh_I-Rh_{II} = 2.379(1), Rh_I-O = 1.998(8), 2.004(7), $Rh_{II}-O$ = 2.025(8), 2.016(8), Rh_I-N = 2.029(9), 2040(9), $Rh_{II}-N$ = 2.06(1), 2.071(9).

$Rh_2(chp)_4(C_3H_4N_2)\cdot 3H_2O$ ($C_3H_4N_2$ = imidazole) is obtained as red-violet crystals from acetonitrile solution by slow evaporation. The origin of the imidazole ligand is uncertain.

The crystals are tetragonal, space group $P4_2/n$-C_{4h}^4, lattice parameters a = 22.372(3), c = 12.457(3) Å, D_{calc} = 1.79 g/cm³, Z = 8. The molecules have the "lantern" structure with the chp ligands orientated in 3:1 fashion and with the imidazole group coordinated to the least sterically hindered axial rhodium site (see figure in original paper). Principal bond distances in Å: Rh_I-Rh_{II} = 2.385(1), Rh_I-O = 2.019(8), $Rh_{II}-O$ = 2.033(7), 2.033(7), 2.024(7), Rh_I-N = 2.054(9), 2.085(9), 2.053(9), $Rh_{II}-N$ = 2.063(9), Rh_{II}-N(imidazole) = 2.129(9).

F. A. Cotton, T. R. Felthouse (Inorg. Chem. **20** [1981] 584/600).

2.9.4 Complexes with Substituted Acetamides

Complexes with Trifluoroacetamide

$CF_3C(O)NH_2$ = trifluoroacetamide

$Rh_2\{NH\cdot C(O)CF_3\}_4$ is prepared by heating $Rh_2(O_2CCH_3)_4$ in molten trifluoroacetamide at 160 to 165°C for 2 h [1] or 140 to 150°C for 5 to 6 h [2] and is crystallised from CH_3OH/H_2O as a light blue solid after removal of excess ligand by vacuum sublimation [1]. A "lantern" structure is proposed, see **Fig. 23**. Chemical or electrochemical oxidation affords the mixed-valence species $[Rh_2\{NHC(O)CF_3\}_4]^+$ [1, 3]. Cyclic voltammograms are reproduced for $Rh_2\{NHC(O)CF_3\}_4$ in CH_3CN [1] and in C_5H_5N/CH_3CN mixtures [3]. Half-wave potentials and electrochemical parameters for the oxidation of $Rh_2\{NHC(O)CF_3\}_4$ in non-aqueous media as CH_3NO_2, PhCN, CH_3CN, C_3H_7CN, $(CH_3)_2CO$, THF and $(CH_3)_2NC(O)H$ have been reported [3]. Electronic spectra (650 to 350 nm in CH_3CN, CH_3OH and $(CH_3)_2CO$ solution [3] and 650 to 350 nm in CH_3CN solution, complex undergoing electrochemical oxidation [1, 3]) are reproduced.

High performance liquid chromatography-mass spectrometry enables 4 isomeric forms of $Rh\{NHC(O)CF_3\}_4$ to be separated. Electronic spectra (700 to 350 nm) and mass spectra are reproduced for the three most abundant isomers [2].

$Rh_2\{NH\cdot C(O)CF_3\}_4py_2$ is obtained as yellow crystals by slow evaporation of a methanol solution containing $Rh_2\{NH\cdot C(O)CF_3\}_4$ (major isomer) and a stoichiometric amount of pyridine [2].

The crystals are orthorhombic, space group Pnma-D_{2h}^{16}, lattice parameters a = 15.523(4), b = 16.613(6), c = 10.407(6) Å, Z = 4, D_{calc} = 2.01 g/cm³. The molecule has the familiar "lantern" structure with axial pyridine ligands; the bridging $NHC(O)CF_3$ ligands are arranged head-to-tail in 2/2 fashion. Bond distances in Å: Rh-Rh = 2.472(3), Rh-N(py) = 2.29, Rh-O = 2.12(1), Rh-N = 2.01(1), 1.98(1), 2.31(1), 2.26(1) [2].

Formation of the pyridine adduct has been investigated spectrophotometrically and the electronic spectrum is reproduced [3]. The thermal decomposition has been studied and a thermogravimetric curve is reproduced [1].

References:

[1] A. M. Dennis, R. A. Howard, D. Lançon, K. M. Kadish, J. L. Bear (J. Chem. Soc. Chem. Commun. **1982** 399/401). – [2] A. M. Dennis, J. D. Korp, I. Bernal, R. A. Howard, J. L. Bear (Inorg. Chem. **22** [1983] 1522/9). – [3] K. M. Kadish, D. Lançon, A. M. Dennis, J. L. Bear (Inorg. Chem. **21** [1982] 2987/92).

Complexes with N-Phenylacetamide

$CH_3C(O)NHPh$ = N-phenylacetamide

$Rh_2\{PhNC(O)CH_3\}_4$. Two of the four possible isomers are prepared by heating a mixture of $Rh_2(O_2CCH_3)_4$ and excess phenylacetamide at 150°C for 48 h and are separated by high-performance liquid chromatography. The first isomer, which forms green crystals, has been characterised by ^1H NMR and is thought to possess a lantern structure with a 3/1 arrangement of the acetamido ligands, see **Fig. 24**. The second isomer, which crystallises as a blue **monohydrate**, has a similar lantern structure but a different (indeterminate) arrangement of the acetamido groups. Electrochemical and spectral data for both isomers have been reported; cyclic voltammograms and time-resolved electronic spectra (1000 to 400 nm) for the complex undergoing progressive electrochemical oxidation are reproduced.

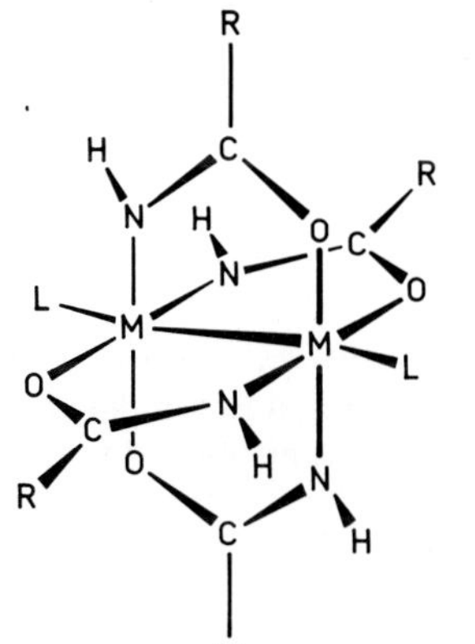

Fig. 23
Structure of $Rh_2\{NH{\cdot}C(O)CF_3\}_4$.

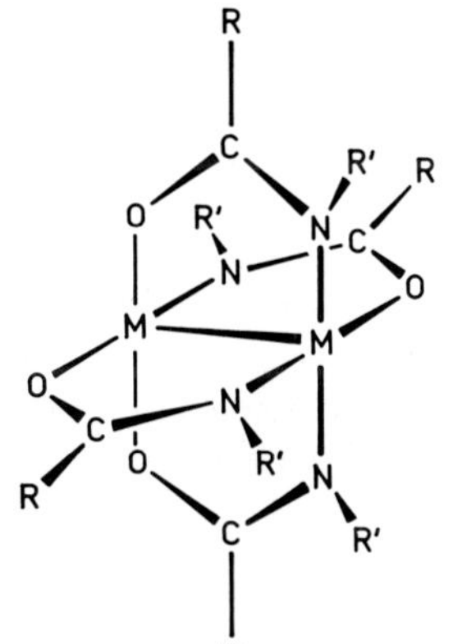

Fig. 24
Structure of $Rh_2\{PhNC(O)CH_3\}_4$.

J. Duncan, T. Malinski, T. P. Zhu, Z. S. Hu, K. M. Kadish, J. L. Bear (J. Am. Chem. Soc. **104** [1982] 5507/9).

2.9.5 Complexes with Anils

Complex with p-Dimethylamino-anil of Phenylglyoxal, [RhCl$_2$L$_2$]Cl

$(CH_3)_2\overset{+}{N}=\!\!\!\left\langle\!\!\!\bigcirc\!\!\!\right\rangle\!\!\!=\!N-CH=\underset{\underset{O^-}{|}}{C}-C_6H_5$ = L

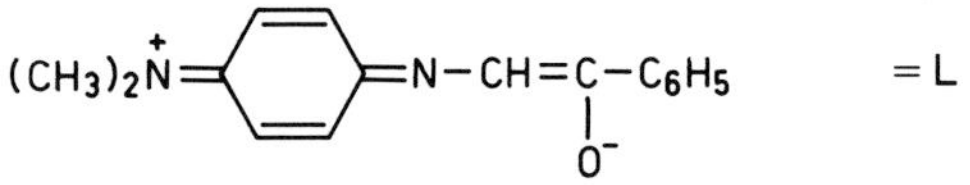

Only a general reference to preparative methods is given. The electronic spectrum has been reported and assignments given, λ_{max} = 739 ($^1A_{1g} \rightarrow {}^3T_{1g}$), 490 ($^1A_{1g} \rightarrow {}^3T_{2g}$), 432 ($^1A_{1g} \rightarrow {}^1T_{1g}$), 320 ($^1A_{1g} \rightarrow {}^1T_{2g}$), 239 nm ($^1A_{1g} \rightarrow a_1 \, {}^1T_{1u}$); L → M charge transfer. Other parameters: 10 Dq = 27 924.5 cm^{-1}, Racah's B = 508.06 cm^{-1}, C = 4803.5 cm^{-1}, LFSE = 74.45 kcal/mol (311.2 kJ/mol). The magnetic moment μ_{eff} = 0.48 B.M. Chelation by the anil ligand through N and O donor atoms is proposed.

R. K. Upadhyay, M. L. Singhal, A. K. Bajpai (Acta Chim. [Budapest] **96** [1978] 19/26).

Complex with p-Diethylamino-anil of Phenylglyoxal

$(C_2H_5)_2\overset{+}{N}=\!\!\!\left\langle\!\!\!\bigcirc\!\!\!\right\rangle\!\!\!=\!N-CH=\underset{\underset{O^-}{|}}{C}-C_6H_5$ = L

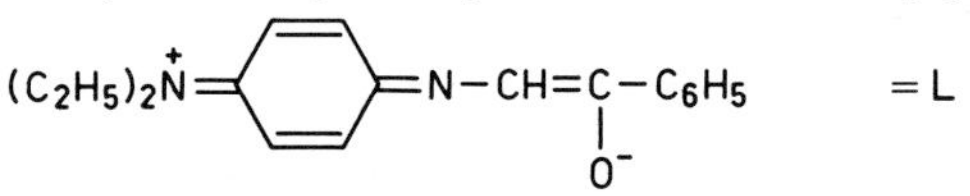

[RhCl$_2$L$_2$]Cl is prepared by mixing RhCl$_3$ and the free ligand in acetone, and deposits from the concentrated solution as pink crystals. The electronic spectrum has been recorded and assignments given, λ_{max} = 695 ($^1A_{1g} \rightarrow {}^3T_{1g}$), 380 ($^1A_{1g} \rightarrow {}^1T_{1g}$), 267 ($^1A_{1g} \rightarrow {}^1T_{2g}$), and 222 nm (charge transfer). Other parameters: 10 Dq = 22 260 cm^{-1}, Racah's B = 700 cm^{-1}, LFSE = 106.5 kJ/mol; magnetic moment μ_{eff} = 0.3 B.M., conductance 119.64 cm$^2 \cdot \Omega^{-1} \cdot$ mol^{-1} (10^{-3} M in DMF solution).

H. S. Verma, R. C. Saxena (Z. Naturforsch. **33b** [1978] 1001/4).

2.9.6 Complexes with Hydroxamic Acids

HON:C(OH)C(OH):NOH = oxaldihydroxamic acid = H$_2$odho
C$_6$H$_5$C(OH):NOH = benzohydroxamic acid = Hbho
o-HO·C$_6$H$_4$C(OH):NOH = salicylhydroxamic acid = Hsho
o-NH$_2$C$_6$H$_4$C(OH):NOH = anthranilohydroxamic acid = Haho

Rh$_2$(odho)$_3$. This apparently binuclear complex is prepared by heating under reflux an acidified (CH$_3$CO$_2$H) aqueous alcoholic solution containing RhCl$_3$ and oxaldihydroxamic acid. It deposits as a brown precipitate, soluble in mineral acids but insoluble in alkali, ammonia solution and common organic solvents [1].

A red-violet precipitate of the same stoichiometry is obtained by dissolving the brown precipitate in concentrated HCl and carefully neutralising the solution with solid sodium acetate. This product is insoluble in water and NaOH solution but soluble in mineral acids, methanol and ethanol [2].

Na[RhCl$_2$(bho)$_2$]·5H$_2$O is prepared by adding an aqueous acidified (CH$_3$CO$_2$H) solution of benzohydroxamic acid to a buffered (pH = 5 to 6, NaO$_2$CCH$_3$) aqueous solution of RhCl$_3$ and heating the mixture for 100 to 110 min when a light chocolate precipitate separates. The complex loses all water of crystallisation at 126°C but does not melt or decompose further below 320°C. It is sparingly soluble in methanol and hot water but insoluble in benzene, diethyl ether, acetone and ethylacetate [2].

Na[RhCl$_2$(sho)$_2$] is prepared by heating an aqueous buffered solution of $RhCl_3$ and salicylhydroxamic acid (pH = 4 to 5, CH_3CO_2H/NaO_2CCH_3) for 7 to 8 h when a light brown precipitate appears. The complex is almost insoluble in water but soluble in mineral acids and highly soluble in methanol and ethanol [2].

Na[RhCl$_2$(aho)$_2$]·3H$_2$O is prepared by heating an aqueous acidified (HCl) solution of $RhCl_3$ and the sodium salt of anthranilohydroxamic acid under reflux for 3 to 4 h, and deposits from the cooled solution as a pale yellowish-red precipitate. The hydrate loses all water molecules and changes to brown-black colour at 128°C. The complex is soluble in mineral acids, methanol and ethanol, but not in water [2].

References:

[1] R. S. Mishra (J. Indian Chem. Soc. **44** [1967] 400). – [2] R. S. Mishra (J. Indian. Chem. Soc. **46** [1969] 1074/7).

2.9.7 Complexes with 2-Pyridylazo and 2-Quinolylazo Dyes

General Literature:

S. Shibata, 2-Pyridylazo Dyes in Analytical Chemistry, in: H. A. Flaschka, A. J. Barnard, Chelates in Analytical Chemistry, Vol. 4, Dekker, New York 1972, pp. 1/232.

2-Pyridylazocresols

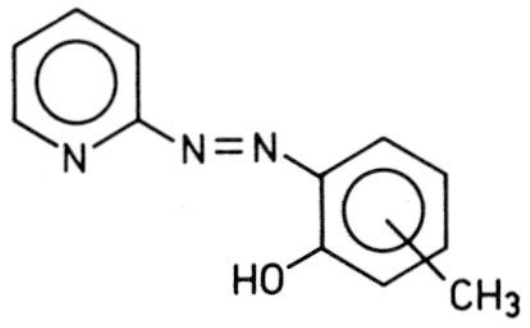

Colour reactions of selected 2-pyridylazocresols with rhodium(III) have been investigated and electronic spectra reported ($\lambda_{max} \approx 620$ to 570 nm) [1].

1-(2-Pyridyl)azo-2-naphthol (PAN)

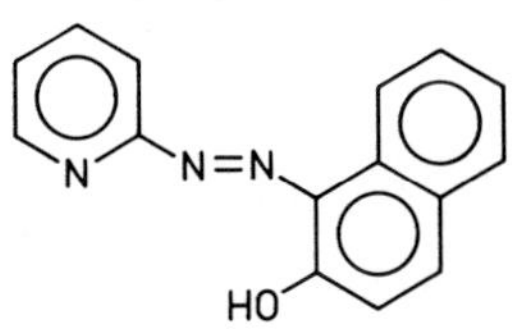

The formation of a 1:1 complex and a green 1:2 complex between PAN and rhodium(III) in buffered hot aqueous alcohol solution (pH ≈ 5, NaO_2CCH_3/CH_3CO_2H) has been reported. The green complex is insoluble in water but extracts into chloroform. The electronic spectrum (680 to 520 nm) is reproduced, λ_{max} is reported to occur at 598 nm. Plots of absorbance versus buffer pH and [PAN]:[Rh] ratio are reproduced [2].

Procedures for the determination of rhodium spectrophotometrically in the presence of iridium [1] and palladium/platinum [3, 4] have been described.

The extraction of Rh^{III}-PAN complexes from aqueous solution into chloroform has been studied. A plot of % rhodium extraction versus pH of aqueous solution is reproduced and data on the back extraction of rhodium from $CHCl_3$ into aqueous solutions at various pH values and in the presence of masking agents (citrate, cyanide, thiourea, thiosulphate and fluoride) have been reported [5]. A spectrophotometric procedure for determination of rhodium has been described [5, 6].

4-(2-Pyridylazo)resorcinol (PAR)

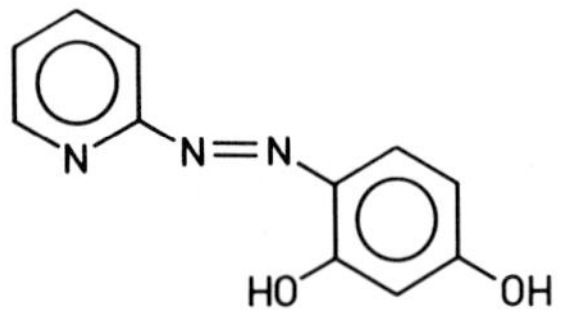

Formation of 1:1 and 1:2 complexes between rhodium(III) and PAR has been described. Molar absorptivities have been determined, equilibrium and stability constants have been computed and electronic spectra (680 to 400 nm) are reproduced [7].

A more detailed study leads to the conclusion that at least four rhodium-PAR species are formed:

Conditions	Colour	λ_{max} in nm
1) Rh:PAR ratio 1:10	pink red	520
2) excess Rh; pH = 2.2 to 4.9	green	570, 430
3) pH = 4.0 to 6.8	—	500

Electronic spectra (680 to 400 nm) of rhodium-PAR solutions at various pH values (1.9 to 6.2), plots of optical density versus pH and plots of optical density versus [Rh^{III}] or [PAR] concentrations are reproduced [8].

1-(2′-Pyridylazo)-2-phenanthrol (PAP)

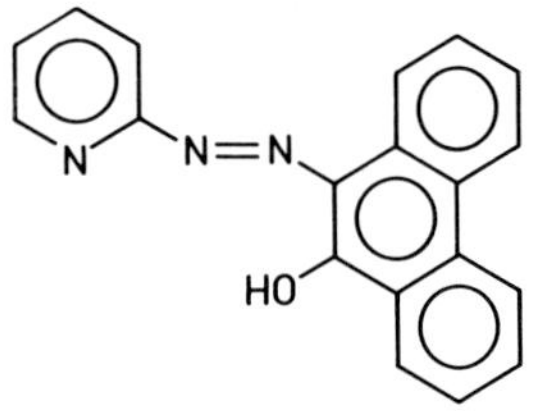

The ligand PAP forms a green, chloroform-soluble complex with rhodium(III) (metal:ligand ratio 1:2) which can be employed in the pH range 4.0 to 8.0 for spectrophotometric determination of rhodium. The electronic spectrum with λ_{max} = 625 and 580 nm has been reported [9].

1-(2-Quinolylazo)-2-acenaphthylenol (QAA)

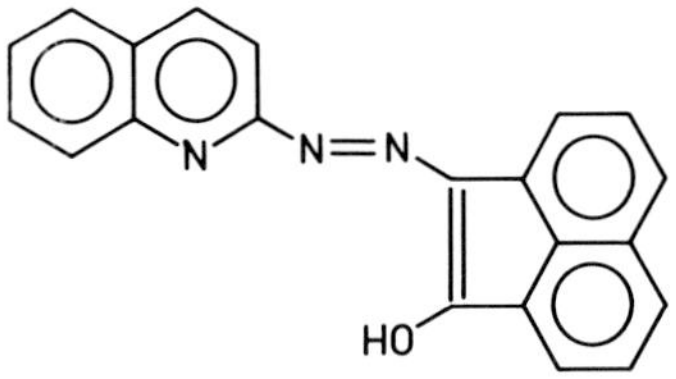

Formation of a reddish complex of rhodium(III) with QAA has been reported, it is insoluble in common organic solvents like CCl_4, $CHCl_3$, C_6H_6, $(C_2H_5)_2O$. No other information is given in the paper [10].

References:

[1] S. I. Gusev, I. N. Glushkova, L. A. Ketova, L. V. Poplevina (Zh. Anal. Khim. **28** [1973] 9/12; J. Anal. Chem. USSR **28** [1973] 5/7). – [2] J. R. Stokely, W. D. Jacobs (Anal. Chem. **35** [1963] 149/52). – [3] V. M. Ivanov, A. I. Busev, V. N. Figurovskaya (Zh. Analit. Khim. **29** [1974] 2260/2; J. Anal. Chem. [USSR] **29** [1974] 1943/5). – [4] V. N. Figurovskaya, A. I. Busev, V. M. Ivanov (Zavodsk. Lab. **39** [1973] 132/3 from C.A. **78** [1973] No. 154554). – [5] M. N. Cheema, I. H. Qureshi, M. Ashraf, I. Hanif (J. Radioanal. Chem. **35** [1977] 311/9).

[6] A. I. Busev, V. G. Grössl, V. M. Ivanov (Anal. Letters **1** [1968] 267/71). – [7] A. I. Busev, V. M. Ivanov, V. G. Grössl (Anal. Letters **1** [1968] 595/602). – [8] A. I. Busev, V. M. Ivanov, V. G. Grössl (Zh. Neorgan. Khim. **13** [1968] 2518/23; Russ. J. Inorg. Chem. **13** [1968] 1301/4). – [9] B. S. Garg, R. P. Singh, A. K. Rishi (Indian J. Chem. A **15** [1977] 367/9). – [10] Y. L. Mehta, B. S. Garg, R. P. Singh (Current Sci. [India] **43** [1974] 11/2).

2.9.8 Complexes with α-Dionemonoximes

α-Benzilmonoxime ($C_6H_5C(O)C(NOH)C_6H_5$ = Hbmo)

Rh(bmo)$_3$ is obtained as a brick red precipitate by adding an ethanolic solution of the free ligand to an aqueous acidic (HCl) solution of $RhCl_3$. The electronic spectrum (1000 to 200 nm), infrared spectrum (4000 to 300 cm^{-1}) and proton NMR spectrum are reproduced in the paper. On the basis of spectroscopic evidence the bmo anion is formulated as an N,O rather than an O,O′ chelate ligand [1].

Acenaphthenequinonemonoxime

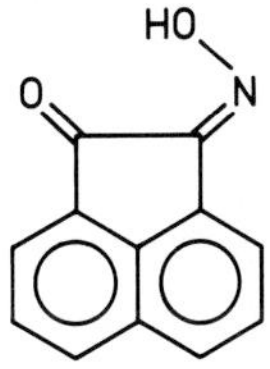

= $C_{12}H_7NO_2$

Acenaphthenequinonemonoxime forms a yellow complex with rhodium(III) in acid solution suitable for use in spectrophotometric determinations of the metal. The complex is extractable into chloroform, exhibits λ_{max} at 390 nm and has been shown by Job's method to have a metal:ligand ratio of 1:3. The electronic spectrum (520 to 370 nm) and Job's curve are reproduced [2].

α-Furilmonoxime ($C_4H_3O \cdot C$: $(NOH)C(O)C_4H_3O$ = Hfmo)

Complex formation between $RhCl_3$/HCl solutions and α-furilmonoxime in the presence and absence of acetate anions has been investigated by spectroscopic methods. Electronic spectra (460 to 320 nm) are reproduced. Equilibria are attained slowly (20 to 25 min at 90 to 95°C) and are pH-dependent. In the presence of acetate anions formation of 1:1 and 1:2 complexes, tentatively formulated as $RhCl_2(fmo)(H_2O)_2$ and $Rh(O_2CCH_3)(fmo)_2(H_2O)$, is observed. Both are yellow and extract into isopentyl alcohol, one is water-soluble, the other is water-insoluble but extracts into chloroform [3].

References:

[1] P. M. Dhadke, B. C. Haldar (J. Indian Chem. Soc. **55** [1978] 18/22). – [2] S. K. Sindhwani, R. P. Singh (Microchem. J. **18** [1973] 686/93). – [3] V. M. Savostina, O. A. Shpigun, T. V. Chebrikova (Zh. Neorgan. Khim. **24** [1979] 2723/7; Russ. J. Inorg. Chem. **24** [1979] 1512/4).

2.9.9 Complexes with Nitrosonaphthols

$Rh(C_{10}H_6NO_2)_3$ ($C_{10}H_7NO_2$ = 1-nitroso-2-naphthol) is prepared by treatment of $RhCl_3$ with excess free ligand in boiling alcohol and acetic acid solution at pH = 4.8 to 5.6 for 15 min and separates on cooling [1]. Electronic spectra (550 to 300 nm) are reproduced [2].

1-Nitroso-2-naphthol-3,6-disulphonic acid (nitroso-R-salt = $M_3[C_{10}H_4O(NO)(SO_3)_2]$). Complex formation between nitroso-R-salt and rhodium(III) under various conditions (pH, concentration, temperature) has been investigated [3, 4, 5]. Electronic spectra are reproduced (620 to 380 nm) [3] and (600 to 420 nm) [4]. Conditions for the spectrophotometric determination of rhodium have been reported [3, 4, 5]. However, there is considerable disagreement concerning conditions required to form coloured rhodium compounds with this reagent and even the colours involved.

$Na_6[Rh\{C_{10}H_4O(NO)(SO_3)_2\}_3]Na_2SO_4 \cdot 11.5H_2O$ is prepared by mixing aqueous solutions of $Rh_2(SO_4)_3$ and nitroso-R-salt, adjusting the pH to 5.5 with NaOH and boiling the mixture for 1 h. The solution formed is concentrated on a steambath to afford red crystals. The electronic spectrum (620 to 380 nm) is reproduced [3].

$[Rh\{C_{10}H_4O(NO)(SO_3)_2\}_2(OH)(H_2O)]^{4-}$, $[Rh\{C_{10}H_4O(NO)(SO_3)_2\}Cl_2(H_2O)_2]^{2-}$, $[Rh\{C_{10}H_4O(NO)(SO_3)_2\}Cl_2(OH)(H_2O)]^{3-}$. These are all postulated as species present in solutions of nitrosyl-R-salt and $RhCl_3$ under different pH conditions [4].

References:

[1] K. Watanabe (Nippon Kagaku Zasshi **77** [1956] 547/50). – [2] K. Watanabe (Nippon Kagaku Zasshi **77** [1956] 1008/11). – [3] O. W. Rollins, M. M. Oldham (Anal. Chem. **43** [1971] 146/8). – [4] L. S. Markova, V. M. Savostina, V. M. Peshkova (Zh. Analit. Khim. **29** [1974] 1378/84; J. Anal. Chem. [USSR] **29** [1974] 1189/93). – [5] S. Nath, R. P. Agarwal (Chim. Anal. [Paris] **47** [1965] 257/61).

2.9.10 Complexes with Miscellaneous Nitrogen-Oxygen Donor Chelates

Hydroxyimino-β-diketones $CH_3C(O)C{:}(NOH)C(O)R$ (R = CH_3, OCH_3, OC_2H_5, NHC_6H_5, $NHC_6H_4CH_3$-p) = HL)

Complexes of the composition **RhL_3** are prepared by stirring a mixture of free ligand and $RhCl_3 \cdot 3H_2O$ in methanol for 2 to 4 days and separate as crystalline precipitates. The hydroxy-imino-β-diketonate ligands are formulated as N,O donors,

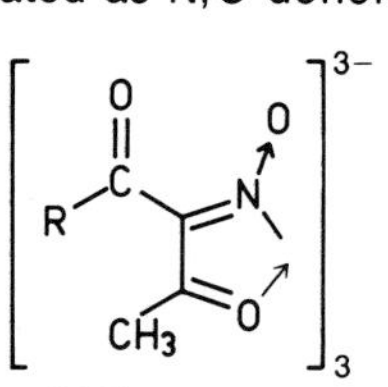

Proton NMR spectra have been reported [1].

Infrared data in cm^{-1} [1]:

R	Colour	ν(CO) (non-coord.)	ν(CO) (coord.)	ν(NO)	ν(CN)
CH_3	brown	1670(vs)	1520(s)	1200(s)	1620(m)
OCH_3	dark violet	1680(vs)	1540(s)	1230(s)	1640(m)

R	Colour	ν(CO) (non-coord.)	ν(CO) (coord.)	ν(NO)	ν(CN)
OC_2H_5	light violet	1685 (vs)	1540 (s)	1195 (s)	1640 (m)
NHC_6H_5	violet	1655 (vs)	1525 (s)	1200 (s)	1600 (m)
$NHC_6H_4CH_3$-p	yellow-brown	1650 (vs)	1520 (s)	1200 (s)	1630 (m)

(v)s = (very) strong, m = medium

Arylalkyltriazene Oxides (ArNH-N=N(O)R; R = CH_3, Ar = C_6H_5, m-$CH_3C_6H_4$, p-$CH_3C_6H_4$; R = C_2H_5, Ar = C_6H_5).

Complexes of the composition **Rh{ArN–N=N(O)R}$_3$** are prepared by refluxing an aqueous ethanolic mixture of $Rh(NO_3)_3$ and free ligand for 45 min, then adjusting the pH to ~4 using $NaHCO_3$. They separate as orange crystals on addition of water to the cooled reaction mixture and are recrystallised from CH_2Cl_2/hexane as shiny orange crystals.

Electronic spectra:

Ar	R	λ_{max} in nm	
C_6H_5	CH_3	363	565
m-$CH_3C_6H_4$	CH_3	314	498
p-$CH_3C_6H_4$	CH_3	318	469
C_6H_5	C_2H_5	357	568

A spectrum (450 to 250 nm) is reproduced for Rh{C_6H_5N–N=N(O)CH_3}$_3$. Proton NMR data have been reported and selected spectra are reproduced. The complexes have been formulated as *mer* tris chelates containing N,O-coordinated triazene oxide ligands [2].

References:

[1] V. Balasubramanian, N. S. Dixit, C. C. Patel (Transition Metal Chem. [Weinheim] **5** [1980] 152/4). – [2] B. Behera, A. Chakravorty (J. Inorg. Nucl. Chem. **31** [1969] 1791/6).

Complexes with 2-Aminoethanol

$RhCl_3(NH_2CH_2CH_2OH)_3$ is prepared by dropwise addition of 2-aminoethanol to a cooled stirred aqueous solution of $RhCl_3 \cdot 3H_2O$ and separates after a few days as fine dark yellow crystals. The electronic spectrum has been reported and assignments given, λ_{max} = 539, 449, 357 (sh), 339, 247 (sh) and 217 nm. Infrared data have been recorded and assigned, the spectrum (4000 to 400 cm^{-1}) is reproduced. The conductance is 17.7 $cm^2 \cdot \Omega^{-1} \cdot mol^{-1}$ for a fresh 10^{-3} M solution and increases to 22.5 $cm^2 \cdot \Omega^{-1} \cdot mol^{-1}$ after 90 min due to hydrolysis of one chloride ligand.

$RhBr_3(NH_2CH_2CH_2OH)_3$ is similarly prepared from $RhBr_3 \cdot 3H_2O$ and forms a light brown solid. The electronic spectrum has been reported and assignments given; λ_{max} (in nm) = 555 (sh), 465, 382 (sh), 316 (sh), 283 (sh), 245, 232 and 203 nm in H_2O solution and 555 (sh), 476, 381 (sh), 316 (sh), 286 (sh) and 247 in 0.5 M KBr. Infrared data have been recorded (4000 to 200 cm^{-1}) and assigned. The conductance is 26.9 $cm^2 \cdot \Omega^{-1} \cdot mol^{-1}$ for a fresh 10^{-3} M solution and increases to 49.8 $cm^2 \cdot \Omega^{-1} \cdot mol^{-1}$ after 90 min due to hydrolysis of one bromide ligand.

$[RhCl_2(NH_2CH_2CH_2OH)_4]Cl$ is prepared by slow addition of aqueous 2-aminoethanol to a stirred aqueous solution of $RhCl_3 \cdot 3H_2O$ containing a few drops of ethanol, and separates as a yellow crystalline precipitate. The electronic spectrum has been reported and assigned; $\lambda_{max} = 500$(sh), 429, 333(sh), 308, 247(sh) and 212 nm. The infrared spectrum (4000 to 400 cm^{-1}) is reproduced and assignments have been given. The conductance is 113 $cm^2 \cdot \Omega^{-1} \cdot mol^{-1}$ for fresh and aged 10^{-3}M solutions. On treatment with nitric acid in aqueous solution the salt is converted to the nitrate form, **$[RhCl_2(NH_2CH_2CH_2OH)_4]NO_3$**. The infrared spectrum (4000 to 400 cm^{-1}) is reproduced and assignments have been given.

$[RhBr_2(NH_2CH_2CH_2OH)_4]Br$ is similarly prepared using $RhBr_3 \cdot 3H_2O$. The electronic spectrum has been reported and assignments given; $\lambda_{max} = 526$(sh), 453, 340(sh), 286, 241 and 202 nm in aqueous solution; 526(sh), 453, 340(sh) and 287 nm in 1 M KBr. Infrared data (4000 to 180 cm^{-1}) have been recorded and assignments given. The conductance is 101 $cm^2 \cdot \Omega^{-1} \cdot mol^{-1}$ for fresh and aged 10^{-3}M solutions. On treatment with nitric acid in aqueous solution the salt is converted to the nitrate form, **$[RhBr_2(NH_2CH_2CH_2OH)_4]NO_3$**.

$[RhCl(NH_2CH_2CH_2OH)_5](ClO_4)_2$ is prepared by addition of 2-aminoethanol to an ice-cold aqueous solution of $RhCl_3 \cdot 3H_2O$, filtration of the reaction mixture and addition of the filtrate to a saturated aqueous $NaClO_4$ solution. It separates as a light yellow crystalline precipitate. The electronic spectrum has been reported and assignments given; $\lambda_{max} = 500$(sh), 435(sh), 341, 278(sh), 236(sh) and 200 nm. The conductance is 234 $cm^2 \cdot \Omega^{-1} \cdot mol^{-1}$ for fresh and aged 10^{-3}M solutions.

$[RhBr(NH_2CH_2CH_2OH)_5]Br_2$ is prepared by heating an aqueous mixture of $RhBr_3 \cdot 3H_2O$ and 2-aminoethanol on a water bath for 5 min, and deposits from the filtered solution as a yellow crystalline precipitate.

M. A. Golubnichaya, I. B. Baranovskii, G. Ya. Mazo, I. F. Golovaneva, R. N. Shchelokov (Zh. Neorgan. Khim. **26** [1981] 147/55; Russ. J. Inorg. Chem. **26** [1981] 77/81).

Other N-O Donor Chelates

Monobenzoylaminopyridine. The extraction of Rh^{3+} ions from aqueous solution into $CHCl_3$ or $C_2H_4Cl_2$ using monobenzoylaminopyridine and halide (Cl^- or Br^-) or thiocyanate anions has been investigated. No other details are given in the abstract [1].

Bis(8-hydroxy-5-quinolyl)methane

$[Rh(C_{19}H_{12}N_2O_2)]_n$ =

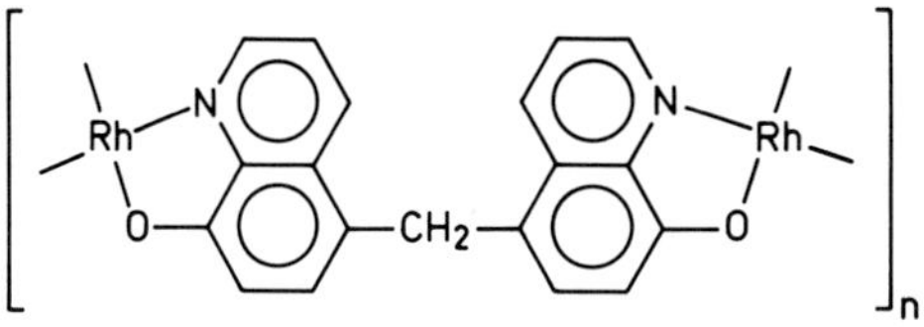

A polymeric species apparently containing rhodium in the oxidation state II has been reported and its ability to catalyse the hydrolysation of olefins has been investigated [2].

8-Amino-7-hydroxy-4-methylcoumarin

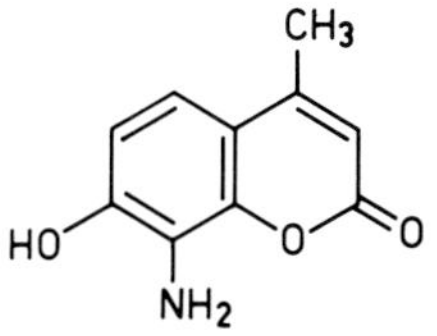

= Hahmc

$Na_2[RhCl_4(ahmc)]H_2O$ is prepared by mixing aqueous $RhCl_3$ and ethanolic Hahmc solutions at pH = 4.5, then raising the pH to 12.5 by dropwise addition of NaOH solution. The brown solid which deposits from the cooled, concentrated solution after 3 to 4 d gives dark chocolate-brown crystals from ethanol. The electronic spectrum has been reported and assignments given; λ_{max} (in nm) = 820 ($^1A_{1g} \rightarrow ^3T_{1g}$), 390 ($^1A_{1g} \rightarrow ^1T_{1g}$), 263 ($^1A_{1g} \rightarrow ^1T_{2g}$), 223 (charge transfer); Racah parameters (B = 772 cm^{-1}, C = 6725 cm^{-1}) and a value for 10 Dq (32370 cm^{-1}) have been measured [3].

Infrared data (4000 to 300 cm^{-1}) have been reported and assignments given. The complex is diamagnetic ($\chi_g = -29.28 \times 10^{-6}$ at 294.2 K) and is a 2:1 electrolyte (conductivity 275 $cm^2 \cdot \Omega^{-1} \cdot mol^{-1}$). The ahmc anion is thought to coordinate through N and O donor atoms [3].

Diaminoacetylurea ($NH_2CH_2CONHCONHCOCH_2NH_2$ = diaminoacetylurea). Thermodynamic parameters ($\Delta G°$, $\Delta H°$ and $\Delta S°$) and stability constants (log K_1, log K_2 and log β) for rhodium(III) complexes with diaminoacetylurea at different ionic strengths (μ = 0.15, 0.125, 0.10 and 0.075 M) and temperatures (25 and 30°C) have been determined [4]. Electronic spectra have been reported [5].

Biuret ($NH_2CONHCONH_2$ = biuret). Potentiometric studies on biuret complexes of rhodium(III) at different ionic strengths (μ = 0.2, 0.15, 0.10 and 0.05 M) have been performed at 25°C. The Calvin-Bjerrum pH titration technique has been applied to determine successive stability constants of 1:1 and 1:2 complexes. The biuret ligand is thought to coordinate as an N,N'-donor chelate [6].

References:

[1] A. T. Pilipenko, V. G. Markhotko, Nguyen-Mong-shinh (Visn. Kiivs'k. Univ. Ser. Khim. No. 11 [1970] 3/5 from C.A. **76** [1972] No. 80568). – [2] V. Vaisarová, J. Hetflejš (Z. Chem. [Leipzig] **14** [1974] 105/6). – [3] D. K. Rastogi (Australian J. Chem. **25** [1972] 729/37). – [4] P. C. Srivastava, S. K. Adhya, B. K. Banerjee (J. Indian Chem. Soc. **57** [1980] 985/90). – [5] P. C. Srivastava, B. K. Banerjee (Technology [Sindri, India] **12** [1975] 139/42).

[6] P. C. Srivastava, B. K. Banerjee (J. Indian Chem. Soc. **56** [1979] 779/81).

Table of Conversion Factors

Following the notation in Landolt-Börnstein [7], values which have been fixed by convention are indicated by a bold-face last digit. The conversion factor between calorie and Joule that is given here is based on the thermochemical calorie, $cal_{th\,ch}$, and is defined as 4.184**0** J/cal. However, for the conversion of the "Internationale Tafelkalorie", cal_{IT}, into Joule, the factor 4.1868 J/cal is to be used [1, p. 147]. For the conversion factor for the British thermal unit, the Steam Table Btu, BTU_{ST}, is used [1, p. 95].

Force	N	dyn	kp
1 N (Newton)	1	10^5	0.1019716
1 dyn	10^{-5}	1	1.019716×10^{-6}
1 kp	9.8066**5**	9.8066**5** $\times 10^5$	1

Pressure	Pa	bar	kp/m^2	at	atm	Torr	lb/in^2
1 Pa (Pascal) = 1 N/m^2	1	10^{-5}	1.019716×10^{-1}	1.019716×10^{-5}	0.986923×10^{-5}	0.750062×10^{-2}	145.0378×10^{-6}
1 bar = 10^6 dyn/cm^2	10^5	1	10.19716×10^3	1.019716	0.986923	750.062	14.50378
1 kp/m^2 = 1 mm H_2O	9.8066**5**	0.98066**5** $\times 10^{-4}$	1	10^{-4}	0.967841×10^{-4}	0.735559×10^{-1}	1.422335×10^{-3}
1 at = 1 kp/cm^2	0.98066**5** $\times 10^5$	0.98066**5**	10^4	1	0.967841	735.559	14.22335
1 atm = 76**0** Torr	1.0132**5** $\times 10^5$	1.0132**5**	1.033227×10^4	1.033227	1	760	14.69595
1 Torr = 1 mm Hg	133.3224	1.333224×10^{-3}	13.59510	1.359510×10^{-3}	1.315789×10^{-3}	1	19.33678×10^{-3}
1 lb/in^2 = 1 psi	6.89476×10^3	68.9476×10^{-3}	703.069	70.3069×10^{-3}	68.0460×10^{-3}	51.7149	1

Work, Energy, Heat	J	kWh	kcal	Btu	MeV
1 J (Joule) = 1 Ws = 1 Nm = 10^7 erg	1	2.778×10^{-7}	2.39006×10^{-4}	9.4781×10^{-4}	6.242×10^{12}
1 kWh	$3.\mathbf{6}\times10^{6}$	1	860.4	3412.14	2.247×10^{19}
1 kcal	4184.**0**	1.1622×10^{-3}	1	3.96566	2.6117×10^{16}
1 Btu (British thermal unit)	1055.06	2.93071×10^{-4}	0.25164	1	6.5858×10^{15}
1 MeV	1.602×10^{-13}	4.450×10^{-20}	3.8289×10^{-17}	1.51840×10^{-16}	1

1 eV/mol = 23.0578 kcal/mol = 96.473 kJ/mol

Power	kW	PS	kp m/s	kcal/s
1 kW = 10^{10} erg/s	1	1.35962	101.972	0.239006
1 PS	0.73550	1	**75**	0.17579
1 kp m/s	$9.8066\mathbf{5}\times10^{-3}$	0.01333	1	2.34384×10^{-3}
1 kcal/s	4.184**0**	5.6886	426.650	1

References:

[1] A. Sacklowski, Die neuen SI-Einheiten, Goldmann, München 1979. (Conversion tables in an appendix.)
[2] International Union of Pure and Applied Chemistry, Manual of Symbols and Terminology for Physicochemical Quantities and Units, Pergamon, London 1979; Pure Appl. Chem. **51** [1979] 1/41.
[3] The International System of Units (SI), National Bureau of Standards Specl. Publ. 330 [1972].
[4] H. Ebert, Physikalisches Taschenbuch, 5th Ed., Vieweg, Wiesbaden 1976.
[5] Kraftwerk Union Information, Technical and Economic Data on Power Engineering, Mülheim/Ruhr 1978.
[6] E. Padelt, H. Laporte, Einheiten und Größenarten der Naturwissenschaften, 3rd Ed., VEB Fachbuchverlag, Leipzig 1976.
[7] Landolt-Börnstein, 6th Ed., Vol. II, Pt. 1, 1971, pp. 1/14.